# Multiplexed Networks for Embedded Systems

# Multiplexed Networks for Embedded Systems

## CAN, LIN, Flexray, Safe-by-Wire...

**Dominique Paret**

Translated by

**Roderick Riesco, MA**

*Member of the Institute of Translation and Interpreting, UK*

John Wiley & Sons, Ltd

Originally published in the French language by Dunod as "Dominique Paret: Réseaux multiplexes pour systemés embarqués. CAN, LAN, FlexyRay, Safe-by-Wire". Copyright © Dunod, Paris 2005.



Email (for orders and customer service enquiries): cs-books@wiley.co.uk
Visit our Home Page on www.wiley.com



***Other Wiley Editorial Offices***

John Wiley & Sons Inc., 111 River Street, Hoboken, NJ 07030, USA

Jossey-Bass, 989 Market Street, San Francisco, CA 94103-1741, USA

Wiley-VCH Verlag GmbH, Boschstr. 12, D-69469 Weinheim, Germany

John Wiley & Sons Australia Ltd, 42 McDougall Street, Milton, Queensland 4064, Australia

John Wiley & Sons (Asia) Pte Ltd, 2 Clementi Loop #02-01, Jin Xing Distripark, Singapore 129809

John Wiley & Sons Canada Ltd, 6045 Freemont Blvd, Mississauga, ONT, Canada L5R 4J3

Wiley also publishes its books in a variety of electronic formats. Some content that appears in print may not be available in electronic books.

Anniversary Logo Design: Richard J. Pacifico

***British Library Cataloguing in Publication Data***

A catalogue record for this book is available from the British Library

ISBN 978-0-470-03416-3 (HB)

Typeset in 10/12 pt Times by Thomson Digital
Printed and bound in Great Britain by Antony Rowe Ltd, Chippenham, Wiltshire
This book is printed on acid-free paper responsibly manufactured from sustainable forestry in which at least two trees are planted for each one used for paper production.

*I dedicate this book to my father, André Paret, who passed away many years ago, and who, wherever he is now, will surely be smiling wickedly to himself when he finds that X-by-Wire solutions are about to be introduced in the car industry in a few years' time. He was a pioneer in mechanical and electronic systems for civil aviation, and, more than fifty years ago, brought such expressions as 'fly-by-wire', 'level 5, 6 and 7 multiple redundancy', 'reliability', 'lambda', etc., into my childhood and teenage years.*

*Now the circle is unbroken. Life is always beginning again!*

DOMINIQUE

# Contents

# Preface

Dear reader,

Multiplexed networks such as CAN, LIN and other types are now a mature industrial sector worldwide, and only a few new systems like X-by-Wire are still awaiting the final touches.

Having worked in this field for many years, I felt the need to draw up a report on these newer topics. Very little basic and application-related information and technical training is currently available for engineers, technicians and students. We trust that this book will at least partially make up for this deficiency.

I have waited a long time before taking up my pen again (rather difficult, with a PC!) to write this new volume. This is because many novel products have appeared, accompanied by their inevitable publicity constantly proclaiming that 'everything is good, everything is fine'. I decided to wait for a while until the dust had cleared and we could begin to see the horizon again; as usual, however, this took some considerable time.

In order not to fall into this trap of exaggerated publicity, let me clarify the situation straight away. At present, the version of LIN known as rev. 2.0 has been stable since the end of September 2003, Safe-by-Wire Plus rev. 2.0 since August 2004 and the final version of FlexRay, rev. 2.1, was officially released in March 2005. After these three births and our sincere congratulations to the happy parents, there is no reason to wait any longer before describing their characteristics.

The aim of this book is therefore to provide the most comprehensive guide at a given moment to this rapidly developing field. This book is not intended to be encyclopaedic, but aims to be a long and thorough technical introduction to this topic (and not just a literal translation of what anybody can find on the Web). It is thorough in the sense that all the 'real' aspects of these applications (principles, components, standards, applications, security, etc.) are dealt with in detail. For newcomers to this field, the book offers a global overview of the concepts and applications of the various buses.

I am aware that this sector is developing rapidly, and the content of this book will have to be updated in about three or four years. In the meantime, I can at least set out the basics and the fundamental principles.

I have also done my best to make this volume educational, enabling the reader to make the connection between theory, technology and economics, etc., at all times.

To complete this preface, I should inform you that another book by the same author is published under the title *Le bus CAN –Applications* [*The CAN bus – Applications*] which complements the present volume by dealing more specifically with the application layers of CAN and the details of their implementation; this should meet the requirements of the vast majority of users. However, if you still have even the shadow of a doubt about any of these matters, you are always welcome to contact me by post or e-mail with your questions.

Meanwhile, I hope you gain both pleasure and benefit from this book; above all, enjoy it, because I wrote it not for myself but for you!

**Very important note.** I must immediately draw readers' attention to the important fact that, in order to cover the topic of multiplexed buses and networks in a comprehensive way, this book describes a large number of technical principles which are patented or subject to licensing and associated rights (bit coding, communication methods, etc.), and which have already been published in official professional technical communications or at public conferences, seminars, etc.

Any application of these principles must comply with the current law in all circumstances.

## How to use this book

You are advised to read the next few lines carefully before moving on to the technical details which are provided in the following chapters.

First of all, you should be aware that this book covers many topics which impact on each other, merge and intersect, making it difficult to devise an overall plan. This led me to opt for an overall educational approach to enable readers to make sense of all these communication principles and emerging new protocols which they will discover and use in the future.

The introduction is designed to whet your appetite with a look at an everyday application, namely the mysteries of the on-board and In-Vehicle Network (IVN) systems of motor vehicles. Of course, everything described in this book is equally relevant to all kinds of industrial applications (control of machine tools and production lines, avionics, automation systems for large buildings, etc.).

## Part A

Chapters 1–5, about CAN, are intended to bring you up to speed with the current state of the CAN protocol, all the possible subdivisions of the physical layers and everything relating to conformity problems. Some of you will already be aware of some of this information, but many complementary details of the physical layers have been added or updated. We cannot discuss these topics without (briefly) mentioning the CAL, CAN Open, OSEK and AUTOSAR application layers and the hardware and software tools needed to support assistance with development, checking, production, maintenance, etc.

In Chapter 6, I describe the limits of CAN in terms of functions and applications, and clarify the distinction between event-triggered and time-triggered communication systems, pointing out all the implications of what are known as real-time security applications. This leads to an introduction to the function and content of protocols such as TTCAN and FlexRay, as well as applications of the X-by-Wire type for the latter; in plain English, this means: good-bye to mechanical systems, from now on everything is 'by-wire'. A good start!

At this point we shall have completed the description of what are known as the high-level protocols, and we shall be able to start on the second major section of this book, concerned with other members of the CAN family and the possible relationships between all these buses.

## Part B

I shall start with a detailed description of the new LIN bus (Chapter 7), its development, its properties, its problems and the way to overcome them. The LIN bus is generally presented by its designers as a sub-type of CAN, where the 'sub' is not pejorative, but purely functional. I shall go on to describe (in Chapter 8) the various constraints and possibilities for gateways between the buses described in Part One and these new arrivals, explaining how the Fail-safe – system basis chip (SBC) and other gateways are designed and constructed.

Finally, to complete this long 'bus trip', I shall describe (with only a few details, to avoid having to provide a universal encyclopaedia) the multiplicity of other buses surrounding those described in the previous two sections in on-board systems like those fitted in motor vehicles. Some experts consider that the motor car will eventually become a safe, reliable means of mobility, a domestic outpost (with audio, video, games, etc.), and an extension of the office! On this topic, I shall describe, in Chapters 9–11, wired and wireless serial links operating inside vehicles, in other words 'internal' systems (Safe-by-Wire Plus, 12C, D2B, CPL, MOST, IEEE 1394, etc.) and outside the vehicles, in other words 'external' systems (remote keyless entry, hands-free passive keyless entry, TPMS (tyre pressure monitoring system), tyre identification, GSM, Bluetooth, Zigbee and other related systems).

# Acknowledgements

The field of multiplexed buses is constantly expanding, thanks to the work of many highly skilled people. I have been lucky enough to encounter many of them on numerous occasions, so it is very difficult for me to acknowledge all of them individually.

I must offer my special thanks to numerous friends at Philips Semiconductors of Nijmegen, The Netherlands, and Hamburg, Germany, with whom I have had the pleasure of working in this area for many years, and, at the risk of some unfairness, more particularly to Hannes Wolff, Matthias Muth, the many 'Hans' and other colleagues from the Netherlands, and the many 'Peters' and other colleagues in Germany.

Finally, I would be ungrateful if I did not also thank the many colleagues in the industry, including car makers and parts manufacturers, whom I regularly encounter at work meetings and at the ISO. They will know who they are from my remarks about the development of this book, as a result of which we all hope that this field of multiplexed buses will see the rapid growth that it deserves.

Lastly, I am extremely grateful to the Marcom team of Philips Semiconductors at Eindhoven, especially Manuella Philipsen, for the numerous documents and photographs which she was kind enough to supply and which so enliven this book and the cover.

Dominique Paret
Meudon, 10 june 2006.

# Part A

# CAN: From Concept to Reality

Let us be clear that we are talking about CAN (controller area network).[1] Yet another new protocol and system! True, but we must realize that it is not an easy matter to bring everybody into agreement, especially when everyone has his own objectives and specific technical requirements, and that major markets are sufficient to justify and optimize every concept (which clearly means reducing the cost!). So here is a new serial protocol to be added to the burgeoning family of local area networks (LANs).

As described by the ISO (International Organization for Standardization) standards, CAN is a 'serial communication protocol which efficiently supports the distribution of commands in real time with a high security level. Its preferred fields generally relate to high bit rate, high transmission reliability network applications operating on a low-cost multiplexed cable principle'. You have been warned...

In this part of the book, I will describe the CAN protocol and some of its main applications.

I should point out that this concept was not developed overnight, but is the fruit of lengthy research and experimentation. Clearly, I could move straight on to a description of its operation, but I think this would deprive us of a depth of knowledge of the problems relating to buses used in local networks.

Indeed, some ideas commonly seem to spring up 'naturally', but often the major breakthroughs in our thinking are due to sequences of apparently unrelated events, which are in fact well organized.

So, before revealing the specifics of CAN as devised by R. Bosch GmbH, I wanted to provide an introductory chapter on the major features of this protocol, in an attempt to trace the thought processes followed by a large number of researchers in order to reach the final form of this project.

This approach is unusual, and although this introduction is not intended to be technical in nature, I advise you to read it carefully, as it will certainly give you a clearer understanding of all the ins and outs of the development of this very special protocol, which have made it so successful in the car industry and in other industrial and professional fields.

[1]Sometimes 'CAN bus' is used, not out of ignorance, but simply to avoid possible confusion, which could easily cause an incorrect classification of the book, in web search engines for example. As my last word on this matter, the normal term in use is simply 'CAN', bearing in mind that the term 'bus' represents only one of the many possible applications topologies.

**Note:** To pay tribute where it is due, I have chosen to base the development of this introduction on a model similar to that used by one of the founders of the CAN bus, Professor Uwe Kiencke of the University of Karlsruhe, in an excellent presentation to the International CAN Conference, sponsored by CAN in Automation, in 1994 (Mainz, Germany) . . . proving that you can be a prophet in your own country!

# 1

# The CAN Bus – general

A bus is always a bus – but there are 'buses' and 'buses'!

In fact, from one bus to the next, the problems to be solved remain the same, but the different characteristics of the proposed fields of application modify the hierarchical order of the parameters to be considered, and result in the development of new concepts in order to find neat solutions to the difficulties encountered. Here is a quick list of the virtually unchanging problems encountered in bus and network applications:

- network access concepts, including, clearly, problems of conflict, arbitration and latency;
- deterministic or probabilistic aspects and their relationship with the concepts of real-time or event-triggered systems;
- the concept of network elasticity ('scalability');
- the security of the data carried, in other words, the strategy for managing errors including their detection, signalling and correction;
- questions of topology, length and bit rate;
- questions of physical media;
- radio-frequency pollution, etc.

## 1.1 Concepts of Bus Access and Arbitration

'Distributed' real-time control systems, based on an operating system located within a single processor, interconnected by a communication network with distributed processors, are currently providing a significant addition to 'parallel' systems.

In addition to the simple exchange of data, the processing must be synchronized, in other words, the execution must follow certain interrelated logical sequences. In these systems, the messages relating to synchronization are generally short. They can be created by any process or event in the system and must be simultaneously receiving, in order to maintain the coherence of parallel processing.

*Multiplexed Networks for Embedded Systems: CAN, LIN, Flexray, Safe-by-Wire...* D. Paret

All stations independently generate messages concerning their respective tasks at random (event-triggered) instants.

The transmission requests contend with each other for access to the bus, leading to latencies[1] which are variable rather than constant.

Let us now examine the different principles of arbitration, which are in the running.

### *1.1.1 CSMA/CD versus CSMA/CA*

#### CSMA/CD

For historic reasons, the Carrier Sensor Multiple Access/Collision Detect (CSMA/CD) arbitration procedure was initially developed to resolve these conflicts.

With this system, when several stations try to access the bus simultaneously when it is at rest, a contention message is detected. The transmission is then halted and all the stations withdraw from the network. After a certain period, different for each station, each station again tries to access the network.

It is well known that these data transfer cancellations during contention theoretically also decrease the carrying capacity of the network. The network may even be totally blocked at peak traffic times, which is unacceptable when the network is to be used for what are known as 'real-time' applications.

#### CSMA/CA

In view of the above problems, another principle (one of several) was carefully investigated. This is known as Carrier Sensor Multiple Access/Collision Avoidance (CSMA/CA).

This device operates with a contention procedure not at the time of the attempt to access the bus, but at the level of the bit itself (bitwise contention – conflict management within the duration of the bit). This principle has been adopted in the CAN (controller area network) protocol, and its operation is described in detail in this part of the book.

In this case, bus access conflicts can be avoided by assigning a level of priority to each message carried.

In the case of contention, the message having the highest priority will always gain access to the bus. Clearly, the latency of the messages will then depend markedly on the priority levels chosen and assigned to each of them.

### *1.1.2 The problem of latency*

In the overall design of a system, in order to take all parameters into account, the latency is generally defined as the time between the instant indicating the request for transmission and the actual start of the action generated by this.

For the time being, and for the sake of simplicity, I shall define the latency of a message ($t_{lat}$) as the time elapsing between the instant indicating a request for transmission and the actual start of the transmission.

[1]See Section 1.3.

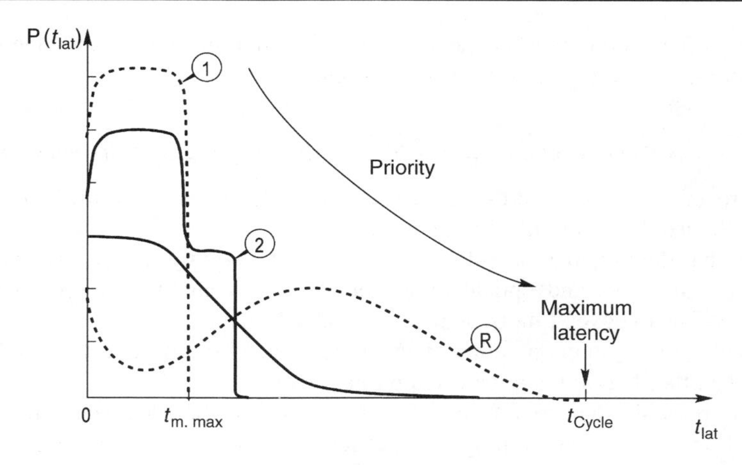


**Figure 1.1**

This concept is widespread and used in statistical analysis, mainly in what are known as 'real-time' systems. The reason is simple: only a few specific messages really need to have guaranteed latency, and then only during peak traffic times. We must therefore consider two kinds of messages:

- $R =$ messages whose latency must be guaranteed,
- $S =$ the rest,

and of course $M = R + S$, the total number of messages.

The curve in Figure 1.1 shows the probability distribution of the latency as a function of the latency, where the transmission request is made once only.

The specific value $t_{\text{cycle}}$ is the time representing a mean activity cycle of the network consisting of $M$ messages having a temporal length $t$ containing $N$ bits. The curves depend on the priority of the messages. The probability distribution of priority '1' returns to 0 immediately after the transfer of the longest message.

### *1.1.3 Bitwise contention*

This CSMA/CA principle, used in the CAN bus (patented since 1975), of the same type as that of the I2C bus, introduces some constraints, both in the representation of the signal in the physical layer and in relation to the maximum geometry '1' of a network operating in this way.

Thus, during the arbitration phase, in order to have higher priority bits in the network erasing those of lower priority, the physical signal on the bus must be

- either dominant,[1] for example, the presence of power, current, light or electromagnetic radiation,
- or recessive,[1] for example, an absence of power.

[1]Increase your Word Power.

By definition, when a dominant bit and a recessive bit are transmitted simultaneously on the bus, the resulting state on the bus must be the dominant state.

### *1.1.4 Initial consequences relating to the bit rate and the length of the network*

Now, here are a few words about the consequences of what I have described above (a part of Chapter 3 will also deal with this thorny question).

We know that the propagation velocity of electromagnetic waves $v_{prop}$ is of the order of 200,000 km s$^{-1}$ in wires and optical fibres (or, expressed another way, each wave takes approximately 5 ns to cover 1 m, or again travels at 200 m μs$^{-1}$).

Theoretically, in a system operating by bit contention, a bit can travel from one end of the network to the other before being detected on its arrival.

Now, it is possible that, a few 'micro-instants' before the bit reaches its destination, the other station, having seen nothing arrive at its terminal, may decide to start transmitting in the other direction. Head-on collision! Disaster looms!

If we call $t_{bus}$ the time taken by the signal to travel the maximum length of the network, the global sum of the outward and return times for the propagation of the signals on the bus is

$$2t_{bus} = 2\frac{l}{v_{prop}}.$$

For example, if $l = 40$ m, then $t_{bus} = 200$ ns.

To enable the station which sent the initial bit to manage the conflicts, the necessary duration of the bit, known as the bit time or $t_{bit}$, must be longer than $t_{bus}$. And for the sake of completeness, we must allow for the time taken (or required) to sample and process the bit at the station where it arrives.

To evaluate the minimum bit time, $t_{bit\text{-}min}$, of the proposed network, it is necessary (Figure 1.2) to allow for

- the outward propagation delays, $t_{out}$,
- the inward propagation delays, $t_{in}$,

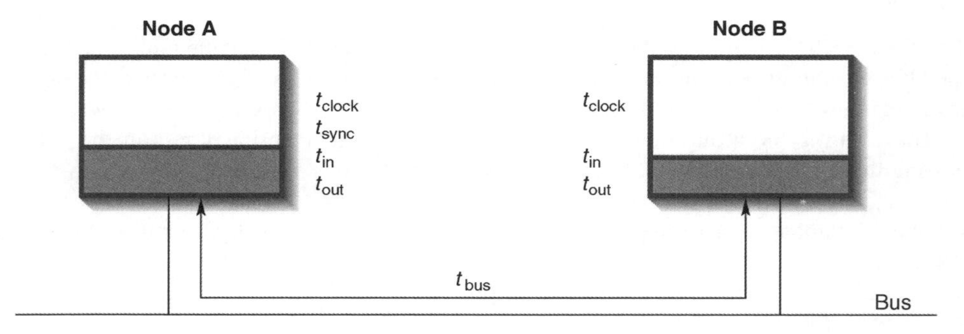


**Figure 1.2**

- the delays due to synchronization, $t_{sync}$,
- the phase differences due to clock tolerances, $t_{clock}$,

giving a total $t_{bit\text{-}min}$ of

$$t_{bit} = 2t_{bus} + 2t_{out} + 2t_{in} + t_{sync} + t_{clock}.$$

**Example.**
With a bit rate of 100 kbit s$^{-1}$, that is a bit time of 10 μs, we can achieve a network length of approximately 900 m.

All these concepts are described in detail in Chapter 3.

### *1.1.5 The concept of elasticity of a system*

The architectural and topological configuration of a distributed system is generally different from one application to another, and, for a single application, it can also develop or be changed over a period, depending on the requirements to be met. To make things clearer, consider the example of the installation of industrial production lines, which, although they always have the same overall appearance, need to be reconfigured from time to time according to the products to be made. Where systems or networks are concerned, we generally use the term 'elasticity' to denote the capacity to withstand a change of configuration with the least possible amount of reprogramming in relation to the data transfer to be provided.

Let us look more closely at the problems associated with the elasticity of a network.

The information received and processed somewhere in a distributed system must be created and transmitted to a station. There is no logical alternative to this. There are two possible cases:

- New information is to be added. Any new information requires a new message transfer and therefore a reprogramming of the communication. In this case, the station which previously sent this specific information element must be reprogrammed according to the new one, whereas the other stations remain unchanged.
- A different situation occurs when a pre-existing specific information element has to be either transmitted from another station or received by additional stations. In this case, the additional receiving station must be reprogrammed to receive the information.

### *1.1.6 Implication of the elasticity of a system for the choice of addressing principle*

Conventional addressing, generally consisting of a 'source' address on the one hand and a 'destination' address on the other hand, cannot provide a system with good structural elasticity. This is because any messages that have to be rerouted will require modifications, even if this is not logically necessary, as mentioned above.

For the CAN concept, in order to provide good elasticity in the system, it was decided to use another addressing principle, based not on the source and destination addresses but on the content of the message itself. This implies two things:

- A message has to be transmitted to all the other stations in the network (the term used for this principle is 'broadcast diffusion').

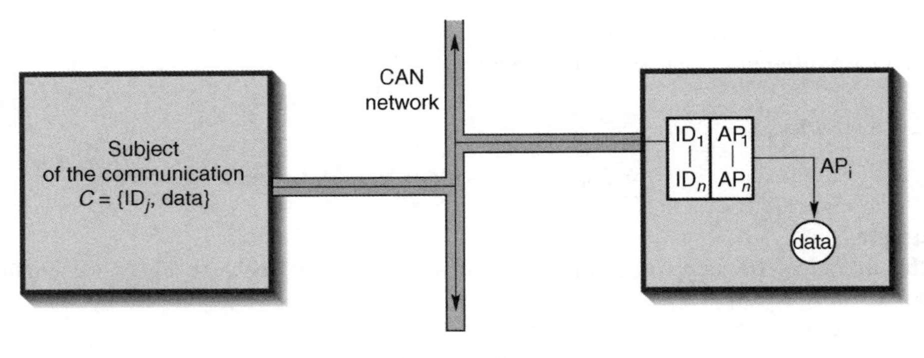


**Figure 1.3**

- The selection processing of the transmitted message is then carried out by what is called 'acceptance filtering' at each station.

For this purpose, the message is labelled with an identifier ID($i$), which is then compared with the list of messages received (or to be received) at each station.

This list contains address pointers AP($i$) towards the communication buffer, enabling the content of the message to be stored (Figure 1.3).

Therefore, *all* the messages are simultaneously received over *all* of the network, and the data consistency is thus guaranteed in distributed control systems.

## 1.2 Error Processing and Management

### *1.2.1 The concept of positive and negative acknowledgements*

The elasticity of systems and identification based on the message content complicate the processing of errors when they occur.

An example of a conventional method of (non-)detection of errors is the return of what is called a 'positive' acknowledgement from the receiving station to the transmitting station, when a message is received correctly. The key information contained within such a positive acknowledgement is generally the address of the receiving station.

In the CAN concept, this idea of a local address completely disappears, and the identifier 'labelling' the message is transmitted to all the participants and received everywhere in the network. This makes it necessary to execute a local error processing task in each of the stations in the network, in the absence of such local message addresses. To achieve this, the CAN protocol concept uses a combination of positive and negative acknowledgements.

The positive acknowledgement ACK+ is expressed as follows:

$$\text{ACK+} = \text{ACK} + (i) \text{ for any } (i).$$

It is sent from all stations ($i$) which have received the message correctly and expresses this (positive) acknowledgement during a single precisely defined time interval (ACK time slot).

The principle described above can also be expressed in two basic forms:

- an optimistic view of the problem, in which we can say that the positive acknowledgement indicates that 'at least one station has received the transmitted message correctly';
- a rather less optimistic (but not totally pessimistic) view: 'The negative acknowledgement of a message must take a form such that it indicates (or will indicate) that there is at least one error in the global system'.

In the latter case, the message indicating the presence of the error must be sent over the network immediately after the detection of the error. This method will ensure that the system can be resynchronized immediately within a single message frame. The latter point is crucial for the security of the applications considered.

### *1.2.2 Error management*

The combination, the redundancy and the analysis of the positive and negative acknowledgements sent from the error processing devices of the different stations are exploited to provide strong presumptions concerning the source of an error (either from the transmitting station or from one of the receiving stations), for example

- The presence of at least one positive acknowledgement sent from a receiver, combined with an error message, signifies that the message has been transmitted correctly at least.
- Conversely, the absence of a positive acknowledgement, together with an error message, indicates that all the receiving stations have detected an error and that there is a strong presumption (or probability) that the error is located in the transmitting station.

### *1.2.3 Error messages*

The error messages used for the CAN concept are of two types (Figure 1.4).

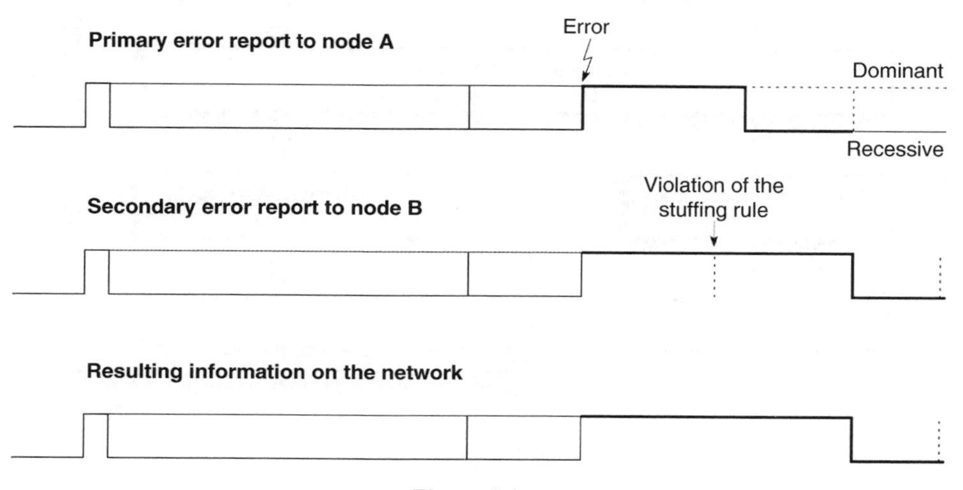


**Figure 1.4**

### Primary error report

A station (or several stations at once) detect(s) an error, causing it (or them) to immediately transmit an error message.

At this point, the error message consists of a sequence of dominant bits and the current message is aborted (not fully processed).

### Secondary error report

Following this cancellation of the message, all the other stations detecting a violation of the format rules transmit an error message on their own behalf, these messages being combined with the first one and thus increasing its duration.

Primary error reports are more probable for a station where the error has occurred than secondary error reports, which are generated in response to the aforementioned error.

### *1.2.4 The concept of an error management strategy*

To manage errors correctly, the ultimate aim is to develop a strategy for the processing of errors. The quality of the processing will depend markedly on whether or not the strategy defined for a given field of applications is proactive.

For the CAN protocol, a strategy called an error (or fault) 'confinement' strategy, shown in outline in graphic form in Figures 1.5 and 1.6, has been defined (and is described in much greater detail in Chapter 2). Here is a brief summary.

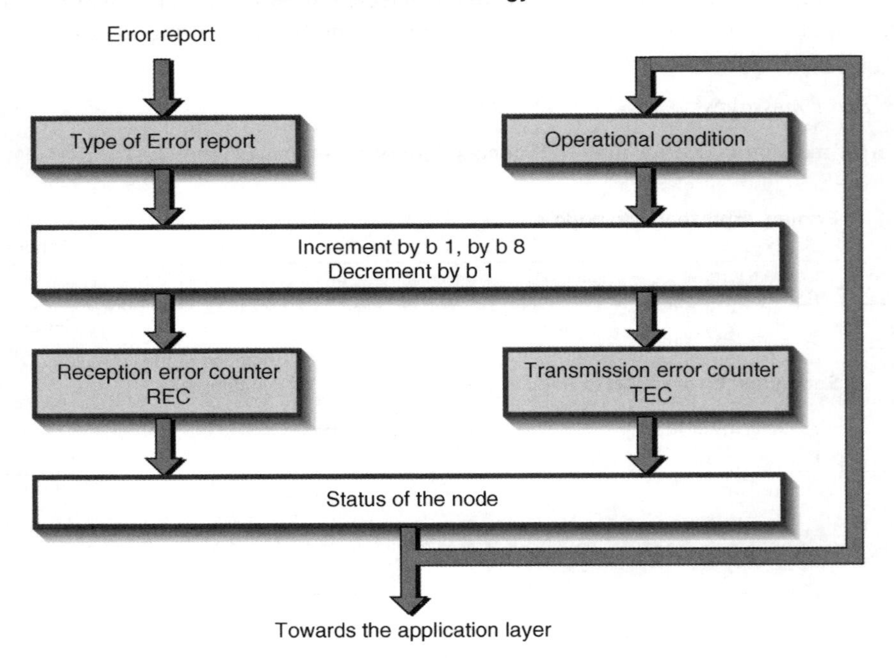


**Figure 1.5**

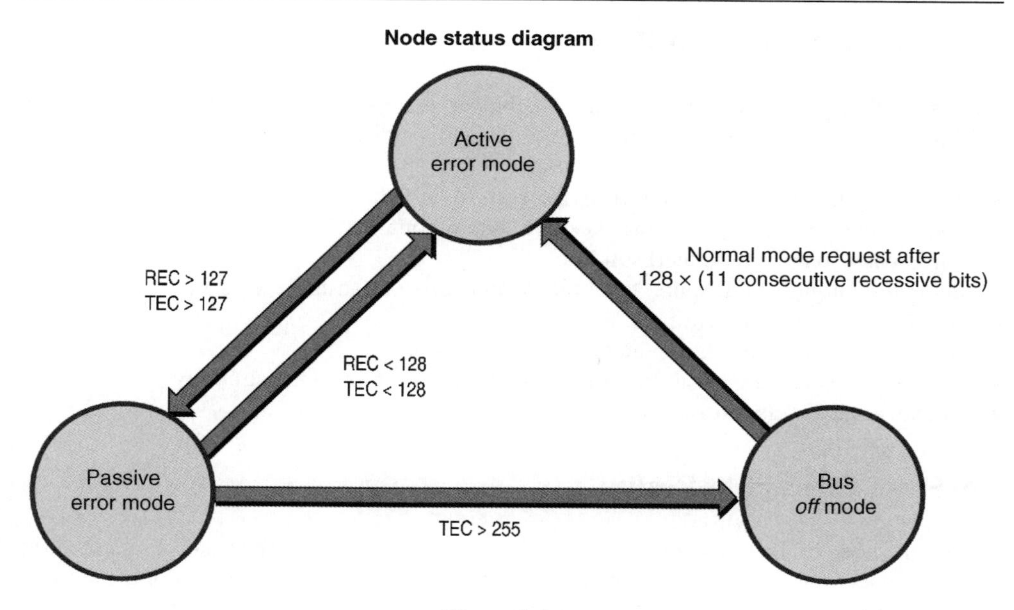


**Figure 1.6**

First, each station has to have two separate error counters: one to record what happens during the transmission of a message and the other to execute a similar task during reception.

Then, according to the type of error report and the operating conditions of the station at the instant in question, the counters are incremented with different weightings according to certain conditions or are decremented.

The purpose of these counters is to record information originating directly or indirectly from all the other stations. Using this device, the counters carry out an averaging operation, giving an approximation of the statistical values of the local quality of the network at a given instant.

To provide more support for this strategy, it was also specified that if too many errors were attributed by such statistical means to a specific station, its state would be transferred from what is called an 'active error' mode to a 'passive error' state, in which it would no longer communicate but would always be active in the management of errors which may occur on the network.

If there are too many transmission errors in the network, the network could be blocked, making all traffic impossible.

To avoid this, it is necessary to specify that, above a certain limit (fixed at 255 for CAN), the station switches to another new state, called *bus off*, during which the station appears to be physically disconnected from the bus so as not to block the bus for any reason.

In this case, its input stage stays in a high impedance state, thus making it possible to observe the signal travelling in the network (monitor function).

This ends the summary of the key points for the study of this new protocol concept, together with the major ideas used to resolve the problems generally raised by the applications.

At this stage, you have the necessary background to study the actual operation of this protocol and of the associated bus.

Now for the work! The bus is waiting, so let us go... (see also *The I2C Bus* [same author, same publisher] in which, at that time, we were only waiting for the... bus!).

## 1.3 Increase Your Word Power

To ensure that we are using the same vocabulary, I now offer for your guidance a few unillustrated extracts from dictionaries:

- *Avoidance*: the fact of avoiding, from the verb 'to avoid'.
- *Confinement*: the act of confining (keeping within confines, limits, edges).
- *Consistency*: keeping together, solidity.
- *Contention*: argument, dispute, from the Latin *contentio* (struggle, fight).
- *Identifier*: 'that which identifies'.
- *Latent*: in a state representing latency (see 'latency').
- *Latency*: time elapsing between a stimulus and the reaction to the stimulus.
- *Recessive*: persisting (still active) in the latent state.

## 1.4 From Concept to Reality

The foundations of the CAN concept were laid, at a given period, on the basis of a state of the market and technology, with an attempt to extrapolate future trends, in respect of the systems and the techniques and technologies to be implemented to make them a reality.

The question that you may well ask is: Why CAN and not another protocol?

So here are some further explanations which will help you to understand; let us begin with a few words about 'site buses'.

### *1.4.1 The site bus market*

For a long time, many companies have been obliged to develop and suggest their own solutions for resolving substantially similar (or related) problems raised by links and communications between systems.

Nearly all of these solutions are known as 'proprietary', and because of the different interests of each of the companies involved, this has led to a significant fragmentation of the market. This diversification of the solutions and the small or medium quantities covered by each of them have therefore, unfortunately, led to a decrease in the quantities of specific components to be developed and produced in order to create a standard on this basis. Among these solutions, many serial buses and links have existed for a long time to resolve similar or virtually similar applications. Regardless of the markets supported (industrial or motor vehicle), the best known names (listed in alphabetical order to avoid offending certain sensitivities) are Batibus, Bitbus, EIB, FIP bus, J1850, LONwork, Profibus, VAN, etc.

Figure 1.7 provides a summary of the numerous buses for covering some or all of these applications.

So why use CAN, which is also a proprietary system?

### *1.4.2 Introduction to CAN*

The size of the motor vehicle market (several million units per year) has led component manufacturers to design integrated circuits handling the CAN protocol. It has also

| | LON | CAN | FIP | Profibus | Bitbus | Interbus |
|---|---|---|---|---|---|---|
| **General** | | | | | | |
| Name of protocol | Local operating network | **Controller area network** | Factory instrumentation protocol | Process field bus | Bitbus | Interbus |
| Fields of application | Industrial domestic automation | Motor vehicles, industrial | Industrial | Industrial | Industrial | Industrial |
| Original designer | Echelon Corp. | **R. Bosch** | Schneider | Profibus Consortium | Intel | Phenix contact |
| Country | United States | **Germany** | France | Germany | United States | Germany |
| Users | LONUSERS | **CIA – CAN in Automation** | FIP-Club International | Profibus Nutzer Organization | | |
| **Specific characteristics of the physical layers** | | | | | | |
| Topology | Bus | **Bus** | Bus | Bus | Bus | Ring |
| Media supported | Diff. pairs RS 485, CPL infrared | Diff. pairs RS 485, Infrared CPL Optical fibres | Diff. pairs RS 485 | RS 485 | RS 485 | RS 485 |
| Max. length (m) | 500 | **40** | | 1200 | 1200 | |
| For a bit rate (Mb) of | 1.25 | 1 | | | | |
| **General characteristics of the protocols** | | | | | | |
| ISO/OSI layers | 2 to 7 | 1 and 2 | 2, 7 | 1, 2, 7 | 1, 2, (5), 7 | 1, 2, 7 |
| Min. bit rate (kbs) Max. bit rate (kbs) | 4.9 1250 | **1000** | 31.25 1000 | 9.6 500 | **62.5** 2400 | 300 300 |
| Max. no. of bytes in message | 249 (+7) | 8(+2) | | 246 (+2) | 128 (+2) | |
| Priorities | 128 | **2048** | | 2 | | |
| Application layers | Lontalk | Numerous | FIP | FMS | RAC | PMS |
| **Electronic components (stand-alone or microcontroller) for protocol handlers** | | | | | | |
| References | 3150, 3120 | [a] | | | 8 × 44, 80C 152 | IPMS |
| Manufacturers | Motorola, Toshiba | [a] | | Siemens | Intel | |

[a] The list is too long to be shown here.

**Figure 1.7**

succeeded in reducing costs significantly, which is not necessarily the case with other buses, which are frequently restricted to smaller scale applications without the benefit of dedicated components or keen prices. Manufacturers have therefore considered the arrival of the CAN bus from another viewpoint (the very special viewpoint of the performance/cost ratio) and have suddenly decided that this type of bus fully satisfies them.

It is surprising what can be achieved by taking a fresh look!

### *1.4.3 The CAN offer: a complete solution*

The strength of the CAN concept, its promotion and its success, lies in the motivation of the people involved in the project to provide the users, at just the right time, with everything they need and cannot easily find elsewhere to deal with the same problems.

From considerable personal experience, I know that this requires a lot of work, but it always brings results when it is well organized. Indeed, all the components for the intelligent development of CAN products have appeared and become available within 2 or 3 years (a period well suited to the market); these components are

- a precise and complete protocol, spelt out clearly;
- the ISO standards for motor vehicle applications;
- competing families of electronic components;
- development of awareness in the industrial market;
- technical literature (articles, books, etc.);
- conferences and congresses for increasing awareness, training, etc.;
- formation of manufacturing groups (CiA, etc.);
- supplementary recommendations for the industry, concerning for example the sockets (CiA) and the application layers (CiA, Honeywell, Allen Bradley, etc.);
- tools for demonstration, evaluation, component and network development, etc.

In short, it all helps!

And now, before we move on to purely technical matters and the applications of the CAN bus, here by way of light relief are a few lines about the history of the protocol.

## 1.5 Historical Context of CAN

A little history always helps us to understand the present, so let us spend some time on this.

Since the early 1980s, many electronic systems have appeared in the car industry. Essentially, this has taken place in three major steps:

- the era when each system was completely independent of the others ... *and everyone lived his own life*;
- a period in which some systems began to communicate with each other ... *and had good neighbourly relations*;
- finally, our own era when everyone has to 'chat' with everyone else, in real time ... *'think global', the world is a big village.*

But let us go back to the beginning.

In those days ... at the start of the 1980s, we had to think about managing future developments. It was in this spirit that, shortly after the appearance of the I2C and D2B 'serial type' buses in the market, many companies concerned with industrial applications (public works, etc.) and some major car manufacturers became interested in communication systems operating (almost) in real time between different microcontrollers, especially for engine control, automatic transmission and antiskid braking systems. For several years, these companies tried to fill the gap by attempting to combine I2C with D2B, in the absence of dedicated solutions.

This was initially done with the aid of conventional UARTs, like those found in ordinary microcontrollers available in the market. Sadly, it soon became clear that although these had very useful properties, they could only reach one part (mainly the passenger area of the vehicle) of the target (the whole vehicle). This form of communication management provided no support, or very poor support, for multimaster communications, and other devices had to be devised and developed. Moreover, their maximum bit rate and the security of the information carried were inadequate. Consequently, there was a gap due to the absence of a bus capable of providing 'fast' multimaster communications, operating over a 'correct' distance and 'insensitive' to its carrier.

In 1983, on the basis of certain performance levels achieved by the I2C asymmetric bus (bitwise arbitration), the D2B symmetric bus (differential pair) and many other factors, the leading German motor components company R. Bosch GmbH took the decision to develop a communication protocol orientated towards 'distributed systems', operating in real time and meeting all the company's requirements. This was the start of the development of CAN. It would be a great untruth to claim that the sole intention was to keep this system under wraps and not use it, when the company was one of the world leaders in motor components.

At this point in the story, let me emphasize the first significant point. This is that the management of R. Bosch GmbH decided to become directly involved in a major way, by cutting through hierarchical relationships and stimulating the development, with the result that the major customers of Bosch were made aware of the state of progress of the project at the start of 1984.

The second major point is that a motor component manufacturer, howsoever large, cannot achieve anything with a concept of this type unless it forms a partnership with universities (particularly the Fachhochscule Braunschweig at Wolfenbüttel, where Professor Wolfhard Lawrenz was later to suggest the name 'controller area network or CAN') and with the creators of known integrated circuits, to make these 'silicon dreams' a reality. Each has its own part to play. These partnerships were established in 1985 with the American giant Intel (useful for ensuring the vital subsequent promotion in the United States) and then, some time later, with Philips SemiConductors, to fill out the European side of the picture. Of course, since that time, many component manufacturers have followed in their footsteps (Siemens, later Infineon, Motorola, later Freescale, NS, TI, MHS, later Temic then Atmel, most of the Far Eastern producers, etc.). In short, the forces were massing.

In the spring of 1986, it was finally time to reveal the new system to the world. The first presentation about CAN was made exclusively to members of the well-known SAE (Society of Automotive Engineers). Where was this? Where else but in the United States, where the car is king, and in a very significant city, Detroit, the cradle and stronghold of the American automobile: if this showed some bias towards the United States (mainly among the car manufacturers, the SAE Trucks and Bus Committee), the Europeans would soon become interested.

Following this, in 1986, everyone converged on the ISO, asking them to. .. set the standards (what else would ISO do?). Any practitioner at the ISO will tell you that the standardization of a protocol requires many years, but forceps can also be useful in helping with the delivery of the little newcomer!

Finally, in the middle of 1987, the reality took shape in the form of the first functional chips, and in 1991 a first top-range vehicle (German) rolled off the production line, complete with five electronic control units (ECUs) and a CAN bus operating at 500 kbit $s^{-1}$.

During the same period, the 'internal' promotions (for motor applications) and 'external' promotions (for industrial applications) were created and actively supported by the SAE and OSEK for the motor industry and also by CAN in Automation (CiA – a group of industrialists, mainly German but subsequently international) for other fields.

So it was the 'engine' of the motor industry that brought this concept to light, but its typically industrial application is not limited to this market. Without ignoring these first applications, therefore, I will try to counteract the general view that CAN is a protocol designed purely for the motor industry, by showing you that it is a highly efficient system for fast local networks.

### *1.5.1 CAN is 20 years old!*

By way of documentation, the table below shows the main stages of development of CAN during its first 20 years of life.

| | |
|---|---|
| 1983 | Start of development of CAN at R. Bosch GmbH. |
| 1985 | V 1.0 specifications of CAN.<br>First relationships established between Bosch and circuit producers. |
| 1986 | Start of standardization work at ISO. |
| 1987 | Introduction of the first prototype of a CAN-integrated circuit. |
| 1989 | Start of the first industrial applications. |
| 1991 | Specifications of the extended CAN 2.0 protocol:<br>• part 2.0A – 11-bit identifier;<br>• part 2.0B – 29-bit identifier.<br>The first vehicle – Mercedes class S – fitted with five units communicating at 500 kbit $s^{-1}$. |
| 1992 | Creation of the CiA (CAN in Automation) user group. |
| 1993 | Creation of the OSEK group.<br>Appearance of the first application layer – CAL – of CiA. |
| 1994 | The first standardization at ISO, called *high and*<br>*PSA (Peugeot Citroën) low speed*, is completed. and Renault join OSEK. |
| 1995 | Task force in the United States with the SAE. |
| 1996 | CAN is applied in most 'engine control systems' of top-range European vehicles.<br>Numerous participants in OSEK |
| 1997 | All the major chip producers offer families of CAN components. The CiA group has 300 member companies. |
| 1998 | New set of ISO standards relating to CAN (diagnostics, conformity, etc.). |
| 1999 | Development phase of time-triggered CAN (TTCAN) networks. |
| 2000 | Explosion of CAN-linked equipment in all motor vehicle and industrial applications. |
| 2001 | Industrial introduction of real-time time-triggered CAN (TTCAN) networks. |
| 2003 | Even the Americans and Japanese use CAN! |
| 2008 | Annual world production forecast: approximately 65–67 million vehicles, with 10–15 CAN nodes per vehicle on average. Do the sums! |

### *1.5.2 The CAN concept in a few words*

It was decided from the outset that CAN should be capable of covering all communication applications found in motor vehicles, in other words, it should carry and multiplex many types of messages, from the fastest to the slowest.

Because of its origin in the car industry, CAN was designed to operate in environments subject to a high level of pollution, primarily due to electromagnetic disturbance. In addition to the transmission reliability provided by an efficient error detection mechanism, CAN has multimaster functionality to increase the possibility of providing fast recovery from errors after their detection. Also, in the case of bus access conflicts, the bitwise arbitration principle adopted makes it possible to provide a 'non-destructive' arbitration method when several stations attempt to send a message simultaneously. This means that, in a contention situation, one station will always gain access to the bus (namely the station sending the highest priority message) and will then finish the communication on its own. Because of this method, none of the bus communication capacity (the bus bandwidth) is lost in the management of bus access conflicts.

The use of this type of non-destructive arbitration and hierarchically ranked messages also makes it a simple matter to meet, in real time, the response times required for the control of systems in which the bus bit rates are not very high.

The disadvantage of this bitwise arbitration method lies in the fact that the maximum length of the network is tied to the chosen bit rate, a high bit rate corresponding to a short-distance network.

In principle, in order to minimize the electromagnetic noise which is mainly due to fast edges of electronic signals during communication on the bus, the communication bit rate should be as low as possible.

See Chapter 2 on the CAN protocol for more details.

### *1.5.3 The market for CAN*

In parallel with the advance of motor applications, a large number of industrial applications – to be fully discussed in this book – have emerged. This success is largely due to the rapid appearance in the market of inexpensive electronic components (integrated circuits) for managing the communication protocol (because the CAN protocol only covers layers 1 and 2 of the ISO/OSI model).

By way of a summary, in the middle of 2004, Figure 1.8 shows the cumulated total of nodes in the market and a comparison with other industrial solutions.

#### **The motor vehicle market**

The motor vehicle market is most important in terms of quantity, simply because of the annual volume of nodes implemented per vehicle produced. Moreover, in future, at least in Europe, the number of CAN nodes on each vehicle will be approximately 5–10 for the engine system of the vehicle, about 10 for the body part, and finally 15, 20, 25 or more for the passenger compartment, depending on the level of equipment and comfort features fitted to vehicles.

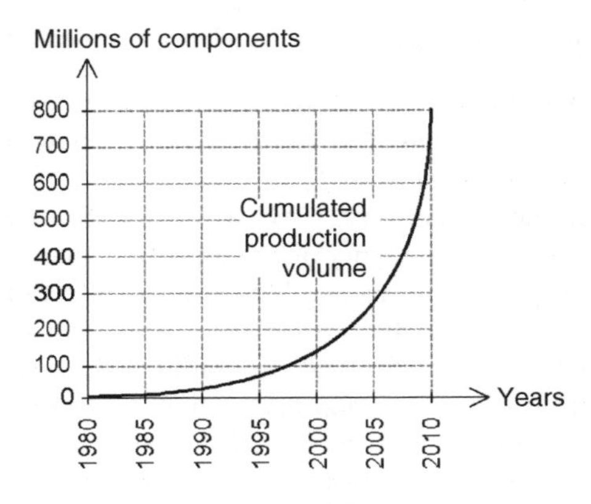


**Figure 1.8**

### The industrial automation market

Although the relative volume of this market is smaller than that described above, it still represents an important amount in absolute terms. Even so, it should be noted that this market was the first to be established on an industrial scale, because of the short lead times required by small and medium businesses to respond to market demand (the development times for industrial applications generally vary from 6 to 12 months, whereas in the motor industry they are of the order of 2–3 years). In 1996, for the last time, the quantity of nodes produced for the automation market exceeded the number for the motor industry market.

Chapters 4 and 5 will describe the wide field of industrial applications which have given strength and life to this concept.

### Application layers and support organizations

Today, for all these markets, there is only one communication protocol: CAN. Regarding the application layers (see the book *Le bus CAN – Applications* [*The CAN Bus – Applications*]), there are many competing proposals, adapted to different application criteria. To gain a clearer idea, let us now examine a list of the main names.

For industrial applications, the principal ones are

- CAL, produced by CAN in Automation,
- CANopen, produced by CAN in Automation,
- DeviceNet, produced by Allen Bradley–Rockwell,
- SDS (smart distributed systems), produced by Honeywell,
- CAN Kingdom, produced by Kvaser,

and, for motor vehicle applications:

- OSEK/VDX, produced by OSEK (open systems and interfaces for distributed electronics in car group),
- J 1939, produced by SAE.

It should also be noted that many development, promotion and standardization organizations (such as CiA – CAN in Automation, ODVA, OSEK, etc.) have sprung up during the same period.

Finally, when a system is devised, it is quite normal for one or more similar competing systems to appear. This is just what happened with the advent of many similar types of bus in the United States, Japan and France (for example, the VAN bus in France, which was devised several years ago by PSA but which had its industrial development halted in 2004). This has been a brief summary, in an intentionally lightweight form, of the historical context of this bus (which is actually a very serious subject). In reading this you would have understood how certain joint technical and marketing approaches can sometimes make a concept into an industrial success story.

## 1.6 Patents, Licences and Certification

To conclude this chapter, a few words in passing concerning patents, licences and standards are offered to you by way of light relief, before the lengthy description of the CAN protocol and its numerous physical implementations.

### *1.6.1 Documents and standards*

Now we come to the matter of standardization: We must comply with certain terms and be accurate in their definition.

#### The original document

The original CAN protocol is described in a document issued by R. Bosch GmbH (the latest edition is dated 1992) and as such it is not an international standard. This document describes the protocol without consideration of all the lower physical layers and all the higher level software application layers. It covers layer 2 of the OSI (Open Systems Interconnect) model in full (LLC and MAC) and part of layer 1 (PLS), giving free rein to all possible applications. On its appearance, one of the aims of this document was to specify layers 1 and 2 of the OSI model without delay, to enable the chip producers to create suitable electronic components as soon as possible to handle the protocol, without prejudicing future applications and while leaving the way open to many new fields.

#### ISO standardization

Over the years, the original CAN documents were filled out and submitted to the ISO so that official international standards could be used as a reference by all those who wished to adopt this protocol. These documents have also attempted to comply where possible with the structure of the OSI communication model. At present (2007), the principal standards relate to motor vehicle applications and are known by the following references:

- ISO 11898-x – road vehicles – interchange of digital information. This is the cornerstone of the CAN standard, consisting of five documents:

- ISO 11898-1 (*data link layer and physical signalling*);
- ISO 11898-2 (*high-speed medium access unit*);
- ISO 11898-3 (*low-speed fault-tolerant medium-dependent interface*);
- ISO 11898-4 (*time-triggered CAN*);
- ISO 11898-5 relates to high-speed CAN and low-power applications.

The highly detailed documents of parts 1 and 2 faithfully reproduce all the text proposed by Bosch (LLC, MAC, PLS), in other words, the former ISO 11898 standard dating from 1993, while supplementing it with much other material mainly concerning the 'media' part of the physical layer (PMA), not included in the original document. The same applies to part 3, which describes certain forms of bus dysfunction in relation to the media recommended in the documents (wires short-circuited, joined to the power supply, wires cut, etc.).

- ISO 11519-x (note that this standard has been superseded by ISO 11898-3):
  - ISO 11519-1 – general and definitions – road vehicles – low-speed serial data communications;
  - ISO 11519-2 – low-speed CAN – road vehicles – low-speed serial data communications.

Many other non-ISO documents are associated with these, especially in relation to the application layers such as CiA, CANopen, SDS, DeviceNet, etc., and in relation to the specific connectivity.

This global set of documents thus enables all designers to develop marketable CAN products, at least as regards the electronic operation of the system (but note that many other minor points of detail are also specified in these documents).

Although they are ISO standards, these documents give no information on the types of connectors recommended or on the higher level application layers. You may say that this is not their role, and you will be (partially) correct. Indeed, in the absence of any law or formal implementation order, anyone is free to obey or disregard the standard. Moreover, regardless of the existence of these standards, car manufacturers are entirely free to do what they want in their vehicles. Clearly, it must be recognized that it is more practical to have a common standard if we wish to use parts from different equipment and component manufacturers. This is another story, largely outside the remit of this book. Briefly, these standards form a good base for future possibilities of interconnection with products from different sources.

For information, no official document mentions the terms 'full CAN' and 'basic CAN' – this should be remembered.

### The CiA

Another event occurred when many companies became interested in using the CAN protocol and wanted to create industrial applications. These companies (major industrial groupings or small or medium businesses) wished to use the bus but found themselves short of resources when times were hard and, realizing that the future lay with CAN, decided to form a group in order to fill some of the gaps in the documents cited above and ensure that they had all the necessary elements for communication. Thus, the CiA (CAN in Automation – international users and manufacturers group) was set up in March 1992, with the following mission (quote): 'to provide technical, product and marketing intelligence information in order to enhance the prestige, improve the continuity and provide a path for advancement of the CAN network'.

First, this group is international rather than purely German, an enormous advantage for the promotion of the protocol. Second, it includes a large number of companies (more than 500) from all sectors, which complement each other and which have the aim of producing CAN products with very good possibilities of interoperation.

In short, it was a priority for all these members to specify all the physical interfaces (cables, connectors, screens, etc.) and higher level application software (level 7 of the OSI/ISO layers). Like a 'little ISO', the participants created technical committees and working groups, leading to the issue of a set of 'CiA recommendations', filling practically all the blank areas left by the ISO.

These CiA recommendations are called CiA draft standards (CiA DS xxx) for the physical part and CAN application layers (CAL) for the software layers. Other more specialized groups have acted in the same way regarding the high-level application levels orientated towards certain fields of application. Examples are Allen Bradley and Honeywell, which have developed the DeviceNet and SDS layers, respectively, for industrial applications (see Chapter 5).

### *1.6.2 Patents*

Many patents have been filed since the development of the CAN protocol and its physical implementations. Some of these have led to epic arguments about their priority, but everything has now been settled. Patents are only one of the external aspects of industrial property protection and are generally only used to make some money, sooner or later, out of the conditions of licence use by friends and competitors in our splendid profession. This has now been done...

The sacrosanct rule of the ISO is that when a standard is officially published – as in the case of CAN – this does not in any way affect the licences and royalties payable to those entitled to them, but these royalties must be 'reasonable' and the licences must be 'non-discriminatory', as expressed by the acronym RAND (reasonable and non-discriminatory). I do not intend to detail the licence fees or royalty payments here, but I can tell you that they are not negligible. You should also know that, as ordinary users, you will certainly not be affected by this. Also, as with many other protocols, it is practically always necessary to use specific integrated circuits to handle CAN. Once again, therefore, it is the component manufacturers (chip producers) who are mainly concerned with licensing problems, not the end users.

### *1.6.3 Problems of conformity*

It is all well and good to have components available for which the manufacturers have paid their licence fees, but one question remains: How can we know if the circuits offered in the market really conform to the standards?

In some situations, this can be a very important question. Consider, for example, the exact moment when you need to brake your car, if the ABS is controlled via the CAN bus. Suppose just for a moment that a bit is not in its proper place at that instant: Who is liable for the accident that may occur as a result? Manufacturers, equipment makers, component producers, the standard?

This underlines the crucial problem of the conformity of protocol handlers, line driver interfaces, etc. with the protocol. For some protocols, there are independent organizations (laboratories) charged with certifying that the product concerned can perform its function or with establishing conformity, approval, a label, etc.

At the present time, as regards the CAN protocol, this type of procedure exists in the form of a document, *CAN Conformance Testing*, reference ISO 16845, which is actually a specific application of the ISO 9646-1 test plan. Before this document existed, Bosch supplied what is commonly called a 'reference model' to its licence holders to overcome the problem. This is a software model providing a battery of tests to be conducted on a protocol handler. If the protocol handler responds correctly to all these tests, it will be considered as conforming to the protocol. This also means that, having passed these tests, the different implementations of protocol handlers will also be able to communicate correctly with each other.

The manufacturer of protocol handlers is therefore responsible for assuring himself, and then ascertaining via an independent certification organization that his product actually passes all the tests proposed by the reference model and the ISO 16845 standard . . . and the users are responsible for making it operate correctly!

### *1.6.4 Certification*

A system can only operate correctly if it is consistent, in other words, if it has a real uniformity of operation. For example, in the present case, all the nodes of a network must react in the same way in normal operation to the same problem, the same fault, etc. There are two possible solutions: either to be the sole owner of a proprietary solution, keeping everything under control by a licensing system, or, for simple market reasons (availability of components, wider range of components, competitive pricing, etc.), to open the solution up to a standard, preferably ISO.

To maintain the technical integrity of the solution, it is a common practice to propose a standard describing in detail the functionality that is to be applied and that is to be provided, but leaving the concrete physical and electronic implementation to the choice of each user. Once a standard has been written, voted on, commented on, corrected and finally issued, we can start work . . . but sometimes the standard may simply be shelved. In principle, compliance with a standard is not compulsory. If so inclined, the government can stipulate its use by passing a law or an order. The same applies to a car manufacturer, who can decide to accept electronic equipment only if it conforms to this or that standard.

But who says that it conforms? This is the sensitive area of the application of a standard! Anyone can say that his product conforms to a standard, if he has made measurements, using his own equipment, to ensure that the values recommended by the standard are matched: but his word is only binding on himself. Unfortunately, this situation, although highly commendable and justifiable, does not satisfy everybody. This is because, on the one hand, the standard is not always very clear and may be subject to different interpretations, and, on the other hand, it is sometimes difficult to measure certain parameters in a reproducible way by the measurement methods in use. To avoid these problems and the ensuing lengthy arguments, it is a very common practice to write a second standard (!), entitled 'Conformance Tests', describing in detail the measurement methods and instruments used for the standard in question. Finally, it is also possible to certify independent organizations for making these measurements so that they can deliver a document (declaration, certificate, approval report, authentication report, etc.) certifying the results of these famous conformity measurements.

To return to the subject of CAN, Bosch was for a long time the only organization capable of giving its support in matters concerning protocol handling, based on its reference model mentioned above. Where the lower 'high speed' layer was concerned, everyone relied on the data sheets of different component manufacturers.

The problem became more critical with the appearance of the LS FT CAN (low-speed fault-tolerant CAN) physical layer. This is because in this case, as described in detail in Chapter 4, all the nodes of the network must signal the same type of fault at the same instant. In view of this, the GIFT (generalized interoperable fault-tolerant CAN transceiver) working group was set up. This group subsequently transferred its work to the ISO, participating strongly in the development of the ISO 11898-3 standard which describes the functional part of LS FT CAN, raising the problems of certification described above. As it takes a long time to convert a document to an ISO standard (of the order of 36–48 months), the GIFT group decided, via a special ICT (International Transceiver Conformance Test) committee, to establish a conformance test document itself, containing more than 100 specific tests, in order to obtain professional validity for this practice. This group also authorized the independent company C&S, managed by Wolfhard Lawrenz and already closely involved in the GIFT and ICT, to conduct these tests.

I will now proceed to examine the more technical matters and explain the operation of the CAN protocol.

# 2

# CAN: Its Protocol, Its Properties, Its Novel Features

In view of the intrinsic power of the CAN (controller area network) protocol and its different possibilities and varied implementations, I have decided to spread its description over several chapters. The key to this presentation is the way I have chosen to divide it.

The description of CAN consists of several parts, which are carefully presented so that they are easier to read and understand than the official documents, which can sometimes be baffling.

This chapter is purely concerned with the protocol, including

- the definition of the protocol,
- the main principles,
- the error processing,
- the rest of the frame,
- CAN 2.0B – description.

Chapters 3 and 4 are entirely devoted to what is generally called the 'physical layer' and its implementation in a medium. Chapter 3 will then give you all the necessary information about the bit structure, synchronization, the medium and the connection to the medium and the problems of propagation. Finally, in Chapter 4, I will discuss the large number of implementations of the physical layer in different media, such as copper wire, optical fibre, current power lines (CPL), etc.

## 2.1 Definitions of the CAN Protocol: 'ISO 11898-1'

Let us start by describing and making a few simple comments on the original document issued by Bosch and then reproduced, with minor changes, in ISO 11898-1.

*Multiplexed Networks for Embedded Systems: CAN, LIN, Flexray, Safe-by-Wire...* D. Paret

In this document, with the aim of ensuring transparency of the product and greater flexibility of implementation, the CAN protocol (providing a 'serial asynchronous link with clock and data time multiplexing') is described and divided essentially according to the standard division into different layers of the ISO/OSI (International Standardization Organization/Open Systems Interconnect) model.

The specification (version 2.0) comprises two parts, called 'A' and 'B':

- Part A describes the most common 'standard' CAN frame. This frame supports 'only' 11 identifier bits, and it will be described in detail in this chapter.
- Part B is intended to describe the CAN frame in its 'extended format'. As it was thought to be insufficient for some applications, the identifier value was changed from 11 to 29 identifier bits.

At the end of the chapter, I will point out the main differences between the type 2.0A and 2.0B frames. Note that the CAN 2.0A and 2.0B versions were designed to provide upward compatibility with any earlier version of the protocol.

As an appetizer, here are the main properties of the protocol structure of the CAN bus:

- hierarchical ranking of messages,
- guaranteed latencies,
- flexible configuration,
- reception of multiple sources with time synchronization,
- multimaster operation,
- error detection and signalling,
- automatic retransmission of corrupted messages when the bus returns to the idle state,
- distinction made between timing errors and permanent non-functionality at a node,
- automatic disconnection of faulty nodes,
- etc.

Figure 2.1 summarizes the division of the ISO/OSI standard and shows very clearly the differences in coverage between the actual content required by the ISO/OSI standard and the content of the CAN protocol reference document. Clearly, instead of covering all seven layers, CAN is concerned only with the whole of layer 2 (the data link layer) and a considerable part of layer 1 (the physical layer). However, this diagram must be supplemented with other information.

### *2.1.1 Content of the different ISO/OSI layers of the CAN bus*

Let us now rapidly review the content of the ISO/OSI layers of the CAN bus.

#### 'Data link' layer – layer 2 – of the OSI/ISO model

The data link layer – layer 2 – is subdivided into two sublayers:

- MAC (medium access control) sublayer

| No. of layer | ISO/OSI model | CAN protocol |
|---|---|---|
| 7 | Application | User specified |
| 6 | Presentation | Blank |
| 5 | Session | Blank |
| 4 | Transport | Blank |
| 3 | Network | Blank |
| 2 | Data link | CAN protocol (with free choice of medium) |
| 1 | Physical | |

**Figure 2.1**

The MAC sublayer is the core of the CAN protocol. Its function is to present messages received from the LLC sublayer, and to accept messages for transmission to the LLC sublayer. The MAC sublayer is responsible for

- message framing,
- arbitration,
- acknowledgement,
- error detection and
- error signalling.

**Important note:** The MAC sublayer is supervised by a management entity, called 'error confinement', which can be described as a self-checking mechanism for distinguishing short-duration faults from permanent failures.

• LLC (logical link control) sublayer
The LLC sublayer is concerned with the following:

- message filtering,
- overload notification,
- the error recovery procedure.

In general (and in theory!), once it has been specified, this transfer layer no longer has to be modified for any given protocol.

**'Physical' layer – layer 1 – of the OSI/ISO model**

The physical layer – layer 1 – comprises three sublayers:

- PLS (physical signalling),
- PMA (physical medium attachment),
- MDI (medium-dependent interface).

The physical layer specifies the way in which the signal is transmitted, and therefore it has the role of ensuring the physical transfer of bits between the different nodes, according to all the properties (electrical, electronic, optical, etc.) of the system. Clearly, the physical layer must be the same for each node within a single network.

Theoretically, this layer is responsible for

- bit representation (coding, timing, etc.),
- bit synchronization,
- definition of the electrical and optical levels of the signals,
- definition of the transmission medium.

**Important note:** The Bosch reference document on the CAN protocol describes only the detailed bit representation (the PLS part) because within the CAN specifications, the characteristics of the driver/receiver of the physical layer are not defined, so that the transport medium and the signal levels can be optimized for any given application.

The operation of such a system is often explained with the aid of layers called 'Object' and 'Transfer' (not shown expressly in the figure), which have the task of processing all the services and functions of the data link layer (layer 2) defined by the ISO/OSI model.

### 'Object' layer

This has the principal task of filtering the messages and processing messages and statuses. Its objectives are therefore

- to find which message has to be transmitted,
- to decide which messages received via the transfer layer are being used,
- to produce an interface with the application layer in relation to the system hardware.

### 'Transfer' layer

The main task of this layer is to deal with the transfer of the protocol, that is

- to manage and control the frame formatting, the bit rate and the temporal conformity of the transfer;
- to arbitrate bus conflicts;
- to verify the absence or presence of errors;
- to signal the different types of error, if any, as well as confinement faults;
- to validate messages;
- to acknowledge the messages.

Besides many other tasks of this layer, it is also responsible for deciding whether the bus is free, so that a new transmission can be started, or if the reception of an incoming message is in progress at that instant.

### 'Application' layer – layer 7 – of the OSI/ISO model

A last major point concerning the reference document and the ISO documents for the CAN protocol: this layer (layer 7) is totally blank. Figure 2.2 clearly shows the functions of each layer described above.

## *2.1.2 Properties of CAN and a small specialized glossary*

Here again, as in previous books (see [*I2C Bus*] or [*Access Bus*]), I cannot claim to have invented this protocol, and it must be admitted that the essential contents of this chapter are a translation (which I trust is at least intelligent, structured and embellished with ample notes!) of the official CAN protocol originated by Bosch and transposed to certain ISO standards.

I have taken great care to include the many details of this protocol, because it is frequently necessary to follow the minutiae of the protocol if we are to understand the operation of CAN properly and make appropriate use of it. So, take your medicine, and be aware that sooner or later this chapter will be your constant bedtime reading – but it is certainly not intended to put you to sleep.

As with most protocols, it is necessary to use a vocabulary appropriate to the matter in hand. So here are the ad hoc terms, and their definitions originating from the ISO standards relating to the CAN protocol (ISO 11898-x, see Chapters 3 and 4), adapted to make this book easier to understand.

**Note:** Although there are slight differences in their meanings, the terms *node, station* and *participant* are used interchangeably to denote an assembly connected to the network, thereby avoiding excessive repetition.

### Sending node (station)

ISO: subassembly connected to a communication network, capable of communicating over the network according to a communication protocol.

In the case of the CAN protocol

- a node (station) generating a message is called the 'sender' of this message,
- a station is called the 'sender' until the bus is idle or until the unit has lost the arbitration.

### Receiving node (station)

In contrast to the preceding case, a node (station) is called the 'receiver' of a message if it is not the 'sender' of a message and if the bus is not free.

### Values of the bus

The bus can have one of the two complementary logical values defined not as '0' and '1' as usual, but as what are called 'dominant' and 'recessive' values.

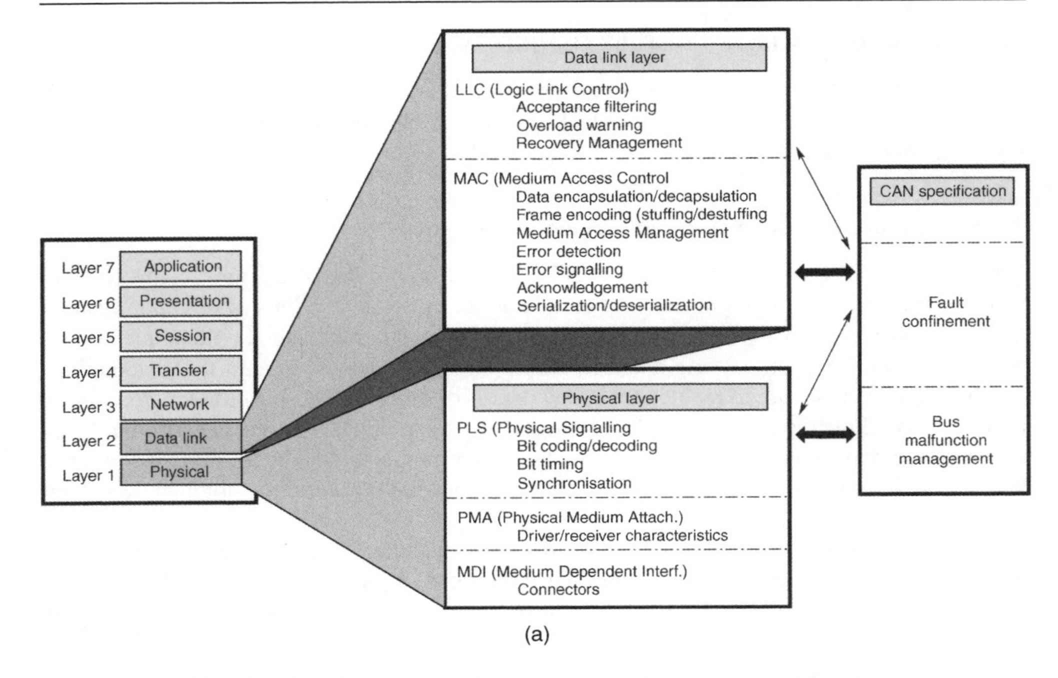
Layer 7 Application
Layer 6 Presentation
Layer 5 Session
Layer 4 Transfer
Layer 3 Network
Layer 2 Data link
Layer 1 Physical
Data link layer
LLC (Logic Link Control)
Acceptance filtering
Overload warning
Recovery Management
MAC (Medium Access Control
Data encapsulation/decapsulation
Frame encoding (stuffing/destuffing
Medium Access Management
Error detection
Error signalling
Acknowledgement
Serialization/deserialization
Physical layer
PLS (Physical Signalling
Bit coding/decoding
Bit timing
Synchronisation
PMA (Physical Medium Attach.)
Driver/receiver characteristics
MDI (Medium Dependent Interf.)
Connectors
CAN specification
Fault confinement
Bus malfunction management

(a)

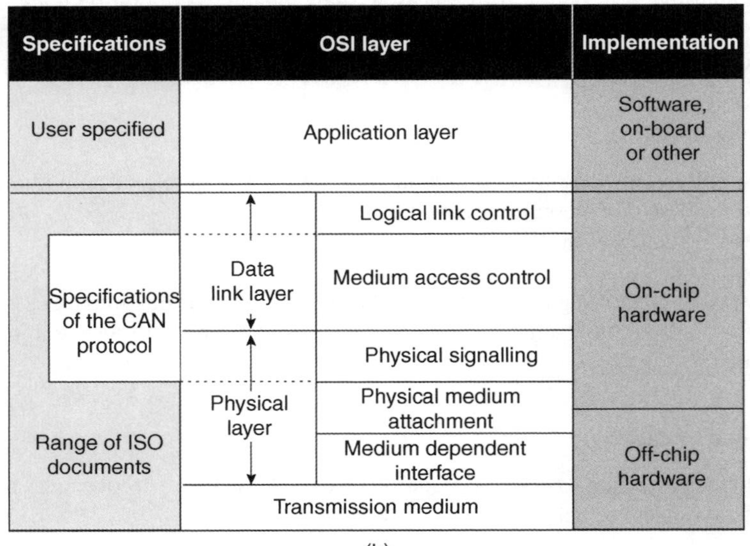
Specifications
OSI layer
Implementation
User specified
Application layer
Software, on-board or other
Logical link control
Data link layer
Medium access control
Specifications of the CAN protocol
On-chip hardware
Physical signalling
Physical layer
Physical medium attachment
Medium dependent interface
Range of ISO documents
Off-chip hardware
Transmission medium

(b)

**Figure 2.2**

In case of a simultaneous transmission of a dominant bit and a recessive bit, the resulting value of the bus will be 'dominant'. You may find this unconventional, not to say nonsensical.

In fact, the CAN protocol does not define (or specify), in any way, the physical medium (copper wire, infrared, radio link, optical fibre, CPL, etc.) in which the CAN physical layer may be implemented. It is therefore difficult to define high and low levels, or '1' and '0', and therefore the writing or specification of an AND-wired element, for example, becomes a mystery – unless we speak of dominant and recessive bits.

Figure 2.3 shows some specific examples of these values and shows how, in conventional electronics for instance (with good old-fashioned gates), we could say that the dominant state for the AND-wired element would be the logical state '0' and the recessive state would be the logical state '1'.

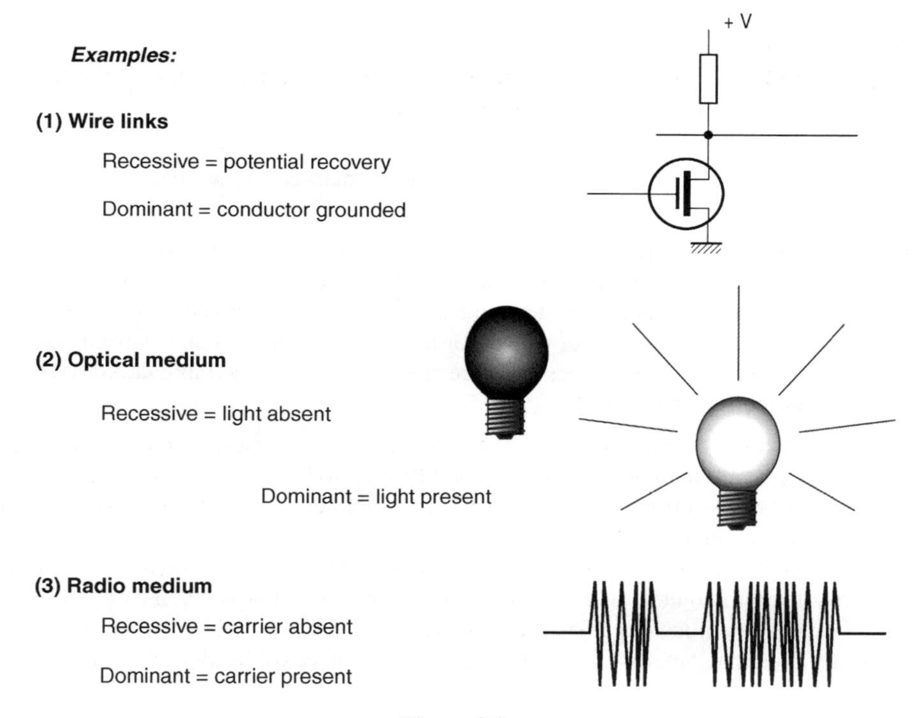


**Figure 2.3**

## Bit rate of the bus

According to ISO, the bit rate of the bus is the number of bits per unit of time during the transmission, regardless of the bit representation.

The CAN bit rate (often improperly called the 'speed') can differ from one system to the next, but must be uniform and constant for any one network. You should therefore pay attention to the internetwork compatibility, because several networks often coexist within a single system, making it necessary to create gateways between them.

### Messages/frames/format

The information carried on the bus is sent in a specified format, and its maximum length is limited (and therefore the maximum time is also limited for a given bit rate).

When the bus is free, any participant can initiate a new transmission.

### Information routing

This point is very specific to the CAN concept. In a CAN network, a node does not have to consider information relating to the configuration of the system it has found its way into. This has a number of important consequences

- Flexibility of the system
  Nodes can be added to the network (or withdrawn), without the need to modify the hardware and/or software of any nodes or application layers.
- Message routing
  For the message routing, the content of a message is 'marked' by an identifier.
- Identifier
  The identifier does not indicate the destination of the message in any way but describes (in the sense of 'giving a general or precise idea of') the content and meaning of the data. This implies that each node is capable of deciding whether or not the message carried on the bus is relevant to it. This function is (or will be, or must be) executed by an electronic message filtering device. In other words, messages are sent out into impenetrable darkness, in case somebody out there is interested!

**Notes:** There is no document (original, ISO/OSI or other) indicating or recommending a particular structure for the message types and architecture to be used. It is useful to know that the protocol gives no directions as to the possible or recommended values to be assigned to the bits forming the identifier.

To provide a degree of flexibility to the design of networks in which numerous participants may need to connect to each other, some corporate groups (such as textile manufacturers) have sometimes organized themselves to draw up 'table' of identifiers and thus curb the potential 'anarchy', which a large range of possible identifiers represent by definition.

**Example of a choice of identifiers.** For guidance, the following lines indicate an example of identifier systems that can be used.

Very frequently, in a CAN network, a node is physically located in a specific place which is reference (for example, in building D, on the 3rd floor, at the north-east corner). Let us suppose that this node has the function, among others, of telling the network about the outside temperature of this place.

If you had to describe this parameter verbally to a friend so that he would understand you clearly, you would say 'the outside temperature at the north-east corner of the 3rd floor of building D' and you would have formally identified this parameter. The identifier associated with the message indicating the value of the parameter (e.g. 12°C) could then be presented in the following form:

- 1st identifier: temp_ext_NE_3D,
- 2nd identifier: D3_NE_temp_ext, in other words, in reverse order, if you so wish.

In fact, in the first case, we have presented one of the general ways of constructing an identifier: this is the standard 'subject/source' description – but we might just as well have used the second form.

The choice of the order of the information and the architecture of the message system will be made when you understand that the arbitration procedure is a function of the weighting of the bits assigned to the identifier.

To return to the first case, we can break this down as follows:

| ID-high part Subjects | | ID-low part Sources | |
|---|---|---|---|
| In 6 bits (64 values) (64 parameters) | | In 5 bits (32 values) (32 nodes) | |
| Principal subject | Sub-subject | Principal source | Sub-source |
| Temperature in 4 bits | Outside in 2 bits | D north–east in 3 bits | 3rd floor in 2 bits |

Many other methods can be used to form the identifiers. They depend on the application.

- Multicast
  According to the ISO, multicast is an address mode in which a simple frame can be used to address a group of nodes simultaneously. The broadcast mode – addressing all the nodes – is a special case of the multicast mode, or vice versa.
  The immediate consequence of the message filtering concept is that all the nodes receive the transmitted message simultaneously and that they can all react immediately to it if they so wish.
  Thus, we have a device which is inherently capable of operating in real time.
- Data consistency
  Because of the structure described above – a message can be received and/or used by one, all or none of the nodes – the data consistency of a CAN network is ensured by the principles of multicast and error processing.
- Priorities and the concept of bus access
  Because of the bitwise arbitration principle (see below), the identifier (via its content) defines a static priority message before access to the bus.
- Data frame, remote frame and remote data
  The data frame is the frame that carries the data.

The remote frame and the data frame associated with it are marked by the same identifier. The term 'remote' (implying a concept of something happening 'at a distance') arises frequently in the CAN protocol. This is because a node, by sending a remote frame, signals to the other nodes in the network that it wishes to receive data from them in the form of a corresponding data frame.

- Multimaster operation
  When the bus is free, any unit can initiate a transmission. The unit whose message has the highest priority (according to the binary content of its identifier) will gain access to the bus and will transmit its message.
- Arbitration
  ISO: Method consisting in assigning the communication medium (signalling bus) to one of the nodes trying to take it over.
  When the bus is free, although all units can start transmitting at once, if two or more units start simultaneously, this creates a bus conflict which is resolved by a 'bitwise' (non-destructive) arbitration throughout the content of the identifier. This arbitration mechanism ensures that there is no loss of time or information.
  Of course, some troublemakers will bring up the case where two messages are started with the same identifiers! In this case, a data frame takes priority over a remote frame. During the arbitration phase, each sender compares the level of the transmitted bit with the level of the bit that it is supposed to be transmitting (using an internal monitoring device of the bus). If these levels are identical, the node continues to send. If a recessive level is sent and a dominant level is observed on the bus, the unit in question loses the arbitration and must close down and send no more bits (look again at the definition of 'recessive' and 'dominant', if necessary).
- Transmission security
  To ensure the quality and security of transmission, numerous signalling, error detection and self-test devices have been implemented, to increase the reliability of information carried by the CAN bus.

  *Error detection.* As regards the error detection, the following measures have been taken:

  - monitoring the bus: The sender checks whether the electrical level that it wished to set on the bus is actually present there;
  - presence of a CRC (cyclic redundancy code or cyclic redundancy check) – message frame check procedure;
  - the bit stuffing method (see below in this chapter for more details).

*Performance of the error detection system.* As all the global errors, all the local errors in the senders, and up to five random errors distributed in a message are detected, the total residual probability of messages affected by errors is less than $4.7 \times 10^{-11}$.

- Error signalling and error recovery time
  All messages affected by error(s) are signalled at each node by a 'flag'. These messages are then considered to be unacceptable and are rejected. The erroneous messages must be retransmitted automatically.

The recovery time (the time elapsing between the moment of detection of the error and the moment when the new message starts) is not more than 29 bits for CAN 2.0A (31 bits for CAN 2.0B), if no other errors are detected.

- Confinement errors
  These are errors in the 'outline', 'frontier' or 'edge' – in other words, in the confines.
  A CAN node must be able to distinguish short-term disturbances from permanent malfunctions. The nodes considered to be defective then change to 'switched off' mode and are (electrically) disconnected from the network. The whole of one section of this chapter will be devoted to the processing of all these errors.
- Connection points
  The topology of the CAN serial communication link is a bus to which a large number of units can be connected.
  The number of units is theoretically unlimited. (There is no direct relationship between the number of possible identifiers and the number of connectable nodes. Please do not confuse 'identifier field possibility' with 'addresses'!) In practice, the maximum number of units in a network will be determined by the delay time (due to propagation phenomena) and/or the electrical loads they create on the bus.
- Single link channel
  The bus consists of a single bidirectional channel which carries the bits. Resynchronization information can be recovered from the carried data. The form in which the channel is implemented (standard wire, optical link, differential pair, etc.) is not determined in the original document of the CAN protocol. For the actual implementation of a network, you should refer to Chapters 4 and 5.
- Acknowledgement
  All the receivers check the consistency of the received message and acknowledge a consistent message, or 'wave their error flags' if the message is inconsistent.
- Switch to sleep mode and wake-up
  To reduce the power consumption of the system, the CAN components can switch to sleep mode, apparently disconnecting their drivers from the bus.

The sleep mode is terminated by an automatic wake-up, triggered as soon as the bus becomes active, or as a result of internal operations in the station itself.

On waking up, the internal activity of the station resumes, although the MAC sublayer waits for the local oscillator of the station to stabilize and to be synchronized with the bus activity (checking for the presence of 11 consecutive recessive bits). On completion of this process, the bus drivers are enabled to reconnect to the bus.

This concludes the simplified introduction to the main terms and intrinsic characteristics of CAN which, as you would have realized, is highly efficient. For this reason, it is generally more suitable for high-level professional applications. I will provide the details of many of the above points subsequently, when we use them for proposed applications.

Now, let us move on to matters which may be somewhat theoretical, but which can at least be observed on an oscilloscope!

### *2.1.3 The general principles – CAN 2.0A standard frame*

As mentioned above, the physical layer described in the reference document does not in any way specify the physical levels (electrical, optical, etc.) which may be present in the bus. To be more specific, let us say that, in the case of applications using wire links, the communication medium is very often in the form of a differential pair (parallel or twisted), and therefore with well-defined electrical levels, defining the logical levels called '0' and '1'.

When the bus is idle, there is no activity and the oscilloscope screen remains hopelessly blank, as in the case of any asynchronous bus. But it is 'Action Stations' when it starts moving! I will start by describing in detail what you can see when 'standard' CAN 2.0A frames travel along the bus.

### *2.1.4 Message transfer*

The message transfer takes place and is controlled with the aid of four types of specific frames and a time interval separating them. These consist of the following:

- data frames, which transport the data from the senders to the receivers;
- remote frames, sent by a unit active on the bus to request the transmission of a data frame whose identifier has the same value as that of the remote frame;
- error frames, transmitted by any of the units present on the bus as soon as an error is detected on the bus;
- overload frames, used to request a supplementary time interval between the preceding and following data frames or remote frames;
- interframes, which separate (in time) the data frames and remote frames from the preceding frames by an interframe space.

All these frames transport information carried on the bus at the lowest level of the physical layer with the aid of 'bits'.

Before going any further, in order to ensure that certain phases of the different frames and the error processing are better understood, it will be useful to examine the distinctive features of the bits and their coding procedure in the CAN protocol.

### *2.1.5 The CAN bit*

I will return to the specific features of the CAN bit when I describe the problems of the physical layer of the CAN bus (in Chapter 3). For the time being, however, to avoid becoming enmeshed in the toils of theory, let us 'predefine' this concept.

There are 'bits' and 'bits'! To define a bit, we must define its duration and the way its content (its value) is coded in this duration. There are many ways of coding a bit (Figure 2.4): some examples are NRZ (non-return to zero) coding, Manchester biphase, diphase, Miller, duo-binary, etc.

#### **NRZ coding**

For the bit stream of the CAN bus frames, the NRZ coding method has been adopted. This means that, for the total duration of the generated bit, its level remains constant, whether it is

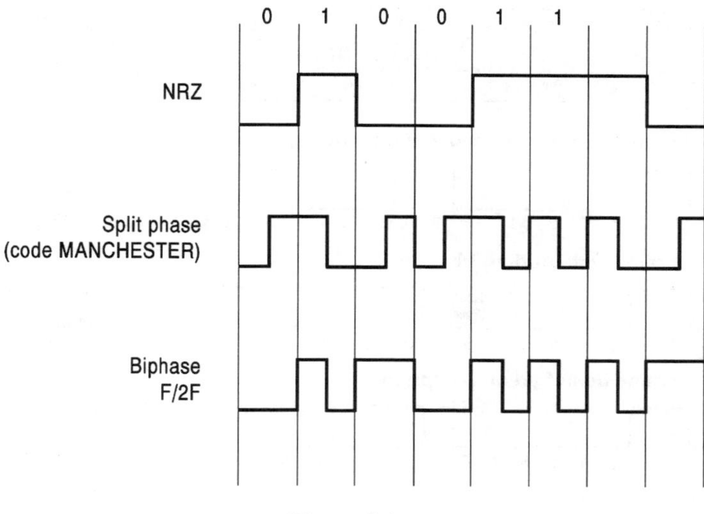


**Figure 2.4**

dominant or recessive. All the frame diagrams shown below in this book will therefore use this representation.

Everybody has their oscilloscope! But alas, you will not find your little darlings there! This is because the bits you have so carefully placed in their frames, as described above, will not appear in this form on the oscilloscope screen at all: after being set up in a row in the transfer register of a controller, they are passed through a special 'mill' before being sent out physically (electrically, optically, etc.) over the communication medium.

A special additional process (one of many) has been established to safeguard the message during its transport over the bus. This is what is known as the 'bit stuffing' method.

## The bit stuffing method

Does this bit stuffing method seem unfamiliar? In fact, it is simple: because the bit is NRZ coded, a particular message may contain many bits having the same value (level) and can make one or more stations think that there may be an anomaly in the network. It was therefore decided to deliberately introduce (at the transmission station), after 5 bits of identical value (either dominant or recessive), a supplementary bit having an intentionally opposite value, to 'break the rhythm', and thus indicate that everything is all right, despite appearances. These bits are called 'stuffing' bits.

Theoretically, this technique slightly protracts the transmission time of a message, but it plays an active part in safeguarding its content during transport. The consequences of this choice are examined in Chapters 3 and 4, where we will study the relationship between the net bit rate and the gross bit rate. On the contrary, by using this method, we introduce into the spectrum of the initial NRZ signal (which by its nature must contain the continuous component) a minimum alternating 'fundamental' which the signal did not previously have

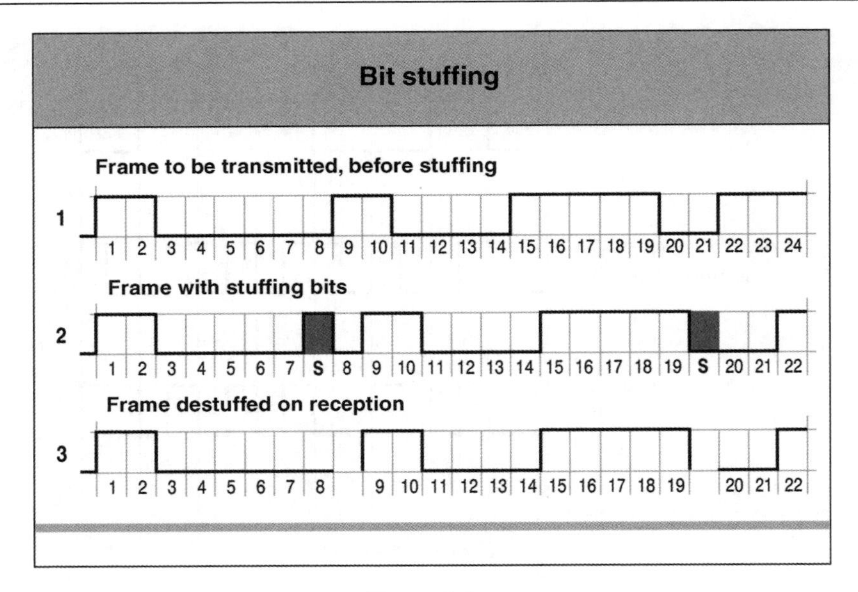


**Figure 2.5**

(Figure 2.5). This stratagem will also enable us to provide links which are electrically isolated by transformers . . . but this is another story to be taken up in Chapter 4.

Clearly, a CAN receiver must be perfectly capable of responding to this stuffing method and will therefore carry out the reverse operation of 'destuffing' on reception, removing these fill bits which were used only for the transport.

The hardware of the integrated circuits makes this bit 'overcoding' method completely transparent to the user, but signals any errors if they happen to occur.

Here is a final important point concerning what are called 'real time' applications. The insertion of stuffing bits during the exchange enables us to create a larger number of transitions in the communication and therefore provides a greater ease of bit synchronization, in spite of the NRZ bit coding, as will be seen in Chapters 3 and 4. An example of bit stuffing coding is shown in Figure 2.6.

## Notes

- As detailed below, only the fragments of frames relating to the start of frame, arbitration field, control field, data field and CRC sequence are coded by the bit stuffing method.
- When the sender detects 5 consecutive bits of identical value in the bit stream for transmission, it automatically inserts a complementary bit into the bit stream being transmitted.
- The remaining bits of the other segments of the data frames or remote frames, such as the CRC delimiter, ACK field and end of frame, have fixed structures and are not stuffed.
- To conclude on this matter, the error frames and overload frames also have fixed structures and therefore are not coded by the bit stuffing method.

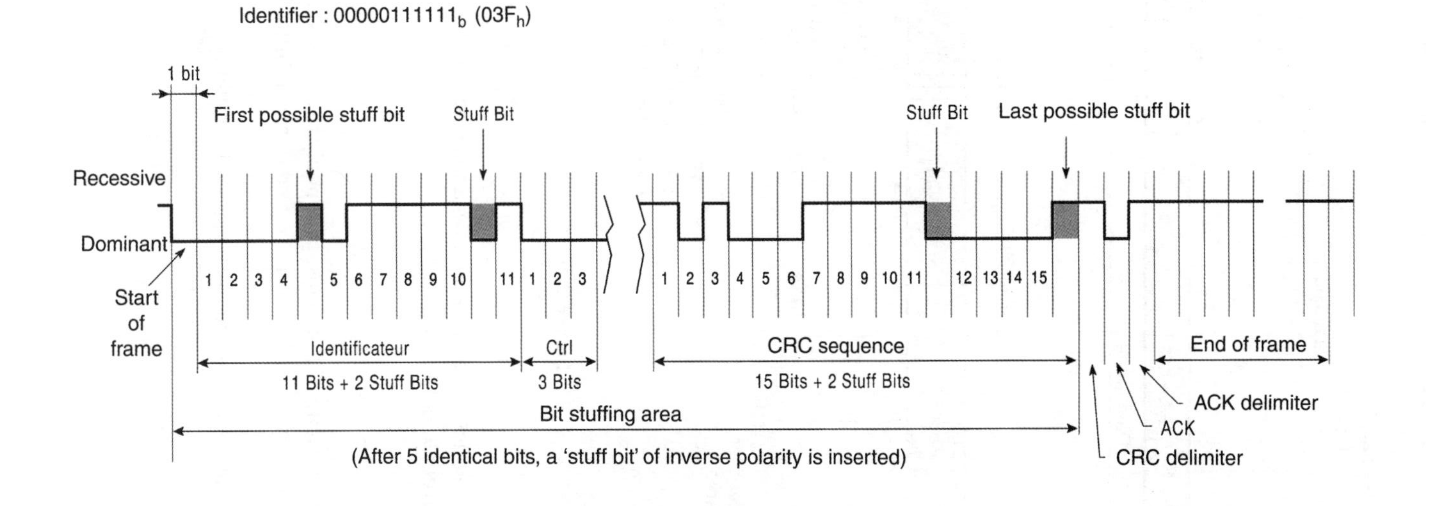


**CAN « Bit Stuffing »**

**Figure 2.6**

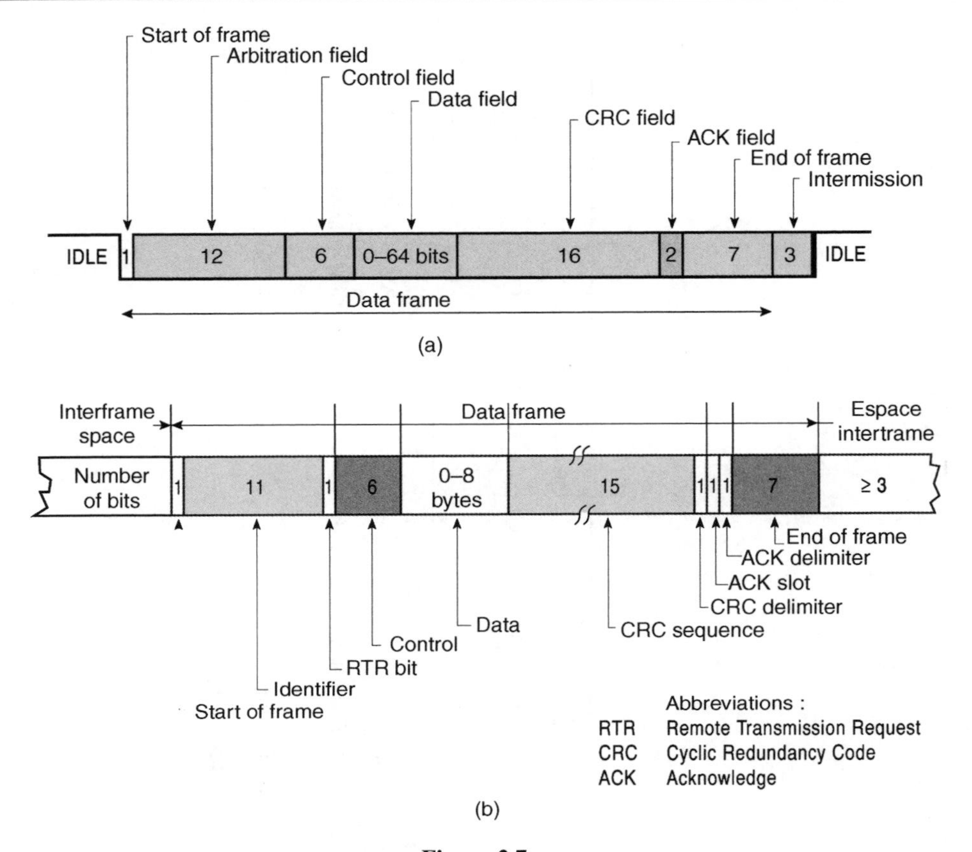


**Figure 2.7**

After this intermission at the 'bit level', let us go on to examine the detail of the content of the different frames that can travel along the bus.

### *2.1.6 Data frame*

First of all, let us examine the structure of the most commonly used CAN 2.0A standard data frame (Figure 2.7a and 2.7b). How and where can we find our own little darlings in all this clutter?

This frame can be broken down into seven main parts called 'fields':

- the start of frame,
- the arbitration field,
- the control field,
- the data field,
- the CRC sequence,

- the ACKnowledgement field,
- the end of frame.

An 8th area, called the interframe space, forms an integral part of the frame.

Now let us rapidly dissect the content of each of the fields to help us understand how the bus works.

### Start of frame

The start of frame of the data frame consists of a single dominant bit, signalling to all receivers the start of an exchange. This exchange can start only if the bus has been idle so far.

All stations must be fully synchronized with each other before the start bit can pass (see Chapter 3 for details of the hardware synchronization mechanism).

### Arbitration field

ISO: Set of bits in the message frame assigned to each message to control arbitration.

The field in which arbitration takes place consists of the identifier bits and the bit immediately following, called the RTR (remote transmission request) (Figure 2.8).

- Identifier
  The length of the identifier is 11 bits, and the bits are transmitted in the order from ID_10 to ID_0, the least significant being ID_0. However, the 7 most significant bits (from ID_10 to ID_4) must not all be recessive.

**Note:** For reasons of compatibility with certain older circuits, the maximum combinations of types are used:

ID = 11 11 11 1*x xx x* (where *x* is an indeterminate value).

In other words, the maximum number of identifiers is

$(2^{11} - 2^{4}) = 2048 - 16 = 2032$ combinations

- The RTR bit
  In a data frame, the RTR bit must be dominant.

### Control field

This consists of 6 bits (Figure 2.9).

- Reserved bits
  The first two bits (sent as dominant in the 2.0A frame) are reserved for later use and ensure future upward compatibility (particularly that of the frame known as the extended CAN 2.0B frame, described at the end of the chapter). CAN controllers must be capable of handling all the combinations of all the bits of the control field.
- Data length
  The last 4 bits of the control field (the DLC, 'data length code' field) indicate the number of bytes contained in the data field.

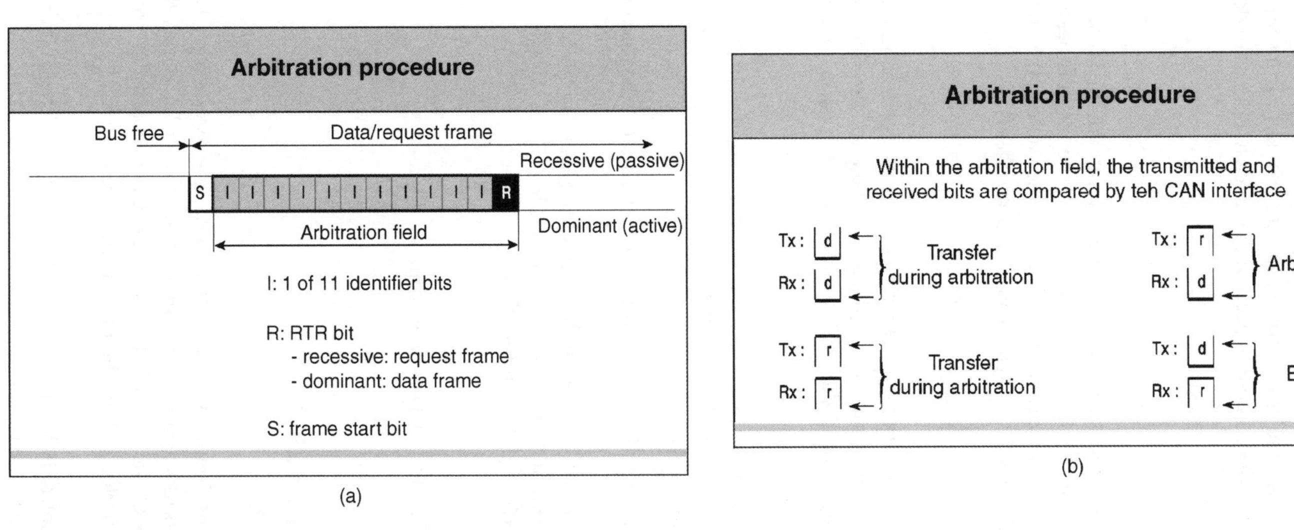

Arbitration procedure
Bus free
Data/request frame
Recessive (passive)
S
I
R
Arbitration field
Dominant (active)
I: 1 of 11 identifier bits
R: RTR bit
- recessive: request frame
- dominant: data frame
S: frame start bit
(a)
Arbitration procedure
Within the arbitration field, the transmitted and
received bits are compared by teh CAN interface
Tx : d
Rx : d
Transfer
during arbitration
Tx : r
Rx : d
Arbitration lost
Tx : r
Rx : r
Transfer
during arbitration
Tx : d
Rx : r
Bit error
(b)


**Figure 2.8**

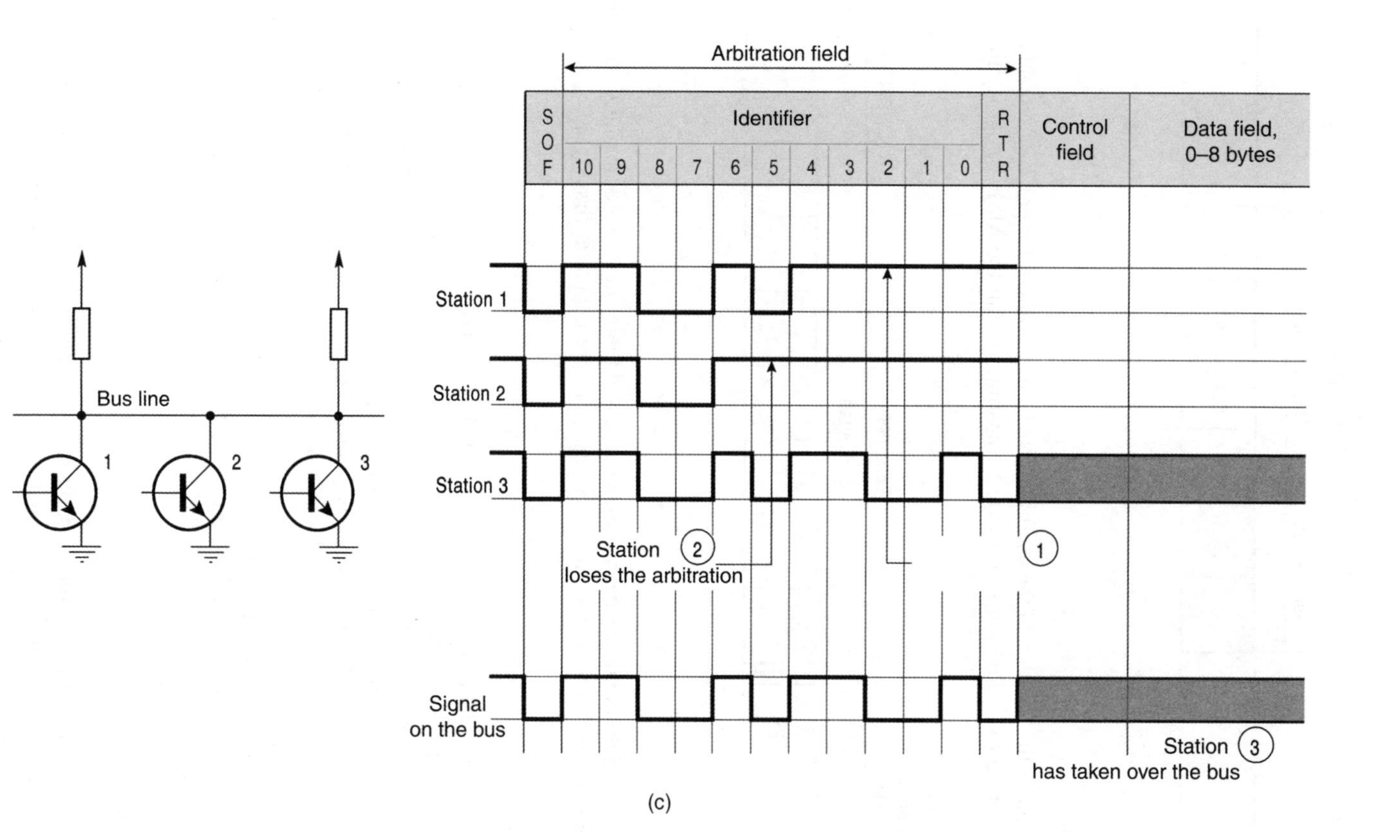


**Figure 2.8** (*Continued*)

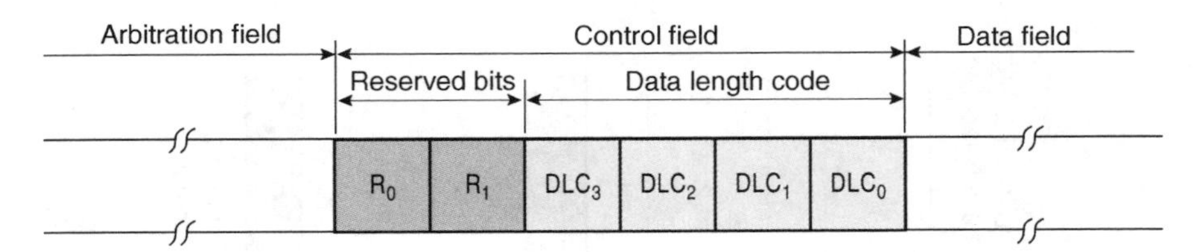


**Figure 2.9**

### Data field

The data field is the location of the useful data to be transmitted (Figure 2.10). It can consist of anything from 0 to a maximum of 8 bytes, transmitted with the MSB (most significant bit) leading.

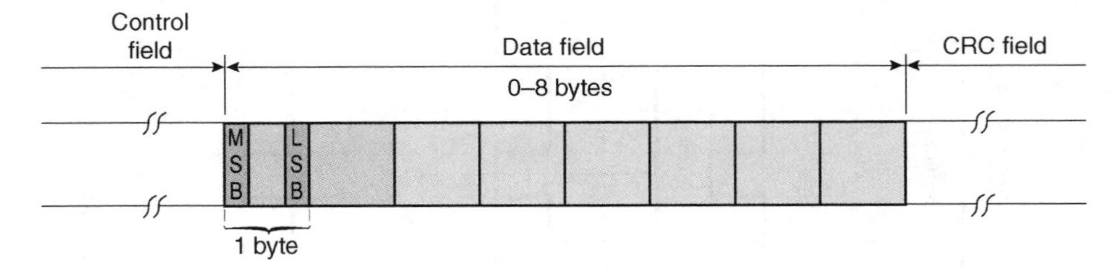


**Figure 2.10**

**Note:** Zero to eight inclusive means nine values, and therefore 4 bits of the DLC to define the number of data elements contained, not 3 bits, as might appear at first sight.

### CRC field

CRC stands for 'cyclic redundancy code'. This field consists of the CRC sequence area followed by a CRC delimiter (Figure 2.11).

To ensure the validity of the transmitted message, all the receivers must strive to verify a CRC sequence generated by the CAN sender, in relation to the content of the transmitted message. The codes used by the CAN bus controllers are (shortened) BCH (Bose–Chaudhuri–Hocquenghem) codes, supplemented with a parity check and having the following attributes:

- maximum length of code: 127 bits;
- maximum number of information digits: 112 bits (a maximum of 83 are used by the CAN bus controller);
- CRC sequence length: up to 15 bits;
- Hamming distance: $d = 6$.

A reminder: The Hamming distance between two binary words (of the same length) is the number of binary elements of the same rank which differ between the two words, and therefore, in the present case, $(d - 1) = 5$, in other words five independent bit errors are 100% detectable within the transmitted code.

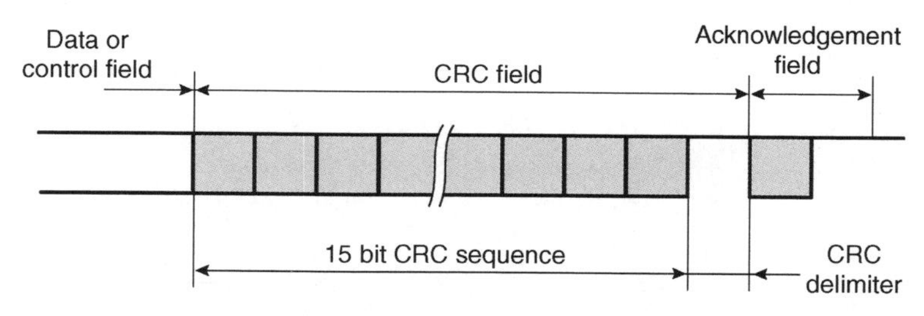


**Figure 2.11**

The CRC sequence is determined (calculated) by the sender of the message, using the following procedure:

(a) The stream of bits (not stuffed or destuffed), consisting of the bits from the start of the frame to the end of the data field (if present), is interpreted as a polynomial $f(x)$ with coefficients 0 and 1 assigned to the actual presence or absence of each bit. The polynomial obtained in this way is then supplemented with zeros for the 15 least significant coefficients.

(b) The above polynomial formed in this way is divided, modulo 2, by the following generating polynomial, including a parity check:

$$g(X) = X^{15} + X^{14} + X^{10} + X^{8} + X^{7} + X^{4} + X^{3} + 1$$

or transcribed into the following binary form:

$$g(X) = 1100010110011001.$$

(c) After the division of the polynomial $f(x)$ by the generating polynomial $g(x)$, the rest of this 'polynomial division' forms the 15-bit CRC sequence. The last-mentioned item is transmitted in the CRC field during the CRC sequence.

The curious (and courageous) readers should consult the example of calculation shown in Figure 2.12a and 2.12b (it works, too!).

*Further information*: A CRC, rather than a checksum, has been adopted for the CAN protocol because cyclic codes have the following properties:

- simplicity of coding and decoding,
- a capacity to detect and correct independent errors or those appearing in packets.

It should also be noted that, out of the known linear codes, it is BCH that has been adopted because, for a given efficiency, it is one of the codes having the greatest correction capacity for independent errors. This capacity is entirely satisfactory because it is very near to the theoretical optimum.

Furthermore, the electronic implementation of the coding and decoding of this code is particularly simple, using shift registers.

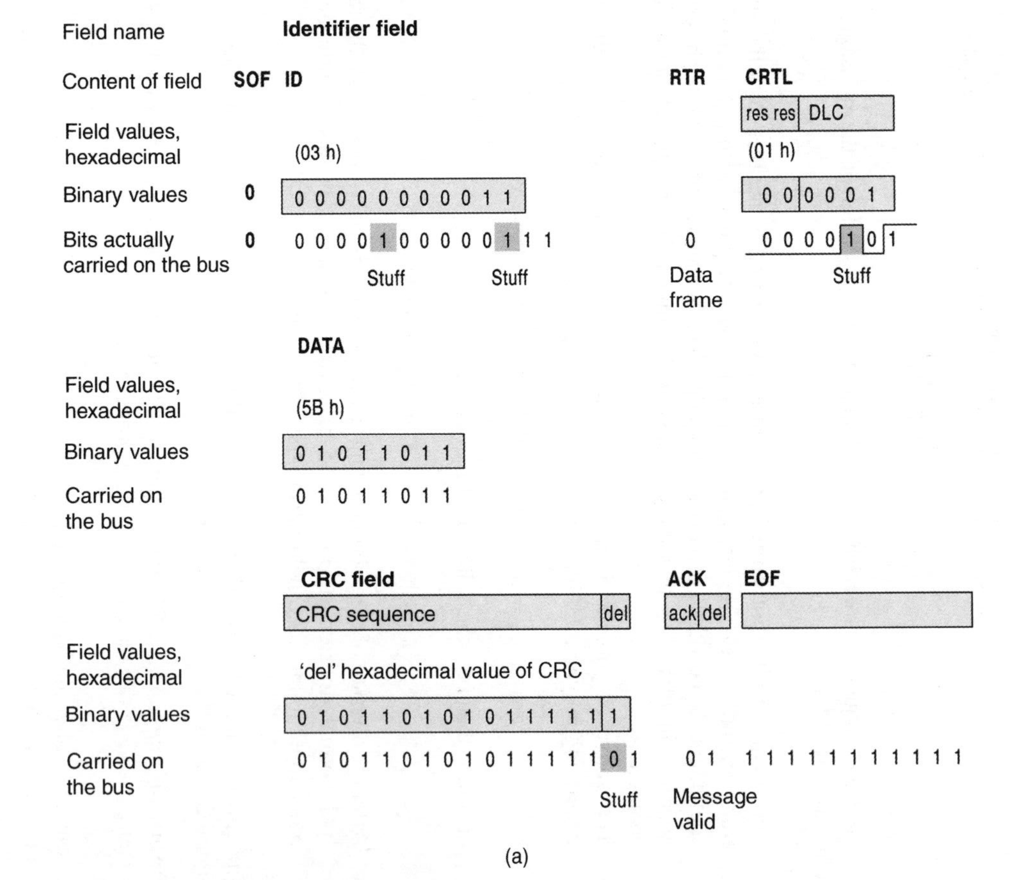

Example of data frame
Field name
Identifier field
Content of field
SOF
ID
RTR
CRTL
res res
DLC
Field values, hexadecimal
(03 h)
(01 h)
Binary values
0
0 0 0 0 0 0 0 0 0 1 1
0 0 0 0 0 1
Bits actually carried on the bus
0
0 0 0 0 1 0 0 0 0 0 1 1 1
0
0 0 0 0 1 0 1
Stuff
Stuff
Data frame
Stuff
DATA
Field values, hexadecimal
(5B h)
Binary values
0 1 0 1 1 0 1 1
Carried on the bus
0 1 0 1 1 0 1 1
CRC field
ACK
EOF
CRC sequence
del
ack
del
Field values, hexadecimal
'del' hexadecimal value of CRC
Binary values
0 1 0 1 1 0 1 0 1 0 1 1 1 1 1 1
Carried on the bus
0 1 0 1 1 0 1 0 1 0 1 1 1 1 1 0 1
0 1
1 1 1 1 1 1 1 1 1 1 1
Stuff
Message valid


(a)

**Figure 2.12**

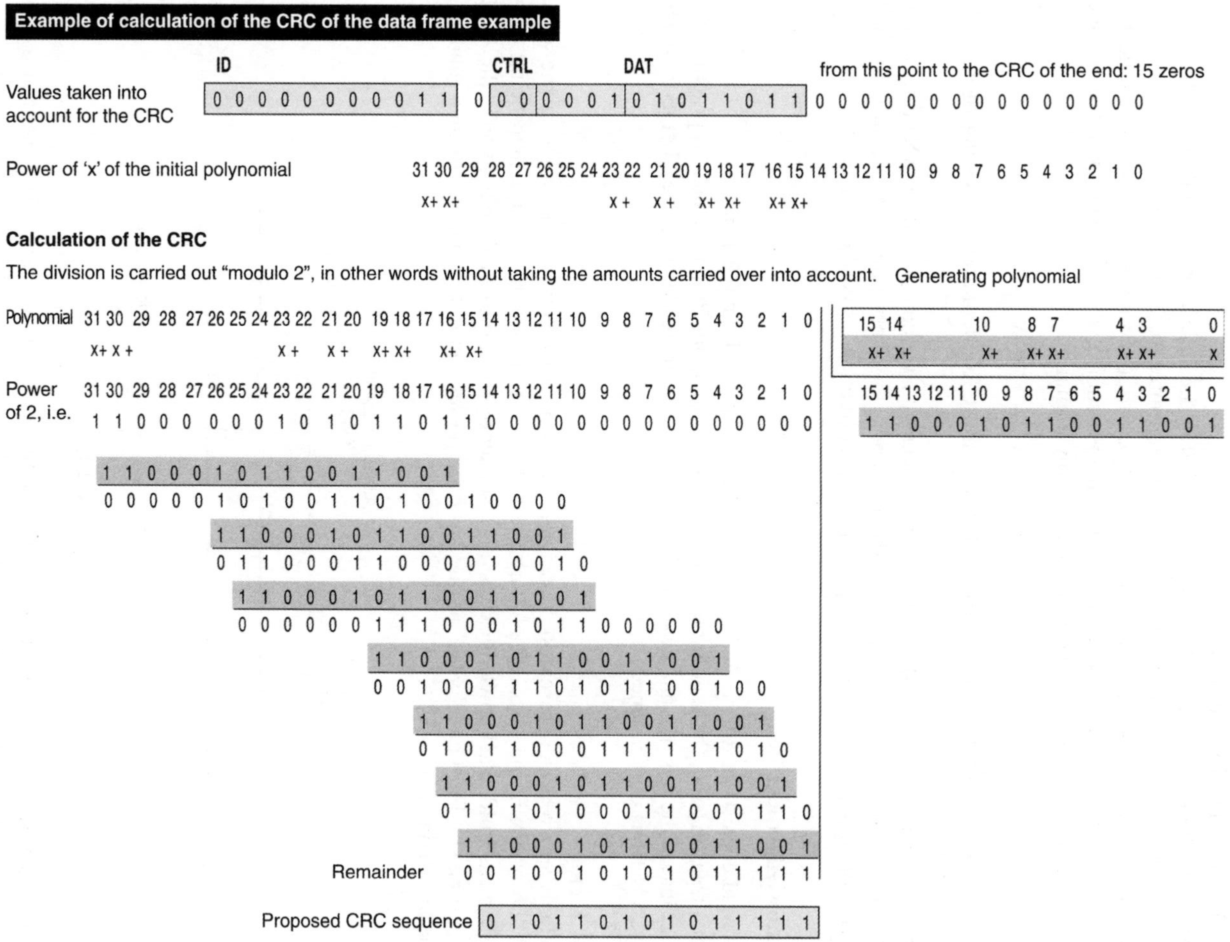

**Example of calculation of the CRC of the data frame example**

```
                        ID                          CTRL      DAT
Values taken into       0 0 0 0 0 0 0 0 0 1 1    0  0 0|0 0 0 1|0 1 0 1 1 0 1 1   0 0 0 0 0 0 0 0 0 0 0 0 0 0 0
account for the CRC                                                               from this point to the CRC of the end: 15 zeros

Power of 'x' of the initial polynomial   31 30 29 28 27 26 25 24 23 22 21 20 19 18 17 16 15 14 13 12 11 10 9 8 7 6 5 4 3 2 1 0
                                         X+ X+                   X +   X +   X+ X+    X+ X+
```

**Calculation of the CRC**

The division is carried out "modulo 2", in other words without taking the amounts carried over into account.

```
Polynomial  31 30 29 28 27 26 25 24 23 22 21 20 19 18 17 16 15 14 13 12 11 10 9 8 7 6 5 4 3 2 1 0
            X+ X +                    X +   X +   X+ X+    X+ X+

Power       31 30 29 28 27 26 25 24 23 22 21 20 19 18 17 16 15 14 13 12 11 10 9 8 7 6 5 4 3 2 1 0
of 2, i.e.   1  1  0  0  0  0  0  0  1  0  1  0  1  1  0  1  1  0  0  0  0  0 0 0 0 0 0 0 0 0 0 0

             1 1 0 0 0 1 0 1 1 0 0 1 1 0 0 1
             0 0 0 0 0 1 0 1 0 0 1 1 0 1 0 0 1 0 0 0 0
                       1 1 0 0 0 1 0 1 1 0 0 1 1 0 0 1
                       0 1 1 0 0 0 1 1 0 0 0 0 1 0 0 1 0
                         1 1 0 0 0 1 0 1 1 0 0 1 1 0 0 1
                         0 0 0 0 0 0 1 1 1 0 0 0 1 0 1 1 0 0 0 0 0 0
                                     1 1 0 0 0 1 0 1 1 0 0 1 1 0 0 1
                                     0 0 1 0 0 1 1 1 0 1 0 1 1 0 0 1 0 0
                                         1 1 0 0 0 1 0 1 1 0 0 1 1 0 0 1
                                         0 1 0 1 1 0 0 0 1 1 1 1 1 1 0 1 0
                                           1 1 0 0 0 1 0 1 1 0 0 1 1 0 0 1
                                           0 1 1 1 0 1 0 0 0 1 1 0 0 0 1 1 0
                                             1 1 0 0 0 1 0 1 1 0 0 1 1 0 0 1
                               Remainder     0 0 1 0 0 1 0 1 0 1 0 1 1 1 1 1

                 Proposed CRC sequence     0 1 0 1 1 0 1 0 1 0 1 1 1 1 1
```

Generating polynomial

```
15 14        10    8 7      4 3       0
X+ X+        X+    X+ X+    X+ X+     X

15 14 13 12 11 10 9 8 7 6 5 4 3 2 1 0
 1  1  0  0  0  1 0 1 1 0 0 1 1 0 0 1
```

(b)

**Figure 2.12** *(Continued)*

Theory and statistical calculations indicate that multiple errors exceeding 6 different bits in the whole content are not detected with a residual error probability of $2^{-15}$ (approximately $3 \times 10^{-5}$) for the CRC part only.

As the start of frame (dominant bit) is included in the code word, no rotation of the code word can be detected in the absence of the CRC delimiter (recessive bit).

When it is received, the message (having been destuffed) comprising the start of frame, the arbitration field, the control field and the data field is subjected to checking with the aid of a CRC constructed around the same generating polynomial.

$$X^{15} + X^{14} + X^{10} + X^{8} + X^{7} + X^{4} + X^{3} + 1.$$

To finish with the CRC field, it then terminates in a CRC delimiter, consisting of a single recessive bit.

A last remark concerning the CRC, its function and its usefulness: In the CAN protocol, the vast concept of CRC is restricted to the detection of errors (and consequently to their subsequent signalling) and is not applied to their correction, as is often the case.

### Acknowledgement field

This field consists of 2 bits: ACK slot and ACK delimiter (Figure 2.13).

In a transmission, the sending unit sends two recessive bits along the bus (in practice, the sender leaves the bus free and switches itself to listening or 'receiver' mode). Let us examine the very serious consequences of what has just been stated.

- ACK slot
  Whenever a receiver present in the network has correctly received a message – in the sense of 'no transmission error' (including the CRC) – the received message then being considered as valid, it acknowledges this to the sender by substituting a dominant bit in the ACK slot time slot for the recessive bit that has been present until this moment. It then sends an 'acknowledgement', ACK.

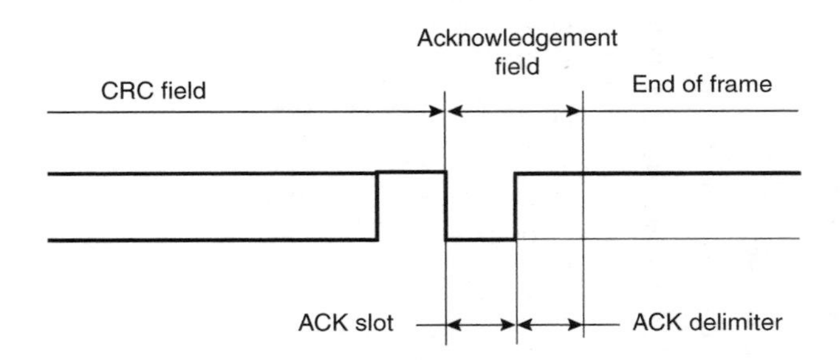


**Figure 2.13**

**Some very important notes:** If the sending station receives an acknowledgement, it means that at least one station in the network has received the message completely and without errors.

Note that the presence of the acknowledgement bit does not in any way indicate that the receiver has accepted or rejected the content of the message (i.e. whether or not it is interested in the content of the message). The only information we can obtain from it is that, at its level, the receiver in question has physically received at its terminal a potentially usable message, which does not contain errors.

All the receivers connected to the network must send an acknowledgement signal over the network if they have not detected any error of any kind. Otherwise the message will be considered invalid, and the receiver will not return an acknowledgement. Moreover (to state the obvious), as the message is invalid, it must contain one or more errors! To comply with the CAN protocol, the receiver must signal the error or errors by transmitting an error frame (see below in this chapter).

We are dealing with a concept designed for operation in what is called a 'shared memory' system. Consequently, the acknowledgement and processing of errors are specified so as to provide all the information in a consistent way across the shared memory.

Theoretically, there is no reason to discriminate between different receivers or messages in the acknowledgement field. In contrast to other protocols (such as I2C), this acknowledgement does not in any way signify that a node is present or not, or even that a node is interested in the message in the network. By way of example, let us briefly review what happens in these two cases.

*Presence or absence of a node.* An important consequence of the above statement is that if a station is disconnected from the bus (by being in 'bus off' mode, for example, or by being deliberately removed physically, i.e. mechanically, from the bus, or again if some of the wires in the bus are cut off), this station will cease to be an integral part of the 'shared memory'.

*Interest (or lack of interest) in the transmitted (and received) messages.* Theoretically, the interest of the network stations in the intrinsic content of the transmitted message is a 'mystery of mysteries'.

In the last two cases of 'identification of a lost node' and 'interest in the message', it is up to the network designers to build additional precautions into their specific application software. It will then be necessary to establish software procedures between the nodes (at the ISO/OSI level 7, the application layer), so that they can acknowledge 'in software terms' their presence or the 'interest' they have in the messages sent and received.

Without examining this subject in detail, it is worth knowing that application software layers for handling these problems (and many others) have already been developed by user groups having the same concerns. Examples that can be mentioned are CAL (CAN application layer) produced by CiA, for general use, and DeviceNet and SDS, produced by Allen Bradley and Honeywell, respectively for rather more specialized use.

- ACK delimiter
  This second bit must (always) be recessive, and therefore when a message has been correctly received by all the stations present in the network, the ACK slot (dominant) bit is surrounded by two recessive bits (CRC delimiter and ACK delimiter).

**End of data frame**

The data frame is terminated by a flag consisting of a sequence of 7 recessive bits, which, it should be noted, is 2 bits longer than the bit stuffing standard length.

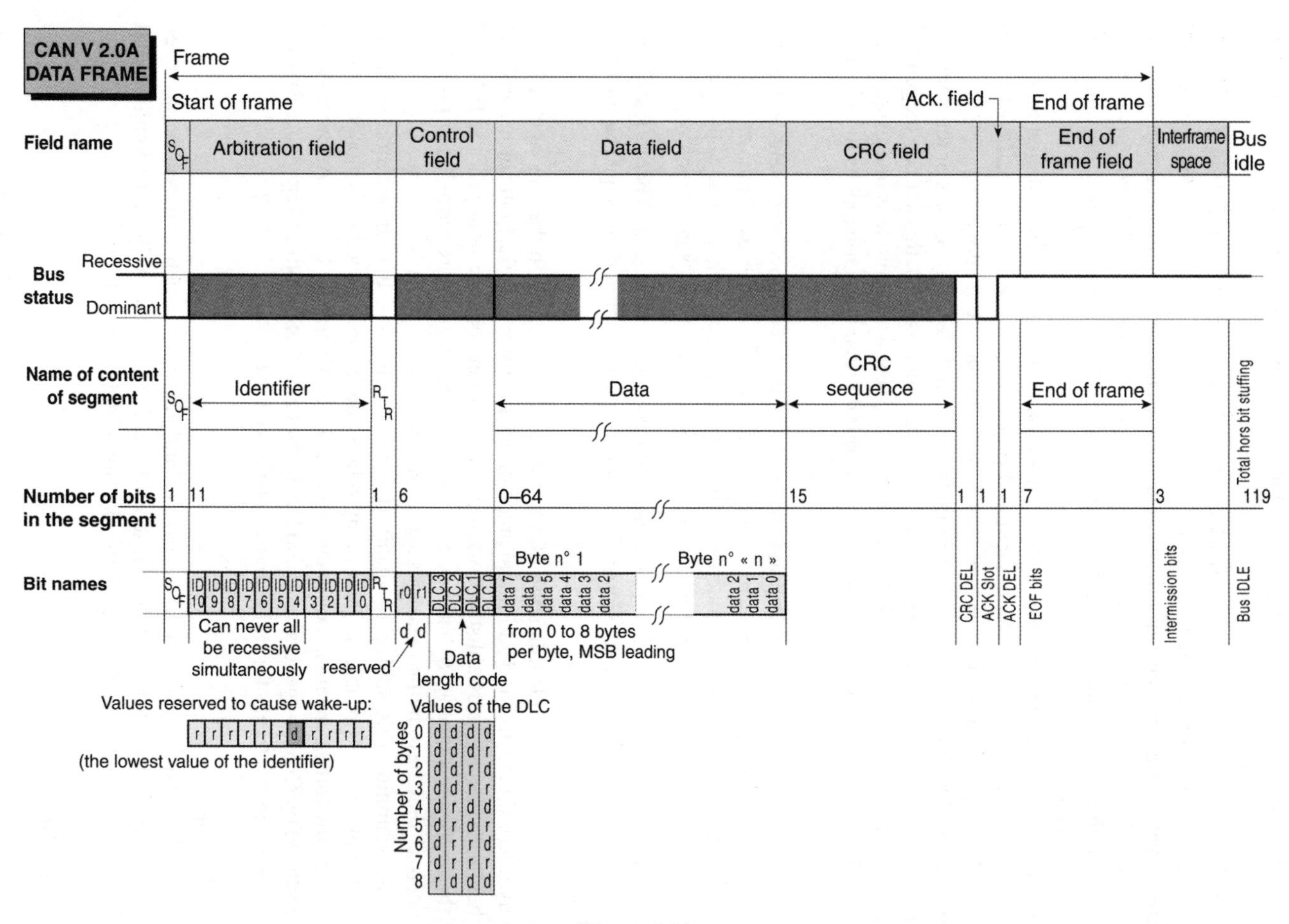
CAN V 2.0A
DATA FRAME
Frame
Start of frame
Ack. field
End of frame
Field name
SOF
Arbitration field
Control field
Data field
CRC field
End of frame field
Interframe space
Bus idle
Bus status
Recessive
Dominant
Name of content of segment
Identifier
RTR
Data
CRC sequence
End of frame
Total hors bit stuffing
Number of bits in the segment
1 11 1 6 0–64 15 1 1 1 7 3 119
Bit names
Byte n° 1
Byte n° « n »
ID 10 ID 9 ID 8 ID 7 ID 6 ID 5 ID 4 ID 3 ID 2 ID 1 ID 0
r0 r1 DLC 3 DLC 2 DLC 1 DLC 0
data 7 data 6 data 5 data 4 data 3 data 2
data 2 data 1 data 0
CRC DEL
ACK Slot
ACK DEL
EOF bits
Intermission bits
Bus IDLE
Can never all be recessive simultaneously
d d
reserved
Data length code
from 0 to 8 bytes per byte, MSB leading
Values reserved to cause wake-up:
r r r r r r d r r r r
(the lowest value of the identifier)
Values of the DLC
Number of bytes
0 d d d d
1 d d d r
2 d d r d
3 d d r r
4 d r d d
5 d r d r
6 d r r d
7 d r r r
8 r d d d

Figure 2.14

This field has a fixed structure, and the bit stuffing logics for coding (in sending) and decoding (in receiving) are disabled during the end of frame field sequence.

## Interframe

Finally, there is an 8th area, called the interframe space, which is detailed below.

As you have been very patient, I can reward you with Figure 2.14 which summarizes and illustrates the essence of the above explanations.

### *2.1.7 Remote frame*

As mentioned above, each station transmits into total darkness, not knowing whether the information sent has been useful to any of the participants. It is also possible that a node may need information of a certain kind, which it does not have, to carry out its allotted task. In this case, a station needing data can initialize the request for transmission of the data concerned from another node, by sending a remote frame.

Let us examine the composition of the remote frame (Figure 2.15). This frame has only six parts, instead of the seven in the preceding frame, which are as follows:

- the start of frame,
- the arbitration field,
- the control field,
- the CRC field,
- the acknowledgement field,
- the end of frame.

This is followed by a 7th area, called the interframe space. In the rest of this description, the remote frame will be compared with the data frame described above, to avoid repetition.

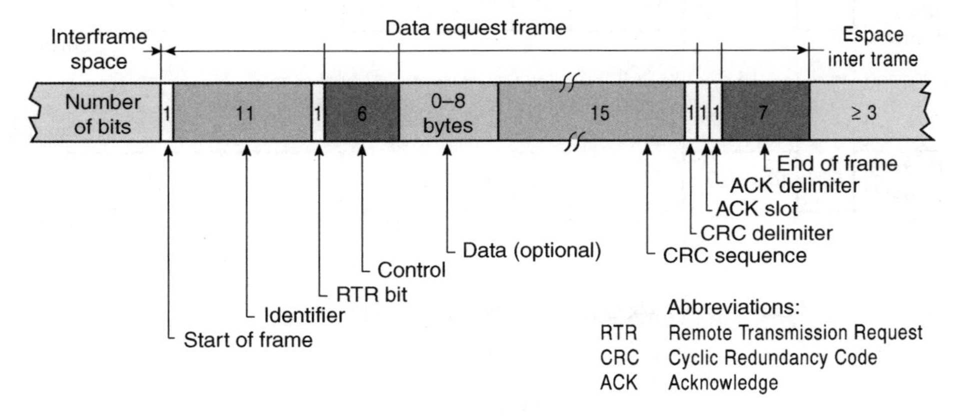


**Figure 2.15**

### Start of frame

The same: see 'data frame' above.

### Arbitration field

The field in which arbitration takes place consists of the identifier bits and the bit immediately following, called the RTR (see Figure 2.16a and 12.6b).

- Identifier
  The same: see 'data frame' above.
- RTR bit

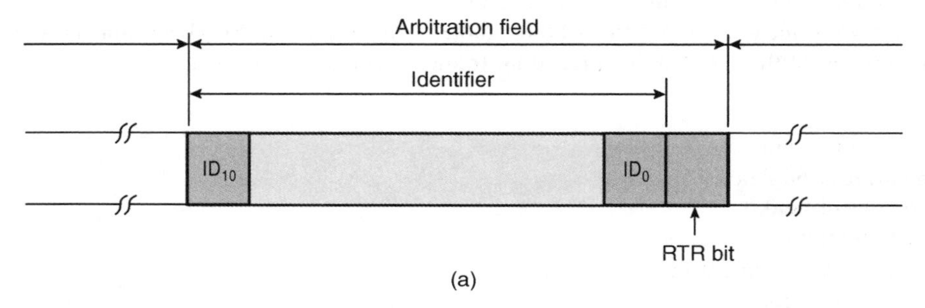


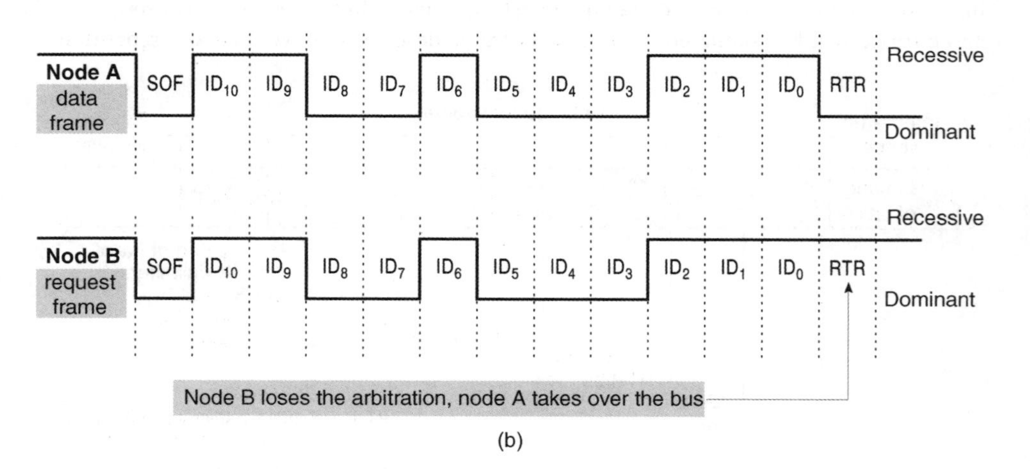


**Figure 2.16**

By contrast with the preceding case, this bit is recessive in a remote frame. Thus, it is this bit that distinguishes a data frame from a remote frame.

**Corollary.** Because the RTR bit is by definition always recessive for a remote frame, we can say that, for a single identifier, a data frame always takes priority over a remote frame (which is reasonable because it already contains the information that the remote frame was supposed to be requesting at that instant).

### Control field

This consists of 6 bits (Figure 2.17).

- Reserved bits
  The same: see 'data frame' above.
- Data length

The last 4 bits indicate the number of bytes in the data field; in other words, those not contained in the remote frame (see below), but which the data frame will be responsible for recovering subsequently.

**Note:** Theoretically, the indication of a data length is unnecessary as the RTR bit is recessive. In fact, it is advisable to specify this field correctly, so that no error occurs on the bus if two microcontrollers start transmitting a remote frame simultaneously.

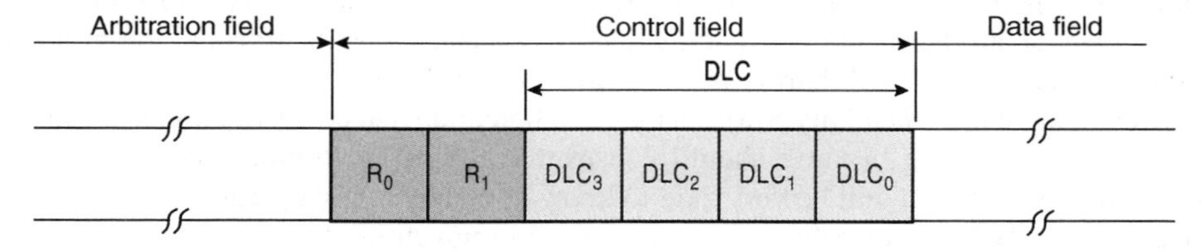


**Figure 2.17**

### Data field

There is no data field, regardless of the message length which may have been declared in the control field.

### CRC, acknowledgement, end of data frame and interframe fields

These fields are identical to those of the data frame.

For a summary of all these, see the representation of an example of a remote frame in Figure 2.18.

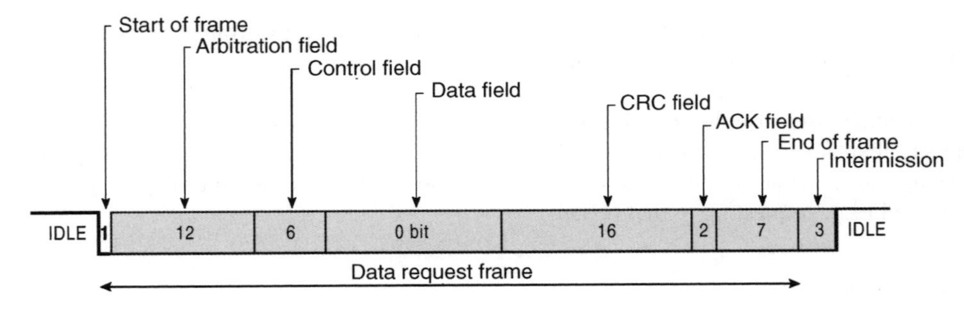


**Figure 2.18**

So far, we have assumed that everything has gone according to plan and the data in the data frames and/or remote frames have travelled without impediment.

Sadly, life is not always like that, and sometimes the frames are disturbed by malicious and badly brought-up parasites. So, in order not to interrupt the story of the transport of the data, I have decided to describe the termination of a frame (the interframes and overload frames) at the end of the chapter and move directly to the hunt for errors.

## 2.2 Errors: Their Intrinsic Properties, Detection and Processing

This is one of the most important parts of the CAN protocol, and it is essential to understand its mechanism (admittedly complicated but impressively efficient!) in order to benefit from all its advantages. This part is often obscured by the fact that the components conforming to the protocol manage all these details inside the circuits (because they conform to the protocol!), and the end user – you – will hardly pay any attention to them.

However, there is tremendous power in the error detection and correction modes, which are well worth investigating because should you ever really be in trouble (and this generally happens quite soon), you will be very glad to know how they really operate. This also avoids the waste of many hours of software development in reinventing features which are already very well implemented in the hardware.

### *2.2.1 General*

In view of the complexity of error processing, it is not as easy as it may seem to present the content of this section without forcing your neurons to develop rather more nodes than there already are in the network! After several exhausting readings of the original protocol (which is admittedly professional) and long reflection and debate with professionals and teachers concerning the educational method for presenting this subject, I have opted for the following plan, which consists of two main parts:

- A general part:
  - the description of the different types of errors that may occur,
  - the global philosophy and strategy of error processing and recovery.

- The details:
  - the error detection methods,
  - the error signalling methods (and frames),
  - the error recovery methods.

### *2.2.2 The different kinds of errors that may occur*

The CAN protocol has many very efficient resources for hunting down parasitic elements. In one of the preceding sections, I have outlined the main sources of errors which will now be examined in detail and placed in a hierarchy.

Many types of errors can occur

- in the physical layer itself (bit errors and bit stuffing errors):
  - the fact that the bit itself is affected by errors (by parasitic effects, for example),
  - a bit stuffing error for any reasons, involuntary (parasites, transmissions, forgotten elements) or sometimes deliberate, as mentioned below in the 'error frame',
  - the fact that all is well at the level of the bit and the frame structure, but that, for example, the frame has not been acknowledged, causing an ACKnowledgement error, or the CRC value is not what was expected;

- in the frame layer, in the sense that its internal structure has been violated (information in the wrong place, etc.), the following errors can occur:
  - CRC delimiter error,
  - ACKnowledgement delimiter error,
  - end of frame error,
  - error delimiter error,
  - overload delimiter error.

In all these examples, the presence of errors will be signalled by an error frame which is generated on the bus to inform those entitled, but, before detailing its structure and mechanism, I must look at some more theory and definitions relating to what are called 'confinement errors'.

### *2.2.3 The concepts of a network 'health bulletin'*

In addition to the fact that errors can happen, it is always instructive to know the type of error and the way they occur: if they happen rarely and are therefore 'not too troublesome', or if they happen very frequently and become very troublesome. In the latter case, the microcontroller or microcontrollers managing the network must be informed when there are 'long lasting' disturbances and when the bus activity has returned to normal. During persistent disturbances, the 'controller' part of the bus switches to what is called 'bus off' mode, and the local part of the CPU can then take the agreed default values. The presence of minor (or brief) disturbances in the bus does not affect the controller part of the bus. For this purpose, the term 'confinement' implies a vast mechanism having the purpose of determining whether a node is:

- not disturbed at all,
- slightly disturbed,

- rather more seriously disturbed by errors,
- too disturbed and must switch to bus off.

Determining all this requires a miracle, no? Well, no!

### *2.2.4 The mechanism for processing confinement errors*

One of the purposes of this mechanism is not only to enable hardware faults and disturbances to be detected, but also, and especially, to locate them so that precise intervention is possible.

The rules for processing confinement errors are described in the CAN protocol in such a way that the microcontrollers closest to the place of occurrence of the error react with the highest priority, and as quickly as possible (in other words, they become 'error passive' or 'bus off'). Consequently, errors can be more easily located, and their effect on the normal activities of the bus can be minimized.

All microcontrollers conforming to the CAN protocol must have two quite separate internal counters, the 'transmit error counter' and the 'receive error counter', with the task of recording (counting) the errors occurring in transmission and reception, respectively. This mechanism has something in common with the game of snakes and ladders: a few steps up, then all the way down again . . . although it is rather more technical!

In fact,

- if the message is correctly transmitted or received, the counter in question is decremented;
- if the message contains errors, the counter in question is incremented.

Also,

- Error counters do not use proportional counting methods.
- An error increments the counter in question by an amount which is much greater than the amount by which the same counter would be decremented if a message had been transmitted or received correctly.

Furthermore,

- Over a long period, this may cause the error counting to increase, even if there are fewer incorrect messages than correct messages. In this case, the level of the error counters reflects the relative frequency of the problems occurring on the bus.
- The ratio between the incrementing and decrementing of the counters – in other words, the weighting of the point counting – depends on the ratio between correct and incorrect messages on the bus. By definition, the protocol sets this value to 8.

For your information, this mechanism is entirely implemented in silicon form in the components.

### *2.2.5 Adding up the score!*

We are reaching the end of this section. Before making a decision, we must add up the score.

Regardless of the exact rules for adding up the scores, shown in detail in Figure 2.19, we can immediately state the consequences of the resulting totals.

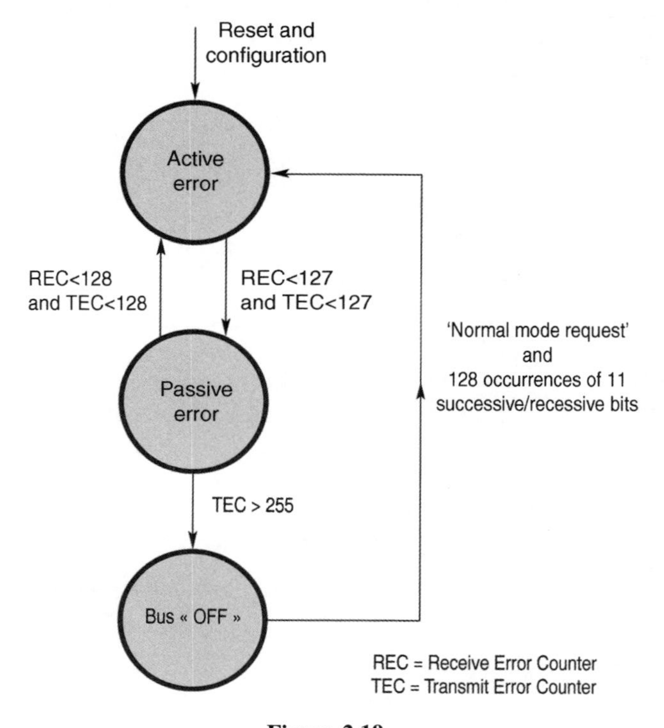


**Figure 2.19**

A graphic illustration of the results and their consequences is provided in Figure 2.20a and 2.20b.

### From zero to 127 inclusive: the error active state

When the values of the two error counters are in the range from 0 to 127, the CAN station (or node) is said to be operating in the 'error active' state (mode).

This means that the node in question will not only continue to receive and send normally, but also transmit an 'active error flag' in the error frame if an error is detected.

**Note:** The protocol recommends that we should'be alarmed' as soon as the counter reaches 96 points (the 'warning limit'), a score indicating a significant accumulation of error conditions. Generally, at this level, an 'error status' or 'error interrupt' break is caused within the CAN microcontroller, generating a request for control of these disturbances.

### From 128 to 255 inclusive: the error passive state

When the values of *one or other* of the two error counters are in the range from 128 to 255, the CAN station (or node) is said to be operating in the 'error passive' state (mode).

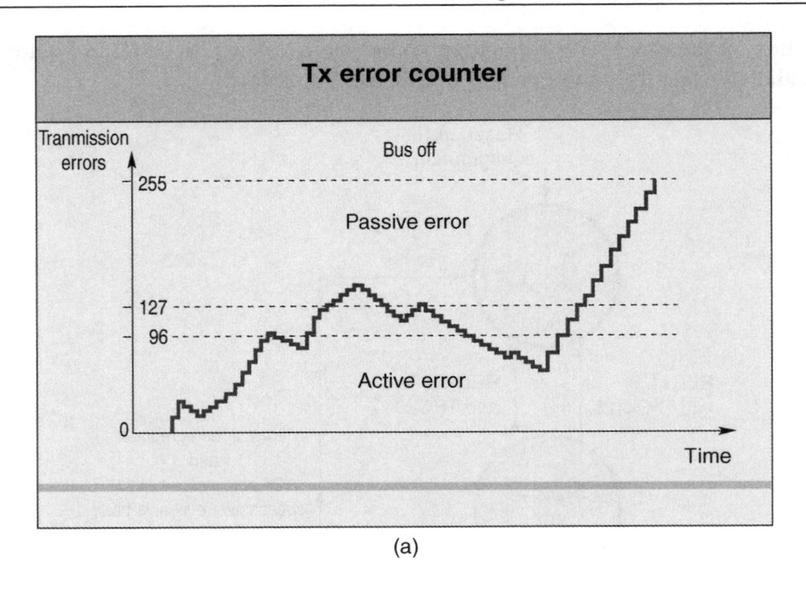


(a)

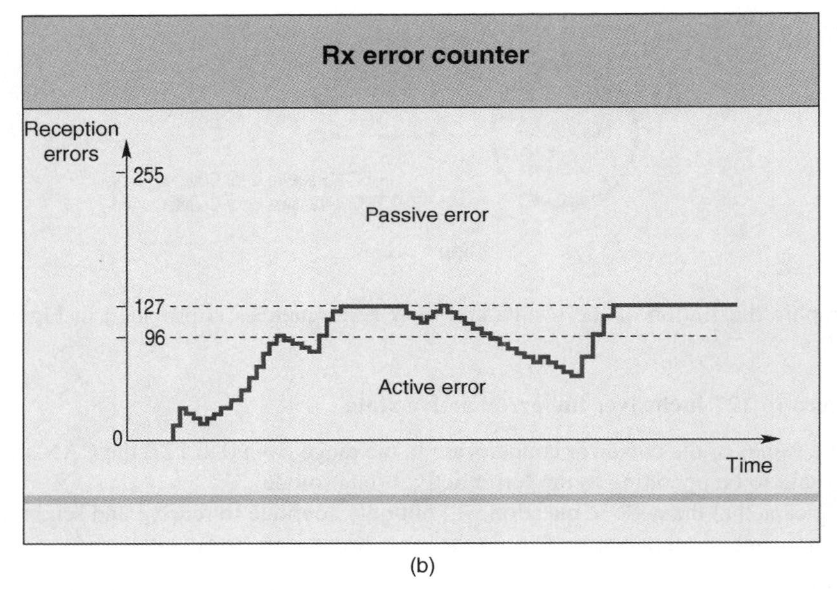


(b)

**Figure 2.20**

This means that the node in question will continue not only to receive and send normally, but also transmit 'passive error flags' only in the error frame if an error is detected.

**Special note:** An error condition causing a node to become error passive obliges the node to send an active error flag. This is the case, for example, when the node has been error active with a counter at 125 for example, and gains another eight counting units on detection of an error, making 133, and therefore switching to error passive.

### Above 255: the 'bus off' state

If the transmission counter value exceeds 255 ($\geq 256$), the CAN station (node) is said to have entered the 'bus off' state (mode). This means that the node in question, becoming tired of being unable to receive or transmit messages successfully, goes off to sulk in a corner and stops receiving and sending normally. In this case, the bus off unit is no longer permitted to have any effect or carry out any activity on the bus. Its bus driver stages must be disconnected electronically from the bus.

Theoretically, the node can remain in this state indefinitely if nobody takes any notice of it, until it decides that its (justified) sulk has gone on long enough and it wants to return to the network, having taken its previous unhappiness into account and resetting its counters to zero.

The protocol, acting as referee, allows a 'bus off' node to become 'error active' again (having reset all its error counters to zero) after it has seen 128 error-free occurrences of 11 recessive bits each on the bus (demonstrating that many messages whose ACK delimiter, end of frame and intermission bits have been carried correctly and the bus has found them to be in good health).

## *2.2.6 Rules for error detection in CAN*

The rules for error detection in CAN are explained in Figure 2.21.
**Note:** A '*warning*' condition is signalled at 96.

### Locating the defective node(s)

In this section, it is important to note how the transmission and reception error counters in a node add up their points locally, according to whether the node is receiving (in the case of an erroneous message received at its terminals) or transmitting (in the case of an error made by itself or the impossibility of sending correctly as it would wish, because of network problems, for example), so that the position of the node in the network is not ideal. Figure 2.22a and 2.22b shows a precise example of what has just been stated. This clearly shows how, regardless of the sending or receiving states, a 'faulty' node always accumulates more points than the other nodes, making it very easy to trace them by the subsequent absence from a network (in the sequence of their switch to error passive mode followed by their disappearance after switching to bus off mode), and consequently to take ad hoc decisions.

### Statistics on the health of the network

Clearly, a knowledge of the state of the internal counters makes it possible to document and provide information on the health of the station and the state of the network in its immediate vicinity. Moreover, because each station has to inform the network of local disasters, it immediately becomes possible for each station to draw up its own network quality statistics, if it has access to the precise content of its counters.

## The rules of CAN

The reception and transmission error counters are modified according to the following rules

Note that more than one rule can be applied during a single message transfer

| | Reception error counter | Transmission error counter |
|---|---|---|
| **Exchange successful** | | |
| **Transmission** After a successful transmission (ACK received and no error up to the end of the frame), the | | decrements by 1 unit unless this was the case |
| **Reception** After successful reception (reception without error up to ACK SLOT and correct transmission of the ACK bit), the | **decrements by 1 unit if it was between 1 and 27**<br>**remain at 0 if it is already 0 if it is >127 it is returned to between 119 and 127** | |
| **when a transmitter :** | | |
| **Detects** a bit error during the transmission of an active error flag or an overload flag | | increments by 8 units |
| **Transmits** an error flag, the | | increments by 8 units |
| unless : (a) the transmitter is error passive and detects an ACKnowledgement ERROR because it has not detected a dominant ACK and does not detect a dominant bit during the transmission of its passive error flag | | no change of state of the counter |
| (b) the transmitter sends an error flag because a stuff error has occurred during arbitration | | no change of state of the counter |
| **quand un RÉCEPTEUR :** | | |
| **Detects** (a) an error, the unless the error is a bit error occurring during and active error flag or an overload flag | **increments by 1 unit** | |
| (b) a 'dominant' bit as the first bit after an error flag has been sent | **increments by 8 units** | |
| (c) a bit error during the transmission of an active error flag or an overload flag | **increments by 8 units** | |

(a)

**Figure 2.21**

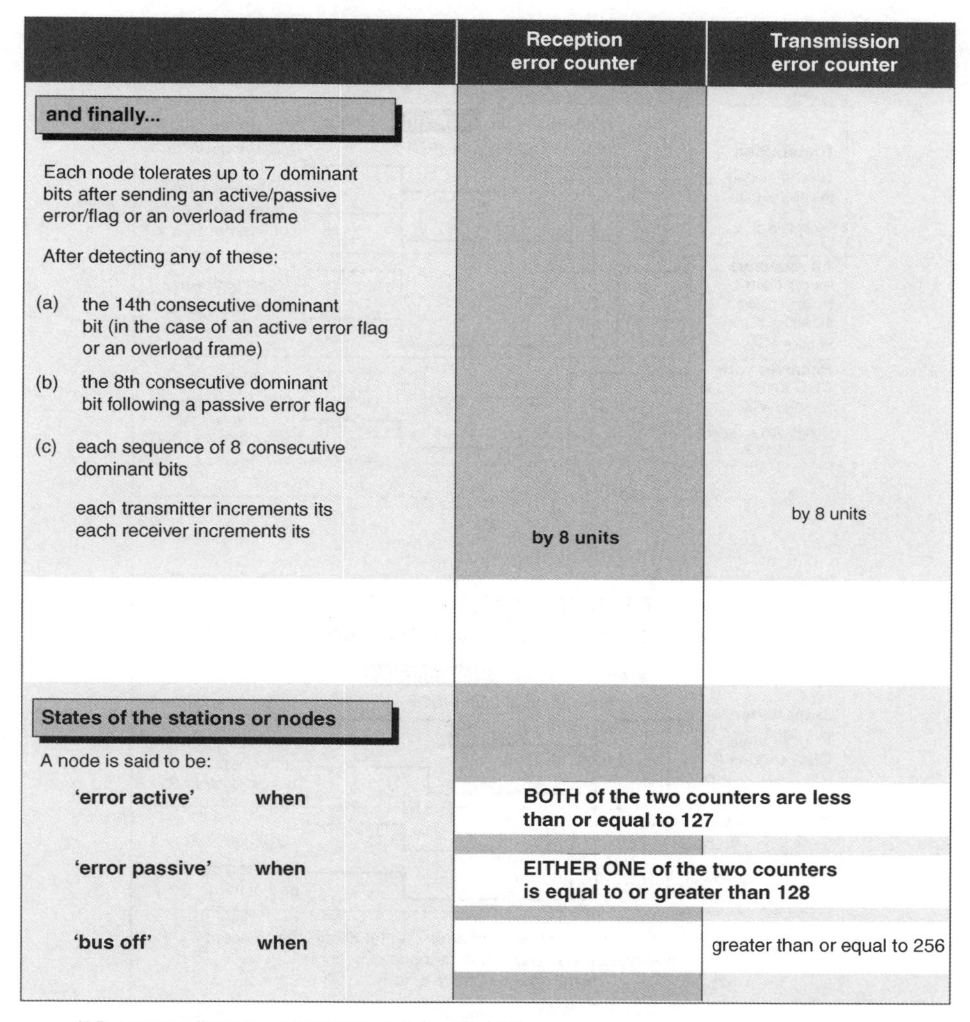

| | Reception error counter | Transmission error counter |
|---|---|---|
| **and finally...** | | |
| Each node tolerates up to 7 dominant bits after sending an active/passive error/flag or an overload frame | | |
| After detecting any of these: | | |
| (a) the 14th consecutive dominant bit (in the case of an active error flag or an overload frame) | | |
| (b) the 8th consecutive dominant bit following a passive error flag | | |
| (c) each sequence of 8 consecutive dominant bits | | |
| each transmitter increments its | | by 8 units |
| each receiver increments its | **by 8 units** | |
| **States of the stations or nodes** | | |
| A node is said to be: | | |
| **'error active' when** | **BOTH of the two counters are less than or equal to 127** | |
| **'error passive' when** | **EITHER ONE of the two counters is equal to or greater than 128** | |
| **'bus off' when** | | greater than or equal to 256 |

N.B. : une condition de « WARNING » est signalée à 96.

(b)

**Figure 2.21** (*Continued*)

In its present version, the CAN protocol does not require the provision of such detailed information to the CAN manager/controller. The only way of gaining an idea of the internal values of the counters at present is to observe the different error flags relating to these values on the bus, via certain relationships. But beware: the deciphering of this information is not a simple matter!

Clearly, it would be essential to have external information available, at the terminals of a component, about interruptions where the values of 96, 128 and 255 are exceeded. And why

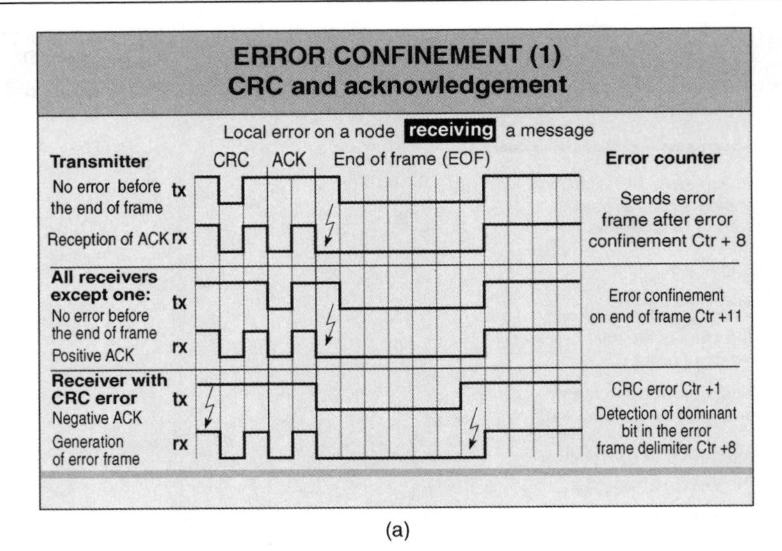


(a)

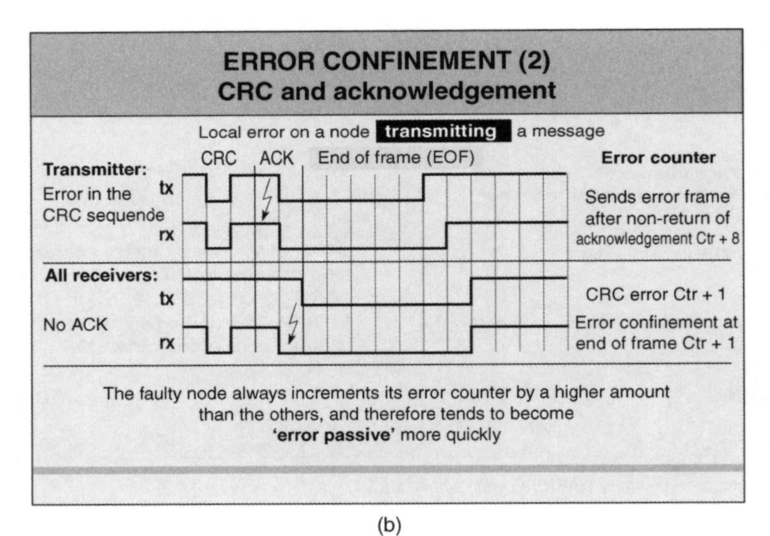


(b)

**Figure 2.22**

not have something even better? Let us dream a little ... How wonderful it would be to have two tiny internal registers, where we could read the contents of the transmission and reception error counters to satisfy all our statistical yearnings! You can be reassured that many CAN components of the current generation include these facilities. Such an enrichment of the protocol, how delightful ...

So much for these flights are fancy; now that the context of the problem is well established in relation to errors and their management, let us examine the method of error detection and signalling and the recovery process.

**Error detection**

Having specified the principles of error management that are to be applied, let us now investigate the mechanisms of error detection.

There are five different types of errors, which are not mutually exclusive.

*In the physical layer (bit errors and bit stuffing errors)*

- Bit error
  A CAN sender checks at all times that the level of the bit that it wishes to send along the bus actually matches the desired level. If it does not match, this is signalled by a bit error, except in the following cases:
  - If a recessive bit is sent in the stream of stuffed bits in the arbitration field or in the ACK slot. This is because, in these time intervals, the recessive bit can be 'squashed' by a dominant bit without any error occurring. In this case, the CAN microcontroller normally interprets this situation not as a bit error but as either a loss of arbitration in the first case or as a possible acknowledge response. In the acknowledge slot, therefore, only the receiving CAN microcontrollers (not the transmitting ones) can recognize a bit error.
  - A transmitter sending a passive error flag and detecting a dominant bit must not interpret it as a bit error.

- Stuff error
  In the start of frame, arbitration, control, data and CRC fields, bit stuffing is present. There are two ways of detecting a bit stuffing error:
  - A disturbance generates more than 5 consecutive bits of the same polarity (note that this disturbance can be deliberate, to create a bit stuffing error in order to signal an anomaly to all participants – see below).
  - A disturbance falsifies one or more of the 5 bits preceding the stuff bit. In this case, the receivers *should* – but do not – recognize a bit stuffing error. The reason for this? Other error detection procedures, such as CRC checks and checks of format violation concerning CAN receivers, will be responsible for signalling it. The same applies to bit errors for CAN transmitters. So, this type of error will not go undetected after all.

*Everything is all right at bit level, and at frame structure level – but what if, for example ...*

- ACKnowledgement error
  This error is detected by a CAN transmitter when it does not see a dominant bit on the bus during the time reserved for the ACKnowledge slot.
- CRC error
  The receiver calculates the CRC in the same way as the transmitter (finding the remainder from the division). Consequently, there is an error if, after calculation and checking, the value of the CRC does not match the expected value (including parity).

*Violation of the frame structure (information in the wrong place, etc.)*

Structure errors relate to

- CRC delimiter errors,
- ACKnowledgement delimiter errors,
- end of frame errors,
- error delimiter error,
- overload delimiter errors.

During the transmission of these fields, an error condition is recognized if a dominant level is detected in place of a recessive level.

***Summary***

The CAN protocol uses the following error detection mechanisms:

- bus monitoring,
- cyclic redundancy check,
- message frame check,
- bit stuffing,
- acknowledgement,
- error signalling.

*Bus monitoring*. The bit sender compares the signal it wishes to send with the signal that it physically observes on the physical line of the bus. If the signal present on the bus is different from the sent signal, the sender then sends an error frame, except in the case of the arbitration phase (during the transmission of the message identifier) and the ACK slot. Thus, the bit errors affecting all the stations present on the bus cannot lead to non-detectable errors, because they would be detected by the sender of the frame.

*Cyclic redundancy code – CRC*. The 15 CRC bits are calculated from each bit of the SOF up to the last data bit. The BCH code used to create the CRC gives a Hamming distance of 6, including a parity check in the stuffed bit sequence.

*Frame check message*. The SOF, RTR, IDE, DLC, and EOF fields and the delimiters must be consistent with the CAN specification. If one of the fixed-format fields of the received frame (except the last bit of the EOF) does not conform to the standard, the receiver sends an error frame and does not accept the received frame.
*Bit stuffing*. A violation of the bit stuffing rule between the SOF and the CRC must be considered as an error.

*Acknowledgement*. The sender of a data frame or remote frame treats an absence of acknowledgement as an error and destroys the EOF field, sending an error frame.

*Error signalling*. Each station detecting an error starts an error frame so that the other stations present in the network see a violation of the bit stuffing rule, or of the fixed format of the delimiters, or of the EOF fields. Note once again that at this moment all the stations seeing a deliberate bit stuffing error respond by also sending an error frame.

Figure 2.23 summarizes the corresponding error detection procedures for the CAN transmission and receiving modes.

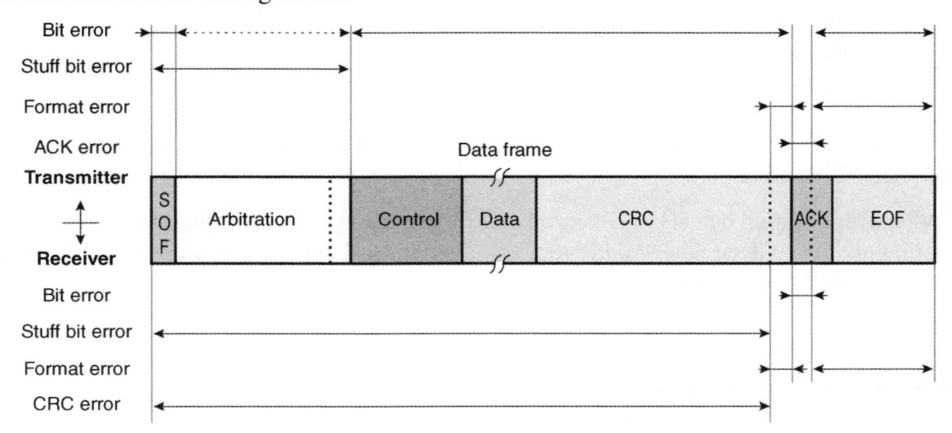


**Figure 2.23**

Now that all the errors have been detected, let us move on to the phases of signalling and informing all our companions on the bus via a dedicated error frame representing the preceding events.

## Error signalling

*The essentials*

When errors have been detected, it generally makes sense to report them to the network, for the following reasons:

- to inform the participants of the network and enable them to take action;
- and in order to inform the other participants, at the same time, of the local state of the station at that instant (with allowance for the state of the station's counters of previously detected errors), and thus to provide some indication of the 'local' quality of the area of the network where the station is located.

For this purpose, a node detecting an error condition is required to signal it by transmitting different error flags according to the instantaneous states of its own error counters:

- for a node in the error passive state, a passive error flag;
- for a node in the error active state, an active error flag.

Before explaining the content of the error frames, I shall now describe the special mechanism for controlling the precise instant of their start.

*Start of error frames*

Here are some important remarks:

Error frames are triggered at different times according to the sources of the errors.

- In the case of
  - a bit error,
  - a bit stuffing error,
  - a structure error,
  - an acknowledgement error,

the transmission of an error flag starts simultaneously with the next bit, from the station which has detected the error.

- Whenever a station detects a CRC error, the transmission of an error flag starts with the bit following the ACKnowledge delimiter, unless an error flag for another error condition has already been started elsewhere in the network.

*2.2.7 Error frame*

Let us examine the composition of the error frame (Figure 2.24). It comprises only two fields:

- an error flag field,
- a field delimiter.

This is followed by the area called the interframe space.

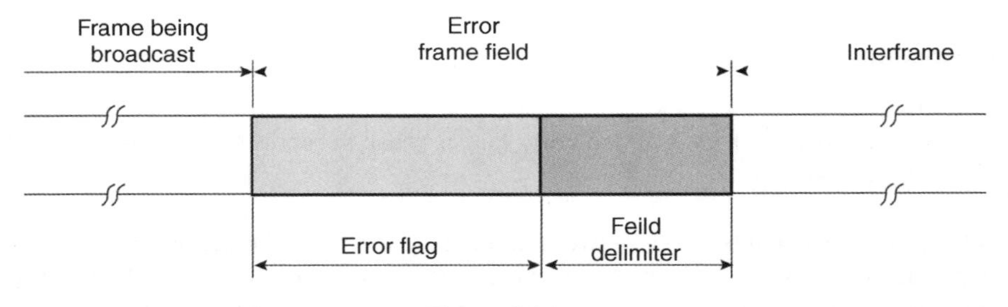


**Figure 2.24**

**The error flag field**

This first field is formed by the superimposition of the error flags to which the different stations present on the bus have contributed. There are two kinds of error flags:

- active error flags,
- passive error flags.

**Note:** For the terms precisely defining what are known as error active and error passive stations, please refer to the preceding sections.

*Active error flag*
An error active station detecting an error condition signals this by transmitting an active error flag. By definition, an active error flag consists of 6 consecutive dominant bits (Figure 2.25).

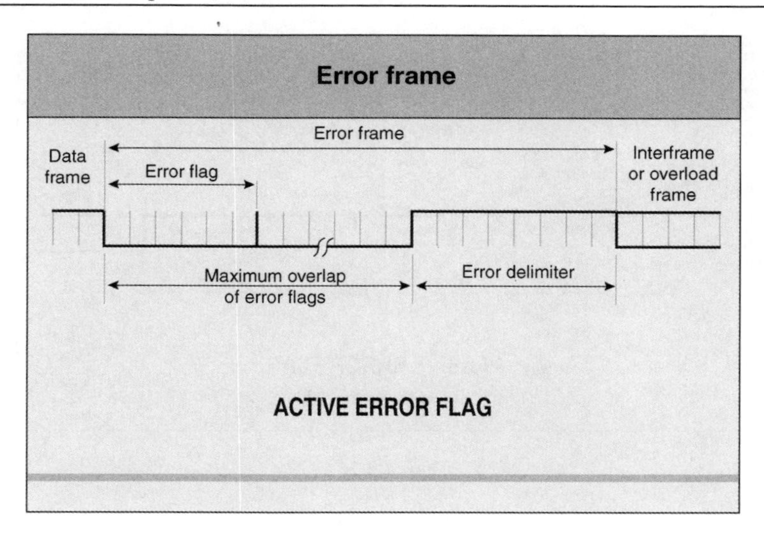


**Figure 2.25**

By their very nature, the active error flags violate the sacrosanct law of bit stuffing applied from the start of frame to the CRC delimiter (see above and below), since 6 consecutive bits are identical, or destroy the fixed and specified format of the ACK field, as well as that of the end of frame field whose bits are normally only recessive.

Consequently, all the other stations present on the network also detect an error condition and also start the transmission of a new error flag (active or passive, according to their own states). This becomes a destructive chain reaction!

The sequence of dominant bits that can be seen on the bus in this case results from the superimposition of the participation of the different error flags which are supplied by all the individual stations (and which each of them can examine – monitor – in real time).

The CAN document allows the total length of this sequence to vary from a minimum of 6 bits (the length of the active error flag) and a maximum of 12 bits, to prevent the bus from being blocked indefinitely.

*Passive error flag*

A station which is said to be error passive and which detects an error condition 'tries' to signal this by transmitting a passive error flag. By definition, a passive error flag consists of 6 successive recessive bits . . . unless these bits are 'squashed' by dominant bits sent from other nodes (Figure 2.26).

According to the principle of recessive bit transmission, a passive error flag cannot interrupt a current message between other different controllers present on the bus, but this type of error flag can be ignored ('squashed') by the other controllers.

Having detected an error condition, a controller in passive error mode waits for 6 consecutive bits of identical polarity and, when it has identified them, interprets them as an error flag.

The error passive station waits for 6 consecutive bits of the same polarity, beginning with the start of the passive error flag. The passive error flag is terminated when these 6 identical bits have been detected.

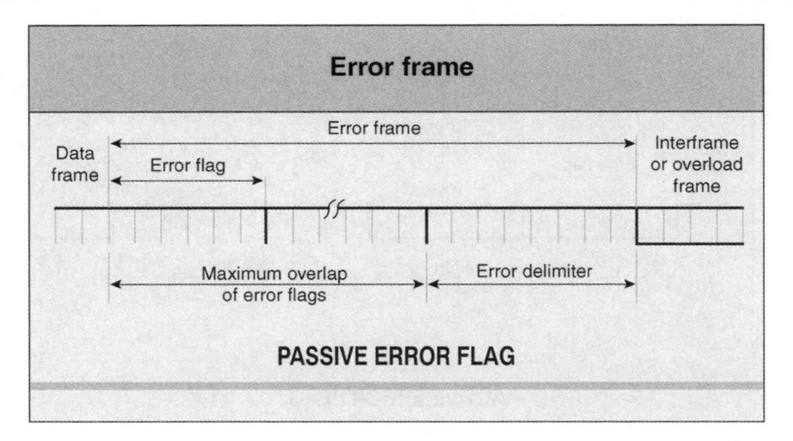


**Figure 2.26**

### Error delimiter

The error delimiter consists of 8 recessive bits.

After the transmission of an error flag, each station 'sends' recessive bits and examines (monitors) the bus, until it detects a change from a dominant level to a recessive level. At this instant, each CAN controller has finished transmitting its error flag and has additionally sent the first of the error delimiter bits. All the CAN controllers are then able to start the sequence of the remaining 7 recessive bits, to finish the construction of the 8 bits of the error delimiter.

After this event and an intermission field, all the 'error active' controllers present on the network can start a transmission simultaneously.

If a detected error is signalled during the transmission of a data frame or a remote frame, the current message is destroyed and a retransmission of the message is initialized.

If a CAN controller recognizes a frame error, a new error frame will be transmitted. Several consecutive error frames can make the CAN controller become 'error passive' and leave the network unblocked.

**Note:** In order to terminate an error frame correctly, a passive error station may need the bus in bus idle mode for a time equivalent to at least 3 bits (if there is a local error at a receiving station in passive error mode). Therefore the bus should not be 100% loaded at all times.

After these long discussions, Figure 2.27 shows some concrete examples of what can happen in different circumstances.

### *2.2.8 Error recovery*

Error recovery is carried out by the automatic retransmission of the disturbed frame, until the frame can be sent correctly and there are no more error messages.

This can continue for ever and ever . . . or almost, because if the errors do happen to persist, the error counters will be incremented and sooner or later the circuit creating (or recognizing) the error will be the first to switch to bus off mode.

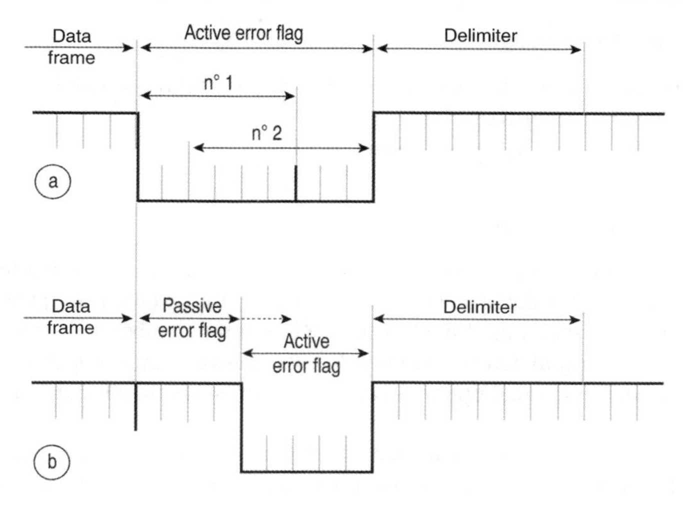


**Figure 2.27**

After the removal of this 'troublemaker', there should be no more errors, and the message should be sent correctly. If the problems persist, then we do the same with another troublemaker. I can see your question coming! What if all of them are troublemakers? In that case, all of them will disappear, one by one, and you will be the only station continuing to communicate over the network: Nobody will reply to you, so there will be no acknowledgement, no error, no error counter, etc. So it will be time for you to take rest as well! And you too will switch to bus off mode ... and everyone will be sleeping (with one eye open)! The bus will then be momentarily 'blocked' because no station will be able to see (with that one open eye) the 128 occurrences of 11 recessive bits to enable it to wake up.

At this point the hero arrives, in the form of the undaunted time-out and his noble and faithful 'time-out control', to wake you with a kiss, like the sleeping beauty! The kiss that wakes you is simply a smack on the power-on-reset switch of the integrated circuit! A bit brutal, admittedly, but very effective!

### *2.2.9 Validity of messages*

The precise instant when a message is considered valid differs according to whether you are the sender or the receiver of a message.

#### **For the sender of the message**

For the sender of the message, the message is considered valid if no error is detected up to the end of the 'end of frame'. If the message has been corrupted, retransmission will automatically follow and will be carried out in according with the rule of message priority.

So that it can compete for bus access with other messages, the retransmission must start as soon as the bus is in idle mode.

### For the receivers of a message

For the receivers of a message, the message is considered valid if no error is detected up to the penultimate bit of the 'end of frame'.

In all other cases, the messages are considered invalid.

## *2.2.10 Error processing performance*

The performance of the error processing in a protocol is not an irrelevant matter. Admittedly, this aspect is not covered in the protocol, but it is an underlying factor, as it has played a part, for example, in the specification of the type of CRC and the number of its bits. Furthermore, this consideration has stimulated (too) many disputes between different protocols for a long time because it relates to the real operation of the networks which, although similar, are often not strictly identical.

To achieve an equitable settlement, teams of academic researchers frequently develop theories backed up with ample calculations, for a thorough justification (or refutation) of the qualities and virtues of one or another device in the protocols.

Being unwilling to start yet another argument, I shall simply list the principal results which are known and accepted by those in the profession.

### Error classes

The residual errors are generally divided into a number of major classes:

- errors that do not disturb the frame length (Figure 2.28). In this class, all the fields of the transmitted frame are interpreted as valid by the receiving station. The fields affected by bit errors can only be the ID, DLC and DATA fields.

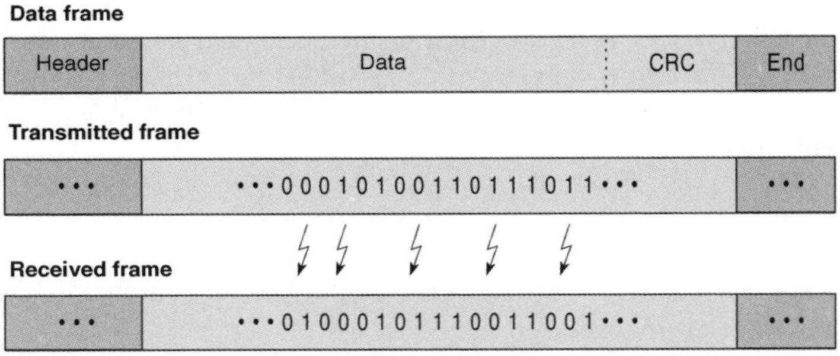

Data frame

Header | Data | CRC | End

Transmitted frame

••• | •••0001010011011101 1••• | •••

Received frame

••• | •••0100010111001 1001••• | •••

**Figure 2.28**

- errors contributing to a change in the predictable length of the frame (Figure 2.29). If the SOF, RTR, IDE or DLC fields are changed by bit errors, the receiver expects a frame having a different length from the frame sent originally, and interprets the incoming frame with a different interpretation from that expected originally.

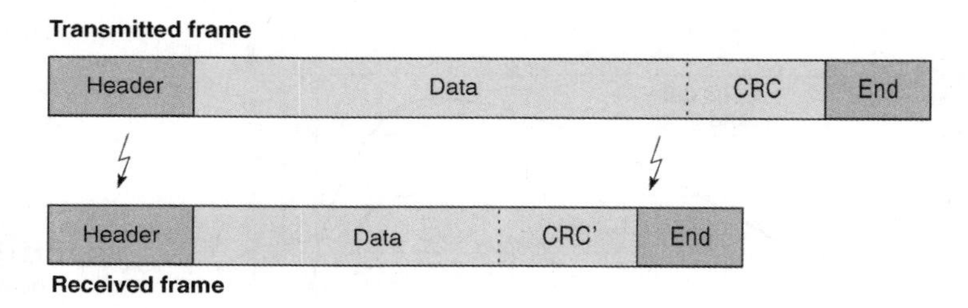


**Figure 2.29**

Depending on the changes produced by the errors, the receiver may be expecting a shorter or a longer frame. However, the probability of a residual error is much lower for increases than for reductions in frame length due to errors.

### Error transformation

Other types of errors can occur. For example, parasites may affect the content of the data without violating the bit stuffing rules, or a pair of erroneous bits may suppress the presence of a stuff bit.

### Analysing the causes of errors

When the errors have been classified and listed, it becomes possible to make a quantitative analysis of all probable cases of errors. The calculations relating to this analysis are largely outside the scope of this book (for further details, you should consult the highly specialized reports on this subject listed in the References section at the end of the book).

Briefly, the analysis consists of

- summarizing the hypotheses and parameters taken into account;
- modelling the transport channel;
- developing the mathematical formula for the probability of residual error;
- the results:
  - the probability of standardized residual error,
  - the contribution of each class of error,
  - the effect of the length of the data field,
  - the effect of the frame format and the number of stations present on the network.

The results are summarized in Figure 2.30.

## 2.3 The Rest of the Frame

### *2.3.1 Overload frame*

The purpose of this frame is to indicate that a station has been overloaded for a certain time. Let us examine its composition (Figure 2.31).

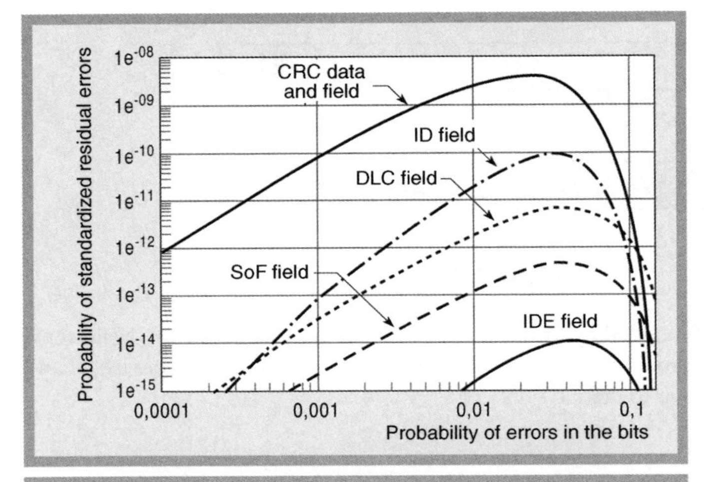


For CAN 2.0A
data = 8 bytes/frame
Number of nodes = 10

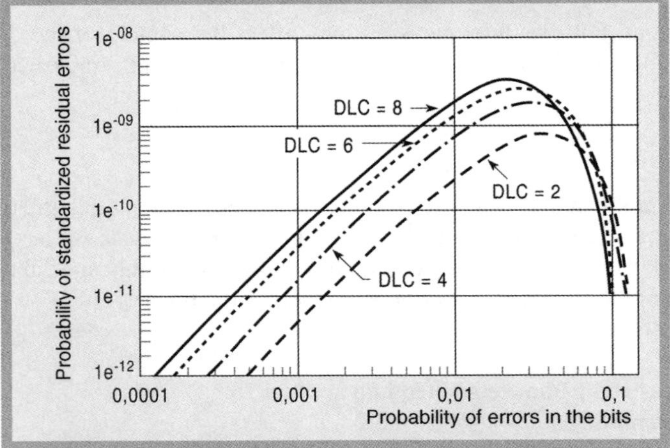


Effect of the DLC

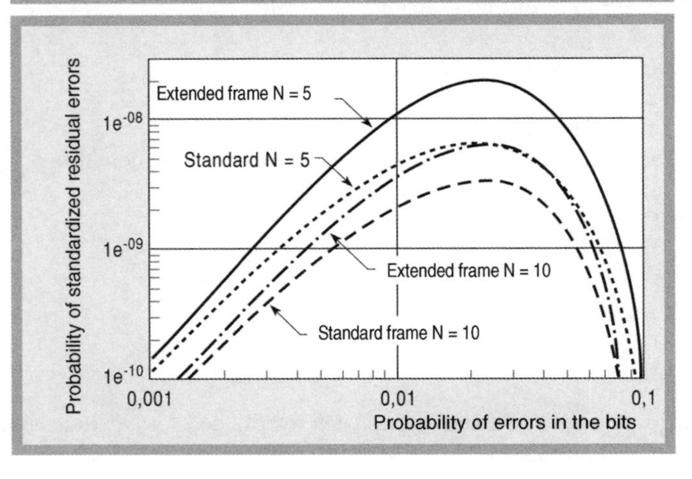


For CAN 2.0A and B
data = 8 bytes/frame
Number of nodes = 5/10

**Figure 2.30**

It comprises only two fields: the overload flag field (OLF) and the field delimiter. As shown in Figure 2.31, it can appear at the end of a frame or of an error delimiter, or at the end of another overload delimiter in place of the start of the interframe.

It is followed by the area called the interframe space, or by another overload frame.

There are two kinds of overload condition leading to the transmission of an overload flag:

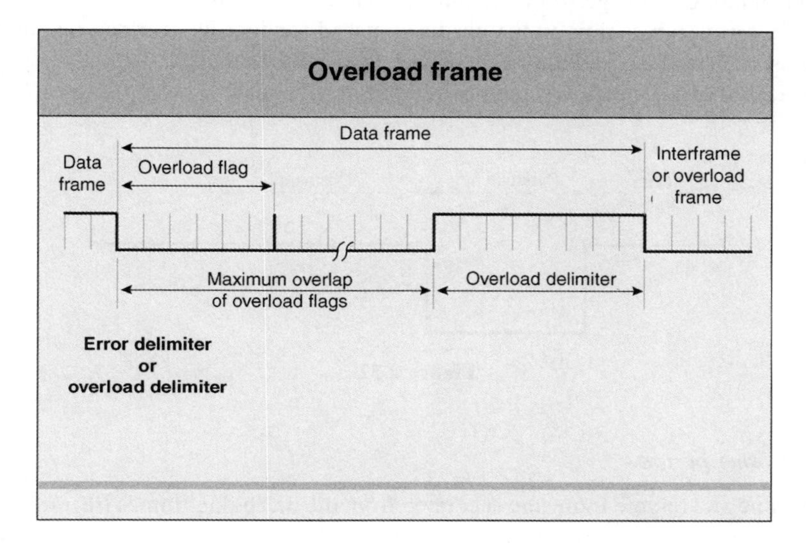


**Figure 2.31**

- the internal conditions of a receiver which require a certain time (a delay) for the acceptance of the next data frame or remote frame. In this case, the start of an overload frame is allowed only in the first time bit of the anticipated intermission;
- the detection of a dominant bit during the intermission phase. In this case, the start of the overload frame occurs immediately after the detection of the dominant bit.

To avoid blocking the bus indefinitely, only two consecutive overload frames can be generated to delay the following data or remote frames.

### Overload flag field

This consists of 6 consecutive dominant bits, like its counterpart the active error flag. The structure of the OLF destroys the defined structure of the intermission field.

All the other stations then detect an overload condition, and each starts to transmit an OLF on its own account. If a dominant bit is detected locally in a node during the 3rd bit of the intermission, the other nodes do not interpret the OLF correctly, but interpret the first of these 6 dominant bits as a start of frame.

Here again, the 6th bit inevitably violates the sacrosanct law of bit stuffing, and therefore all the other stations detect an error condition and each one starts to transmit an error flag on its own account. Once again, we have a destructive chain reaction!

### Overload delimiter

By definition, the overload delimiter consists of 8 consecutive recessive bits, like its counterpart the error delimiter. They also have the same structure.

After the transmission of an overload flag, the station examines the bus until it detects a transition signalling a change from a dominant to a recessive bit.

At this instant, each station on the bus has finished sending its overload flag, and all the stations start transmitting 7 additional recessive bits simultaneously.

Figure 2.32 shows a concrete example of what may happen.

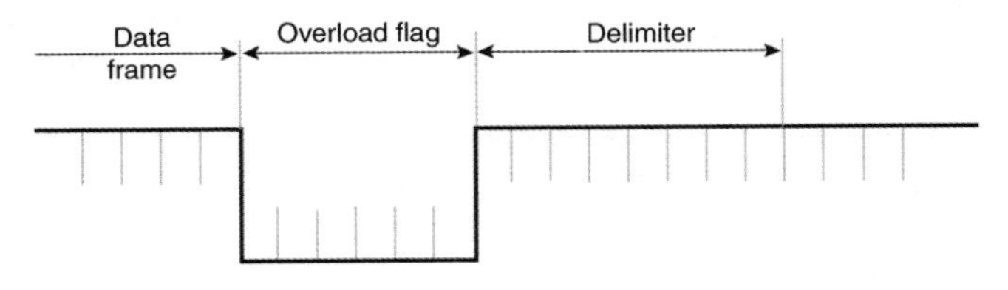


**Figure 2.32**

## *2.3.2 Interframe period*

The data frame and remote frame are separated from the preceding frames (regardless of their type: data, remote, error or overload frame) by a field of bits called the interframe space. However, the overload frame and error frame are not preceded by an interframe space, and multiple overload frames are not separated by an interframe space (see all the above figures and note the details of these operating phases of the bus).

The interframe space consists of two or three fields, as appropriate (Figure 2.33):

- the intermission bit field,
- the bus idle bit field and
- for passive error stations which have been transmitters of the preceding message, a suspend transmission bit field.

### Intermission field

This consists of 3 recessive bits.

During the intermission, no station is allowed to start transmitting the data frame or the remote frame. The only possible action is the signalling of an overload condition.

### Bus idle field

The period for which the bus is idle (at rest, waiting) can be chosen arbitrarily. During this time, the bus (the line) is 'free', and any station having data to transmit can access the bus.

A message which was waiting to be transmitted during the preceding transmission can then start with the first bit following the intermission.

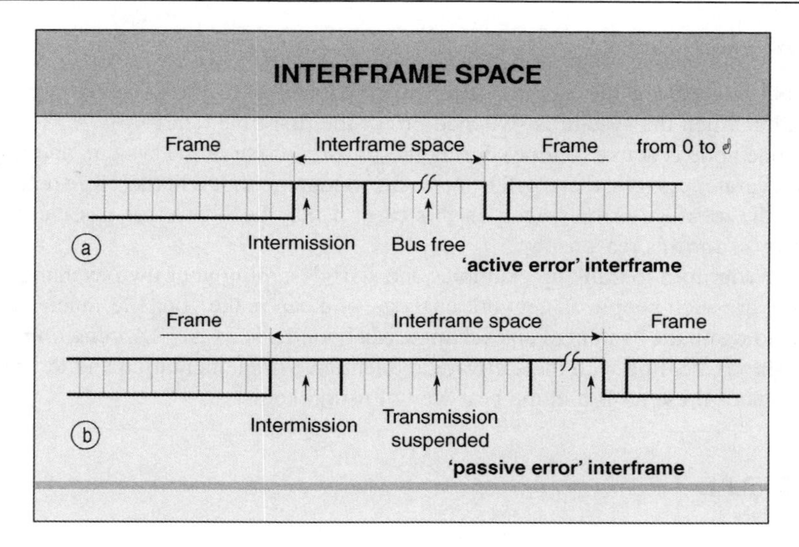


**Figure 2.33**

The detection of a dominant bit on the bus during the intermission is interpreted as the start of frame of a new transmission.

**Suspend transmission field**

After an error passive station has transmitted a message, it sends 8 recessive bits following the intermission field, before starting the transmission of another message or recognizing that the bus is 'free'. If a transmission (from another station) starts in the same time interval, the station becomes a receiver of this message.

*2.3.4 Sleep mode and wake-up*

To reduce power consumption, a participant in the CAN network can switch to sleep mode; in other words it can cease to have any internal activity, having disconnected its bus drivers from the line.

The sleep mode is terminated by a wake-up caused by any activity on the bus or by the internal conditions of the local system.

On waking up, the internal activity of the circuit is restarted (although the transfer layer is waiting for the stabilization of the system oscillator) and it then waits until it has become synchronized with the bus operations (checking the 11 recessive bits), before the bus drivers are switched back to 'on the bus' mode.

This means that it is possible to lose one or more messages during the wake-up phase of a participant, although as these messages are not acknowledged it will clearly be necessary to resend them.

To overcome this problem and wake up the sleeping nodes of a system, a special wake-up message has been reserved for those who may be interested. This message contains the identifier of lowest priority, i.e. 'rrr rrrd rrrr', where 'r' is recessive and 'd' is dominant.

### *2.3.5 Start-up/wake-up*

A final note concerning the above section is concerned with the wake-up not in normal operation, but when the system is switched on for the first time (start-up).

If only one node is active (online) during the start-up phase of the system, and if this node transmits several messages, there will be no acknowledgement. It will therefore detect an error and repeat the message or messages. In this case, it may become error passive, but cannot become bus off for this reason.

Now it is your turn to think up examples and start designing your own exchanges! Do not worry, there are such people as network analysts who can make your life much easier when you are deciding on the content of an exchange. My book *Le bus CAN – Applications* (D. Paret, Éditions Dunod), dealing with the software applications, implementation and testing, reveals all the secrets of these tools and the best way of using them.

## 2.4 CAN 2.0B

### *2.4.1 Introduction*

After the commercial launch of the CAN protocol concept (version 1.2), it soon became clear that the 11-bit identifier field of the 'standard' frame could cause problems in some very specific cases of applications. To provide more user-friendly systems, it was important to devise what came to be called an 'extended' frame, comprising a larger identifier field (29 bits). To achieve this it was necessary to modify the data frame and remote frame formats. This was done with the provision of upward compatibility and with the name of the first 'standard' frame changed from 1.2 to 2.0A, while its new extended version was called 2.0B.

### *2.4.2 Differences between CAN 2.0A and CAN 2.0B*

To avoid unnecessary repetition of lengthy material, I shall use the 'differential method' to indicate the differences between the CAN 2.0A and the CAN 2.0B protocols.

#### CAN 2.0B and the ISO

The description of the CAN 2.0B protocol reproduces the ISO/OSI reference model. You should refer to the start of the chapter if you need to refresh your memory concerning the applications and functions of the different layers.

#### Frames

The concepts of the data, remote, error and overload frames are the same.

#### Frame format

The main differences between the structures of the frame formats lie in the number of bits making up the identifier field and the significance of the 'reserved' bits (Figure 2.34):

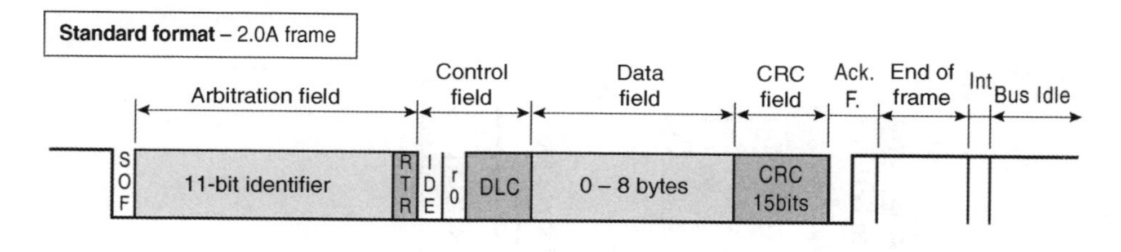


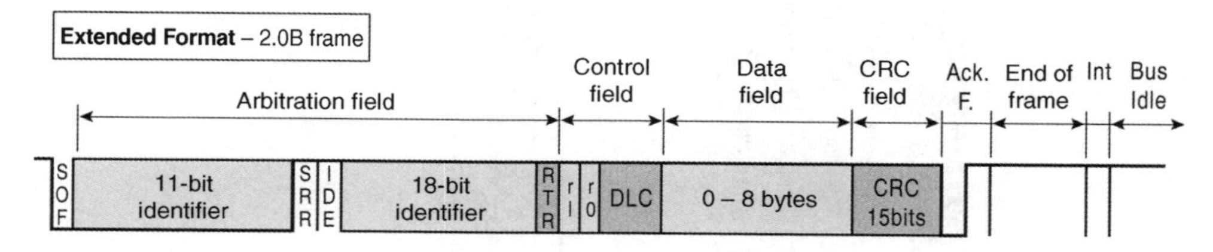


**Figure 2.34**

- 'standard format', CAN 2.0A, 11-bit identifier,
- 'extended format', CAN 2.0B, 29-bit identifier.

Note that the extended format has been constructed in such a way that both formats can coexist in the same network.

### Identifiers and arbitration field

The 2.0A and 2.0B arbitration fields are inherently different (Figure 2.35a and 2.35b).

- In CAN 2.0A, this field consists of 11 bits, renamed ID_28 to ID_18 in the CAN 2.0B frame, and the RTR bit. This (high) part of the identifier field is called the Base ID and provides the 'base priority' of the extended frame.

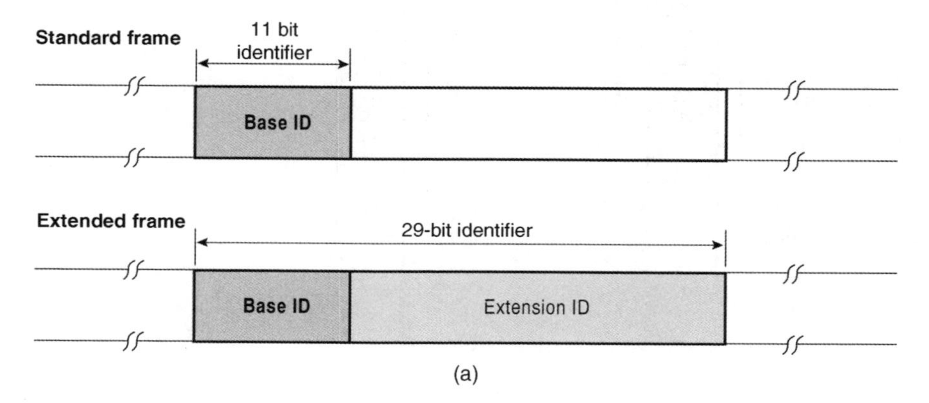


**Figure 2.35**

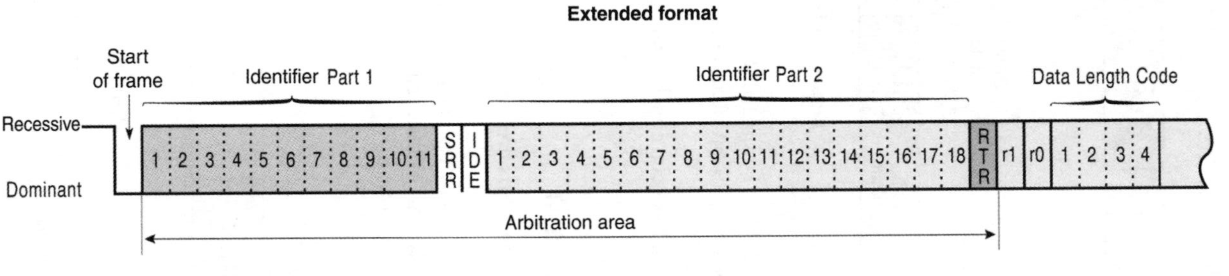


| TX | Bus (RX) | Action |
|---|---|---|
| Recessive | Recessive | Arbitration procedure |
| Dominant | Dominant | Arbitration procedure |
| Recessive | Dominant | Arbitration lost |
| Dominant | Recessive | Bit error |

**CAN arbitration**

(b)

**Figure 2.35** (*Continued*)

- In CAN 2.0B, this field consists of 29 bits, called ID_29 to ID_00, and 3 bits: The former RTR (which has now moved house) and two newcomers, SRR and IDE, extracted from the reserve bits of the CAN 2.0A frame of the control field (they had been waiting a long time for this moment).

The proper names of these three bits are

- RTR: remote transmission request bit,
- SSR: substitute remote request bit,
- IDE: identifier extension bit.

**Notes:** You should carefully note the relative positions and the names of the RTR, IDE and SSR bits in the standard frame (2.0A) and the extended frame (2.0B).

If you are very sharp-sighted, you may again notice the appearance of two new reserved bits (r0 and r1) which have not yet been defined in the new CAN 2.0B frame. You can never take too many precautions!

Now let us look at their more intimate features. Figures 2.36 and 2.37 summarize their functions and actions.

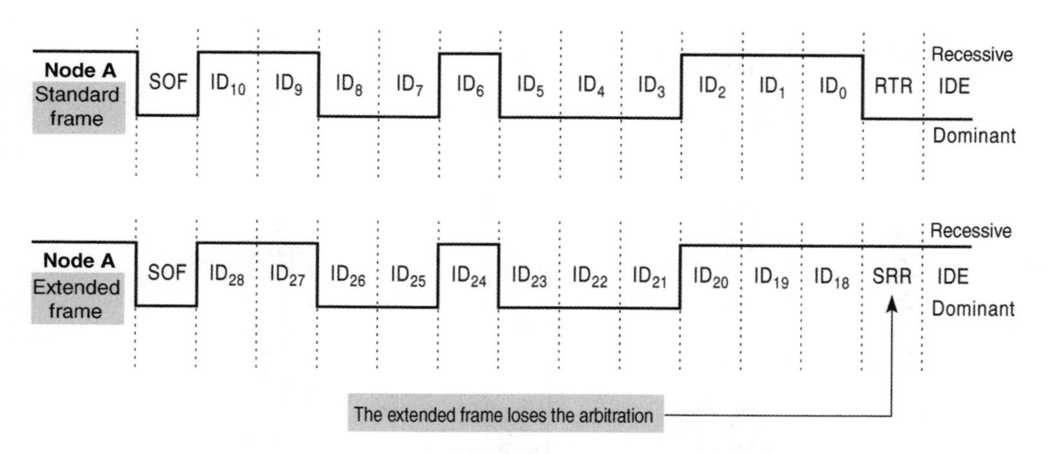


**Figure 2.36**

## Control field

Although still consisting of 6 bits, the control field has a new structure, following changes of the bit names between 2.0A and 2.0B frames.

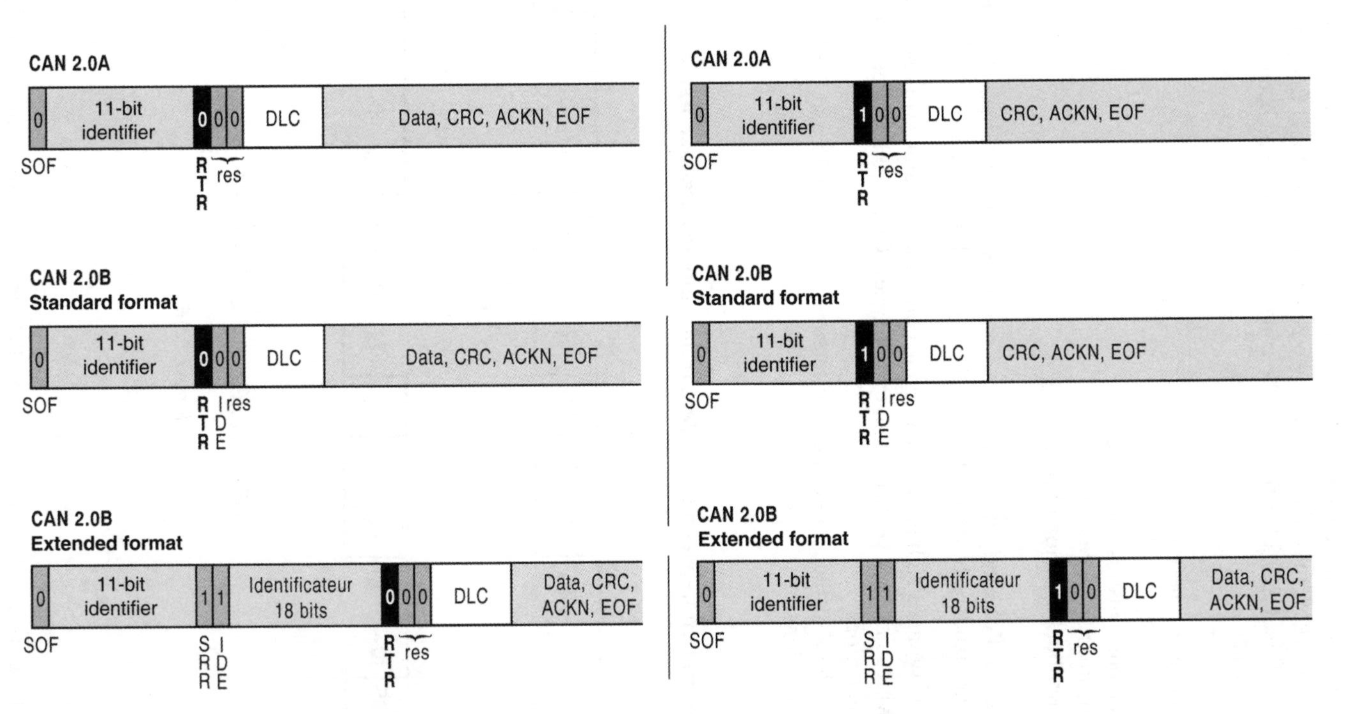
Data frames
Request frames
CAN 2.0A
0
11-bit identifier
0 0 0
DLC
Data, CRC, ACKN, EOF
SOF
RTR
res
CAN 2.0B
Standard format
IDE
CAN 2.0B
Extended format
1 1
Identificateur 18 bits
SRR
1 0 0
CRC, ACKN, EOF

Figure 2.37

### Obligations of a CAN 2.0B controller

By definition, in the official specification, a controller supporting the extended format of part 2.0B must support the standard 2.0A format without restrictions. To avoid any possible misunderstanding, let me quote from the original specification:

'The controllers are considered to conform to the specification of Part B of version 2 of the protocol if they have at least the following properties [author's note: in accordance with the points mentioned in the preceding paragraphs]:

- the controller supports the standard format;
- the controller can receive messages in extended format. This implies that the extended frames cannot be destroyed because of their formats alone. However, it is not necessary for the extended format to be supported by new controllers'.

End of quote, without comments!

Now that this is clear in your mind, let us look at the problems of cohabitation and compatibility of CAN 2.0A and CAN 2.0B circuits in the same network.

### Compatibility of CAN 2.0A and CAN 2.0 B

To conclude this long chapter, I shall now offer some remarks concerning the compatibility between the 2.0A and 2.0B versions of the CAN protocol, in relation to the production of the electronic components.

For specific application-related reasons concerning the extension of the possibilities of the network, users quite frequently want the same bus to carry both 2.0A standard frames and 2.0B extended format frames. This gives rise to a well-known problem: namely, how to find certain network components (not necessarily obsolete) which operate solely and strictly according to part 2.0A, and which automatically generate an error frame on the appearance of a 2.0B extended frame, whose structure is unknown to them in respect of the service bits. This can clearly lead to disaster ...

At this point, we must introduce certain concepts such as 'CAN 2.0A active, 2.0B passive'. This can be broken down into a number of classes of components.

In the case of CAN 2.0A components, we can find, for example

- 2.0A active only,
- 2.0A active and 2.0B passive (no error frames generated on the appearance of a 2.0B frame).

For the CAN 2.0B components, the protocol requires that all the 2.0A and 2.0B frames should be recognized and processed automatically when they are received. In this case, where transmission is concerned, the user is responsible for choosing type 2.0A or 2.0B.

For guidance, Figure 2.38 indicates the privileged application fields of CAN 2.0A and 2.0B.

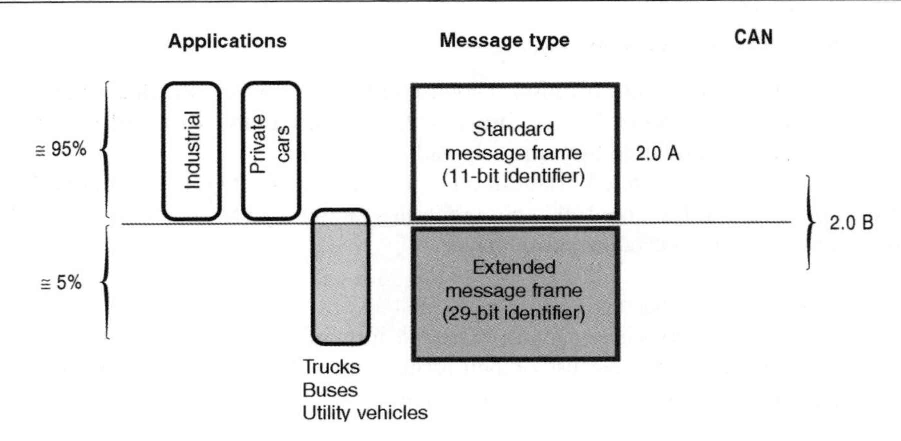


**Figure 2.38**

This concludes (for the time being) the description of the major features of the CAN bus. Admittedly, it is all rather complicated, but sometimes we have to tackle material like this if we are subsequently to understand the integrated circuit operation.

I will now go on to examine the physical layer for the design of the systems, after which we can look at some examples of implementation.

# 3

# The CAN Physical Layer

In the preceding chapters, I described the general principles of the CAN (controller area network) protocol (mainly dealing with OSI/ISO layer 2). As you would certainly have noticed, I have not yet paid much attention to the physical layer along which the CAN frames must travel sooner or later, and its many applications. To fill this gap, I propose to set out a little more of the theory on the physical layer and the problems raised by different types of medium (OSI/ISO layer 1 and layer 0). To make this topic easier to grasp, I have broken down the description into two chapters as follows:

- Chapter 3
  - the physical layer described in the reference document and in the ISO (International Standards Organization) standard,
  - the properties of the bit,
  - the types and structures of networks,
  - the problems of signal propagation, distance and bit rate.
- Chapter 4
  - the different mediums that can be used,
  - the line protection (bus failure management),
  - circuit-to-circuit compatibility,
  - compliance with pollution standards.

## 3.1 Introduction

At this point in the book, I am assuming that you now have a good knowledge of the protocol, the frame management and all the special features of these systems. All we have to do now is step on the bus while it is moving (a rather dangerous thing to do ...!) In short, you know the general structure of the protocol ..., but there are still many specific 'low-level' questions (in the sense of the OSI/ISO layers, of course) that need to be mentioned before we can finally get the bus started in its hardware form:

*Multiplexed Networks for Embedded Systems: CAN, LIN, Flexray, Safe-by-Wire...* D. Paret

- What is the actual speed of the bit?
- At what moment can we say that it is a '1' or a '0'?
- What is the best physical medium for the protocol transport?
- Are ordinary wires good enough for carrying the data?
- Are optical fibres a 'must'?
- What is the optimum bit rate for a given application?
- What maximum operating distance can we expect?
- Are there one or more standard socket(s)?
- How do we 'drive' the line?
- How do we protect the line?

In short, I will raise all the standard questions concerning the physical layer and the related problems of bit rate, choice of appropriate physical interface, standard sockets and the eternal debate over bit rate/distance/medium.

### *3.1.1 The CAN and ISO physical layer*

CAN users often depict the physical layer in an (over-)simplified way, talking about its 'line' aspect only, because this is the feature we first notice. We will be able to identify numerous problems if you are willing to stop thinking of yourself as a standard user for a while and carefully dissect all the aspects of the CAN line.

For this purpose, I shall return to a more academic style of presentation: The ISO has issued documents (see below) describing the standardization of the data link layers and physical layers for specific applications (those for motor vehicles), which can easily be applied or transposed to many other fields. These documents describe the CAN architecture in terms of hierarchical layers, according to the definitions of the 'Open Systems Interconnection (OSI) base reference model', and include specifications relating to the following lower layers:

- Data link layer (layer 2), DLL,
- Logical link control sublayer, LLC,
- Medium access control sublayer, MAC, described in the preceding chapters, and
- Physical layer (layer 1), PL:
  - Physical signalling sublayer, PLS (bit coding, bit timing, synchronization),
  - Physical medium attachment sublayer, PMA (characteristics of the command stages (drivers) and reception stages),
  - Medium dependent interface sublayer, MDI (connectors);
- Transmission medium:
  - bus cable,
  - bus termination networks.

**Notes:** The ISO documents described below (entitled 'CAN high speed' and 'Fault tolerant low speed') differ from each other in respect of the bit rates, PMA, MDI and transmission medium.

The layers above the Data link layer are not covered by the ISO documents. They have to be developed and specified by the project designer.

### *3.1.2 Basic and specific qualities required in the physical support to be used*

Although the type of physical support used to carry the CAN frames is not explicitly defined, but left open in the CAN protocol (for example: wire, RF, IR, optical fibres, visible light, arm signals, semaphore, telegraph or maybe smoke signals – see Figure 3.1), the chosen medium must conform to the CAN structure and architecture by

- being capable of representing bits having dominant and recessive states on the transmission medium;
- being in the recessive stage when a node sends a recessive bit or when no nodes are transmitting;
- supporting the 'wired *and*' function, simply as a result of its implementation;
- being in the dominant state if any of the nodes sends a dominant bit, thus suppressing the recessive bit.

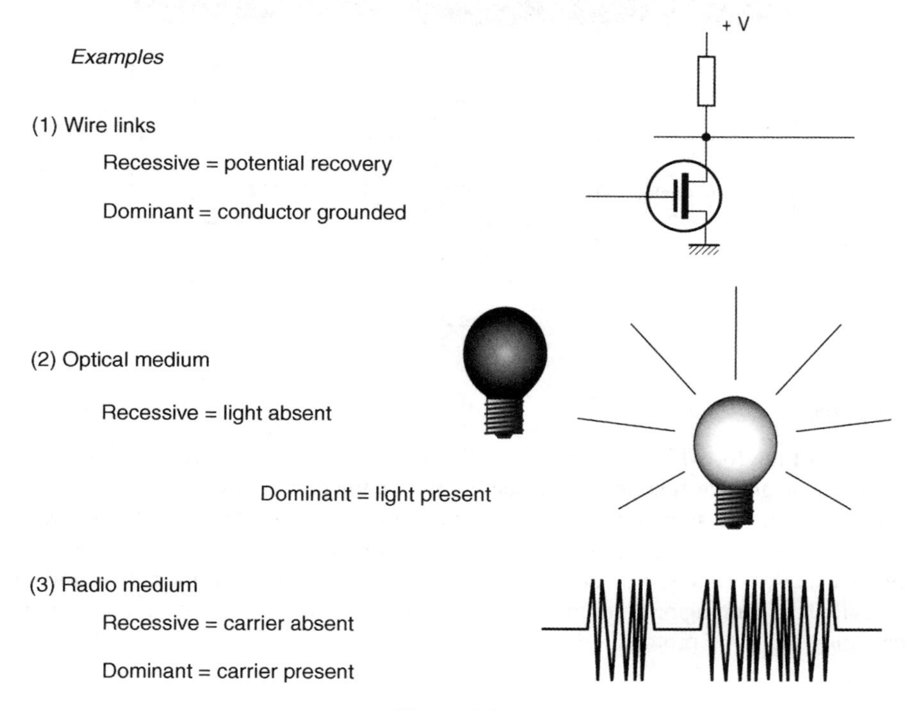


**Figure 3.1**

After these general features which are fundamental to the operation of CAN, let us now examine the details of the 'CAN bit'.

## 3.2 The 'CAN Bit'

We have now arrived in the part of the OSI world concerned with the 'specification of the physical signalling sublayer' (PLS). This long section will lead us into a consideration of the problems of the bit (coding, duration, synchronization, etc.) and the 'gross' and 'net' bit rates of the network.

### *3.2.1 Coding of the CAN bit*

As mentioned above, the bit is coded according to the non return to zero (NRZ) principle – see Figure 3.2.

**Note:** The bit stuffing method described in Chapter 2 does not in any way alter the bit coding type itself!

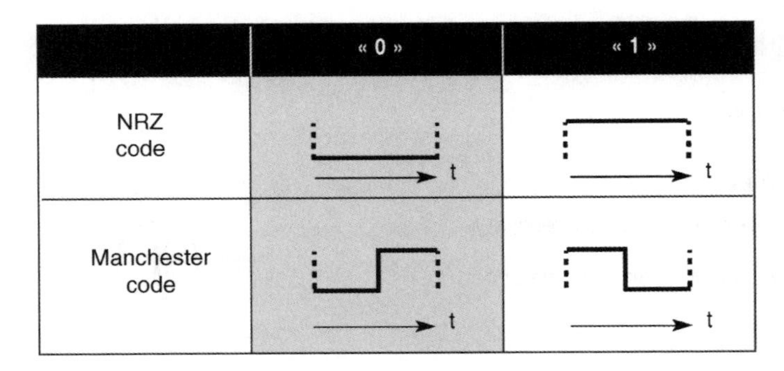


**Figure 3.2**

### *3.2.2 Bit time*

The period of time for which the bit is actually present on the bus is called the bit time, $t_b$.

The bus management functions – carried out within the period of duration of the bit ($t_b$, or 'bit time frame'), and including the synchronization constraints of the stations (ECU, i.e. electronic central units) present on the network, compensation of propagation delays, definition of the exact position of the sample point – are defined by an electronic entity produced in what is called a bit timing programmable logic contained in the integrated circuits responsible for managing the CAN protocol.

### *3.2.3 Nominal bit time*

The nominal bit time is identical to the nominal duration of a bit. By its nature, this value can only be an ideal and theoretical one. It is the value that the system designers wishe to assign nominally to the bit time of their network. Each station in the network must be designed so that it 'nominally' has this bit time $t_b$, but, sadly for the designer, because of the real components which make up each station (and especially because of the details of their oscillators, tolerances, temperature and time fluctuations, etc.), the instantaneous value of the bit will never be the nominal value at any instant ... something to be borne in mind at all times.

The nominal bit time is normally constructed electronically from a specified number of 'system' cycle pulses, based on what should form the internal clock of each station. By definition, it is equal to

$$\text{Nominal bit time} = \frac{1}{\text{Nominal bit rate}}$$

The official CAN protocol document indicates that the nominal bit time is divided into a number of time segments that cannot overlap. In chronological order, these are (Figure 3.3) the following:

- the synchronization segment,
- the propagation time segment,
- segment 1, called 'phase buffer 1',
- segment 2, called 'phase buffer 2'.

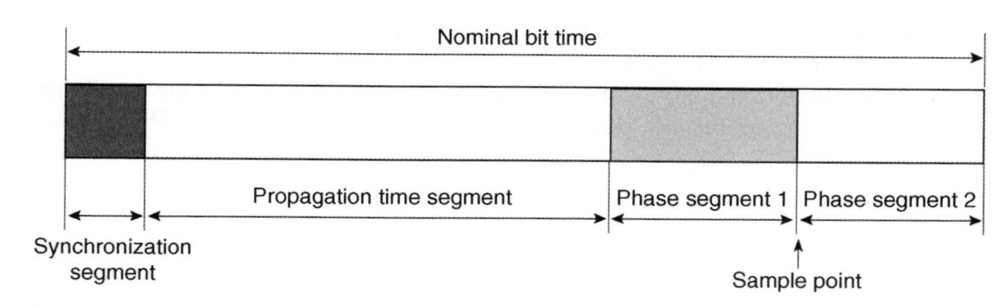


**Figure 3.3**

Now for their definition.

### Synchronization segment

The 'synchronization segment' is the part of the bit time used to synchronize the various nodes present on the bus. By definition, the leading edge of an incoming bit should appear within this time segment.

**Important note:** The word 'should' is justified by the fact that the nominal bit time is equal to the time for which the bit should last, but, as the bit coding is of the NRZ type, it is possible that at the moment when the presence (in time) of an edge is expected, it does not exist (in physical terms) because the signal (bit) present on the bus continues to have its former electrical value and no physical transition signal is produced. Regardless of this, it must be considered that the start of a new bit is probably occurring at this instant.

#### The propagation time segment

This part of the nominal bit time is designed to allow for and compensate the delay time(s) due to the physical phenomenon of propagation of the signal on the medium used and the types of topology used to develop the geometrical configuration (topology) of networks.

To make matters clearer, and enable CAN to operate correctly, I shall simply state that this value is approximately a time equal to twice (outgoing and return) the sum of the propagation times present due to the medium and the delays caused by the input and output stages of the participants which are also connected to the bus (see the section on propagation time below).

#### Phase buffer segments 1 and 2

This pair of phase buffers[1] (phase buffer segments 1 and 2) is used to compensate for the variations, errors or changes in the phase and/or position of the edges of the signals which may occur during an exchange on the medium (due to jitter, variation of thresholds of the level detectors of the input comparators of the integrated circuits, integration and deformation of the signals along the line forming the bus, etc.).

These two segments are designed to be either extended or shortened by the resynchronization mechanism, as described below.

So much for the terminology of these segments. Now let us examine the details of the factors and physical phenomena which make their definition necessary.

### *3.2.4 Sampling, sample point and bit processing*

There is no point in carrying data unless they are to be correctly received, read and interpreted. This requires a considerable number of processing operations on the arrival of a bit, to ensure its integrity. Given these preliminary considerations, we must now define the precise instant when the incoming bit on the bus is sampled and is consequently read and interpreted and has its value defined.

This instant – the 'sample point' – must occur not too soon and not too late during the period of duration of the bit. The choice (and/or the correct position in time) of the sample point is essential for the correct operation, and therefore the success, of a transmission.

Why? The answer is simple; in theory

- on one hand, in order to overcome the constraints of the fundamental principles of the protocol (arbitration and acknowledgement phases), the problems due to the deformation of the signal representing the bit and delays due to the phenomena of signal propagation on the

[1]The term 'phase buffers' is to be interpreted in a sense physically equivalent to the mechanical buffers between railway trucks, whose function is to compensate and absorb the jolting caused by instantaneous differences in speed between the trucks.

medium, the value of the bit should be read and interpreted as late as possible in the bit time to give us the maximum confidence in this value;
- on the other hand, we must not wait for the very last moment, for at least two reasons: firstly, because all the tolerances allowed for in the establishment of the bit time must be taken into account, and, secondly, because of the time required for calculating the value of the bit when it is sampled.

In conclusion, the bit sampling should take place towards the end of the period of duration of the bit.

The position of this precise instant – the 'sample point' – forms, by definition, the exact boundary between phase buffer segment 1 and phase buffer segment 2.

To achieve this, in the design of a project, it is important to remember the other important points listed below, in relation to the precision of the bit sampling instant.

- A start of frame (SOF) obliges all the controllers present on the bus to carry out a 'hardware synchronization' on the first change from a recessive to a dominant level (see below).
- It is also necessary to allow for any arbitration phases during which several controllers can transmit simultaneously.
- To avoid the incorrect positioning (in time) of the sample point of a bit, and thus the possibility of erroneous reading of its value, it is also necessary to include additional synchronization time buffers located on either side of this instant (phase buffers 1 and 2). The main causes of incorrect sampling are
  - incorrect synchronization due to brief pulses (spikes);
  - small frequency variations in the oscillator(s) of each of the CAN microcontrollers of the network, giving rise to phase errors.

The phase 2 time segment (having the same form as phase segment 1 of the CAN protocol) has the task of forming

- a time buffer segment positioned after the sampling point, with the purpose of participating in any necessary (re)synchronization of the bit,
- a reserve time, commonly called the 'information processing time'. This time interval, located and starting immediately after the sample point (i.e. immediately at the start of the phase 2 time segment), is necessary for the calculation of the value of the bit.

**Note on the bit calculation.** To improve the quality of the capture and validation of the bit (in the presence of parasites, an arbitration phase if any, etc.), it has always been permissible to carry out digital filtering by sampling the bit value several times following the sample point instant.

For information, the SJA 1000 circuit (described below) can be programmed, if this is desired, to carry out a 'multiple sampling' of the bit value an odd number of times (three times in succession, in this case), in order to determine the bit value by means of a majority logic. This is what is generally called a 'protocol enhancement'.

Let us now move on to the precise details and evaluation of these time segments.

## 3.3 Nominal Bit Time

Each of these segments plays its part in determining the quality, the bit rate, the length, etc., of a CAN network, and they interact in numerous ways, making it difficult to provide a simple clear description. At the end of this chapter, you will find a table with a simple summary of all the following sections . . . but it will seem clear to you only if you have read your way through all these sections. The theme of the first part of my explanations will be limited to

- showing how the bit time is made up,
- describing the role of the different segments,
- explaining how they are interwoven,
- and, finally, describing a way of determining their values in an example of application.

### *3.3.1 Constructing the bit time*

Everything I have said will be meaningless if we do not associate these segments with time values and corresponding relative magnitudes.

The nominal bit time of a CAN network is fixed and specified for a project, and each station must be able to construct its own nominal bit time locally using its own resources. Each local division of nominal bit time can therefore be structurally different from one station to another. Clearly, it is preferable for the 'project manager', responsible for the network design, to harmonize all the time divisions of all the individual nominal bit times.

To avoid a nameless cacophony, the protocol obliges us to comply with certain rules for constructing the bit time, which we must not depart from. These include, in particular, the definition of the base for the construction of the bit time of a local station as the smallest reference time element usable by this station, called the 'minimum time quantum', and also the definition of some of its derivatives which will now be described.

#### Minimum time quantum of a station

The minimum time quantum is generally derived from the station's clock (for example, a signal emitted from the quartz oscillator to the microcontroller controlling the functionality of the node). We can then move on to the next step which is the definition of the time quantum of the (nominal) bit time.

#### Time quantum of the bit time

The time quantum is a fixed unit of time, derived and obtained from the minimum time quantum from which the CAN bit is to be constructed. The actual construction of this time unit is carried out with the aid of programmable (pre-)dividers (by integers $m$), whose division range must extend from 1 to 32.

Starting from a minimum time quantum (for example, that of the station's clock), the time quantum can have a length of

$$\text{Time quantum} = m \times \text{Minimum time quantum}$$

where $m$ is the division factor of the (pre-)divider.

### *3.3.2 Dividing the bit time*

Once this has been established, the following table shows the values defined by the CAN protocol.

| | |
|---|---|
| Synchronization segment | 1 time quantum (this value is fixed) |
| Propagation time segment | from 1 to 8 time quanta |
| Segment 1 of the phase buffer | from 1 to 8 time quanta |
| Information processing time | ≤2 time quanta |
| Segment 2 of the phase buffer | equal to not more than the value of segment 1 + Information processing time |

The total number of time quanta contained in a bit time which can be chosen by programming cannot be less than 8 or more than 25 (see examples in Figure 3.4).

At this level, the nominal bit time architecture is fully defined.

**Note:** To provide the best adaptation of the performance of the electronic components to the parameters of a CAN network, the values of these segments are generally programmable by means of special registers.

In the specific case of the SJA 1000 protocol management integrated circuit (used by way of example in Chapter 5 for purely technical reasons), its manufacturer considered it appropriate to define a new time 'segment' – time segment 1 (TSEG 1) – characterizing the set of segments compensating for the 'propagation times' and the synchronization buffer 'phase

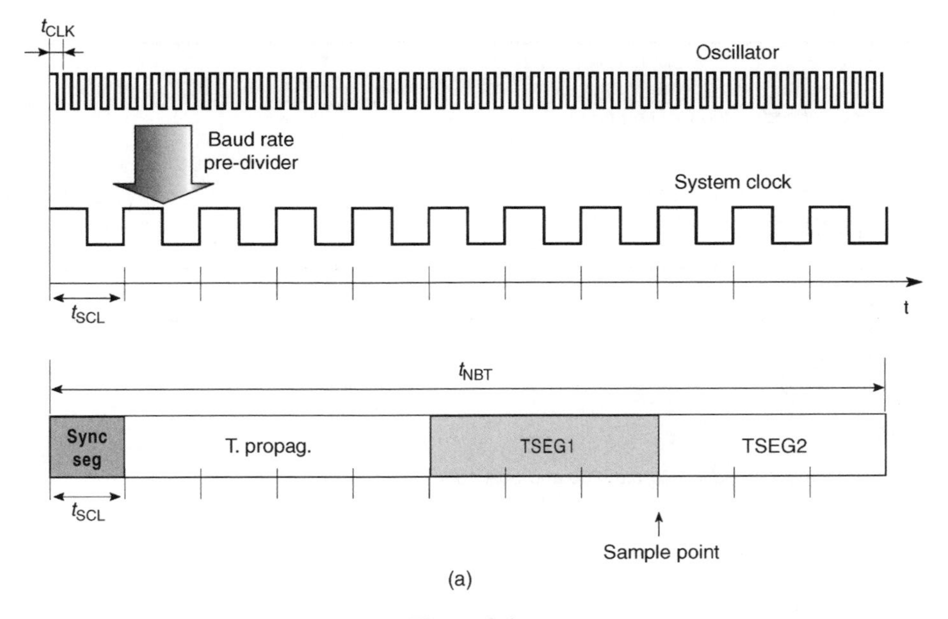


**Figure 3.4**

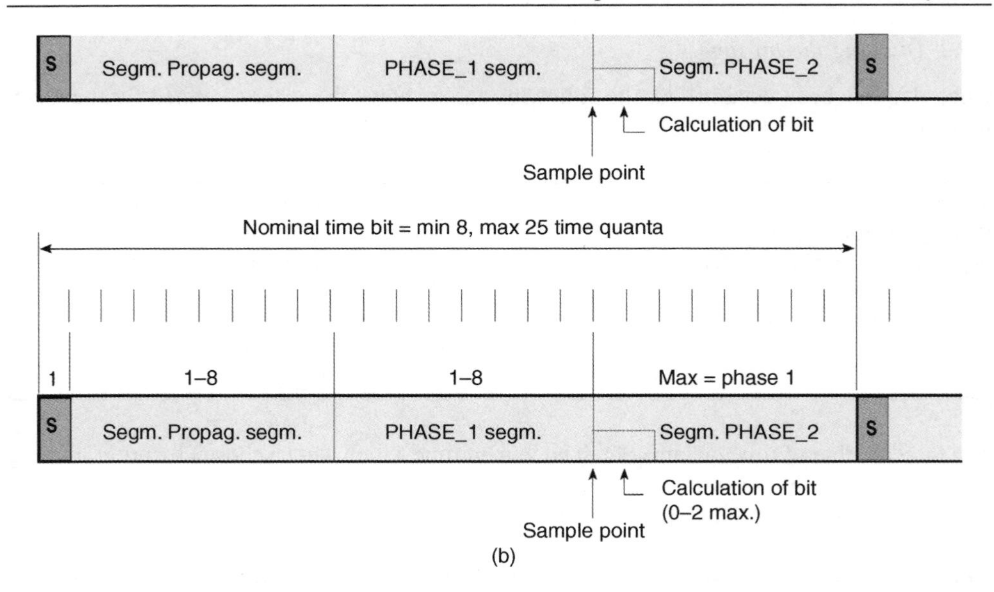


**Figure 3.4** *(Continued)*

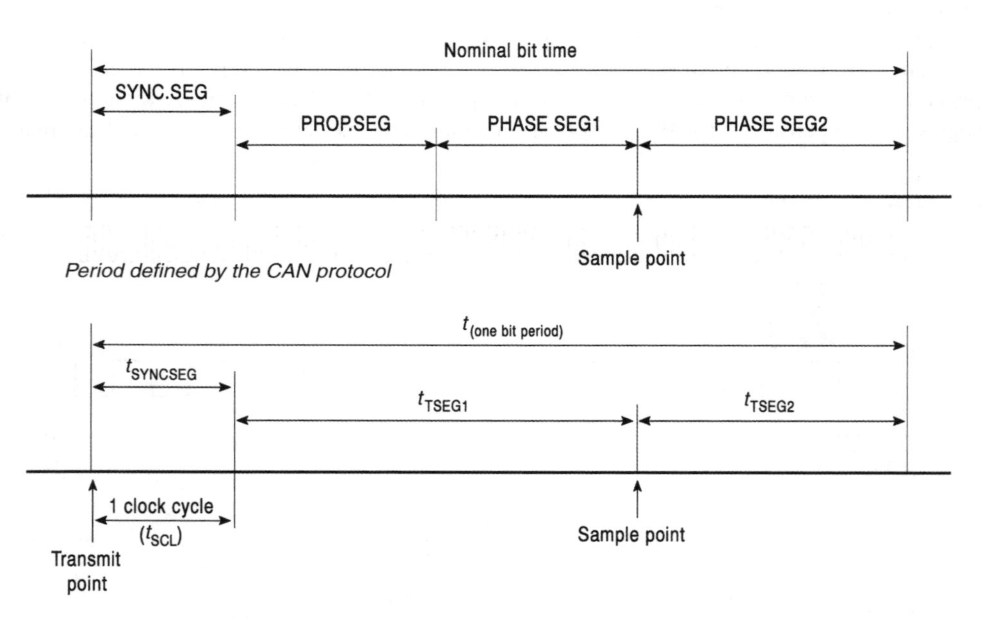


**Figure 3.5**

segment 1' of the CAN protocol (Figure 3.5). This practically amounts to the same thing where the use and general application of CAN are concerned.

A last remark before leaving this subject: It is important for the oscillators of the different ECUs to be 'coordinated' or 'harmonized', so as to provide the system with a time quantum specified in the best possible way.

### *3.3.3 The sequence of events*

Our four segments are now in place, and we need only to define the number of quanta assigned to the propagation segments and to phase buffers 1 and 2. As soon as this is known, it will be easy to deduce the duration of the nominal bit time and therefore the gross bit rate of the network. But there is a lot of work to do first!

This is because, at this level, the set of elements making up the network (medium, components, topology, etc.) all act in combination, and it is vital to classify the parameters rigorously and act in an orderly and methodical way. Let us look at the different segments.

#### Synchronization segment

The duration of this segment is no problem; it is permanently fixed at 1 quantum.

#### The so-called propagation segment

The parameters used to determine the value of this segment (regardless of the clock tolerances of the different stations) are

- the medium (and its choice),
- the propagation speed of the signal on this medium,
- the maximum length of the network (medium),
- the topology of the network,
- the electrical input/output parameters of the line interface components,
- the quality of the received signals, etc.

All these parameters are based on concepts of the medium, topology, etc., which I have grouped under the same heading, namely problems relating to the phenomenon of propagation.

#### Buffer segments 1 and 2

The parameters used to determine the value of this segment (regardless of the media, lengths, etc. mentioned above) are

- the nominal values of the clock frequencies of each node,
- the tolerances of these clocks,

- the temperature and time fluctuations of these clocks,
- the worst cases, etc.

All these parameters affect the capacity to successfully synchronize or resynchronize the stream of incoming bits as a function of the tolerances, fluctuations, etc., which I will cover in the section relating to what are called synchronization problems.

## 3.4 CAN and Signal Propagation

Whole volumes have been written on network theory. My aim is not to reproduce these but to cast some light on the properties of the CAN protocol in its applications to local networks.

The concept of a network implies structure, topology and medium. We cannot diverge from these major standards and the commentaries on them.

### *3.4.1 Network type, topology and structure*

Regardless of the types of medium used to create networks, their topologies are largely dictated by the possibilities of the data transport protocol. In the case we have been describing over many pages, the topology can be varied, and I will give some examples.

#### Bus

The 'bus' is the easiest topology to implement and the most widespread for this kind of protocol. But beware:There is more than one kind of 'bus'! Theoretically, all stations should be connected to the bus by the shortest route (Figure 3.6).

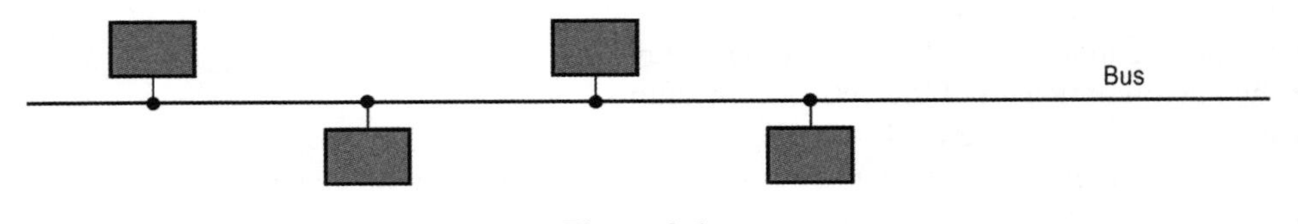


**Figure 3.6**

Another bus, the 'backbone' type (Figure 3.7), is very similar to the previous one, but differs slightly, especially as regards its electrical properties at high speeds. Indeed, everyone will tell you that problems may arise unless you are careful and that you should think in terms of electrical lines with allowance for 'stubs' connected to these lines. In plain English, this means that there is a boundary, a bit woolly perhaps but nonetheless real, between the low speeds where these phenomena can be considered non-existent, and the faster speeds where we start to enter the field of transmission lines proper. This is the eternal question of the relationship between the wavelengths of the signals carried and the distances to be covered. A well-worn argument!

For guidance, for a bit rate (NRZ coded 'square' signal wave) of 1 Mbit $s^{-1}$ (i.e. an equivalent frequency of 500 kHz for square signals), propagated for example in a differential

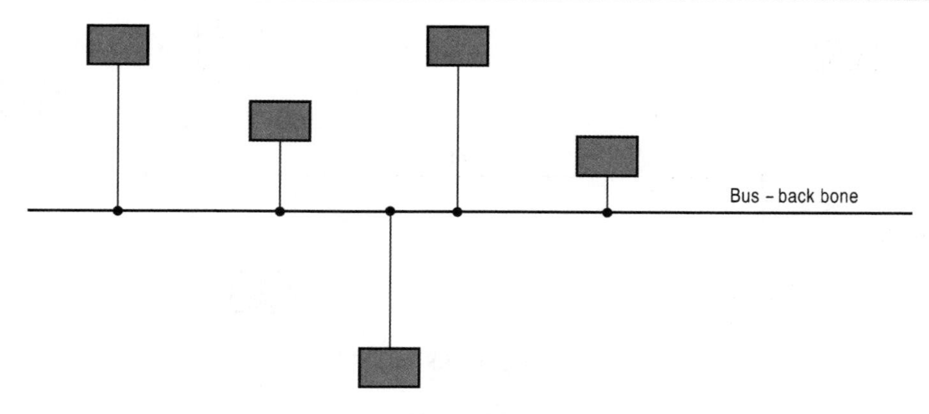


**Figure 3.7**

pair at 200,000 km s$^{-1}$, the fundamental (sinusoidal) frequency of the wave would be a wavelength of

$$\lambda = \nu T$$
$$\lambda = 2 \times 10^8 \times \frac{1}{5 \times 10^5} = 400\,\text{m}$$

As the NRZ square wave is formed from odd harmonics, the other wavelengths of the base signal would be 400/3 m, 400/5 m and 400/7 m, i.e. approximately 133 m, 80 m and 56 m, which is often very short!

As if by chance, some wise thinkers have smoothed the path by establishing specific recommendations for the CAN low speed mode (less than 125 kbit s$^{-1}$) and CAN high speed mode (more than 125 kbit s$^{-1}$ and up to 1 Mbit s$^{-1}$).

The interesting question (in technical and economic terms) is the one you are entitled to ask if you wish to operate on the boundary of the two modes: a kind of 'medium speed' (a non-standard term, invented for this occasion) of approximately 250 kbit s$^{-1}$. In this case, without wishing to penalize the costs, the problem is as complex as that of the high speed mode, as it is essential to think in 'transmission line' terms and consider the characteristic impedances, the impedance matching and mismatching, the rebound, the standing wave rates, the voltage nodes and antinodes, and the terminations (their impedances and physical positions), and to consider the responses to digital signals in the line created in this form (impulse responses, under/overshoot, eye diagram) and consequently the protection of the transported data.

Note, however, that the 'heavy artillery' of the error processing systems described above is very helpful where data transport protection is concerned.

### Star

Often, a network or partial network takes the topological form of a star, as in Figure 3.8. This configuration poses no structural problems.

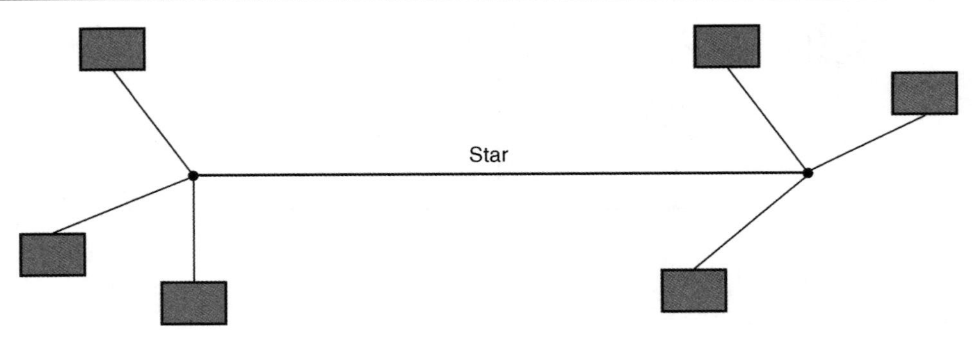


**Figure 3.8**

## Ring

When systems are required to have greater operating security and/or reliability in case of mechanical failure of the network (for example, when cables may be cut), the bus between two or more stations is frequently doubled, or split (in physical terms) (Figure 3.9). Sooner or later, this results in a 'ring' topology (of the 'token' type, for example, leaving aside any question of the protocol design philosophy).

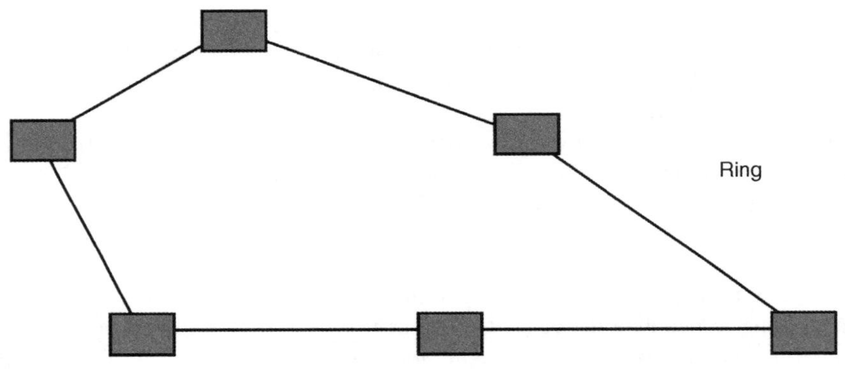


**Figure 3.9**

In the case of the CAN protocol, this particular topology again makes it necessary to examine carefully the estimation and management of propagation time, especially for the resolution of the problem of the 'bit' arbitration phase.

### *3.4.2 Propagation time*

If a line exists, there is also a phenomenon of signal propagation on the line, and therefore a propagation speed, therefore a delay due to the time taken for the propagation of the signal on the medium, a characteristic impedance, and therefore an impedance match (and mismatch, of course), therefore nodes and antinodes, standing waves, line-termination rebound and possible collisions on the bus between incoming signals and reflected signals, etc. Let us examine the propagation times:

We must make a few preliminary assumptions. We will consider that the following are known or defined:

- the choice of medium (for reasons of cost, EMC compatibility, etc.),
- the propagation speed of the signal in the chosen medium,
- the length of the medium (for application-specific reasons),
- the internal delay times of the transmission and reception circuit components.

Note that the values of these parameters are generally easily accessible from data sheets for the various elements concerned.

Now that these parameters are known, let us move on to the core of the problem: the determination of the value of the 'propagation segment' to be assigned to a node.

### *3.4.3 Estimating the value of the propagation segment*

Figure 3.10 summarizes the characteristic example of the phenomena which we shall examine in the worst case, taking into account the paths followed, the signal propagation delays and the tolerances and fluctuations of the oscillators of each participant.

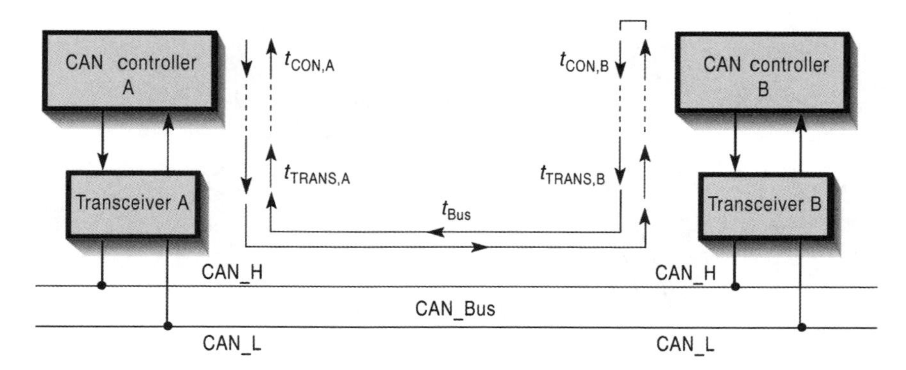


**Figure 3.10**

In a single network, the longest distance between two CAN controllers determines the maximum 'outgoing' propagation delay time.

Beyond this, to ensure the correct operation of the protocol during the arbitration and acknowledgement phases, a longer time interval must be provided to compensate for the total propagation delay times having an effect on an exchange on the network.

An example will make this clearer.

(a) A recessive bit is sent by node A, for example.
(b) The signal from node A is propagated towards node B.
(c) Some 'micro-instants' before the signal reaches its destination, node B, which has not yet noticed the presence of the incoming bit from A, starts to send a dominant bit, for example.

(d) The signal sent from B is propagated towards A.
(e) The collision occurs at a point very near B and makes the state of the line dominant (*and* wired) at this point.
(f) The resulting signal (dominant) is then propagated towards A, taking a certain time, approximately equal to the time taken to travel from A, to arrive.
(g) Station A, receiving a dominant bit while it is sending a recessive bit, can then determine, according to its instantaneous state, whether this represents a loss of arbitration for it (if the bit that it sent formed part of the identifier) or a bit error, etc.

In fact, the duration of the bit sent by station A must therefore be equal to or greater than the global sum of all these times, in order to provide this type of mechanism and ensure the correct operation of the protocol. We could say, although not very correctly, that this total time represents the maximum necessary 'propagation time' taken by the signal to travel from one controller to another and back again. (To be more precise, we would have to allow for the time required for the processing and calculation by each station at the ends of the path.)

Two characteristic examples which may occur are shown in Figure 3.11.

- the case of two CAN controllers on a simple bus during an arbitration phase;
- the case where two nodes are present on a ring network.

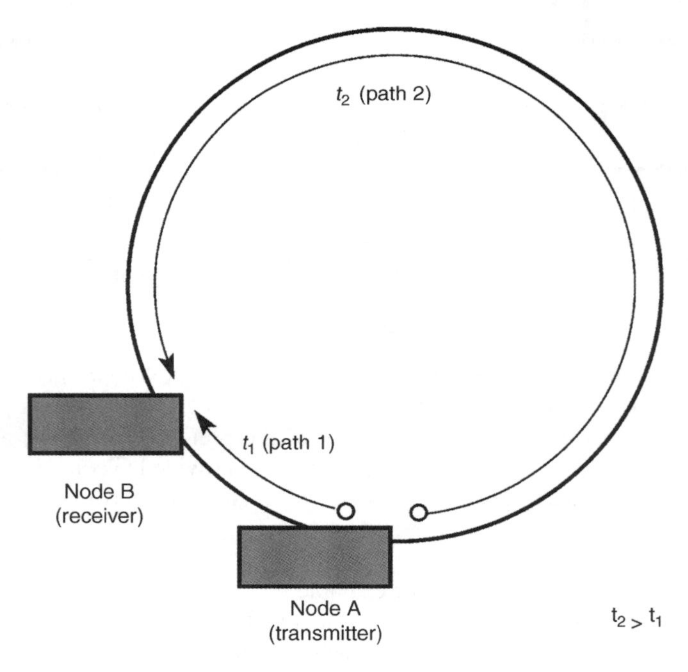


**Figure 3.11**

### *3.4.4 Precise definition of the propagation segment time*

The above explanations make it clear that, for a given medium, the time interval to be assigned to the CAN bit propagation segment plays a considerable part in the evaluation of the maximum length of the network and vice versa.

To precisely define the minimum period to be assigned to the propagation segment (segTprop), we must consider the totality of the time contributions of all the elements in the network (Figure 3.12).

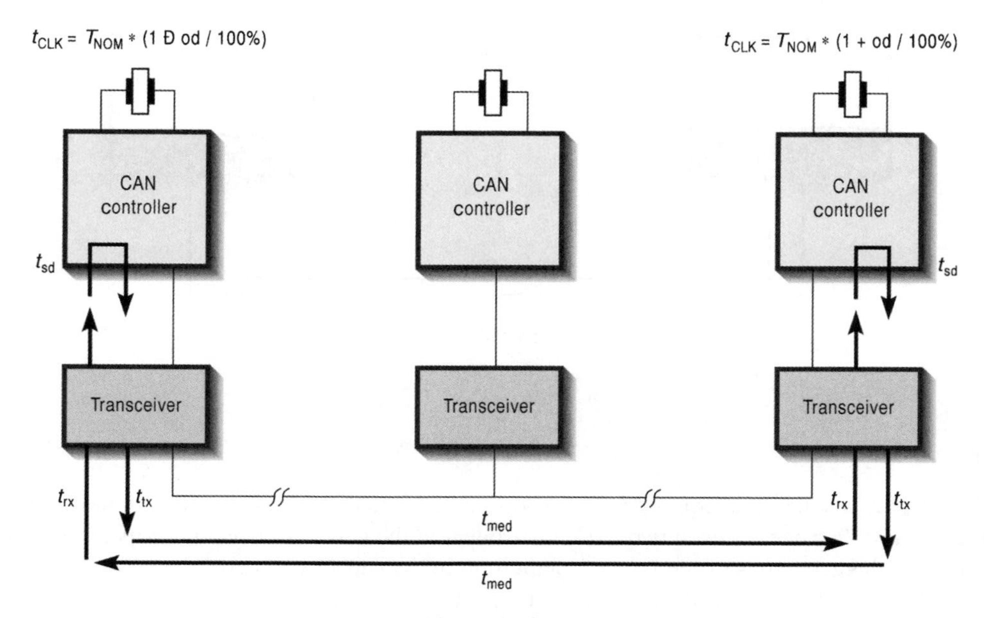


**Figure 3.12**

Following the signal from its output from the transmitter to its actual arrival in the input stage of the receiving station, we find, in succession

- the delay time required by the transmission controller to output the signal on its terminals;
- the delay time required by the transmission interface to generate the signal on the medium;
- the time taken to transport the signal along the medium;
- the time taken by the reception interface for transferring the signal to the reception controller;
- the time taken by the reception controller to process the incoming signal.

These parameters are frequently divided into two classes: those relating more specifically to the physical properties of the medium used ($T_{med}$) and those concerned with the electronic characteristics found in the network ($T_{elec}$).

The sum of these two components ($T_{res}$) characterizes the total time taken by a bit to actually move from one station to the other, i.e. (Figure 3.13)

$$T_{res} = T_{med} + T_{elec}$$

where $T_{elec}$ = sum of the delay times $T_{xxxx}$

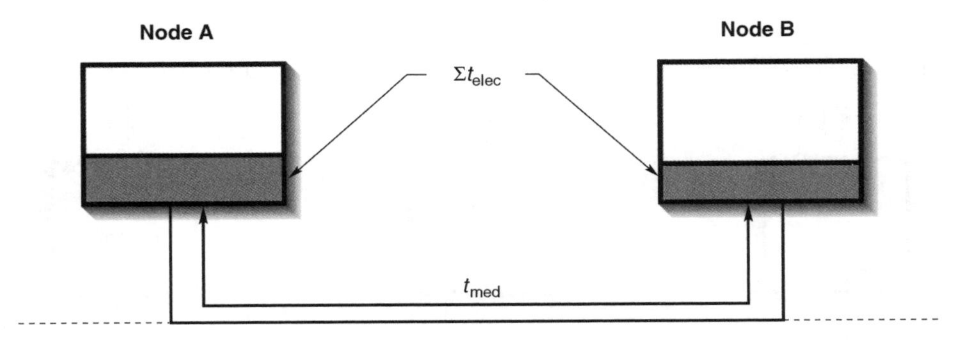


**Figure 3.13**

### $T_{med}$

$T_{med}$ represents the time required for transporting the signal along the medium, in the case of the greatest distance between two CAN controllers in the network. This time depends on the physical characteristics of the type of medium used, its capacity to carry the signal more or less rapidly and consequently its intrinsic propagation speed.

If we call this speed $v_{prop}$, and if the length of the medium is $L$, this time will be

$$T_{med} = \frac{L}{v_{prop}}$$

In the standard case of a wire medium (a differential pair, for example), the propagation speed is substantially equal to 0.2 m ns$^{-1}$, i.e. 200 m μs$^{-1}$ (or, to put it another way, the medium has a propagation time of 5 ns m$^{-1}$).

### $T_{elec}$

This time interval is made up of three elements.

- **$T_{sd}$**

$T_{sd}$ represents the sum of the delay times due to the signal processing carried out in the output stages (on departure) and input stages (on arrival) of the CAN controllers of the departure and arrival stations.

• $\boldsymbol{T_{tx}}$ **and** $\boldsymbol{T_{rx}}$

$T_{tx}$ represents the delay time required by the transmission interface to generate the signal on the bus. $T_{rx}$ represents the time taken by the reception line interface to transfer the signal to the reception controller.

The manufacturers of the line interface components (the line drivers) state the exact values of $T_{tx}$ and $T_{rx}$ in their data sheets, when the incoming signals are 'clean' (properly square). In fact, this is the exception. Let us take a closer look at what happens all the time in reality.

• $\boldsymbol{T_{qual\_sign}}$

Very commonly, the medium used consists of differential pairs (because they are very good at rejecting injected common-mode signals). At the receiver, the input interface of an integrated circuit generally consists of a differential amplifier, provided, followed or associated with a comparator. Of course, this comparator has its own threshold and hysteresis characteristics. These elements, plotted as voltages, must be translated into 'equivalent times' (Figure 3.14) according to the shape and quality of the incoming signals, which are generally very different from the idealized, perfect, beautifully square signals shown in all the text books.

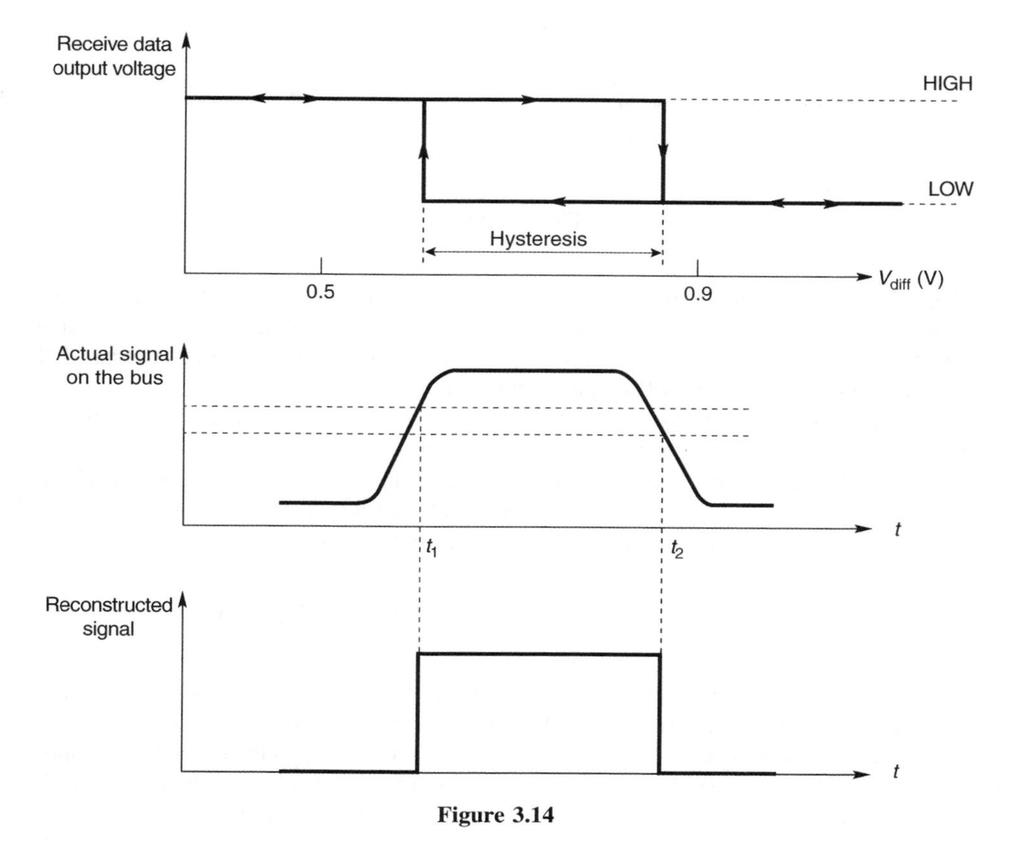


**Figure 3.14**

They are more likely to look like those shown in Figure 3.15, for example as a result of the effects of the inductances and capacities distributed along the line.
Note: the asymmetry between the 'rise time' and the 'decay time', deliberately exaggerated in the figure, is intended to remind you of the presence of two problems which are sometimes completely distinct.

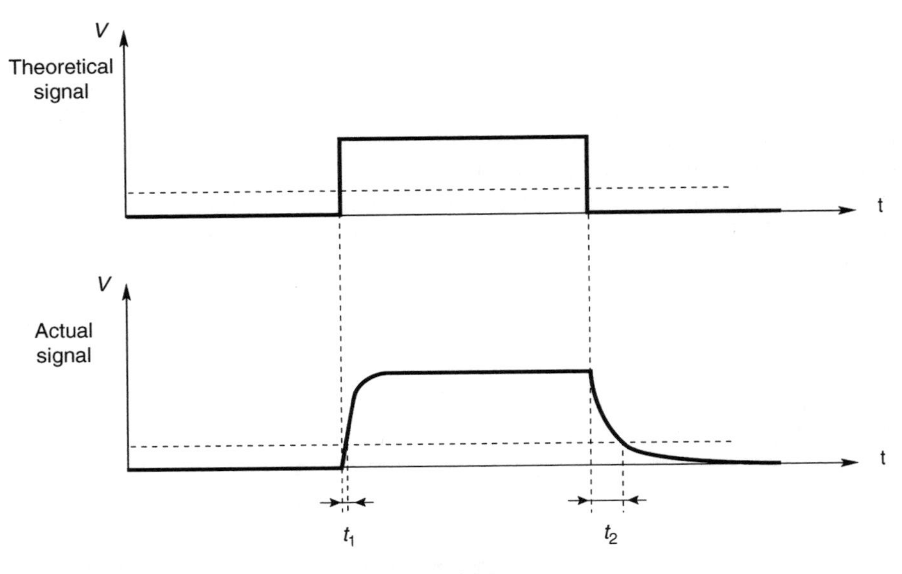


**Figure 3.15**

Clearly, the unfortunate input comparator will do its utmost to retrieve its desired signals, by changing the electrical levels as best it can, but, as shown in the figure, with delay times which depend neither on the pure propagation nor on the intrinsic characteristics of the comparators but on the quality of the incoming signals.

These new electrical 'delay times' ($T_{\text{qual\_sign}}$) can then be considered equivalent to propagation or supplementary distances of the medium.

In view of all these factors, the overall value of $T_{\text{elec}}$ can be fairly well estimated at about a 100 ns. Thus, we have reached the end of this calculation and at last we have the required minimum total value for the duration of the 'propagation time segment':

$$(seg_{\text{T\_prop}})\text{min} \geq 2(T_{\text{med}} + T_{\text{sd}} + T_{\text{tx}} + T_{\text{rx}} + T_{\text{qual\_sign}})$$

The factor '2' is due to the outward and return travel of the signal and allows for the characteristics of the protocol in respect of arbitration and acknowledgement.

When this value has been determined, and given the elementary value of the time quantum defined previously, it is easy to determine the minimum number of time quanta required to construct the 'propagation segment' (from 1 to 8) of the nominal bit time.

### *3.4.5 Corollaries: relations between the medium, bit rate and length of the network*

The same problem is frequently posed in the following two corollary forms:

- What is the maximum distance that can be covered by CAN for a bit rate *X* and a medium Y?
- What is the maximum bit rate that can be used for a network of length *X* using a medium Y?

To answer these, we must return to the principles stated above, but present them differently.

#### Maximum distance

For a given bit rate and medium, CAN makes it clear that the distance between two nodes is largely dependent on

- the intrinsic propagation speed of the physical medium used to construct the line;
- the delay introduced by the output stage of the transmitting station;
- the delay introduced by the input stage of the receiving station;
- the desired nominal bit rate (and therefore the maximum value of the minimum duration of the propagation segment);
- the precise instant when the signal is sampled and the way in which the value of the bit is measured during its physical presence;
- the frequencies and tolerances of the various oscillators of the CAN controllers (microcontrollers or 'stand-alone' devices) present on the network;
- the quality of the signals, etc.

This can be summarized in the following alternative form: 'The maximum distance of a CAN network is primarily determined by the time characteristics of the medium used and the consequences of the principle adopted for the operation of the arbitration procedure defined in the non-destructive protocol (at the bit level) present on the line (and see all the preceding sections)'.

#### Relationship between bit rate and maximum length for a given medium

Let us return to Figure 3.12. As we can see, if the CAN protocol is to operate correctly, what we must take into account is not the travel time of the signal on the network ($T_{res}$), but approximately twice this value, to allow for the fact that the sender of a message must be able to receive return data during the bit time (in the case of the arbitration phase, for example). The value called the 'propagation time segment' of the nominal bit time of a CAN network must therefore be greater than (or at least equal to) approximately twice $T_{res}$:

$$seg_{T\_prog} \geq 2\,T_{res}$$

The propagation time segment represents only a part of the nominal bit time ($T_{bit}$). If $x$ is the percentage represented by this segment, we can write

$$seg_{T\_prog} = x \cdot T_{bit}$$

or alternatively

$$seg_{\text{T\_prop}} = \frac{x}{\text{Baud rate}}$$

Transferring the values found above into the preceding equations, we obtain the equation relating the length of a network to the chosen bit rate, assuming that the values of $T_{\text{elec}}$ and the speed of the medium are known (and therefore that the medium has been chosen)and assuming a value of $x$ (chosen or predetermined), thus

$$L \leq (0.2\,\text{m ns}^{-1})\left[\frac{x}{2\ \text{Baud rate}} - T_{\text{elec}}\right]$$

The curve in Figure 3.16 shows the graphic appearance of this equation for the following values: $T_{\text{elec}} = 100$ ns, $v_{\text{prop}} = 0.2$ m ns$^{-1}$ (wire and optical medium), $x = 0.66 = 66\%$.

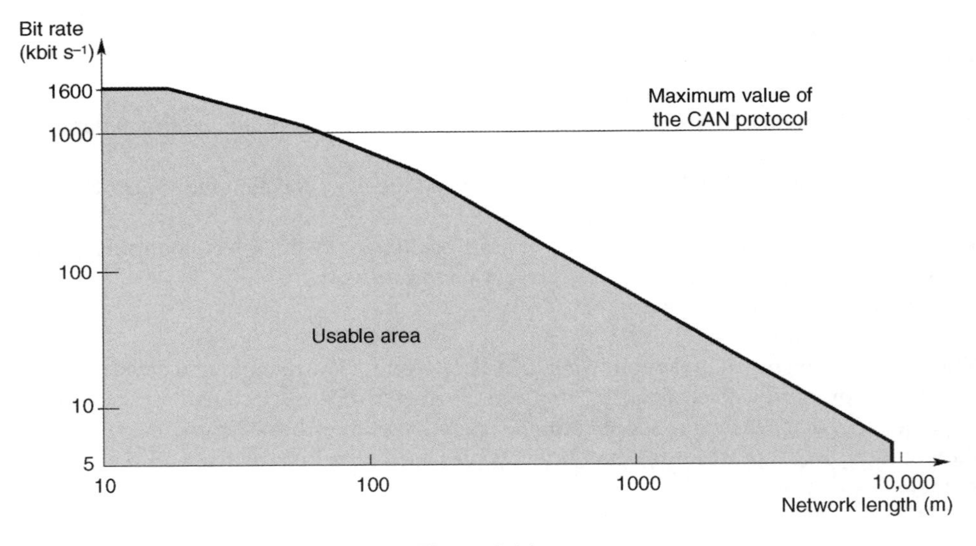


**Figure 3.16**

Using this curve, we can quickly identify the performances we can expect from a network. Clearly, the various parameters must be calculated more precisely for a detailed investigation of a network.

Perhaps you do not like wire pairs? Perhaps optical fibres terrify you? Would you rather use a different medium? No problem . . . You now have all the necessary ingredients for your own calculations. Each to its own!

**Example.**

Now we have all the necessary ingredients for quantifying a network, in respect of the propagation times and compensation based on appropriate electronic circuits in the circuitry of the CAN microcontrollers. As we cannot deal in abstractions forever, it is time to choose a real example, such as the SJA 1000 circuit (of course, this example could be adapted for other components). This component has internal registers (BTR0 and BTR1) whose contents can be used to manipulate the values to be assigned to the different segments of the bit time. Figure 3.17 shows the parameter definitions for these components.

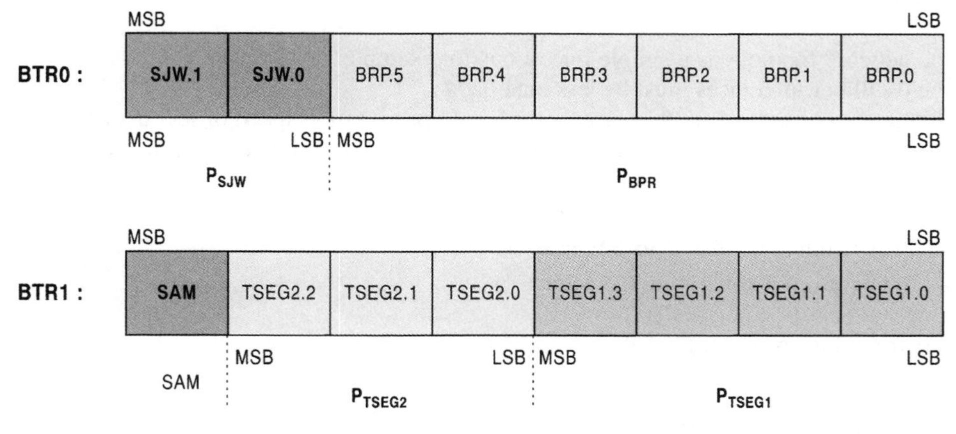


**Figure 3.17**

With all this technology in place, we can at last go on to examine the values to be entered in the registers, and calculate the maximum propagation delay time that can be supported by our network, and, if the propagation time of the medium (cable) is known, the maximum length that can be supported by our network.

By way of example, so that you can have an immediate idea of various orders of magnitude, we obtain the table of results shown in Figure 3.18 (into which, in addition to the speed/distance relation, I have directly introduced the values to be loaded into the BTR0 and BTR1

| Bit rate | Maximum distance | Bus timing | |
|---|---|---|---|
| | | BTR0 | BTR1 |
| 1.6 Mbit s$^{-1}$ | 10 m | 00h | 11h |
| 1 Mbit s$^{-1}$ | 40 m | 00h | 14h |
| 500 kbit s$^{-1}$ | 130 m | 00h | 1Ch |
| 250 kbit s$^{-1}$ | 270 m | 01h | 1Ch |
| 125 kbit s$^{-1}$ | 530 m | 03h | 1Ch |
| 100 kbit s$^{-1}$ | 620 m | 43h | 2Fh |
| 50 kbit s$^{-1}$ | 1.3 km | 47h | 2Fh |
| 20 kbit s$^{-1}$ | 3.3 km | 53h | 2Fh |
| 10 kbit s$^{-1}$ | 6.7 km | 67h | 2Fh |
| 5 kbit s$^{-1}$ | 10 km | 7Fh | 7Fh |

**Figure 3.18**

registers, to achieve these bit rates when a 16 MHz quartz oscillator is used). This table will be meaningless unless we carefully state in the caption, as here, the parameters included in the calculations, namely

- the network topology is a simple line (a bus for example, not a ring);
- the oscillator tolerances must be less than 0.1%;
- the line (a twisted pair in this case) must have a propagation speed of less than 5 ns $m^{-1}$;
- the sum of the delays introduced by all the electronic circuits of the transmitter and receiver must not exceed
  - 70 ns (at 1.6 Mbit $s^{-1}$),
  - 90 ns (from 250 kbit $s^{-1}$ to 1 Mbit $s^{-1}$),
  - 300 ns (from 5 kbit $s^{-1}$ to 125 Mbit $s^{-1}$).

Apart from this, everything is OK.

## 3.5 Bit Synchronization

The term 'synchronization' as used in the CAN protocol has many implications, due on the one hand to the large distances that may be covered by the bus, and consequently the significant delays due to the propagation times of the signals carried, and on the other hand to the accuracies, fluctuations and tolerances of the frequencies of the oscillators of each station.

The following paragraphs will attempt to detail all the subtleties of these procedures. In fact, their whole purpose is to define the moment when the edge of the bit has arrived (or should have arrived) with respect to the moment of its expected arrival, so that the station can react and sample the signal accordingly.

So keep your spirits up and pay careful attention, for this is a complicated matter!

### *3.5.1 Some general considerations and important notes*

Before starting, let us examine some points which are rather important for a good understanding of the subject. In order to grasp the mechanisms and operation of the CAN bus 'synchronization' system, we must thoroughly digest the following information.

- Concerning the internal clocks of each node

Generally, each node has an internal oscillator which belongs to it and which operates continuously. After division, the clock frequencies of these oscillators enable us to find the corresponding frequency and duration of the (unique) bit rate chosen for the CAN network in question, together with all the parameters of its individual nominal bit time (see or review the preceding section).

The clock frequencies of all these nodes do not necessarily have exactly the same values (for the same nominal and theoretical bit rate of the CAN bus), either because of the strict values and tolerances, or because of the fact that the initial clocks of the different nodes are not absolutely identical (one node operating at 12 MHz, another at 16 MHz, a last one at

11.0592 MHz, for their own functional reasons), the internal (pre-)divisions not yielding exactly the same values.

Assuming that all the nodes have been made to have the same nominal clock value and the same division factors, the local clocks will never be strictly identical from one node to another, only because of the tolerances of the quartz oscillators, or, for example, because of the fluctuations due to the fact that each node operates in an environment having a substantially different temperature.

Even assuming that all the nodes can be (or are) timed by an identical and unique clock signal, all the timing signals would be 'synchronous' but would not necessarily be strictly 'isochronous', as they would not necessarily have the same time phase, because of the distance between the nodes and the propagation phenomenon.

• . . . consequently,
Each node (via the component managing the protocol) has all its own resources to enable it to calculate its own bit time. This is because each circuit has its own internal clock and can therefore construct the division (. . ., phase segment 1, . . .) of 'its' own individual nominal bit time at the user's command and with the aid of its internal registers. The designer is therefore responsible for defining and choosing these (and his) values (or for accepting the default values offered by the manufacturer of the component managing the protocol).

• everything has a beginning – even a CAN transmission!
There is always a point when the network is first switched on – a first 'power on' – and therefore first ' power on resets' of the components of the stations, which are accompanied by their inevitable initialization sequences and, before the first exchange on the bus commences, a certain time interval when the bus is inactive (in idle mode).

Having recalled all these points, we can soon begin our work, once we have dealt with one more matter: Why should we want each node, with its autonomously operating clock, to construct its own nominal bit time when it is switched on? In fact, all its internal pre-dividers are in place (pre-programmed), and the node is waiting only for the network to start-up so that it can synchronize the start of its first nominal bit time with the first signal edge presented to it (normally this will be a recessive-to-dominant edge signalling the SOF bit of the very first frame).

Now that you are fully up-to-speed, let us examine the fine detail of the synchronization mechanism.

### *3.5.2 Bit synchronization*

As a general rule, in asynchronous transmission such as that used for CAN (with no wire dedicated to a permanent clock signal during a transmission), the recovery of the value of a bit at the receiving station frequently raises major problems, due to the lack of regular timing information. These problems are due to a number of features of the CAN protocol.

The first feature of the CAN protocol is that the CAN bit is NRZ coded, and it is inherently difficult to recover transitions from one bit to the next when they are sometimes structurally absent!

What must we do to be sure of knowing whether a bit having the same value as the preceding bit is actually arriving? Invent a local clock! This is what has been done by specifying the local construction of a nominal bit time at each station. Thus, we need to synchronize only the local clock with the edges of the incoming bits (if any), and pray that everything will run properly from then on; in other words, hope that there is not too much clock slippage (due to tolerances, etc.) between the message source and the receiving station, otherwise we could be facing disaster by taking the value of one bit for another because of the accumulated shifts! What can we do in this case?

- Resynchronize with the incoming bit stream? This is impossible, because there may be numerous bits with the same value.
- Resynchronize from time to time with the edges present during the exchange, and try to keep the local clocks in time with that of the source transmitter for as long as possible?

In this context, the CAN protocol has a second interesting feature. It arises from the bit stuffing coding principle adopted for transmitting the messages on the medium. In principle, in the greater part of the transmitted CAN frame, not more than 5 consecutive bits (see the note below) can have the same physical (electrical, optical, etc.) value on the medium, so there will be at least one transition for every 5 bits. This feature makes it possible to design a much more reliable mechanism for resynchronizing a unit bit within a bit stream in a frame, so that it becomes easier to be independent of possible clock slippage.

**Note:** To conclude this section and provide a more precise definition, note that the text describing the CAN protocol indicates that the maximum distance between two transitions to be taken into account for correct resynchronization is 29 bits (and not 5, as stated in our simplified outline above). This is because there can be long stretches of the CAN frame which include no bit stuffing and which must be taken into account, for example ends of frames followed by interframes during which more than 10 successive identical (recessive) bits followed by an overload frame must be identified.

It should be pointed out that the value of 29 bits enables us to allow for all problems of accumulation of errors and/or clock or phase slippage, etc., due to the tolerances of the components (quartz oscillators, etc.).

### *3.5.3 Implementation of CAN bit synchronization*

The CAN protocol describes two modes of synchronization: hardware synchronization and resynchronization. The text requires component manufacturers to construct these synchronization devices in the electronic form of state machines. The purpose of these circuits is to compare the position of the leading edge of the incoming signal with the start of the current bit timing of the station, and to adapt the duration of the latter value if necessary, by a hardware synchronization and/or resynchronization mechanism.

#### Hardware synchronization

We say that there is a 'hard synchronization' mechanism when the leading edge (transition) of the incoming bit is located – or is forced to be located – within the synchronization segment of the nominal bit time that has just started.

In short, the new bit time restarts with the synchronization segment in hard synchronization (Figure 3.19). This is in fact the equivalent of the well-known direct synchronization systems of oscilloscopes. (See the operating rules in the following sections.)

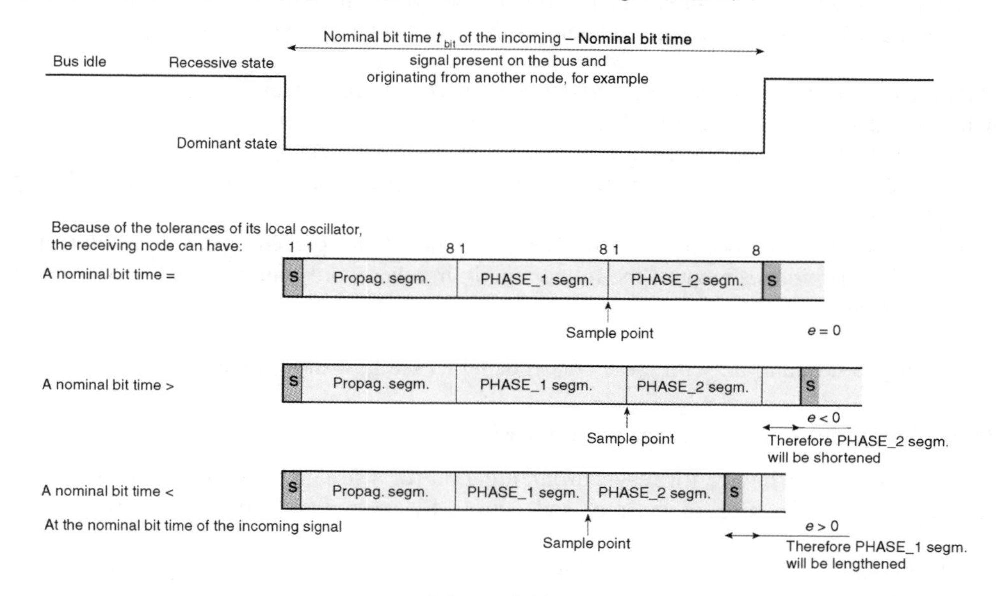


**Figure 3.19**

### Bit resynchronization

'Bit resynchronization' is a mechanism which can be used in the transmission of an isolated bit or a stream of bits of a message to compensate for

- either the variations of individual (instantaneous) frequencies of the CAN controllers on the network;
- or the changes in length of the nominal bit times introduced by switching from one transmitting station to another, for example during arbitration phases when the edges do not arrive at strictly the same instant, because of the different propagation times (or distances) of the different candidates in the arbitration procedure.

It is also possible (and sometimes desirable), for the purposes of 'real time' operation, to resynchronize the bit stream by acting dynamically, at any instant, on the values of the time intervals PHASE_SEG1 and PHASE_SEG2 as explained below.

The resynchronization mode exploits various new features of CAN, namely

- phase errors and their associated signs,
- phase jumps and the widths of phase jumps.

Let us examine these two concepts.

**Phase error of a signal transition (or edge)**

The phase error $e$ of an edge is given by the relative discrepancy in position between the edge of the incoming bit and the predicted position of the synchronization segment of the bit time (see Figure 3.19 again). The measurement of the phase error is characterized by

- its intensity (its magnitude), measured and expressed in time quanta;
- its sign, defined as follows (see Figure 3.19 again):
  - $e = 0$ if the edge occurs within the synchronization segment,
  - $e > 0$ if the edge occurs before the sample point of the current bit, in other words during phase segment 1,
  - $e < 0$ if the edge occurs after the sample point of the preceding bit and before the synchronization segment of the following bit, in other words during phase segment 2 of the preceding bit.

Now that you are familiar with these concepts, let us see how they are applied.

**Phase jump and resynchronization jump width**

In order to design a device for resynchronizing a bit (or a stream of bits), it is important to start by defining its area of validity and action, by immediately defining a number of constraints.

The maximum value of this constraint is called the resynchronization jump width (RJW), whose value (expressed in time quanta) can be programmed between 1 and the minimum value of the range of values included in (4, number of quanta of phase segment 1).

Now let us look at the actual effects of resynchronization. They are determined by the relation between the amplitude of the phase error $e$ which causes the resynchronization phenomenon and that of the programmed value of RJW. There are two distinct cases:

$$|e| \leq \text{RJW}$$

In this case, if $e$ is positive or negative, but less than or equal to RJW, the effect will be the same as that of a hard synchronization (see above, and Figure 3.20).

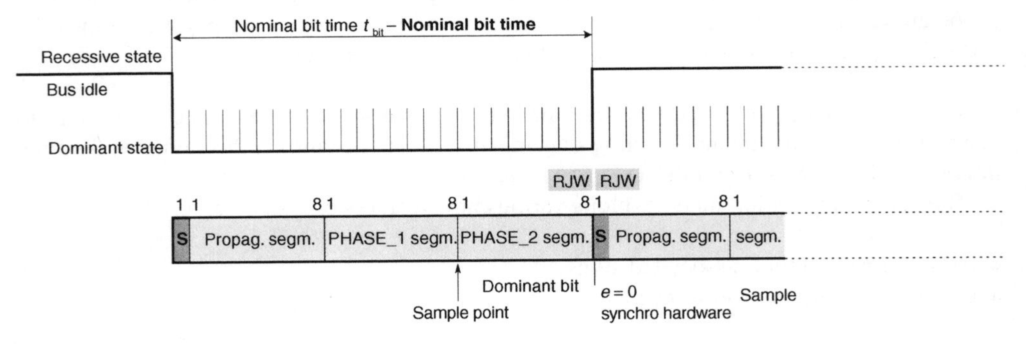


**Figure 3.20**

$$|e| > \text{RJW}$$

If $e$ is positive and greater than the RJW, there will be a 'resynchronization' of the nominal bit time by elongation of phase segment 1, which in this case will be increased by a quantity equal to the RJW (Figure 3.21).

If $e$ is negative and greater than the RJW, there will be a 'resynchronization' of the nominal bit time by a shortening of phase segment 2, which in this case will be decreased by a quantity equal to the RJW (Figure 3.21).

As shown above, the effect of the resynchronization operation makes it possible to reset the sample point to the most suitable moment of the incoming bit, to measure it with greater accuracy.

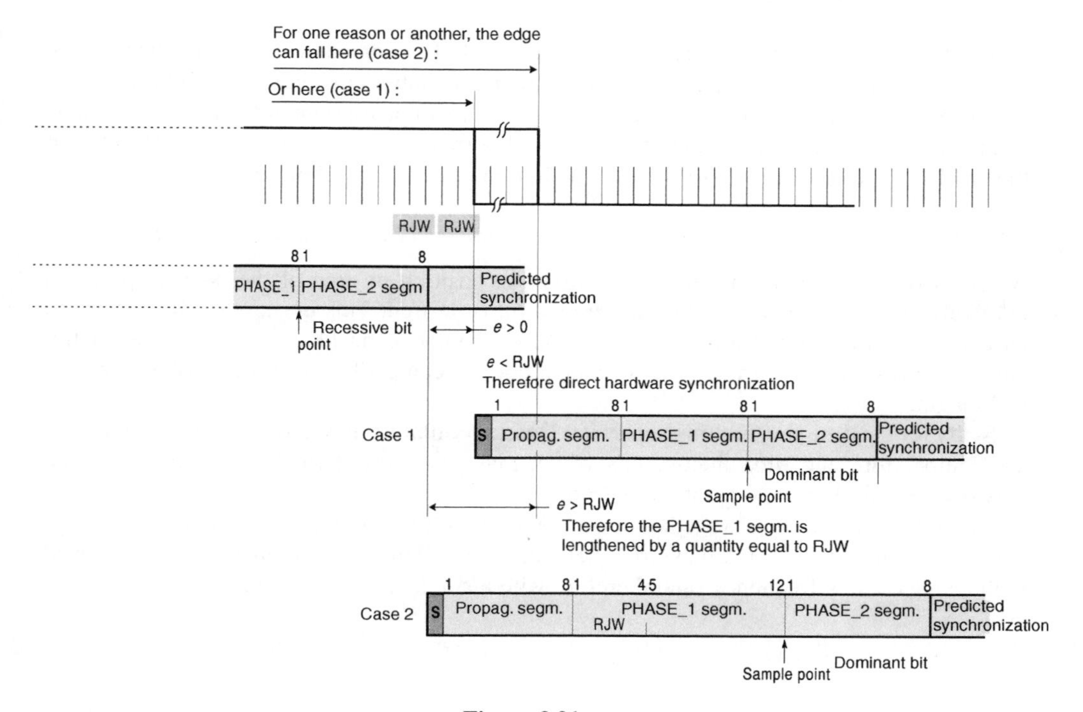


**Figure 3.21**

### *3.5.4 Operating rules for the two synchronization modes*

The CAN protocol indicates that the hardware synchronization mode and the resynchronization mode must obey the following rules:

- Only one type of synchronization is permitted within a nominal bit time (hardware synchronization or resynchronization).

- When the bus is inactive (in the idle state), the hardware synchronization is carried out whenever there is a transition from the recessive to the dominant state (this actually corresponds to the start of a frame).
- A transition (signal edge) is used to ensure its synchronization if (and only if) the value read on the bus immediately after the transition is different from that detected at the preceding 'sample point' (the previously read bus value).
- All other transitions (from recessive to dominant, and possibly from dominant to recessive in the case of low bit rates) meeting the conditions of Rules 1 and 2 will be used for resynchronization, with the exception of the case in which a node transmitting a dominant bit does not carry out a resynchronization as the result of a transition from recessive to dominant having (or causing) a positive phase error, if only recessive to dominant transitions are used for the resynchronization.

The latter exception ensures that the delay time of the command driver and that of the comparator input do not cause a permanent (and/or cumulative) increase in the bit time.

To complete and conclude the explanations of the use of phase segments 1 and 2, let us now examine some remarks concerning the effect of the frequency fluctuations and tolerances of the clocks of the different stations on the definition of the nominal bit time.

### *3.5.5 Effect of the precision, fluctuation and tolerances of the oscillator frequencies*

As you would have noted in the last section, the position in time of the 'sample point' is relatively critical if a good quality of the read bit is desired. This means that, if we wish to ensure a high level of uniformity in a network, all the nodes must have clocks whose quality and tolerances are of the same order of magnitude, all compatible with the topology, bit rate and length.

Sadly for the designer, nothing is perfect: Quartz oscillators and resonators have tolerances and temperature and time fluctuations which must be allowed for by participants in the network, so that they can operate even in the worst case.

Bringing all these values together under the heading 'oscillator drifts', symbolized by 'od' below, we can state that in a network where all the participants are nominally synchronized with a single clock frequency (and therefore with a single nominal bit time), we find

$$[T_{\text{clk min}} = T_{\text{nom}}(1 - \text{od}/100\%)] < T_{\text{clk nom}} < [T_{\text{nom}}(1 + \text{od}/100\%) = T_{\text{clk max}}]$$

We can also express this in the following form: To enable a network to operate correctly, all the participants must operate with oscillators having very strictly defined performances and tolerances.

The conclusion is simple: Use quartz oscillators! But this raises the economic problem of the cost of a node when its task is a simple one, which is often the case with certain 'passive' input/output devices. In these cases, it would be preferable to use ceramic resonators, although their tolerances are higher (of the order of 1.5%). For this purpose, some recommendations are made to integrated circuit manufacturers in the CAN protocol, in order to avoid the effects of these larger tolerances. By following these recommendations, the bus can be made to operate correctly with a bit rate of 125 kbit $s^{-1}$ and oscillators with a tolerance of 1.58% can be used.

To conclude on this subject, I must point out some important recommendations for operation:

- If a network has to use different protocol handlers and it is not known whether or not they conform to these recommendations, there is only one answer: We must use quartz oscillators.
- The node (the station) having the least severe constraints for its oscillator will determine the precision required for all the other nodes.
- Ceramic resonators can be used only if all the other nodes in the network use the enhanced protocol.

### *3.5.6 Quick estimation of RJW as a function of the maximum tolerance on the frequency of the oscillator of a node*

The CAN protocol specifies that it must be possible to resynchronize in the presence of a maximum of 29 successive recessive bits. This means that the cumulative frequency shifts (or slippages) of the oscillators between the message sender and the other network participants must be calculated so that the sample point is always in the correct place with respect to the incoming signal, and so that the value of the bit as detected, read and measured is always correct. The value of $T_{\text{clk min}}$ must be taken into account for one node and $T_{\text{clk max}}$ for the other node right up to the final bit in all cases, including the worst case.

To estimate the value of RJW to be loaded into the microcontroller registers, some additional values must be defined. First of all, it is necessary to know that the value of RJW acts in the same way in the two phase segments 1 and 2 and that its task is to compensate for the tolerances of the different clocks (Figure 3.22). Statistically, therefore, we can write that, for one bit,

$$T_{\text{bit}} \frac{x}{100} = \text{RJW}$$

where RJW is the resynchronization phase jump width, $T_{\text{bit}}$ is the nominal bit time and $x$ is the maximum tolerance of the oscillator of a node.

In fact, where the worst case is considered, this value must be doubled, as the clock of one station can be instantaneously at $(T_{\text{clock}} - x/100)$ whereas the clock of another station is at $(T_{\text{clock}} + x/100)$ and therefore $2(T_{\text{bit}}(x/100)) = \text{RJW}_{\text{max}}$, for the duration of a bit.

However, we must allow for the possibility of the presence of 29 successive bits having the same value. So, taking into account the possible cumulative effect of the delays

$$29 \times 2\left(T_{\text{bit}} \frac{x}{100}\right) = \text{RJW}_{\text{max\_max}}$$

This last value is therefore the one that must be loaded into the nominal bit time to give the network a chance to operate correctly at all times. We also know that the structure of the nominal bit time will always have the form

$$T_{\text{bit}} = 1\,t_{\text{q}} + \text{seg}_{\text{T\_prop}} + \text{seg}_{\text{phase\_1}} + \text{seg}_{\text{phase\_2}}$$

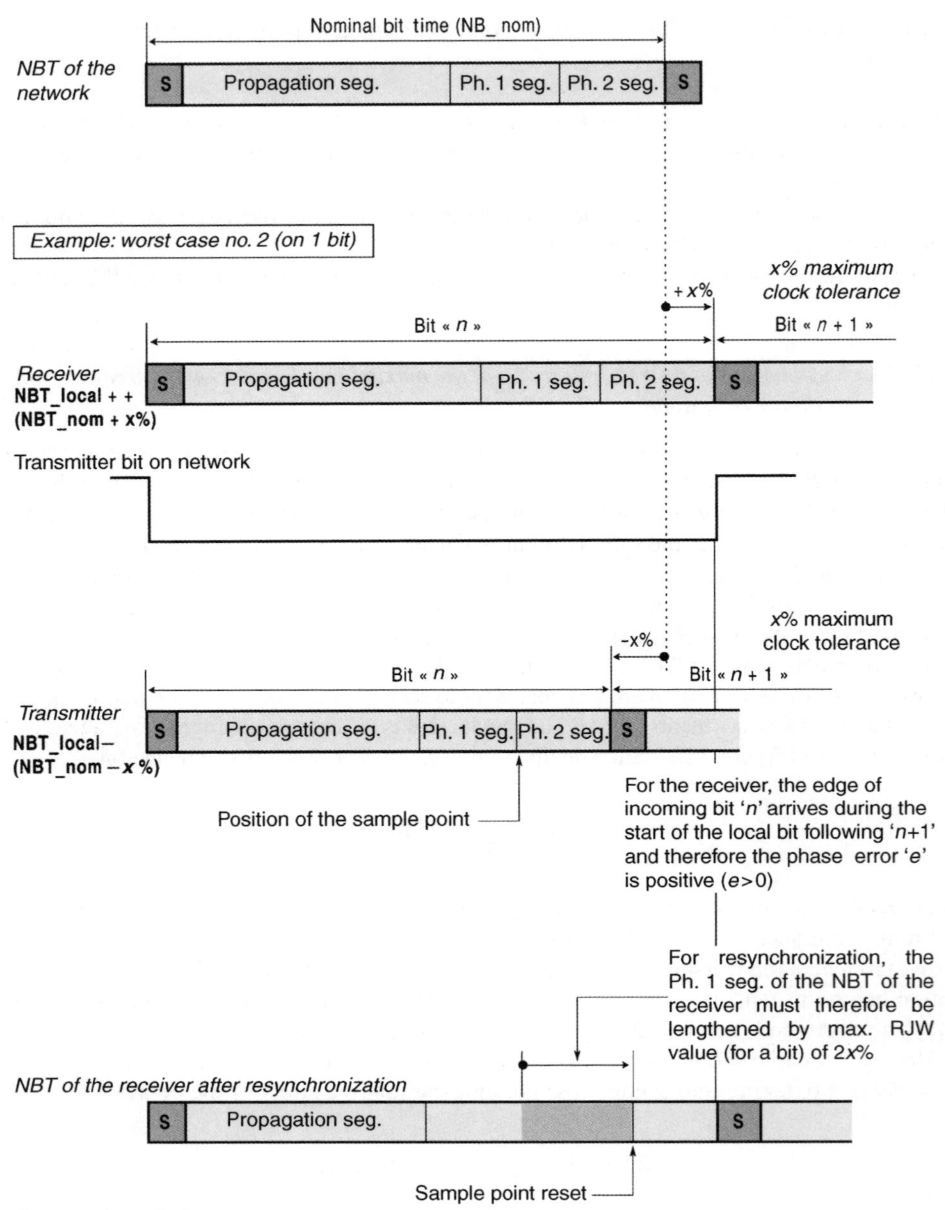


By way of conclusion

As the CAN bit is NRZ coded and the protocol states that there can be a sequence of 29 identical bits (recessive), Ph. 1 seg. and Ph. 2 seg. must have a minimum value of (RJW_max_max) = (29x2.x%), to allow for the accumulation of phase errors.

(a)

**Figure 3.22**

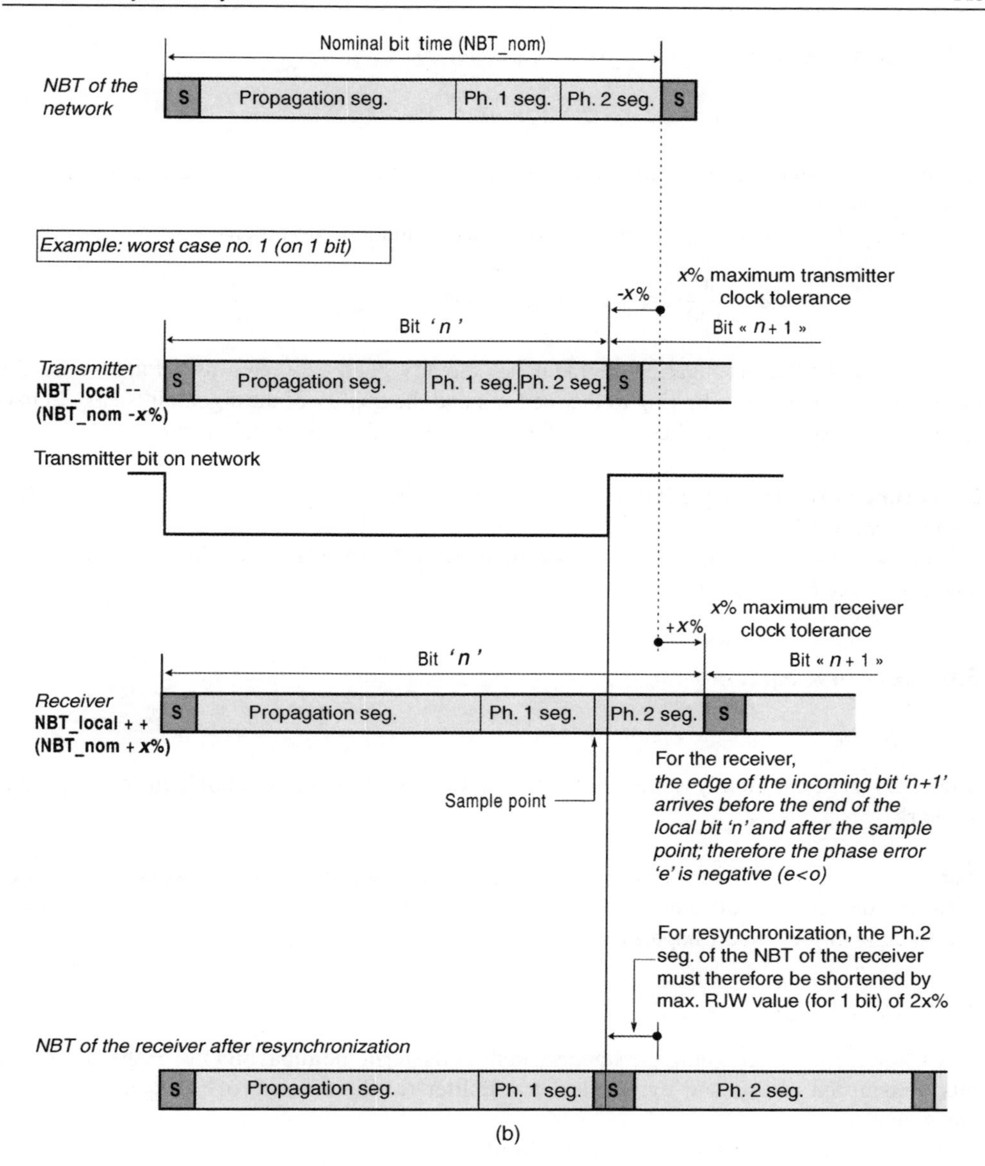


(b)

**Figure 3.22** *(Continued)*

and that, depending on the sign and value of the phase error at a given instant, either phase segment 1 can be elongated or phase segment 2 can be shortened (but not both at once!).

Theoretically, phase segment 1 can have a zero value (obligatory minimum of 1 $t_q$) because it cannot be elongated. However, phase segment 2 can only be decreased and must be at least equal to the value of $RJW_{max\_max}$.

We can then estimate the value of $T_{\text{bit\_min}}$:

$$T_{\text{bit\_min}} = 1t_q + \text{seg}_{\text{\_T\_prop}} + 1t_q + \text{RJW}_{\text{max\_max}}$$

where $t_q$ is the time quantum (the synchronization segment must always be equal to at least 1 time quantum).

Carrying the value of $T_{\text{bit\_min}}$ into the preceding equation, we obtain

$$58 \times \frac{x}{100}(2t_q + seg_{\text{Tprop}}) = \text{RJW}_{\text{max\_max}}\left(1 - 58\frac{x}{100}\right)$$

As the value of the propagation segment has already been estimated as a function of the medium, the components linking to the medium and the quality of the signal, it is easy to find the value of $\text{RJW}_{\text{max\_max}}$ to be used for the calculation of the nominal bit time.

**Important note:** This equation is valid only if $1 - 58\frac{x}{100}$ is positive, i.e. if $x$ is less than approximately 1.7.

In conclusion, the sum of the tolerances supported by CAN in relation to the clocks is approximately 1.7% maximum.

## 3.6 Network Speed

### *3.6.1 Bit rate*

The CAN protocol clearly states that the bit rate must be fixed and uniform for a specific network.

**Note:** An application very commonly consists of a number of CAN networks and/or sub-networks operating at different speeds, and CAN 'gateways' must be created to move from one to another (see also Chapter 8).

### *3.6.2 Gross bit rate*

In relation to bit transmission, the standard defines the term 'nominal bit rate' as the number of bits transmitted per second by an ideal transmitter (in the absence of a resynchronization mechanism).

This gross bit rate can be as high as 1 Mbit $s^{-1}$.

**Note:** Two classes of speeds have also been formally standardized by ISO:

- the mode called 'low speed', where the maximum speed is limited to 125 kbit $s^{-1}$ (inclusive);
- the mode called 'high speed', from 125 kbit $s^{-1}$ up to 1 Mbit $s^{-1}$ maximum.

The details of these two modes will be examined later.

### 3.6.3 Net bit rate

To estimate the net bit rate (number of useful bits carried per second), we must draw up a list of the complete contents of a frame (a data frame for example), allowing for its distinctive features in respect of the number of bytes carried by it and the number of associated stuff bits that it contains.

As the CAN protocol supports versions 2.0A and 2.0B, it is useful to examine the two types of frames.

Here is the estimate, in terms of number of bits, of a standard frame (2.0A), in sequence:

- 1 start bit – SOF,
- 11 identification bits,
- 1 RTR bit,
- 6 control bits,
- 64 data bits (maximum),
- 15 CRC bits,
- 1 CRC delimiter bit,
- 19 stuff bits (maximum, in the worst case),
- 1 ACK slot bit,
- 1 ACK delimiter bit,
- 7 end of frame bits,
- 3 interframe space bits,

i.e. a maximum of 130 bits (not including the overload frame, etc.).

Here is the estimate, in terms of number of bits, of an extended frame (2.0B):

- 1 start bit – SOF,
- 11 identification bits,
- 1 SRR bit,
- 1 IDE bit,
- 18 identification bits,
- 1 RTR bit,
- 6 control bits,
- 64 data bits (maximum),
- 15 CRC bits,
- 1 CRC delimiter bit,
- 23 stuff bits (maximum, in the worst case),
- 1 ACK slot bit,
- 1 ACK delimiter bit,
- 7 end of frame bits,
- 3 interframe space bits,

i.e. a maximum of 154 bits (not including the overload frame, etc.).

### *3.6.4 Dispense with the trimmings, and what is left?*

Let us take the example of the extended frame (CAN 2.0B).

For a (gross) bit rate of 1 Mbit $s^{-1}$ (1 bit = 1 μs), the extended frame lasts for a maximum of 154 μs for a maximum of 64 μs of useful data, giving a net bit rate of approximately $64/154 = 0.145$ (to carry 8 bytes, with the values of the identifier and data bits, etc., carefully specified to maximize the number of stuff bits and so on).

As we have associated 154 bits at 1 Mbit $s^{-1}$ with the eight useful data elements of the message, these would have been carried at a net bit rate of approximately 415 kbit $s^{-1}$.

This calculation is rather simplified but has the merit of quickly yielding an order of magnitude of the net bit rate; it is worth using with a few additional adjustments for real applications, allowing for the real number of data carried per frame, the possible arbitration and error losses, etc.

The details of these calculations lie outside the scope of this book, but Figure 3.23 provides a better idea of the net bit rates obtained as a function of the numbers of bytes carried in operation at 1 Mbit $s^{-1}$, subject to some minor adjustment for the effect of their contents in the presence or absence of stuff bits.

*CAN net bit rate*
*(in kilobits per second)*

For a gross bit rate of 1 Mb $s^{-1}$, assuming that the number of useful bits for a message (a CAN object) is as follows :

* for CAN 2.0A: 11 (ID) + 1 (RTR) + 4 (DLC) + number of bits of data transmitted
* for CAN 2.0B: 29 (ID) + 1 (RTR) + 4 (DLC) + number of bits of data transmitted

Example:

The complete CAN 2.0A frame (Including the stuff bits) contains about 138 bits for 8 bytes of data, i.e. a net bit rate of: (11 + 1 + 4 + 64) / 138 = (80 / 138) x 1 Mb $s^{-1}$ = 579 kb $s^{-1}$

Net bit rate (in kilobits per second)

| Number of bits transmitted per CAN frame | CAN 2.0A standard frame | CAN 2.0B extended frame |
|---|---|---|
| 1 | 72.1 | 61.1 |
| 2 | 144.1 | 122.1 |
| 3 | 216.2 | 183.2 |
| 4 | 288.3 | 244.3 |
| 5 | 360.4 | 305.3 |
| 6 | 432.4 | 366.4 |
| 7 | 504.5 | 427.5 |
| 8 | 576.6 | 488.5 |

**Figure 3.23**

### *3.6.5 Minimum value of the ratio between the net and gross bit rate*

Figure 3.24 provides another illustration of the principle of bit stuffing found in an 'ordinary' message where a stuff bit is introduced after every 5 bits having the same value.

Figure 3.25 indicates the area of a data frame (CAN 2.0A) in which the stuff bits can be present.

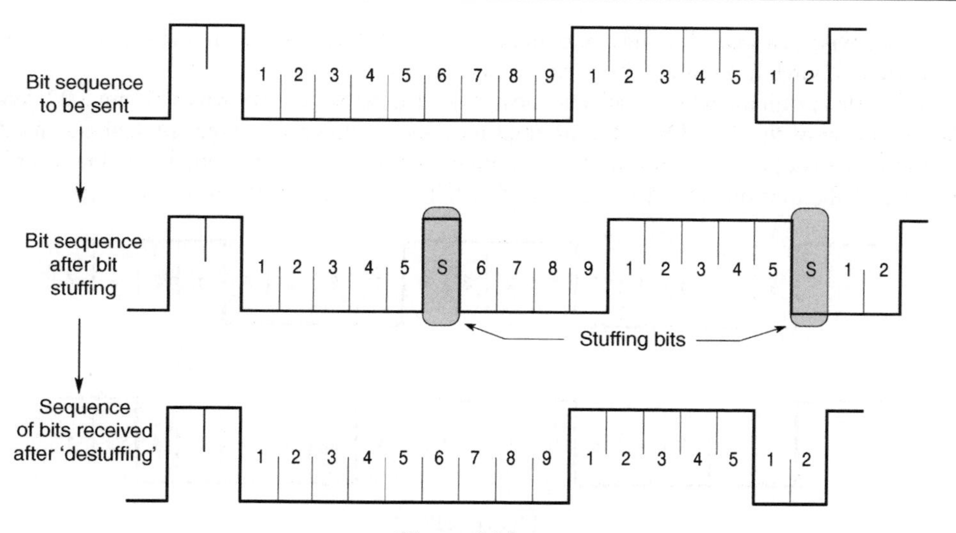


**Figure 3.24**

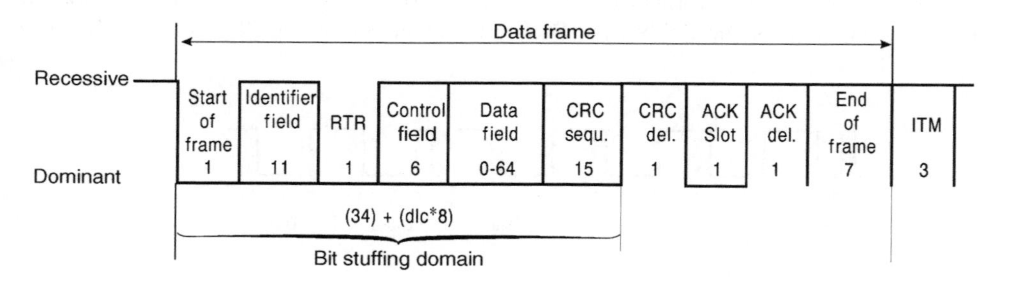


**Figure 3.25**

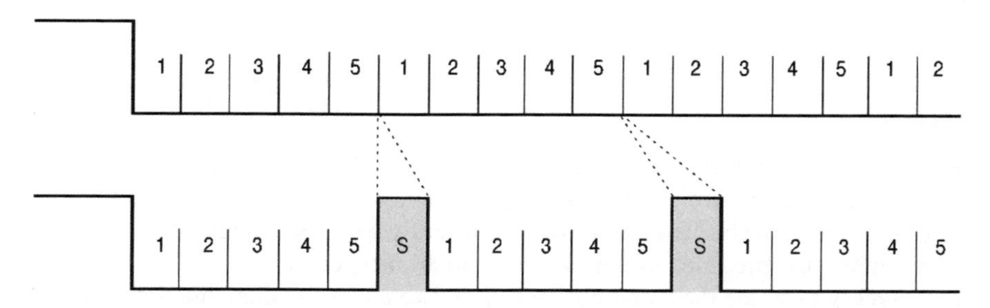


**Figure 3.26**

If we envisage a signal having a maximum number of stuff bits, we might initially think of a message as shown in Figure 3.26, giving a ratio of 5:1.

Sadly, this is not the worst case! The most awkward message is shown in Figure 3.27: With the exception of the first 5 bits, the original message contains sequences of 4 identical bits. This makes it necessary to insert stuff bits more frequently (for every sequence of 4 useful bits, because of the introduction of the first stuff bit). The ratio thus falls from 5:1 to 4:1!

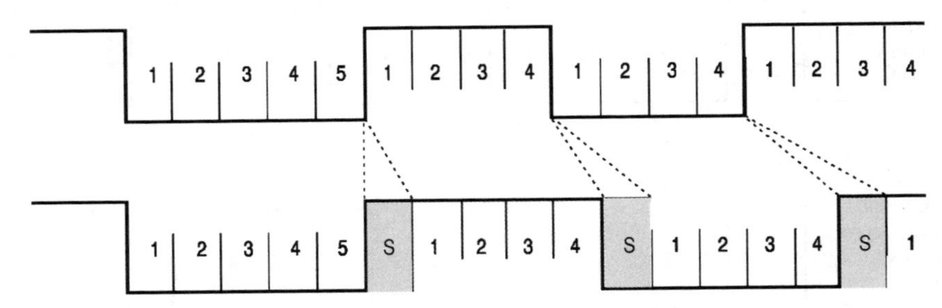


**Figure 3.27**

So you must be cautious in your calculations and when creating messages that you want to be 'fast'. For information, the message containing the maximum number of stuff bits (allowing for the RT bit, the DLC, the calculation of its CRC, etc.) is shown in Figure 3.28.

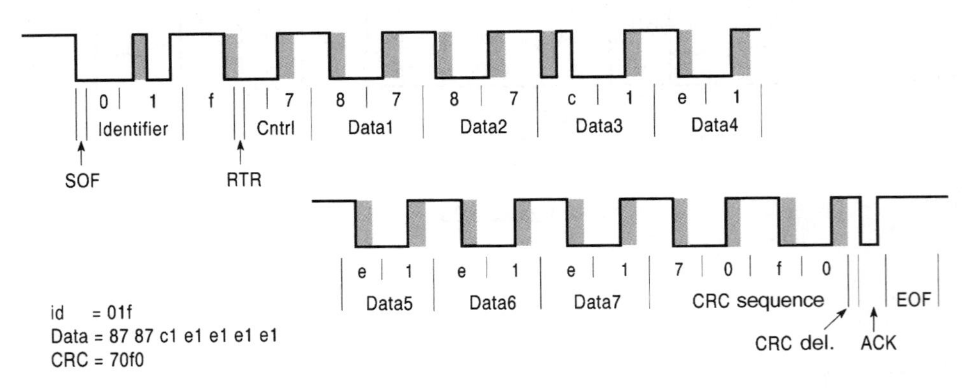


**Figure 3.28**

### *3.6.6 Latency*

The latency is the time interval between the instant of initialization (transmission request) and the actual start of transmission (including all parameters). Its value depends on a wide range of conditions present on the bus and the messages currently being exchanged (the number of data bytes transmitted, the presence or absence of error frames, etc.).

Let us quickly examine the worst possible case, and try to calculate the time for which we must wait to access the network if it is not subject to disturbance (no error frame).

In the CAN protocol, the length of a message can be expressed as follows:

| Frame type | Standard CAN 2.0A | Extended CAN 2.0B |
|---|---|---|
| Number of bits of the frame subject to bit stuffing | $n = 1 + 11 + 1 + 6$ $+ 8\text{DLC} + 15$ $= 34 + 8\text{DLC}$ | $n = 1 + 11 + 1 + 1+$ $18 + 1 + 6 + 8\ \text{DLC} + 15$ $= 54{+}8\ \ \text{DLC}$ |
| Number of bits of the frame not subject to bit stuffing | $1 + 1 + 1 + 7 = 10$ | $1 + 1 + 1 + 7 = 10$ |
| Number of interframe bits not subject to bit stuffing | 3 | 3 |
| Total length of the CAN frame | | |
| Without any stuff bits | $34 + 8\text{DLC} + 10 + 3$ | $54 + 8\text{DLC} + 10 + 3$ |
| i.e. for 8 data bytes | $111 + 3 = 114$ | $128 + 3 = 131$ |
| With stuff bits | The maximum number of stuff bits is $s_{max} = (n - 1)/4$ (a ratio of 4/1 in the worst case, review the previous section if necessary). | |
| For 8 bytes carried | $(34 + 8 \times 8 - 1)/4$ | $(54 + 8 \times 8 - 1)/4$ |
| i.e. | 24 | 29 |
| Maximum length of the frame | | |
| For 8 data bytes | $114 + 24$ | $131 + 29$ |
| i.e. | 138 | 160 |

In this case (an extreme worst case, and very unrealistic because this most unusual message with a maximum of stuff bits is highly improbable, as you know that it can exist and you will do everything possible to avoid it), the maximum latencies guaranteed by the protocol for the identifier of highest priority would be 137 and 159 bit times for CAN 2.0A and 2.0B, respectively.

**Note:** You would certainly have noticed that this value includes a bit time of less than the maximum length of the longest frame that can (statistically) exist in a transmission. If you were very unlucky, you would have tried to start an exchange immediately after the first SOF bit of the message already being transmitted (therefore one bit less) and you would therefore be in a state of 'latency'. This state would not exist if you had started your message at exactly the same time as the current message, as you would have inevitably won the bus arbitration.

Surprising? No, because for the calculation of a latency, it is implicitly assumed that the message you wish to transmit has a higher priority than all the messages that can be transmitted in this time zone of the operation of the system. For the complete calculation of the latencies of the network, the whole system must be subjected to a statistical analysis, which lies outside the remit of this introductory book on CAN.

### *3.6.7 Conclusion and summary of this section*

We have reached the end of these important sections describing the CAN bit, the nominal bit times and the resynchronization mechanism. A thorough knowledge and mastery of these concepts is essential for the definition of the set 'medium type/maximum length/maximum bit rate and tolerances of components' for a given network.

The properties, the qualities and the resulting compromises are summarized in the definition (and/or choice) of the nominal bit time and the choice of the relative values of the specific division into the different time segments which make it up.

To be as concise as possible (Figure 3.29):

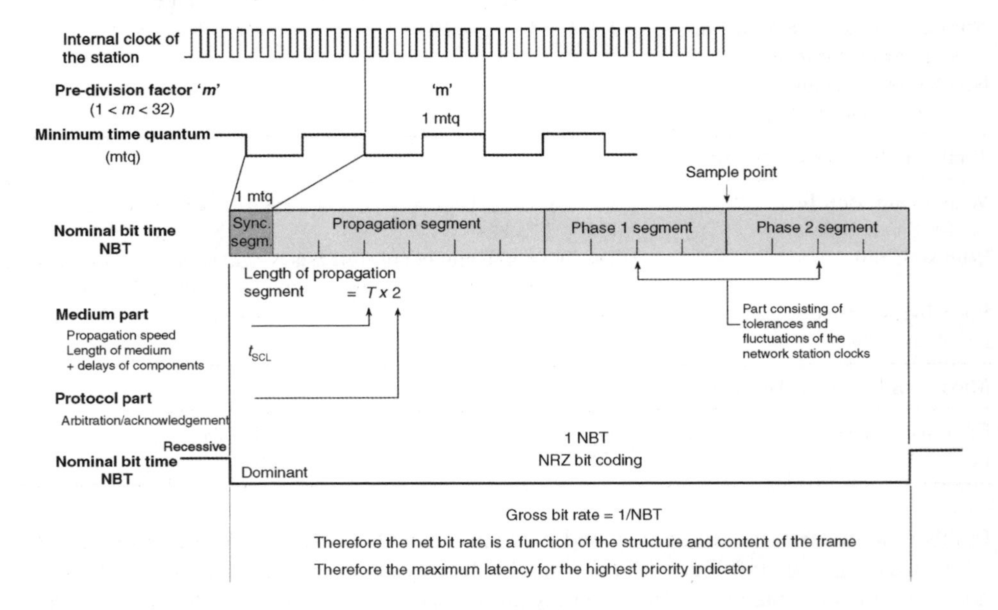


**Figure 3.29**

- The choice of a type of medium implies a knowledge of its signal propagation speed.
- The desired length of the medium and the time performances of the components for coupling to the medium are used to calculate the minimum time required for the 'propagation time segment' of the nominal bit time (roughly twice the time taken to travel from one end of the network to the other).
- The tolerances and fluctuations of the oscillators of the various participants can be used to estimate the maximum values of the RJW, and consequently the minimum values of phase segments 1 and 2 of each participant.
- The set of the minimum propagation segments, increased by phase segments 1 and 2, provides an initial approximation of the minimum nominal bit time, and therefore of the maximum 'gross bit rate' which the planned CAN network can theoretically support (despite any attenuation of the signal due to the medium used, preventing it operating correctly, there must still be at least some of the signals left!).
- Depending on the type of messaging system required (number of bytes carried per frame, etc.), this gross bit rate can be used to estimate the net bit rate (average and maximum), and consequently the maximum latency of a message having a given priority (choice of identifiers).

## An example summarizing Chapter 3

| | | |
|---|---|---|
| Nominal clock frequency of a node | $F_{clock\_nom}$ | = 10 MHz |
| i.e. a nominal period | $T_{clock\_nom}$ | = 100 ns |
| Maximum tolerance of network clocks | $\%F_{clock}$ | = 0.5% |
| i.e. a possible time fluctuation | $\%T_{clock}$ | = 0.5 ns |
| Minimum time quantum of the node | — | = 100 ns |
| Chosen value of the pre-divider $m$ $(1 < m < 32)$ for the node in question | m | = 10 |
| Time quantum of the network | $t_q$ | $= 100 \times 10 = 1000$ ns = 1 μs |
| Medium chosen for the planned application | — | = wire, differential pair |
| Propagation speed of the medium | $v_{prop}$ | $= 200,000$ km s$^{-1}$ |
| Desired length of the network | L | = 600 m |
| Propagation time of the medium | $T_{prop\_med}$ | = 3 μs |
| Sum of 'electrical' times ($T_{sd}$, $T_{tx}$, $T_{rx}$, $T_{qual\_sign}$) | $T_{elec}$ | = 100 ns = 0.1 μs |
| Minimum necessary value of the 'propagation time segment' | — | $= 2 \times (3 + 0.1) = 6.2$ μs |
| Minimum number of time quanta necessary | $1 \ll 8$ | = 6,2/1 = 6,2... i.e. = 7 quanta |
| Accepted value of the 'propagation segment' | $seg_{T\_prop}$ | = 7 μs |
| Minimum value of $RJW_{max\_max}$ (see formula in text) | $RJW_{max\_max}$ | = 3.39 μs |
| Number of quanta accepted for phase seg. 2 | $seg_{phase\_2}$ | = 4 min., 4 accepted, i.e. 4 μs |
| Number of quanta accepted for phase seg. 1 | $seg_{phase\_1}$ | = 1 min., 4 accepted, i.e. 4 μs |
| Length of the nominal bit time | $1\ t_q + seg_{T\_prop} + seg_1 + seg_2$ | $= 1 + 7 + 4 + 4 = 16$ μs |
| Gross bit rate of the network in the above conditions | — | = 62.5 kbit s$^{-1}$ |

Following this interlude, and now that the major principles are in place, let us move on to the matters of the topologies and types of medium that can be used to carry CAN frames.

# 4

# Medium, Implementation and Physical Layers in CAN

The preceding chapters have hopefully brought us up to date with the nature and the operation, considered in its strictest sense, of the CAN (controller area network) protocol – part 1 of ISO 11898.

As I have mentioned several times, the original specification describing the protocol was drawn up to meet the requirements of multimaster communication applications, for real-time operation with multiplexed wiring. If your memory serves you well, you will recall that the designer's official specifications cover the whole data link layer and the bit specification of the OSI/ISO (International Standardization Organization/Open Systems Interconnect) communication model, while leaving the user free to choose the network topology and the type of transport medium to be used.

Rather evasively, therefore, I have avoided giving any electrical values for the signals in the diagrams in the preceding chapters, simply using the terms 'recessive' and 'dominant' for the values of the bits.

This model does not cover the actual structure of the physical layer, and, to avoid having a plethora of different solutions, many companies and organizations have worked on the problem of specifying the electrical properties required in a node at the interface with a transmission medium. After all, regardless of all other considerations, the system has to work and has to be constructed in the real world from real elements.

The aim of this chapter is to describe most of the subdivisions and hidden subtleties of the commonest implementations of the CAN physical layer in relation to the media used, together with the associated problems of conformity and radio interference. You should note that the physical implementations of the CAN physical layer are constantly developing. This is not a sign of instability in the concept, but simply an indication that the CAN solution is very much a living solution, with more and more numerous applications.

This chapter will therefore deal with parts 2 and 3 of ISO 11898.

This brings us face to face with the second fundamental problem of CAN, namely the construction of the physical layer, the lowest layer in the OSI model. It is time now to deal with

*Multiplexed Networks for Embedded Systems: CAN, LIN, Flexray, Safe-by-Wire...* D. Paret

the concrete realities, so we will examine the conventional problems of media, their characteristics, the lengths of networks, parasitic elements, insulation, etc.

The first part of this chapter will deal with the following major points:

- the various possible transmission media,
- the different modes of coupling to a CAN network,
- their performance,

whereas the second part examines

- compliance of networks with pollution standards (electromagnetic compatibility or EMC),
- the line protection (bus failure management),

and finally, we will rapidly survey some further problems relating to compatibility with other circuits.

## 4.1 The Range of Media and the Types of Coupling to the Network

The media or physical systems for CAN frame transmission can be constructed using various well-known technologies. In this field, we will encounter the conventional media, with their performance, characteristics, qualities and defects. Each of these potentially implies certain specific forms of coupling to the network (for example, the electrical or optical isolation of certain stations in the network). The list of media and their different coupling modes is summarized below, divided into

- wire media of all kinds:
  - single-wire, simple and asymmetrical;
  - two-wire:
    - ○ in differential mode, in the form of parallel or twisted pairs (screened or unscreened);
    - ○ with galvanic isolation using optocouplers or transformers;
    - ○ repeaters;
    - ○ current power lines (CPL):
      - ■ on specialized lines;
      - ■ on power lines:
        - □ alternating current type (220 V, 50 Hz, etc.);
        - □ direct current type (12 V/24 V battery, etc.);
        - □ ASK and FSK modules;
- optical media using optical fibres;
- electromagnetic media:
  - ASK and FSK modulated radio-frequency waves;
  - infrared waves.

At present, wire media are most widely used for carrying CAN frames, and in order to deal with the greatest possible number of questions relating to this matter in general, I shall examine the details of a large number of possible applications of this type of medium.

### 4.1.1 Single-wire medium

The single-wire medium is the simplest medium available for supporting the CAN protocol. When I described the composition of the basic data frame, you would have noted that its content could be carried by a single data line – in other words, a single 'wire', a single 'strand' or a single 'fibre' – plus, of course, the ground return in the case of a wire system. So now we just need to specify the electrical levels representing the recessive and dominant states of this medium.

Figure 4.1 shows an example of a simple single-wire implementation of this kind, which is asymmetrical with respect to ground, designed with the use of bipolar transistors whose wiring structure implements the electrical function of an 'AND-wired' circuit and in which the dominant level is the low electrical level (ground) and the recessive state is the 'high' level (the + of the power supply, using pull-up resistors). In this case, the stages of coupling to the single-wire line are provided using the 'open collector (or drain)' connection (as found, for the same reasons, in the well-known I2C bus).

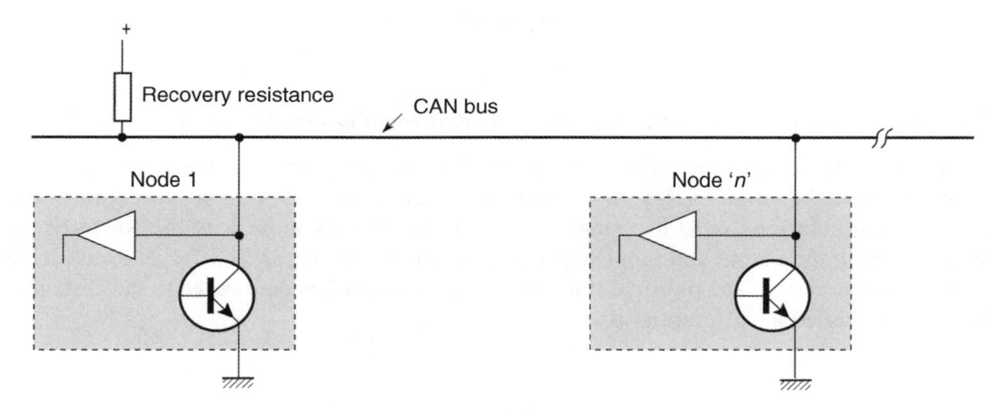


**Figure 4.1**

For implementing the physical part of the CAN network connection, this configuration is by far the cheapest and still conforms perfectly to the protocol (as regards the bit-wise arbitration rules, the possibility of re-reading bits during transmission, etc.). Along with its low cost, however, this architecture has the same faults as those of the bus I2C bus because of its asymmetrical structure with respect to ground and consequently the intrinsic protection of the set of signals (identifiers, data, CRC, etc.) against parasitic elements while they are carried on this type of medium.

To avoid these nuisances, we can consider various solutions. One example is the provision of line buffering to reduce its impedance (Figure 4.2).

This solution solves many problems, but it can be rather critical for high-quality applications for many typical CAN applications (with high speeds and/or long distances), despite the useful presence of systems for error detection and signalling and automatic resending of disturbed frames. It is true that, if we are to achieve the ultimate aim of CAN (connecting systems which may be up to several hundred metres apart), we must think about eliminating parasitic elements.

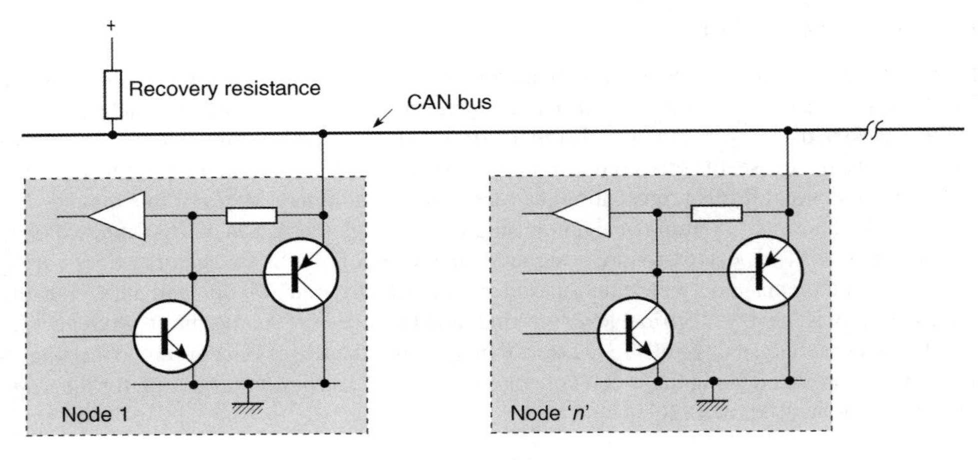


**Figure 4.2**

### *4.1.2 Twin-wire medium using a differential pair: advantages of the differential mode*

For eliminating parasitic elements originating from external sources, and therefore considered to have the same 'energy' at two points very close to each other in space, one of the conventional solutions is to use stages operating in what is known as the 'differential mode', in both the output and input stage configurations (see Figure 4.3 for an example of such a configuration). The physical line, consisting of two wires connected to these stages, is generally called a 'differential pair'.

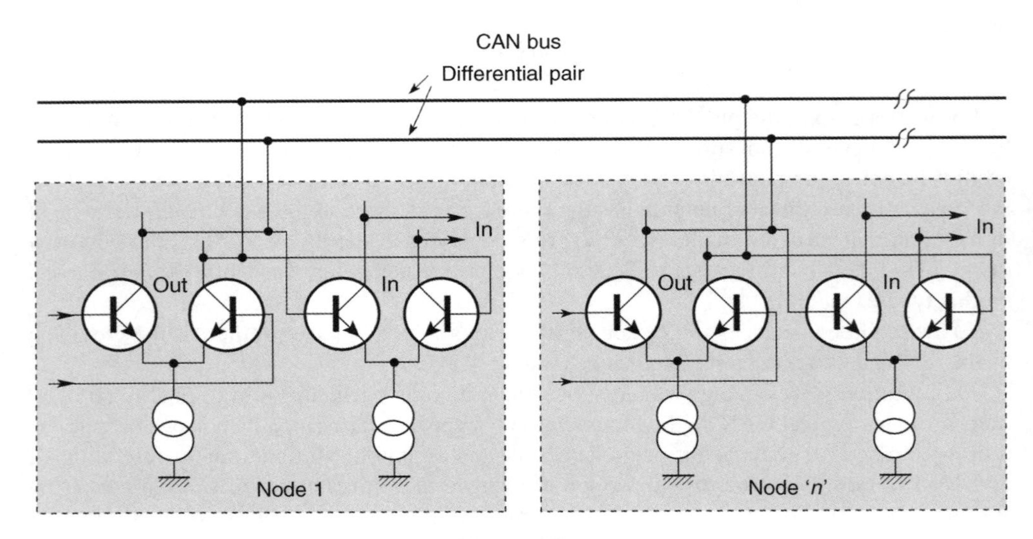


**Figure 4.3**

It is easy to show that the electrical nature of circuits operating in this differential mode is such that they have the structural property of providing a considerable 'common mode' rejection (rejection of a signal which appears simultaneously on the differential inputs and has the same phase and amplitude at a given instant and which is therefore likely to be an incoming parasitic signal). The downsides of this method are that, first, we have to physically install two wires to transmit the active signal (the CAN frame), and, second, in order to avoid complex problems related to returns from the input and output amplifier components, it is considered appropriate to physically separate two entities, namely the transmission output Tx from the reception input part Rx.

### Practical implementation of differential lines

Two major designs are generally used: single parallel differential pairs and twisted differential pairs, including a variant in the form of screened twisted pairs.

*Single parallel differential pairs*
Figure 4.4 shows an example of an implementation of such a pair, which has the structural qualities outlined above but can pose a number of problems due to the electromagnetic pollution it causes when frames are in transit. To overcome this, we must move on to the next paragraph.

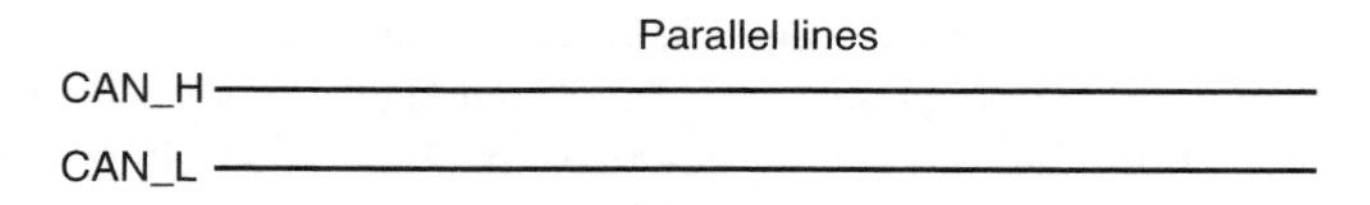


**Figure 4.4**

*Twisted differential pairs*
These differential pairs are twisted around themselves (Figure 4.5). Consequently, when 'differential' signals are emitted (with identical amplitudes and symmetrical and opposite shapes on the two wires), the twisted structure greatly reduces the radiated radio interference (by cancelling out the two fields produced on each strand). For reasons which will be detailed below, these pairs are often used to form the physical layer of a CAN bus.

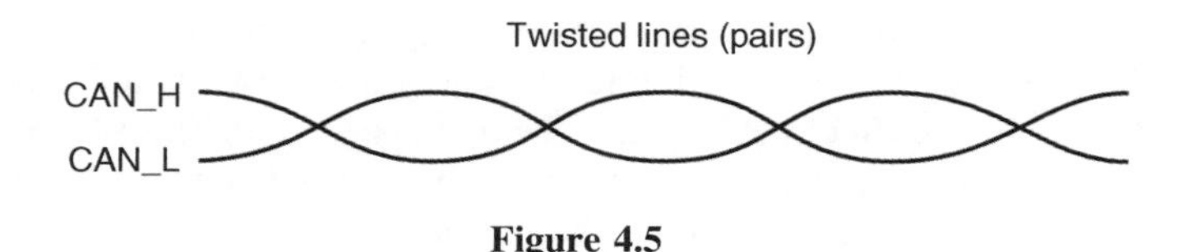


**Figure 4.5**

Unfortunately, for many reasons, the signals emitted from integrated circuits are rarely exactly equal in amplitude or of opposite sign at all times (mainly during the transitional phases of bits, in changes from recessive to dominant and vice versa).

So what can we do to solve this problem, you may ask? The answer is to go on to the next paragraph!

*Twisted and screened differential pairs*

The final (or finest) point – or almost!

Twisted pairs are used to reduce most of the interference, and the finishing touch is to add a screen around the pairs (Figure 4.6), the end(s) of the screen being connected to ground.

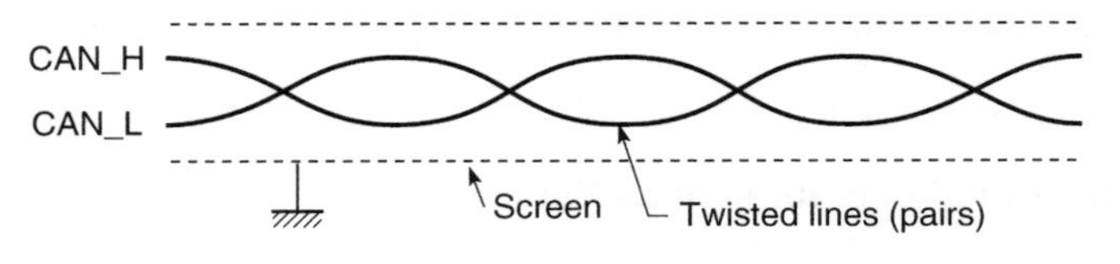


**Figure 4.6**

You would be surprised to discover how many problems lurk behind the concluding part of the last sentence – 'the end(s) of the screen being connected to ground'. Which end? Both ends? What to? How?

Usually, the screen of the differential pair raises more problems than it solves, especially in the case of long distances and a poor ground return – so beware! It is often better to do without one – and this will save money too.

For these important reasons, some manufacturers of line drivers for differential pairs guarantee an excellent degree of symmetry in the output stages. They also offer devices for controlling the slope of the rise times and decay times of signals travelling along the lines, to reduce the asymmetry of the signals carried and thus enable simple unscreened twisted pairs to be used in most cases.

*Wire solutions, CAN and ISO standards*

This concludes our brief overview of the 'theory' of differential pairs. We could now go on to look at another type of medium, but this would mean missing out much of the history of CAN. In fact, as this type of medium is very widespread in industry, an ad hoc working group of the ISO has been studying the standardization of this particular medium, with the aim of precisely defining certain characteristics in the context of the highly specific applications of multiplexed networks in the car industry.

Three ISO standards clearly specify how CAN is to be related to this type of differential pair wire medium (but do not provide a mechanical specification of the connector). They divide the qualities and types of connection to the medium into two major groups: the low-speed mode (ISO 11519 and ISO 11989-3) and the high-speed mode (ISO 11989-2). Leaving aside the matters of speed, extended frames, etc., these standards mainly differ in their proposals for the electrical specifications of the physical layers. They define the operation and electrical characteristics of the transmission medium, i.e. the physical layer, but do not cover the mechanical aspects (forms of connector) or any of the problems of the other layers (such as layers 2, 6 or 7).

Note that the appearance of these standards in time did not follow the logical order of the increase of speed in CAN. To clarify this point, the CAN applications were designed as follows:

- initially in CAN high-speed mode (125 kHz at 1 Mbit $s^{-1}$) – ISO 11989-2;
- followed by the CAN low-speed mode (from 10 to 125 kHz) and its variants:
  - in a former version, ISO 11519,
  - in the current and most widespread version, including the management and correction of certain faults on the bus – *fault-tolerant low-speed CAN* (FT LS CAN) – *ISO 11898-3*,
  - and finally, an economical solution, dispensing with differential pairs and standardization, and using only one wire!

To follow this development, the sequence of the text below keeps to this order of appearance in the market.

## 4.2 High-Speed CAN, from 125 kbit $s^{-1}$ to 1 Mbit $s^{-1}$: ISO 11898-2

We shall now examine the different implementations of the physical layer of CAN in the case of applications running at high speed, in other words, at bit rates from 125 kbit $s^{-1}$ to 1 Mbit $s^{-1}$.

### *4.2.1 Simple differential wire pair*

This is certainly the most widespread implementation of the physical layer, being used both in motor vehicle applications and in industry. It is based on compliance with the electrical and time values described in the ISO standard 11898-2 for high-speed CAN, these values being simply those which have existed ever since the creation of CAN.

Figure 4.7 shows the electrical levels adopted for the dominant (nominally +2 V) and recessive (0 V) states, the upper part showing the nominal signals between the two wires of the

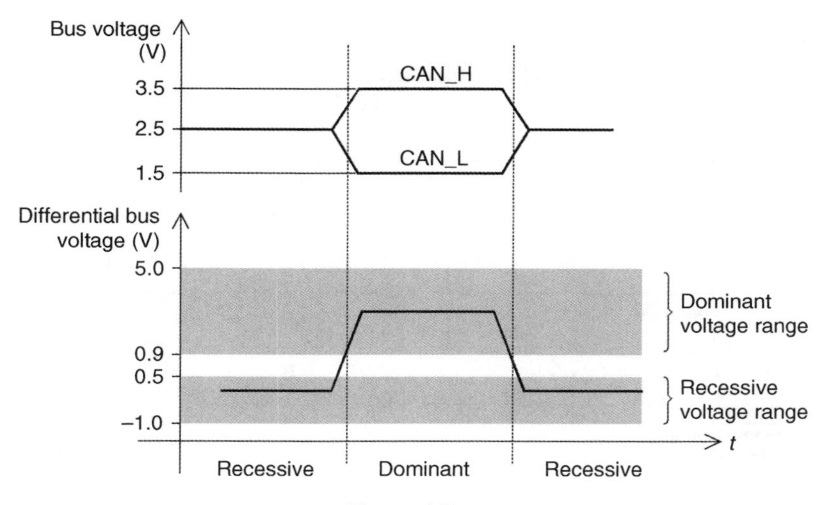


**Figure 4.7**

bus, named CAN_H and CAN_L for this purpose, and the lower part showing a differential representation of the voltage CAN_H – CAN_L (dominant state = +2 V nominally and recessive state = 0 V), including the tolerances.

The principal characteristics described in the high-speed CAN standard are

- a speed from 125 kbit $s^{-1}$ to 1 Mbit $s^{-1}$;
- a network length of up to 40 m at the maximum speed of 1 Mbit $s^{-1}$;
- a number of network participants in the range from 2 to 30 nodes;
- the physical layer is a differential pair (twisted) with ground return;
- a characteristic line impedance of 120 Ω;
- a wire resistance of 70 mΩ $m^{-1}$;
- a nominal signal propagation time of 5 ns $m^{-1}$ on the bus;
- an output current of more than 25 mA supplied by the transmitter;
- a line protected from short circuits (from 3 to +16 V, or 32 V);
- a common mode range from – 2 to +7 V, simple power supply under 5 V, etc.

The conventional architecture for a high-speed CAN node implementation is shown in Figure 4.8. In this architecture, the 'microcontroller + CAN protocol handler' unit very often forms only one of the units within a microcontroller with an integrated CAN controller.

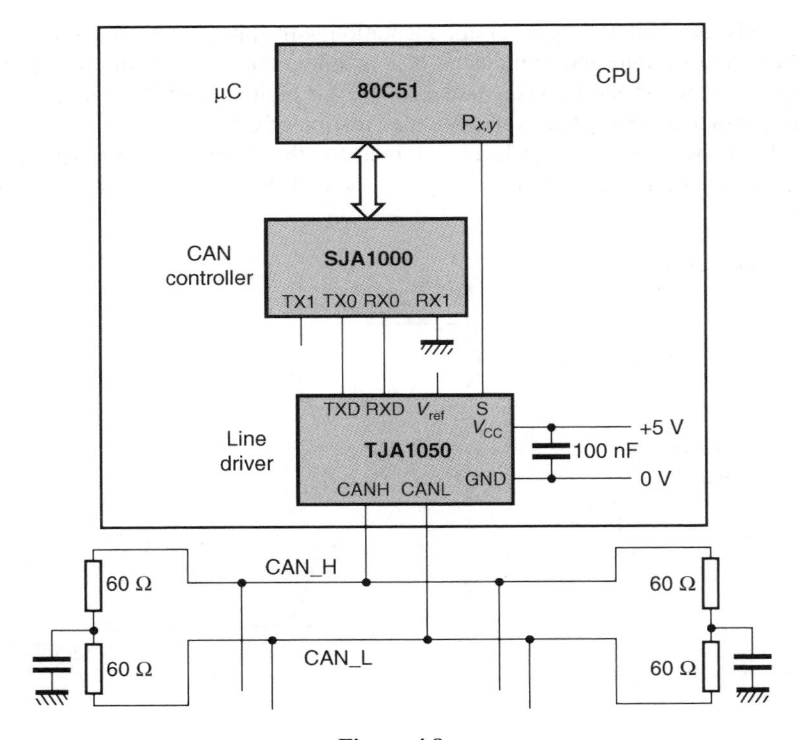


**Figure 4.8**

Because of the different technologies used (with microcontrollers operating at 5 or 3.3 V and line drivers having to withstand up to 60 or 80 V in certain conditions), the line drivers are hardly ever integrated into the microcontroller.

To return to technical matters, the implementation of the high-speed CAN physical layer raises three main technical problems.

The first is that higher speeds lead to more pollution, especially as regards radio-frequency interference. One of the main problems in this implementation therefore relates to the art and method of limiting EMC pollution (again at the lowest cost) and all its associated difficulties. I will deal with this matter in a special paragraph at the end of this chapter.

The second problem concerns the maximum number of nodes that can be connected to the CAN network. In these types of networks with ever-increasing populations of nodes (provided at the outset or to cover potential options which can be implemented), all these random factors must be allowed for.

Figure 4.9 shows the equivalent diagram that should be studied to determine the values of the differential PDs found at line terminations, as a function of the number of nodes connected in parallel to the bus, the total ohmic resistance of the connecting wires and the minimum and maximum currents that can be delivered by the line driver, in order to estimate whether or not the input voltage of the most remote driver will be greater than the threshold for achieving the value required by the standard.

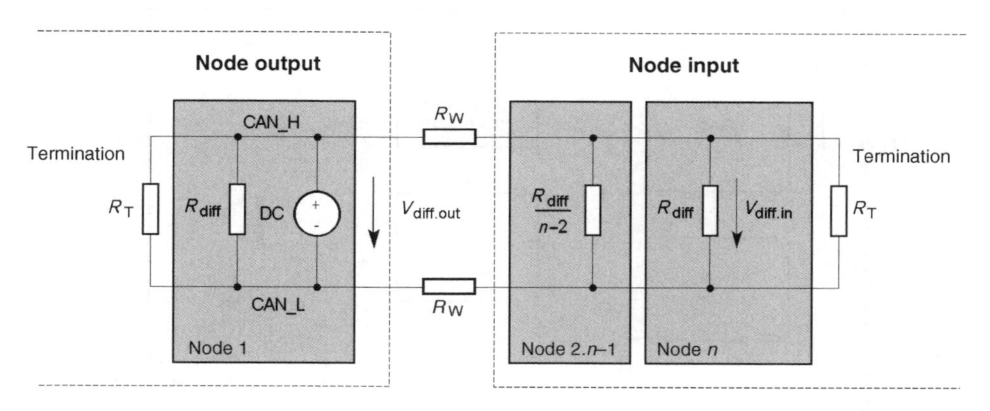


**Figure 4.9**

The third and final point relates to the power consumption of the line driver stages, as an increasing number of nodes remain live when a vehicle or a system is halted. As they are becoming more numerous, each of them must have a very low consumption when in its 'sleep' state, or else the batteries will rapidly run down at inconvenient times (during long holiday periods, or where cars are kept in manufacturers' parking lots before sale, etc.).

Otherwise, we only have to deal with small everyday problems of design (the shrinking surface area of printed circuits), weight (to allow for vibration), cost, multiple sources of supply, availability and the like, in other words, the old routine problems.

*4.2.2 Example of components for the ISO 11898-2 HS CAN layer*

In view of the annual volume of consumption of these parts, most of the world's component manufacturers have a strong presence in the market, each trying to find a niche based on subtle minor differences enabling its components to be used for special applications.

Here again, in theory, the interoperability of this wide range of components in a single network is appropriate and requires conformity tests. In fact, the number of electrical and time values described in ISO 11898-2 is rather limited, and the conformity of the guaranteed values of component data sheets with these specifications is very often relied on in the marketplace.

To provide some specific details, drawn from the wide range of possibilities in the market, I have selected three examples, including the 82C250, an 'ancestor', but still useful, and three newcomers, TJA 1050, 1040 and 1041, which seem to have a few differences between them, and yet . . .

**PCA 82C250**

The PCA 82C250 (Figure 4.10), well known in the market for many years, fully satisfies the functional requirements. Depending on the desired speed, it may or may not be necessary to add a double choke to reduce the EMC (see also the special paragraph on EMC in this chapter).

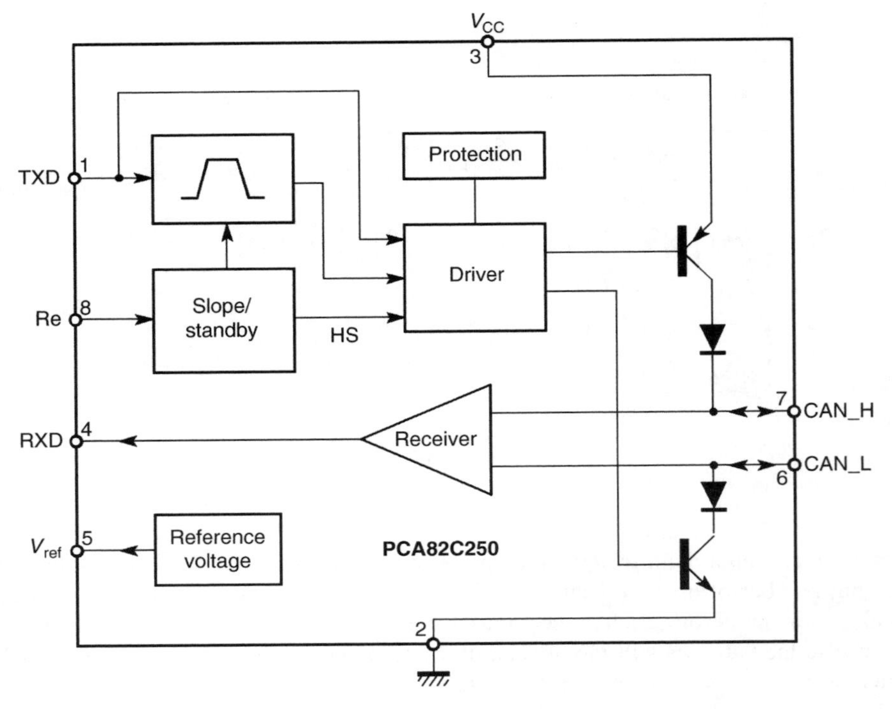


**Figure 4.10**

As shown in the block diagram, this circuit also has a terminal for adjusting the slope of the output signals sent to the CAN bus to limit the use of this choke.

### TJA 1050

A product of newer technology, the TJA 1050 was designed to succeed the first, in a more specialized technology known as SOI (silicon on insulator) (Figure 4.11), considerably reducing the EMC problems and providing much better immunity. In each of these areas, the gain is of the order of 20–25 dB compared with the circuit described in the preceding section (see also the special paragraph on EMC at the end of this chapter), making it unnecessary to use the choke mentioned in the preceding section. However, some designers still like to use a 'belt and braces' approach, providing a filter coil even when it is not essential in a standard network architecture ... and then they complain of the high cost of their solution!

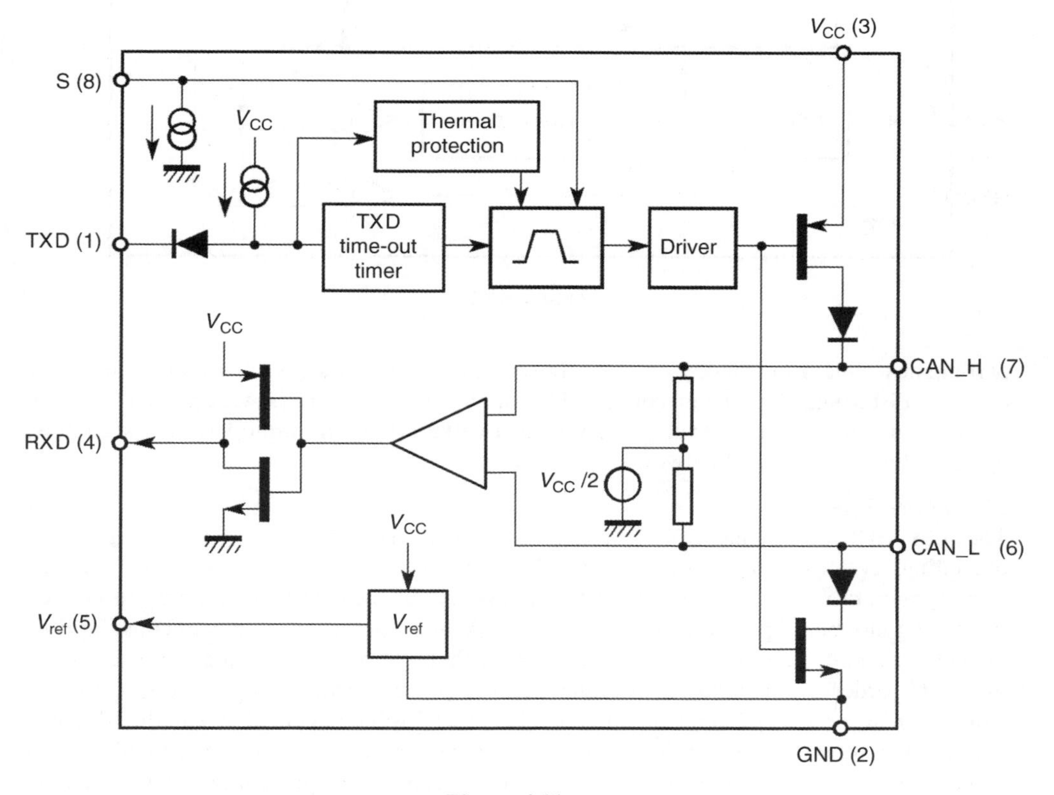


**Figure 4.11**

Note that the first two circuits described do not have terminals delivering a mid-voltage $V/2$, called the 'split voltage' $V_{split}$, whose usefulness will be demonstrated later.

### TJA 1040

The TJA 1040 (Figure 4.12), based on the same technology as the TJA 1050, is designed to have very low power consumption, making it possible to create a CAN network in which some

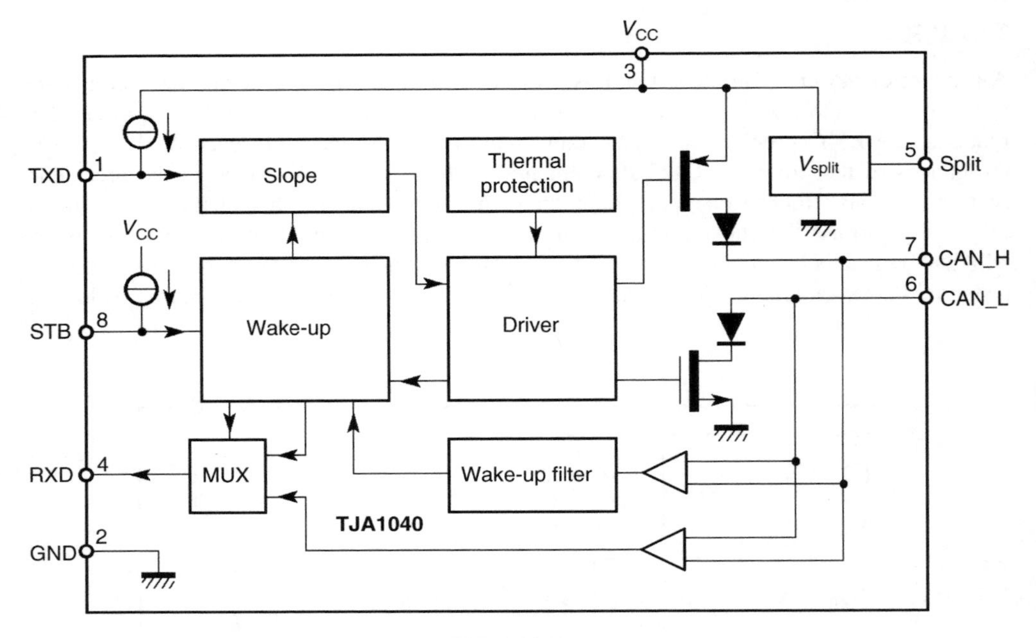


**Figure 4.12**

participants remain in standby or sleep mode, whereas others are switched off completely (without disturbing the bus, of course). This application structure is widely used in motor vehicles and on-board industrial environments. I will return to this point later in a section on what are known as 'partial networks'.

*The voltage $V_{split}$*

The TJA 1040 circuit has a $V_{\text{split}}$ pin that has just been mentioned in passing in the preceding sections. Let us now consider this functionality that is very important for the correct operation of a high-speed CAN network. As mentioned above, there are now an increasing number of participants in each network, and, for reasons of power consumption, the network controller often sends some or all of them 'to bed', according to the operating phase, by ordering them to sleep either deeply (power down) or lightly (sleep mode). Nothing new or complicated about that. The real problem occurs when the sleeping participants are woken up. If this happens in a firm and decisive way (by a local reset or wake-up using special pins), the problem is often resolved by providing several wake-up instants in advance, so that the participant is ready when the CAN communication frames start to travel along the network. However, if the wake-up is done 'softly', simply by the travel of a frame along the network, the line driver has to wake up very rapidly and be ready to operate without delay, to not to miss reading the content of the frame being transmitted. This gives rise to well-known problems of controlling transient phenomena during the establishment of supply voltages, etc.

If no special device is provided, then, when the circuit is resupplied with power, the mean value of the supply voltage (corresponding by definition to the value of the recessive bit in high-speed CAN) is not established immediately, as there is a common-mode jump at the start of each message (Figure 4.13), generally causing incorrect interpretation of the first bits of the frame (identifiers, etc.), and therefore the generation of error frames, retransmission of frames, etc. Furthermore, at this precise instant, because of the high electrical asymmetry of the signals, this whole process plays an important part in EMC radiation.

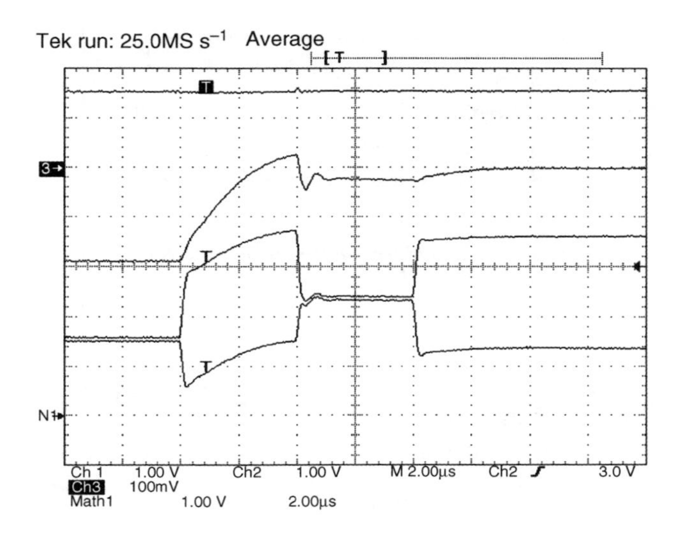


**Figure 4.13**

If a voltage $V_{split} = V/2$, drawn from the integrated circuit, is provided, these phenomena are eliminated (see Figure 4.14 in which the common-mode jump has disappeared), permitting an instantaneous start of the correct decoding of the message sent on the bus. Thus, the presence of numerous nodes in the network implies the use of high-speed CAN to manage the numerous messages and also implies that some nodes must be put into sleep mode to reduce power consumption, with fast wake-up and the provision of a $V_{split}$ voltage.

The conventional diagram of the application of a high-speed CAN line driver is thus developed into that shown in Figure 4.15.

### *4.2.3 Differential pair with diagnosis of the CAN HS physical layer*

This type of implementation is an enhancement of those described above. At the time of writing (2006), this implementation has been standardized and is increasingly common in the market, as although its components cannot provide for the repair of electromechanical bus failures or operation in degraded mode, they can at least control low-consumption modes and signal the presence of disturbances in the physical layer. This is a first step towards 'fault-tolerant high-speed CAN', which at present is both difficult to implement and not in demand in the market. Here again, for the same reasons as for FT LS CAN (see below), the GIFT group has resumed its work (in fact, it has never really stopped and should really change its name because the system is not truly fault tolerant!), in order to define all the functional

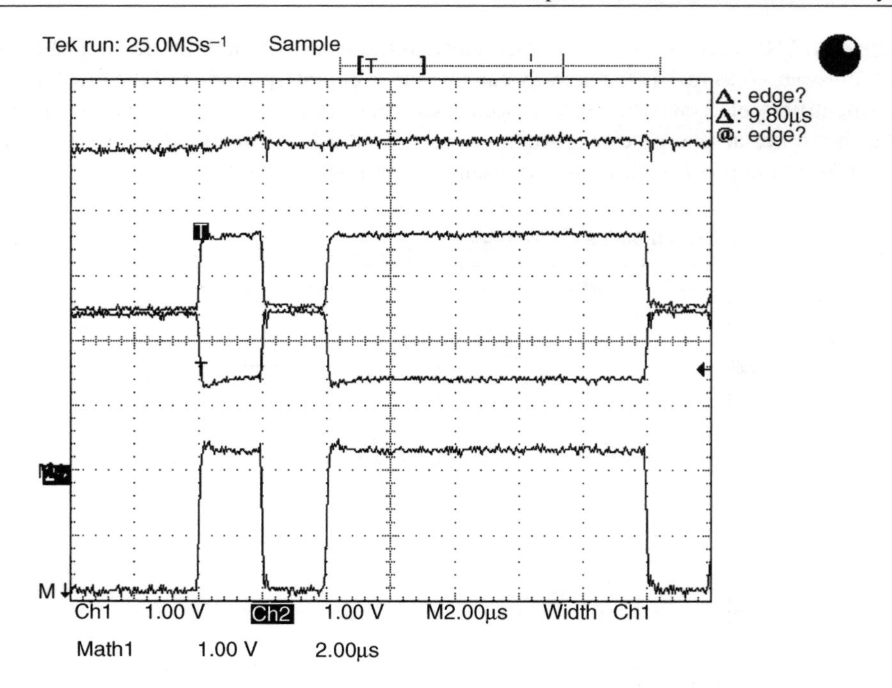


Figure 4.14

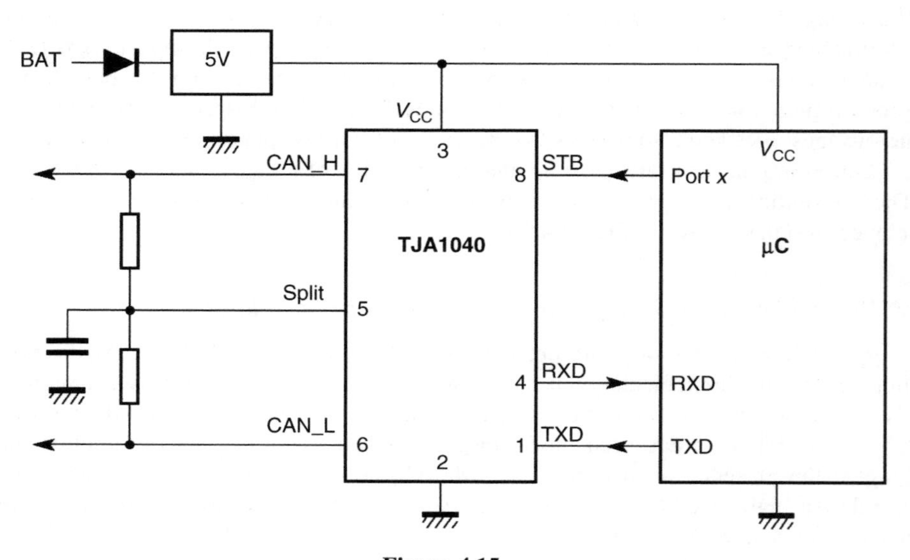


Figure 4.15

interoperability implied by this. A 'golden chip' (reference circuit), the Philips/NXP Semiconductors TJA 1041, is being used as the model for drafting the ISO 11898-5 standard, which will also take into account the appearance of multiplexed solutions operating at 42 V under battery power.

### 4.2.4 *Example of components for the ISO 11898-5 HS CAN layer: TJA 1041*

Clearly, this circuit has more pins than its predecessors, as it has to communicate with a microcontroller in order to send to the microcontroller the diagnosis it has made and allow the microcontroller to decide what use to make of the data. The block diagram of this circuit is shown in Figure 4.16.

Its pin layout is also compatible with the simpler circuits (see Figure 4.17 and note the compatibility of the pins with standard HS CAN drivers). You should also note that the pin

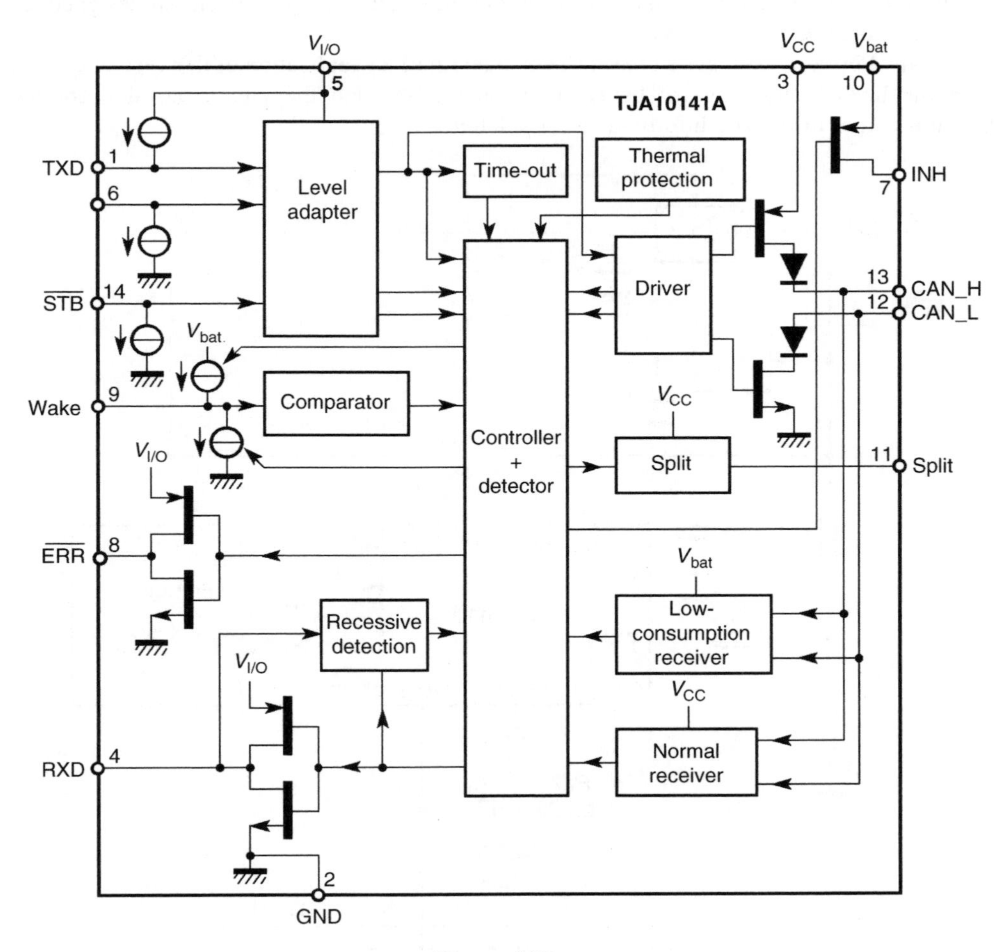


Figure 4.16

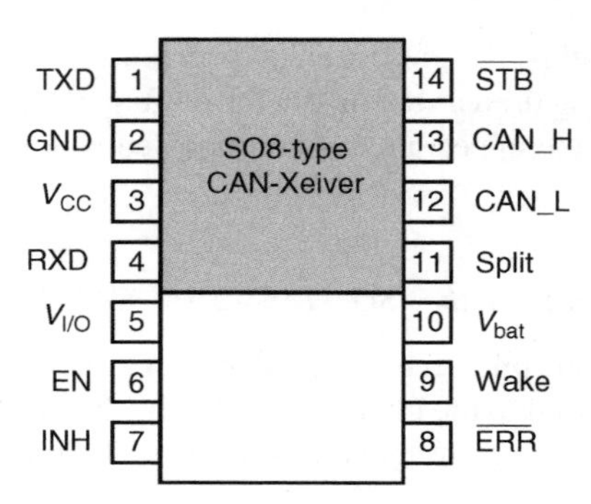


**Figure 4.17**

layout of the circuit enables it to be physically implemented in a printed circuit designed for possible double implementation.

Finally, Figure 4.18 shows a conventional diagram of an application of this type of circuit, with the links to be provided between the line driver and the microcontroller, for the transmission of diagnostic information to the latter.

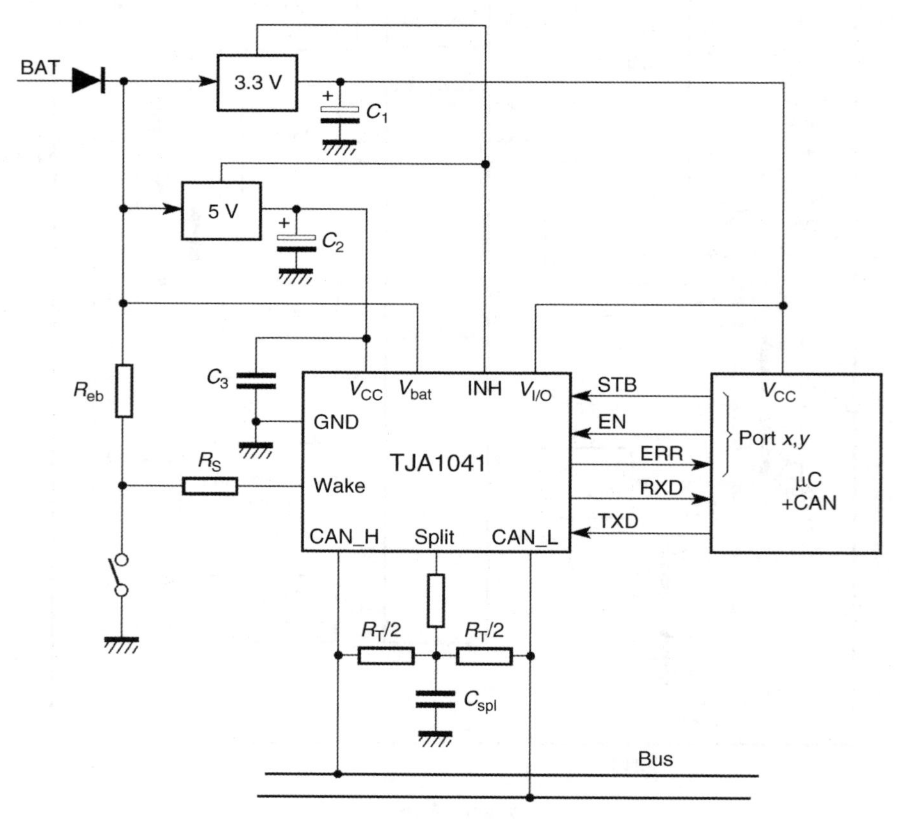


**Figure 4.18**

### 4.2.5 *Partial HS CAN networks*

Yet another problem for CAN components to solve! In many systems, some of the nodes often have to remain live for functional reasons (with the supply voltage present and the node either operational or not – for specialists, this is what is called 'clamp 30', from the generic name of the connector used for this purpose in all vehicles in the market), whereas the others, because of the pressing need to save power, are deliberately disconnected from the supply (i.e. after the contact called 'clamp 15') (Figure 4.19).

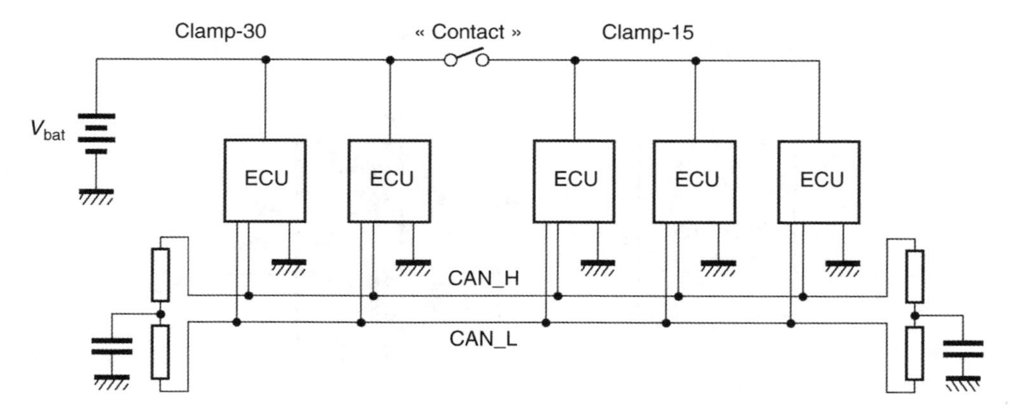


**Figure 4.19**

They can also be put into sleep mode – in the form of a doze (standby, using a few hundred μA) or a deep sleep (sleep mode or power down mode, using a few tens of μA) and will always be ready to wake up when necessary. Such systems are called 'partial network' systems (Figure 4.20).

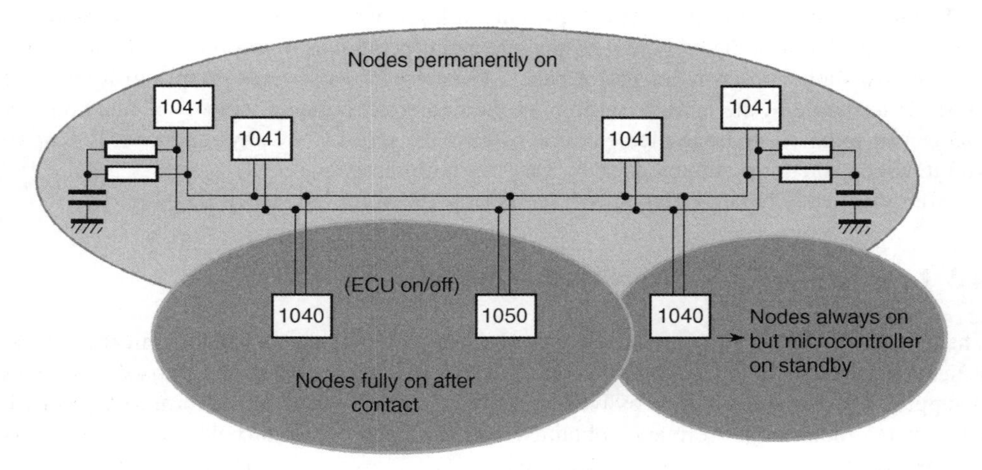


**Figure 4.20**

The key point is that the nodes which are in sleep mode or switched off do not disturb the ones which remain awake. As the bus wires are same for all stations, the 'snoring' of some of them must not disturb the operation of the others. Generally, there is no easy answer to this, because the inactive nodes must be in a state which makes them 'transparent' for the correct operation of the bus, and yet they must not be switched back on by the signals travelling along the bus while they are asleep, and must not send any reverse currents along the network, as sometimes happens (Figure 4.21).

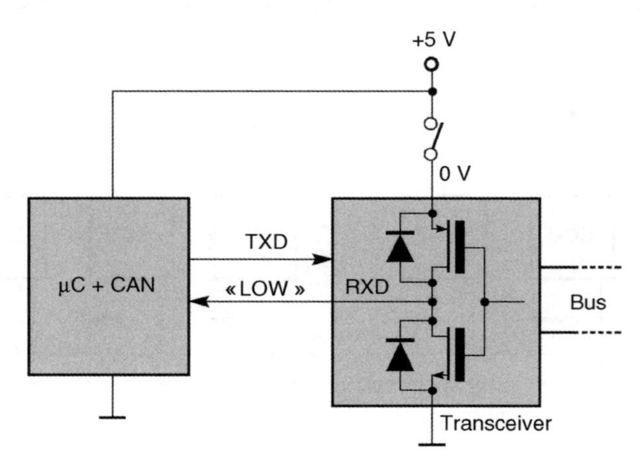


**Figure 4.21**

Similarly, should the sleepers be awakened with a trumpet blast or a tender kiss? In electronic terms, this equates to a wake-up using a dedicated pin on the integrated circuit, for applying a closely specified local wake-up signal, or using a remote wake-up procedure triggered by the start of activity in the network. Clearly, we must be able to handle both processes!

To finish with the subject of the components of high-speed CAN line drivers, without any obtrusive special pleading on behalf of any one manufacturer, as most products in the market are roughly similar in terms of performance, Figure 4.22 shows an example of a range of products available in the market, setting out the numerous possible variants of functionality that can be required in the implementation of the high-speed CAN physical layer. It is up to you to choose the most suitable product for your requirements.

After this heady burst of high speed, let us slow down for low-speed CAN.

## 4.3 Low-Speed CAN, from 10 to 125 kbit $s^{-1}$

The low-speed CAN mode is defined as the mode in which the network communication speed is in the range from 10 to 125 kbit $s^{-1}$. Theoretically, any physical layer can be used, as long as it supports the dominant and recessive levels, and there is no need to conform to a standard. Clearly, the radiation performance obtained will vary, and conformity with EMC standards will be easier or harder, according to the type of physical layer used. To help you to choose, the

| | **PCA82C250/51** | **TJA1050** | **TJA1040** | **TJA1041/1041A** |
|---|---|---|---|---|
| ISO 11898-2 | Yes | Yes | Yes | Yes |
| Standby mode with remote wake-up | Yes | No | Yes | Yes |
| Sleep mode with remote wake-up | No | No | No | Yes |
| Local wake-up | Yes | — | Yes | Yes |
| Consumption in standby/sleep mode | 100 μA | 5 mA | 10 μA | 20 μA (for the whole node) |
| Invisible when switched off | No | Approx. | Yes | Approx. |
| Listen-only mode | No | Yes | No | Yes |
| TXD dominant timer | No | Yes | Yes | Yes |
| Enhanced bus clamping protection (RXD failures) | No | No | No | Yes |
| Bus diagnostics | No | No | No | Yes |
| Overload protected | Yes | Yes | Yes | Yes |
| Split voltage pin | No | No | Yes | Yes |
| μC compatibility | 5 V | 3.0–5 V | 3.0–5 V | 2.8–5 V |
| Casing options | DIL8, SO8, bare chip | SO8, HVSON8 bare chip | SO8, bare chip | SO14, bare chip |
| Principal applications | Nodes always live | Main nodes in large networks, without standby | Partial networks or nodes always live | Partial networks or nodes always live |

**Figure 4.22**

following paragraphs will deal with examples commonly used by the profession in industrial applications or motor vehicles.

### *4.3.1 Symmetrical – on differential pair*

Again, I shall start by trying to make the CAN physical layer operate with a differential pair. There are various possible cases.

### As in ISO 11898-2 high-speed mode ... but slower!

Why not take the simple approach? As CAN generally operates in high-speed mode, let us just reduce its speed below 125 kbit $s^{-1}$. After all, it is allowed!

Many industrial users have done this without problems ... and from their point of view, they were quite right to do so. Effectively, you make no changes, you just reduce the speed. Job done! The bit time increases, and so does the communication distance, the slopes of the signals decrease, pollution is reduced ... If we ignore any malfunctions due to electromechanical incidents occurring along the network, everyone will be happy.

Sometimes, when required by the application (because of very severe EMC radiation or interference problems, for example) low-speed CAN (below 125 kbit $s^{-1}$) is used on a differential pair but over short distances at which no mechanical disturbances of the bus can occur. In this case, the LS FT solution (see below) is not justifiable. This is why we often encounter low-speed CAN applications using the physical layer of the ISO 11989-2 (high-speed) standard for the implementation of this kind of network. In this case, the conventional high-speed CAN components are used.

### In accordance with ISO 15519

Yes, for many years there has been a standard defining the electrical levels to be used with differential pairs in low-speed CAN mode. It is called ISO 15519-2. Sadly (or happily?) nobody uses it any more! This is because this standard, dating from the early 1990s and describing the implementation of CAN in low-speed mode on differential pairs, has been rarely used, if ever, for simple practical reasons. Because of the slower speed, resulting in longer communication distances and a higher probability of electromechanical disturbances in the network, many users soon wanted enhancements in the fault-tolerant performance of the physical layer of the low-speed CAN mode. It was therefore rapidly replaced with the one described below.

### With a fault-tolerant solution (FT LS CAN) – ISO 11898-3

FT LS CAN has been mentioned briefly in my list of solutions, but it deserves a few additional paragraphs. In fact, many mysteries are hidden behind this name.

Because of the low speed of the LS CAN network, implying a long bit time, it is possible to construct networks with a considerable bus length (as much as 5–10 km). It will soon be found that as the length of a network increases, it is more likely that it will suffer from, and/or produce mechanically on the bus, various kinds of electromechanical disturbances due to short circuits and/or open circuits (wires isolated or cut).

*Prevention, detection and management of line faults*
For communication over a CAN network to be reliable, it must be capable of being maintained even in very severe conditions of physical disturbance. For this purpose, the internal configurations of the CAN interfaces of the various components must be capable of resolving almost all ordinary (and extraordinary) disturbances while continuing to transmit, even in 'degraded

mode' (but transmitting nonetheless!), and signalling and/or indicating any faults that occur. Clearly, many industrial companies have followed this philosophy and provide abundant products of such kind. The resulting additions and improvements made to the original CAN protocol, in the form of FT LS CAN, now represent one of the most important 'hardware enrichments' of CAN.

The principle of FT LS CAN is simple. If there are any faults on the bus (Figure 4.23), they must be detected, signalled and repaired as well as possible.

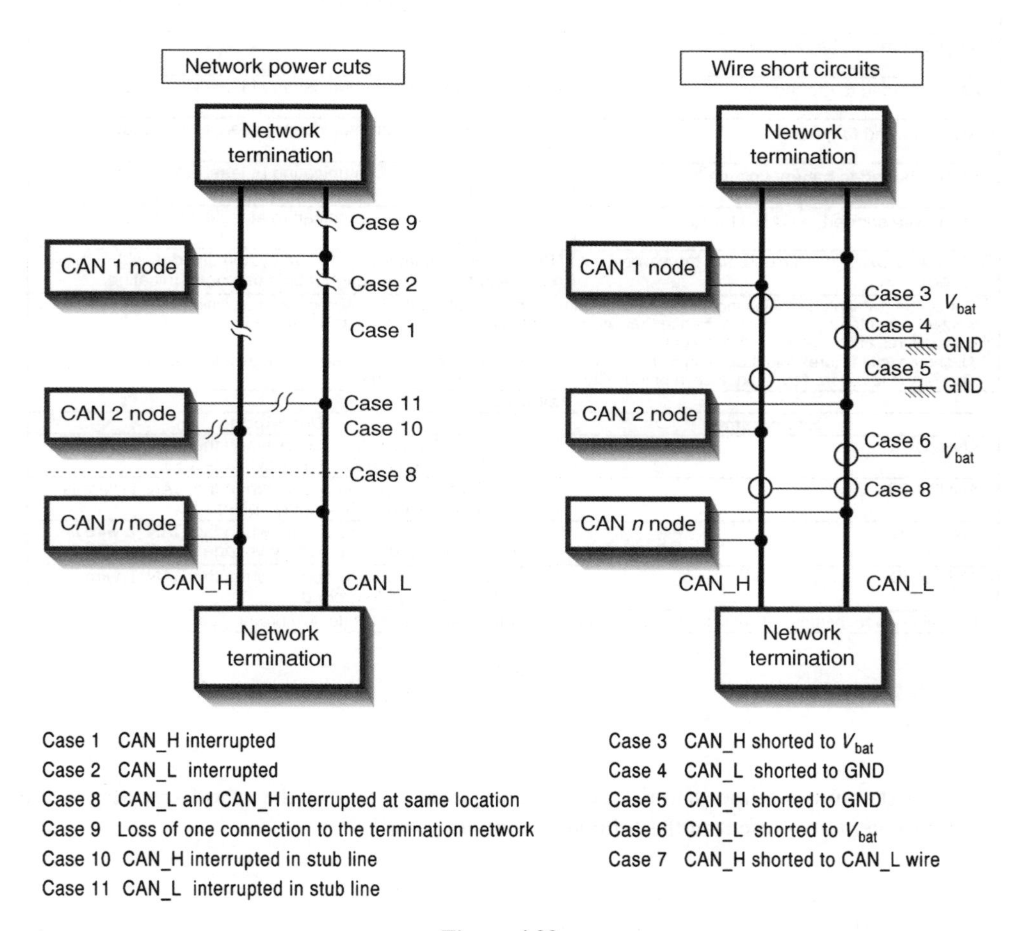


**Figure 4.23**

Figures 4.24 and 4.25 summarize part of the list given in ISO 11898-3 for two types of families of disturbances and disruptions (bus failures) that a CAN line must be able to withstand.

| Description of bus failure | Behaviour of the network |
|---|---|
| One node becomes disconnected from the bus[1] | The remaining nodes continue communication. |
| One node loses power[2] | The remaining nodes continue communicating at least with reduced signal to noise ratio. |
| One node loses ground[2] | The remaining nodes continue communicating at least with reduced signal to noise ratio. |
| Open and short failures | All nodes continue communicating at least with reduced signal to noise ratio. |
| CAN_L interrupted | All nodes continue communicating at least with reduced signal to noise ratio. |
| CAN_H interrupted | All nodes continue communicating at least with reduced signal to noise ratio. |
| CAN_L shorted to battery voltage[3] | All nodes continue communicating at least with reduced signal to noise ratio. |
| CAN_H shorted to ground[3] | All nodes continue communicating at least with reduced signal to noise ratio. |
| CAN_L shorted to ground[3] | All nodes continue communicating at least with reduced signal to noise ratio. |
| CAN_H shorted to battery voltage[3] | All nodes continue communicating at least with reduced signal to noise ratio. |
| CAN_L wire shorted to CAN_H wire[4] | All nodes continue communicating at least with reduced signal to noise ratio. |
| CAN_L and CAN_H interrupted at the same location[1] | No operation within the complete system. Nodes within the remaining subsystems might continue communicating. |

[1] Due to the distributed termination concept these failures shall not affect the remaining communication and ate not detectable by a transceiver device. Hence they are not treated and are not part of this standard.
[2] Both failures are treated together as power failures.
[3] All short circuit failures might occur in coincidence with a ground shift (seen between two nodes) in a range of +/–1.5V.
[4] This failure is covered by the detection of the failure 'CAN_L shorted to ground'.

**Failure events**

| Event name[1] | Description |
|---|---|
| CANH2UBAT | Failure that typically occurs when the CAN_H wire is short circuited to the battery voltage $V_{bat.}$ |
| CANH2VCC | Failure that typically occurs when the CAN_H wire is short circuited to the supply voltage $V_{CC.}$ |
| CANL2UBAT | Failure that typically occurs when the CAN_L wire is short circuited to the battery voltage $V_{bat.}$ |
| CANL2GND | Failure that typically occurs when the CAN_L wire is short circuited to ground. |

[1] The failure event names may occur with the indices N (for normal mode) and LP (for low power mode).

**Figure 4.24**

The first family of line disturbances or disruptions is that of the 'short circuits' to the different potentials present in the following configuration:

- case 3: CAN_H short-circuited to $V_{bat}$,
- case 4: CAN_L short-circuited to ground,
- case 5: CAN_H short-circuited to ground,
- case 6: CAN_L short-circuited to $V_{bat}$,
- case 7: CAN_L and CAN_H short-circuited to each other.

The second family is that of the 'open circuits':

- case 1: CAN_H interrupted,
- case 2: CAN_L interrupted,

| 4 • Médium, implémentation et couches physiques CAN | 4.3 CAN à bas débit (*low speed CAN*) de 10 à 125 kbits/s |
|---|---|

| FAILURE | DESCRIPTION | TERMINATION CANH (RTH) | TERMINATION CANL (RTL) | CANH DRIVER | CANL DRIVER | RECEIVER MODE |
|---|---|---|---|---|---|---|
| 1 | CANH wire interrupted | on | on | on | on | differential |
| 2 | CANL wire interrupted | on | on | on | on | differential |
| 3 | CANH short-circuited to battery | weak; note 1 | on | off | on | CANL |
| 3a | CANH short-circuited to $V_{CC}$ | weak; note 1 | on | off | on | CANL |
| 4 | CANL short-circuited to ground | on | weak; note 2 | on | off | CANH |
| 5 | CANH short-circuited to ground | on | on | on | on | differential |
| 6 | CANL short-circuited to battery | on | weak; note 2 | on | off | CANH |
| 6a | CANL short-circuited to $V_{CC}$ | on | on | on | on | differential |
| 7 | CANL and CANH mutually short-circuited | on | weak; note 2 | on | off | CANH |

**Figure 4.25**

- case 8: CAN_L and CAN_H interrupted at the same location,
- case 10: CAN_H interrupted towards a node,
- case 11: CAN_L interrupted towards a node,
- and finally, case 9: CAN_L or CAN_H interrupted at the start of the line.

**Note:** In the above figures, you would have noticed the well-known case of a CAN station whose power supply ground return wire is disconnected. Many other cases are recorded by users in the motor vehicle and industrial markets but do not appear in the ISO list, and it should also be possible to resolve these (by means of bus failure management) in order to make the link reliable.

In short, this again raises the problem of failure detection, signalling and management, and the action to be taken to resolve the problem. Everyone knows the different measures which can be taken routinely (or otherwise) in case of difficulty. Here is a short list.

*Fault detection*

This consists of designing a device for monitoring the faults that may occur and detecting them (for example, by detecting an output level not matching the desired level). Figure 4.24 shows some examples of disturbances (short circuit, open circuit, etc.). Theoretically, an engineer of average skill should have no difficulty in resolving the problem, but ... In fact, it is not a simple matter to implement these fault monitoring and detection devices without degrading the quality of the signals carried on the bus in ordinary bus operation.

*Fault signalling*
It is all very well to detect a fault, but it is selfish to keep this information to yourself! It is common politeness to let the neighbours know about it, by signalling the presence of a fault (by modifying an error flag bit, for example). Evidently, the ultimate aim is to spell out to the other participants in the network the nature of the fault (for example, 'short circuit between the CAN wire and ground'). To do this, it is necessary to draw up the most exhaustive list of faults which may be encountered, by using a table of error types, corresponding to specific messages or interruptions. Some or all of this list may or may not be described in a standard. At this level, we can say that an error has been detected and signalled. The diagnostic phase is practically complete. Here again, there is nothing new under the sun!

Sometimes we may stop here, and the 'application' (software or hardware) will be responsible for taking over to assure some of the operating modes (degraded mode, restricted mode, survival mode, fallback mode, etc.). Yes – I did write 'some', not 'all'! In fact, numerous strategies can be used, according to circumstances and requirements (always allowing for the cost involved).

*Correction/repair*
When an operating fault has been signalled, do we want to correct it partially or completely? Two of the most commonly used strategies are described here.

*Complete repair.* The luxury version! The emergency plan is deployed, and everything is done to make the necessary replacements wherever they are needed. The system therefore has internal redundancy. This is applicable to high-level security systems (see the description below of the principle of the FlexRay concept of the X-by-Wire family).

*Operation in degraded mode.* Generally, operation in what is called 'degraded' mode is tolerated: this consists in the toleration of a few minor imperfections which do not disturb, or only slightly disturb, the global operation of the network. To make this clearer, let us take the example where one of the two wires of the differential pair of the CAN bus is short-circuited to ground. Here, action is taken in response: the system switches to transmission mode on the remaining wire, thus in an asymmetrical single-medium topology with respect to ground. Clearly, in degraded mode the bus will no longer operate as a pure differential pair and will be less resistant to parasitic signals, but the messages can continue to travel on the remaining wire. Similarly, in the case where two wires of the differential pair are short-circuited together for any reason, by a mechanical assembly on which we are unable to act in any way, this will also operate in asymmetric single-wire mode with respect to ground.

Admittedly, the network radiates more and is more susceptible to parasitic signals because of its poorer common mode immunity. Statistically, as the parasitic signals are less successfully rejected, there is a higher probability of transmission errors on the bus, leading to more frequent error signalling by CAN, leading to error recovery, leading to a slightly lower net communication rate of the network . . . but, in the last analysis, because of the CAN protocol, the message will get through – otherwise it is not worthy of the name CAN! But . . . clearly, if anyone is concerned about the urgency of the message to be transmitted, this could cause problems.

This technique is therefore called 'fault tolerant', meaning that it tolerates faults in the physical layer. We also say that we are carrying out 'bus failure management'. In short, this works very well.

The last two hypotheses are used in ISO 11898-3 according to the cases of disturbance detected.

*Fault recovery*
The concept of repair during the failure phase implies that the system can also detect the instant at which the incident ceases to be present, so that it can reinstall itself automatically and operate again as it did before the incident.

When the fault has disappeared (for example, if the short circuit was intermittent or if a repairman has removed the short circuit from the bus), the system automatically restarts at its own pace with all its initial performance levels. Thus, it has completely recovered its operation.

*Electrical levels of the FT LS CAN physical layer*
The electrical levels adopted for the *FT LS CAN* physical layer are shown in Figure 4.26.

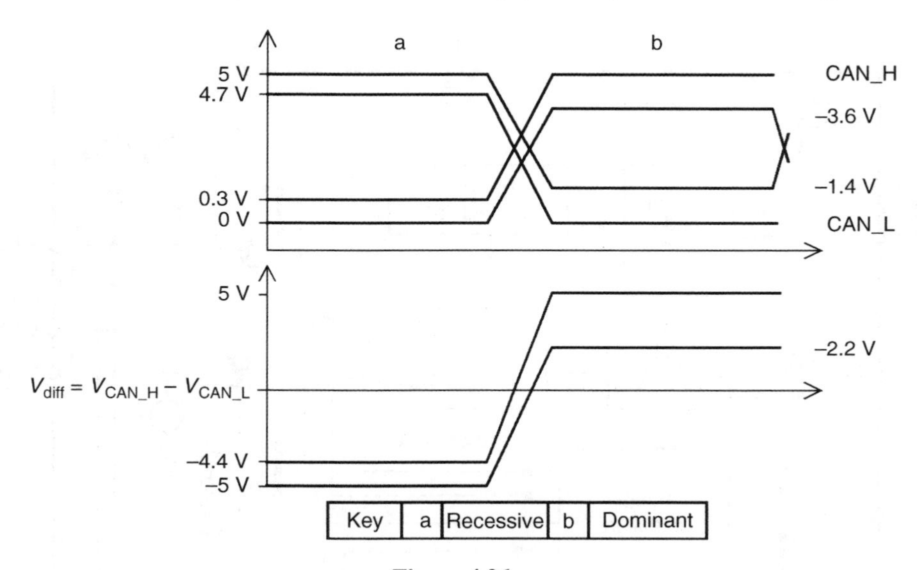


**Figure 4.26**

As shown in the figure, they are very different from those of HS CAN. This deliberate choice is mainly due to the fact that, as the maximum speed is lower, in the presence of radio-frequency disturbances, it is possible to impart more swing to the amplitude of the signals present on the line, and thus to have a better signal to noise ratio and fewer transmission errors.

The main characteristics of FT low speed CAN are as follows:

- maximum speed, 125 kbit $s^{-1}$,
- line length up to $x$ metres, determined solely by the capacitive load of the bus,
- from 2 to 20 nodes,
- differential (twisted) pair with ground return,
- a single termination network for each wire of the bus,
- an output current of more than 1 mA supplied by the transmitter;

- a simple design of the bus termination (connection) circuit,
- a line protected from short circuits (from −6 to +16 V (32 V)),
- a simple 5 V power supply,
- low EMC radiation, etc.

*Example of components for the ISO 11898-3 FT LS CAN layer*
Figure 4.27 shows the block diagram of the Philips/NXP Semiconductors TJA 1054A circuit, not because I am obsessed with this company's components, but because this component has been the 'golden device' forming the base for the development of ISO 11898-3 which is described a little further on. There are many other components produced by other manufacturers (ST, Freescale, Infineon, etc.) at the present time, but beware: not all of them strictly conform to ISO 11898-3.

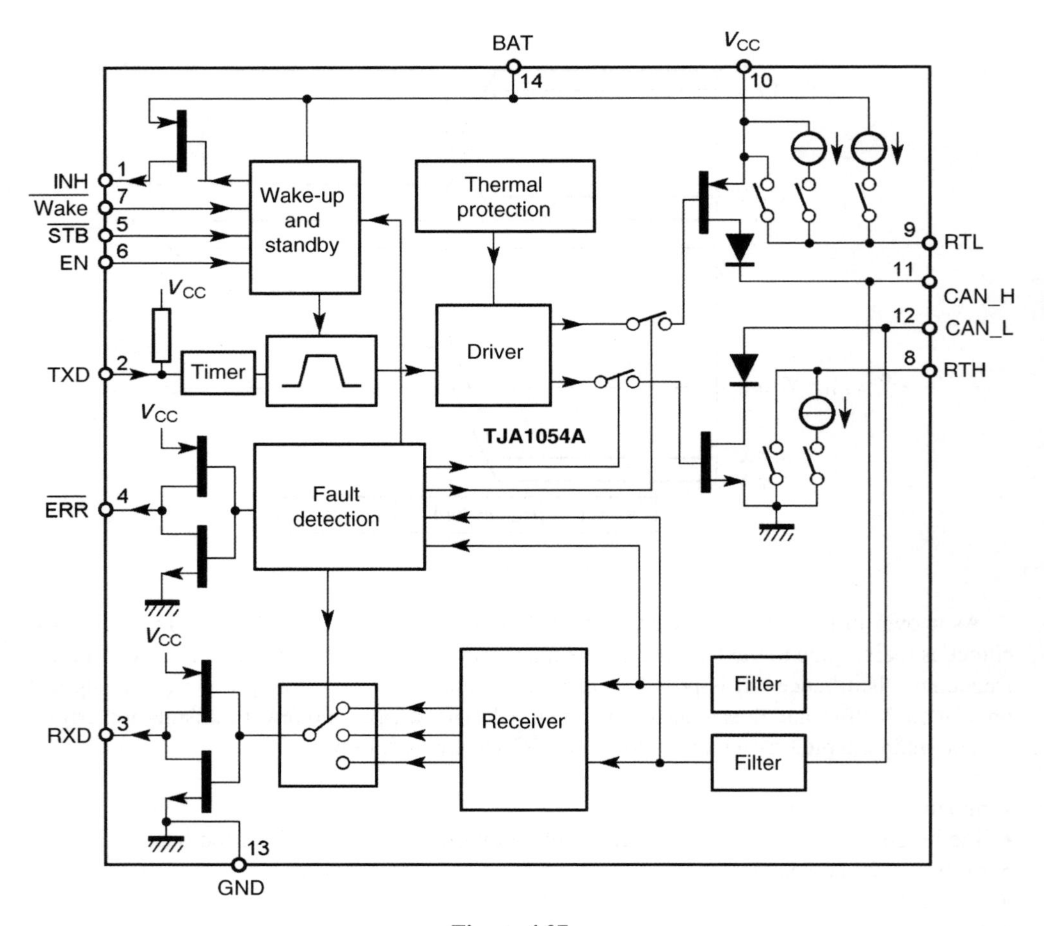


Figure 4.27

This is all very well, and the intention is very good . . . only there is a 'But' with a capital 'B'.

*Problems of compatibility and interoperability*
The main problem to be overcome in all systems of this type (fault tolerant) is that of maintaining the functional integrity – the consistency – of the network, especially in relation to the quality of the diagnosis of the faults which may occur. This is because each participant in the network must detect the same fault in an identical way, at the same instant, and must signal it in the same way and recover its performance at the same instant in order to safeguard its structural integrity: and this is far from easy!

Clearly, it is (more or less) easy to imagine resolving this problem within a single range of components produced by a single manufacturer (in respect of tolerances, temperature variations, etc.). However, this philosophy has to be extended to all similar products from different suppliers, in order to meet the commercial demand for 'double-sourcing' procurement, which is so strong among the purchasing departments of equipment manufacturers.

The different manufacturers of integrated circuits have therefore had to align themselves in order to meet their customers' requirements. In this context, car manufacturers and some equipment and component manufacturers decided, some time ago, to set up an ad hoc working group on this subject, known as the GIFT/ICT (*generalized interoperable fault-tolerant CAN transceiver/international transceiver conformance test*), to ensure this convergence of outlook and to produce a document on this highly critical subject of functional interoperability in FT LS CAN. This document, based on the Philips Semiconductors TJA 1054 golden device, was published some years later by ISO, under the official standard title ISO 11898-3 – *road vehicles* – CAN – *part 3: low-speed, fault-tolerant, medium-dependent interface* (from 40 to 125 kbit $s^{-1}$).

It describes the following main points in detail:

- the medium-dependent interface (MDI),
- the physical medium attachment (PMA),
- the electrical levels on the bus,
- the topology of the network,
- the events of faults in the buses and power supplies,
- fault detection and management.

*The problem of conformity*
The concept of a standard implies conformity with this standard, with all the related conformity tests. Theoretically, another standard called 'Conformance Tests' should also be drawn up within the ISO, and independent test laboratories or authorities should be appointed for certifying the conformity, or non-conformity, with this standard. As this new ISO development phase is also very lengthy, the GIFT group has also created its own test plan (ICT – Test Specification V1.2), containing not less than 120 tests to be passed, in order to get down to its work more quickly. A mark of 120/120 is required to obtain the coveted 'conformity authentication' document. By way of example, the authentication document for the TJA 1054A is shown in Figures 4.28 and 4.29. So if you have any doubts on this point concerning a component in the market, you will know that it is up to you to do the checking!

**Fachhochschule**
- University of Applied Sciences.
c&s communication & systems group
Prof. Dr. Ing. W. Lawrenz
- Director c&s -

Salzdahlumer Strasse 46/48
D-38302 Wolfenbuttel

Philips
Fault Tolerant Low Speed
Transceiver TJA1054AT

**Authentication on**
**CAN Transceiver**
**Conformance**
Test Specification done by GIFT ICT group, derived from ISO 11898-3 WD current Transceiver Component Specification

c&s group is a subdivision of the Fachhochschule Wolfenbuttel. As such c&s is worldwide recognized as a neutral expert in testing of communication systems such as CAN Transceivers, CAN, CAN Software Drivers, (CAN) Network Management.

Herewith c&s group is proud to confirm thet the following tests on the subsequently specified device implementations have been performed by c&s resulting in the findings given below.

**c&s Conformance Test Results**

| | |
|---|---|
| Transceiver | **Philips** |
| Component/Part Number | **TJA 1064AT N1A6R2 HnH 0122** |
| Date of Tests | **08/2002** |
| Version of Test Specification | **Version 1.2 www.cs.group.de · GIFT ICT - public** |

Tests Results

| | |
|---|---|
| • Homogeneous Network 40 Nodes | **Passed** |
| Double Failure Recovery Ratio | 120 of 120 passed |
| • Heterogeneous Network 40 Nodes | **Passed** |
| Mix of 2: 10P / 30 DUT | |
| Double Failure Recovery Ratio | 120 of 120 passed |
| • Heterogeneous Network 40 Nodes | **Passed** |
| Mix of 6: 10P / 8M / 8s / 8I / 16 DUT | |
| Double Failure Recovery Ratio | |
| Mix of 4: 11P / 7M / 7I / 16 DUT | 120 of 120 passed |
| • Heterogeneous Network 2 Nodes | **Passed** |
| • Further Observations | **None** |

i.A. Lawrenz

WolfenbUttel, 05.09.02 LA. Lawrenz, Director c&s

Abbreviations to identify components: P = Philips, M = Motorola, S = STMicroelectronics, I = Infineon, DUT = Device Under Test

**Figure 4.28**

## *4.3.2 Asymmetric, on one wire: single-wire (SW) CAN or one-wire CAN*

### General

In view of all this, leaving aside any standardization of the physical layer, a single-wire solution (plus ground, of course), called one-wire CAN, was devised a few years ago for simple economic reasons. Clearly, in this case, the line carrying the working signal is asymmetric with respect to ground and has the technical defects associated with its economic qualities.

**Fault Tolerant Low Speed Transceiver Testing**
**Test List v1.2 – 07.10.02**

| Device Name | Test Start | Test End |
|---|---|---|
| TJA 1054 AT – N1A6R2 HnH 0122 | December 2001 | September 2002 |

| Test Nr. | Headline | Result |
|---|---|---|
| | | |
| *7* | ***Tests in Homogeneous Network with 40 Nodes*** | |
| **7.3.1** | **Variation of Power Supply** | |
| 7.3.1.1 | Power Supply = 12V | E/Pass |
| 7.3.1.2 | Power Supply = 6.5V | E/Pass |
| 7.3.1.3 | Power Supply = 27V | E/Pass |
| **7.3.2** | **Variation of GND Shift** | E/Pass |
| 7.3.3 | Variation of Op. Mode | |
| 7.3.3.1 | Transition: [Normal] to [Low Power] | E/Pass |
| 7.3.3.2 | Transition: [Low Power] to [Normal] via Wake Pin | E/Pass |
| 7.3.3.3 | Transition: [Low Power] to [Normal] via Bus | E/Pass |
| **7.3.4** | **Variation of Failure** | |
| 7.3.4.1 | CL_Vx(up)@Rx | E/Pass |
| 7.3.4.2 | CL_Vx(down)@Rx | E/Pass |
| 7.3.4.3 | CH_Vx(up)@Rx | E/Pass |
| 7.3.4.4 | CH_Vx(down)@Rx | E/Pass |
| 7.3.4.5 | CL_CH@Rx(up) | E/Pass |
| 7.3.4.6 | CL_CH@Rx(down) | E/Pass |
| 7.3.4.7 | CL_OW@Rx(up) | E/Pass |
| 7.3.4.8 | CL_OW@Rx(down) | E/Pass |
| 7.3.4.9 | CH_OW@Rx(up) | E/Pass |
| 7.3.4.10 | CH_OW@Rx(down) | E/Pass |
| 7.3.4.11 | Loss of GND | E/Pass |
| 7.3.4.12 | Loss of VBat | E/Pass |
| **7.3.4.13** | **Double** | 120/120 passed |
| 7.3.4.13.1 | [CH_OW + CL_CH] : remove CL_CH | E/Pass |
| 7.3.4.13.2 | [CH_OW + CL_CH] : remove CH_OW | E/Pass |
| 7.3.4.13.3 | [CL_CH + CH_OW] : remove CL_CH | E/Pass |
| 7.3.4.13.4 | [CL_CH + CH_OW] : remove CH_OW | E/Pass |
| 7.3.4.13.5 | [CH_VBat + CL_CH] : remove CL_CH | E/Pass |
| 7.3.4.13.6 | [CH_VBat + CL_CH] : remove CH_VBat | E/Pass |
| 7.3.4.13.7 | [CL_CH + CH_VBat] : remove CL_CH] | E/Pass |
| 7.3.4.13.8 | [CL_CH + CH_VBat] : remove CH_VBat | E/Pass |
| 7.3.4.13.9 | [CH_VBat+ CH_OW] : remove CH_O | E/Pass |
| 7.3.4.13.10 | [CH_VBat + CH_OW] : remove CH_VBat | E/Pass |
| 7.3.4.13.11 | [CH_OW + CH_VBat] : remove CH_OW | E/Pass |
| 7.3.4.13.12 | [CH_OW + CH_VBat] : remove CH_VBat | E/Pass |
| 7.3.4.13.13 | [CH_Vcc + CL_CH] : remove CL_CH | E/Pass |
| 7.3.4.13.14 | [CH_Vcc + CL_CH] : remove CH_Vcc | E/Pass |
| 7.3.4.13.15 | [CL_CH + CH_Vcc] : remove CL_CH | E/Pass |
| 7.3.4.13.16 | [CL_CH + CH_Vcc] : remove CH_Vcc | E/Pass |
| 7.3.4.13.17 | [CH_Vcc + CH_OW] : remove CH_OW | E/Pass |
| 7.3.4.13.18 | [CH_Vcc + CH_OW] : remove CH_Vcc | E/Pass |
| 7.3.4.13.19 | [CH_OW + CH_Vcc] : remove CH_OW | E/Pass |
| 7.3.4.13.20 | [CH_OW + CH_Vcc] : remove CH_Vcc | E/Pass |
| 7.3.4.13.21 | [CH_GND + CL_CH] : remove CL_CH | E/Pass |
| 7.3.4.13.22 | [CH_GND + CL_CH] : remove CH_GND | E/Pass |
| 7.3.4.13.23 | [CL_CH + CH_GND] : remove CL_CH | E/Pass |
| 7.3.4.13.24 | [CL_CH + CH_GND] : remove CH_GND | E/Pass |
| 7.3.4.13.25 | [CH_GND + CH_OW] : remove CH_OW | E/Pass |
| 7.3.4.13.26 | [CH_GND + CH_OW] : remove CH_GND | E/Pass |
| 7.3.4.13.27 | [CH_OW + CH_GND] : remove CH_OW | E/Pass |
| 7.3.4.13.28 | [CH_OW + CH_GND] : remove CH_GND | E/Pass |
| 7.3.4.13.29 | [CL_OW + CL_CH] : remove CL_CH | E/Pass |
| 7.3.4.13.30 | [CL_OW + CL_CH] : remove CL_OW | E/Pass |
| 7.3.4.13.31 | [CL_CH + CL_OW] : remove CL_CH | E/Pass |
| 7.3.4.13.32 | [CL_CH + CL_OW] : remove CL_OW | E/Pass |
| 7.3.4.13.33 | [CL_OW + CH_OW] : remove CH_OW | E/Pass |
| 7.3.4.13.34 | [CL_OW + CH_OW] : remove CL_OW | E/Pass |

**Figure 4.29**

In order to have a 'reasonable' immunity to noise (i.e. parasitic signals) (Figure 4.30), it is preferable not to operate at 5 V, but to use signals of higher amplitude (for example at 12 V, supplying the node directly at the battery supply voltage).

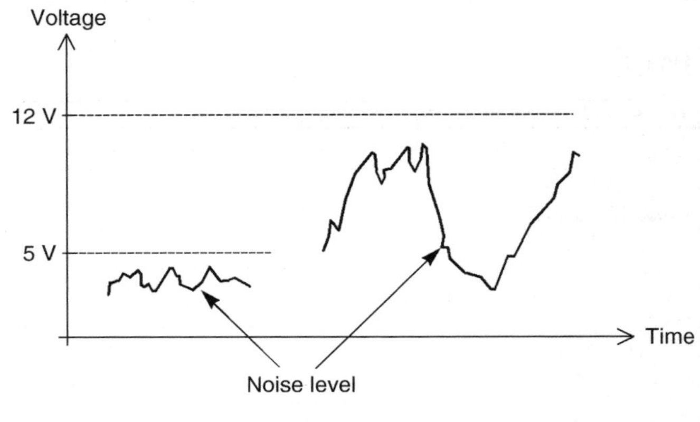


**Figure 4.30**

It is also clear that, for an identical transmission bit rate and similar rise and decay times, the RF radiation produced by the signal travelling on the 'hot' communication wire will be much greater than that of a signal travelling along a differential pair. To be more specific, as a first approximation, measurements show that for an identical radiation (Figure 4.31) to that of a CAN link formed on a differential pair operating at 500 kbit $s^{-1}$, we can expect operation at a maximum of approximately 20 kbit $s^{-1}$ in one-wire CAN, without having to use methods of controlling signal slopes with the aid of a few small filter components and thus altering the rise and decay times and the real length of the bit time.

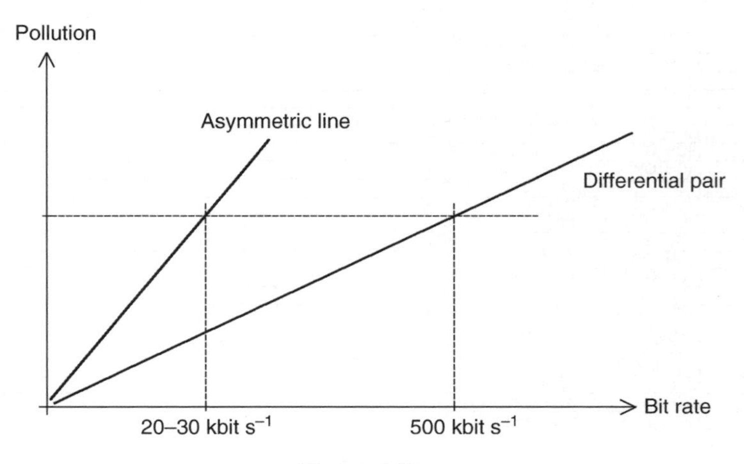


**Figure 4.31**

This provides a general idea of the limits of the usefulness of this solution in terms of performance/cost. In my description of the LIN (local interconnect network) protocol, I will examine all the implications of these words in detail.

Furthermore, when only one transmission wire is used, the problems arising from failures of the 'short circuit' and 'open circuit' types are theoretically insoluble. This immediately indicates that this type of physical implementation is only to be used for non-critical (non-secure) solutions intended to use and comply with the CAN protocol and benefit from its advantages, while accepting the low speeds, without any device for managing the physical disruption of the CAN line, where the cost is to be constrained (for example, for the control of the rear screen heater system in motor vehicle applications).

As you will see in due course, the one-wire CAN physical layer and the LIN physical layer are very similar, but this purely CAN solution has the (major) advantage of maintaining the consistency and performance (in error management and retransmission of frames, for example) of the CAN protocol throughout the network, thus avoiding the need for gateways between protocols or between a protocol and a 'sub-protocol', such as LIN which will be discussed in Chapter 7.

**Example of components for the single-wire CAN layer**

Not many one-wire CAN components are available in the market. They are only offered by Freescale (formerly Motorola) and Philips Semiconductors, as far as I know.

Figure 4.32 shows the block diagram of the Philips Semiconductors AU 5790 integrated circuit, in which we can see the (essential) presence of a detector of ground disconnection from the integrated circuit or from the module, in order to resolve the classic problem of failure of an inexpensive module which is fixed to the vehicle chassis by only one screw, which corrodes after a few years and fails to provide a contact – or, even worse, only provides an intermittent contact – leading to the total blockage of the network concerned.

This circuit also has the characteristic of operating at up to 100 kbit $s^{-1}$. Clearly, this will cause more pollution than at 20 or 40 kbit $s^{-1}$, but it is very useful in the factory production of vehicles, for reducing the time taken to load the programs into the CPUs via the various internal networks of the vehicles.

**Example.**

For information, a production line turns out 400,000 vehicles per year, i.e. approximately 2000 vehicles per day, 200 days per year, i.e. 20 vehicles per hour for 10 working hours each day, in other words, one vehicle every 3 min. If we have to load 500 kB into the various Flash/E2PROM memories of the components in less than 2 min, this means that the mean global net bit rate must be 500 kB/120, i.e. $4.3 \times 8 = 34.7$ kbit $s^{-1}$ (i.e. a time of approximately 0.03 ms per bit or 0.24 ms per byte for writing, re-reading and checking that the bit or byte in question has been input correctly ... not very long!).

*4.3.3 Ground shift*

For many reasons, there are often potential differences and shifts between the earths of the different modules participating in a network. This is due, for example, to the flow of current over long distances between the modules installed on a factory production line or to the

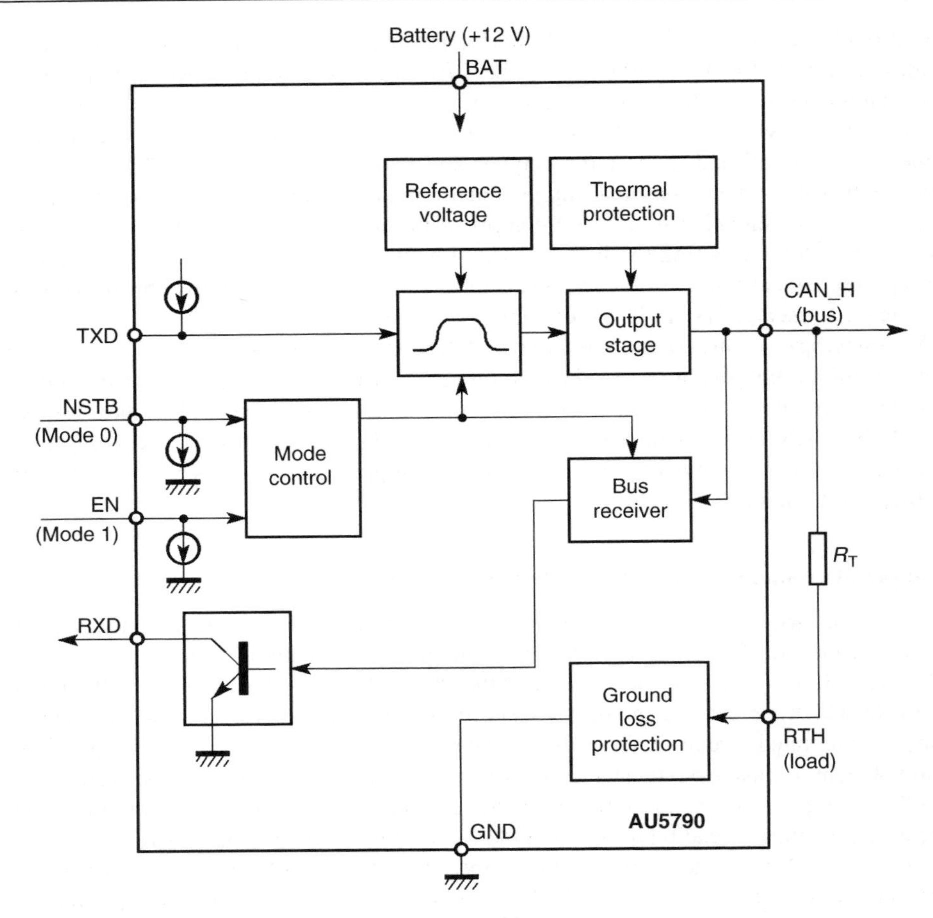


**Figure 4.32**

large instantaneous variations of current within a node or a poor ground return of the module due to a faulty electrical ground connection (corrosion or bad contact), causing what is commonly known as a 'ground shift'. In technical terms, this means that each node can have different single ended voltages on the CAN_H and CAN_L lines of the bus with respect to their own local ground levels, whereas the differential voltage CAN_H – CAN_L between the wires of the bus remains unchanged. To give a specific example, in the case of the TJA 1050 circuit mentioned above, when the CAN_H line voltage is −12 V and the CAN_L voltage is −12 V, the differential input variation only varies from 0.5 to 0.9 V. This very small variation indicates a strong common mode rejection. Note that ISO 11898 only requires a range of −2 to +7 V. See Figure 4.33 for an illustration of all these points.

The first figure shows the case in which the ground potential of the transmitting node 2 is much higher than the ground potential of the receiving node 1 (the figure shows that if the CAN_H voltage reaches the maximum value of +12 V, the maximum difference in ground level between ground_1 and ground_2 will be 8 V).

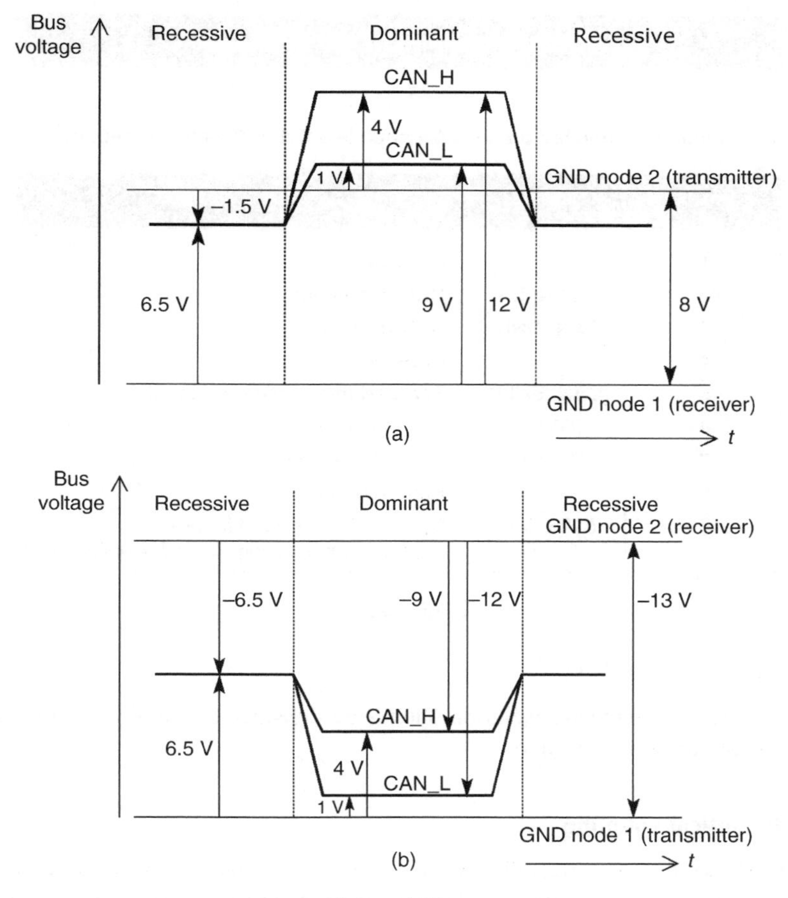


**Figure 4.33**

The second figure shows the case in which the ground potential of the transmitting node 1 is much lower than the ground potential of the receiving node 2. For reasons similar to those explained in the previous paragraph (with CAN_L max = −12 V in this case), the figure shows that the maximum ground shift can be −13 V.

### *4.3.4 CAN – CiA connector for wire links*

As a supplement to the ISO recommendations, intended to harmonize the problems of mechanical connections between the elements of a single network and to avoid the proliferation of systems incompatible with each other, the CiA (CAN in Automation) group recommends, in its note for industrial applications, a preferred type of mechanical connector (sub D 9) and a specific pin assignment (Figure 4.34).

Whether or not you use this is up to you, but bear in mind that many industrial companies (especially in the automation field) follow this recommendation for their connections to CAN networks.

You will also note that this connector allows the provision of a screen pair if necessary.

**Pin assignment of the CAN CiA**

**The connector provided on the CAN node is a Sub D 9 male connector.**

| Pin | Signals | Description |
|---|---|---|
| 1 | | Reserved |
| 2 | **CAN_L** | **CAN_L bus line** |
| 3 | **CAN_GND** | **CAN ground** |
| 4 | | Reserved |
| 5 | **(CAN_SHLD** | **For CAN line screen (optional)** |
| 6 | (GND)) | Optional ground |
| 7 | **CAN_H** | **CAN_H bus line** |
| 8 | | Reserved (error line) |
| 9 | (CAN_V+) | Optional line reserved for positive Supply (remote supply, optical coupling, etc.) |

**Figure 4.34**

### *4.3.5 Isolated CAN wire links*

Two major classes of galvanic isolation should be considered, namely the use of optical coupling and isolation transformers.

#### Isolation by optical coupling

Figure 4.35 shows how the preceding versions can be changed to isolated versions, using simple optical couplers while retaining the same type of medium. This solution is very useful in the case of applications in which the ground return voltage has an annoying tendency to 'fade' with distance, or if there is a high probability of having large instantaneous ground shifts.

**Note:** As mentioned previously, the propagation time of a signal on a wire bus is approximately $5\,\text{ns}\,\text{m}^{-1}$. Optical couplers require their own delay times, generally of the order of 20 ns to 1 μs, to operate correctly. If standard optical couplers with a delay of approximately 100 ns are used, their connection in the circuits is equivalent to the introduction of an additional 40 m of electrical cable. In fact, the signal passes through optical couplers four times, and therefore the return trip of the signal takes 400 ns (Figure 4.36).

#### Isolation by transformer

The CAN protocol permits wire links galvanically isolated by transformer, although the value of its bit is coded in NRZ (non-return to zero), a coding whose spectrum includes the

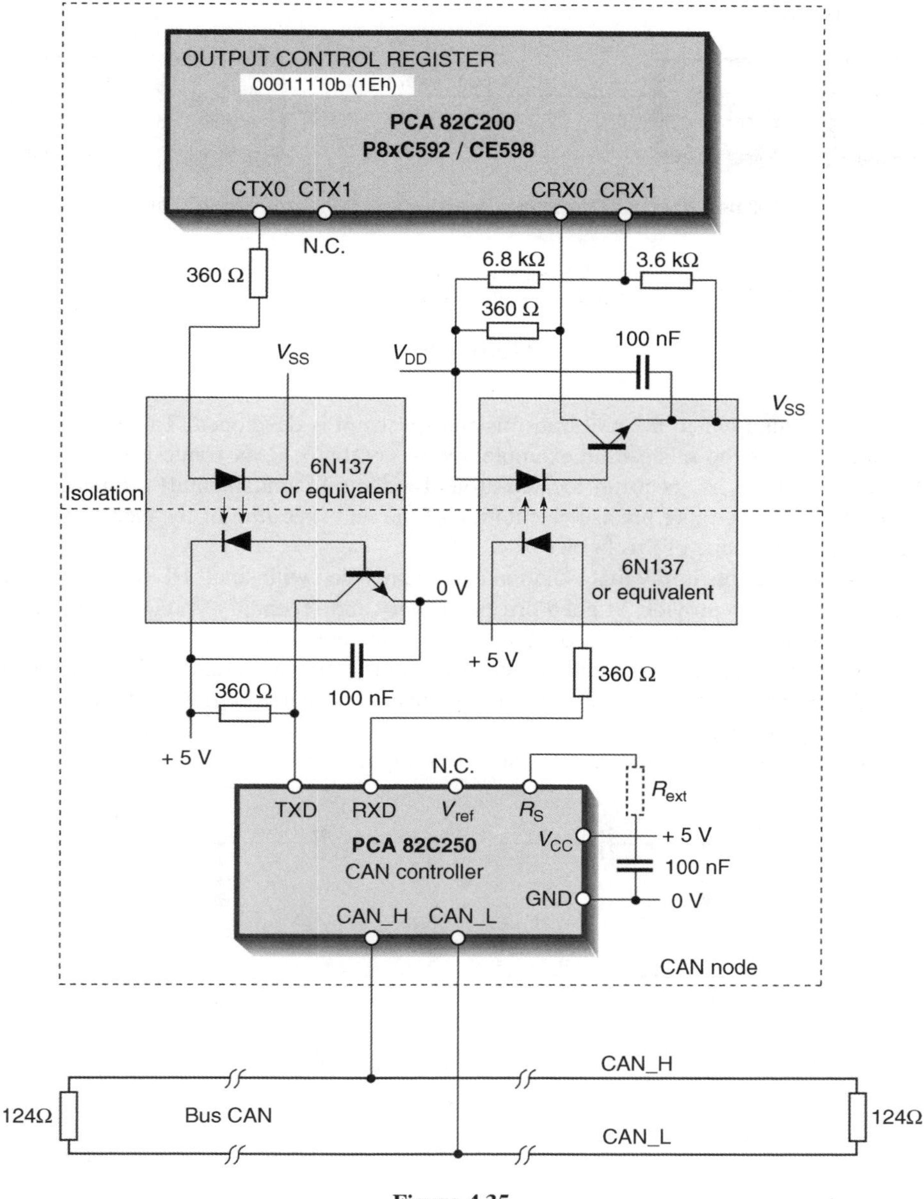


**Figure 4.35**

continuous component by definition. This is due to the fact that most of the transmitted frame is encoded according to a complementary bit stuffing principle (review Chapter 2 if necessary). As mentioned previously, this bit stuffing method, which disrupts the regularity of the value of the transmitted bits once every 5 bits, makes it possible to reconstruct an apparent

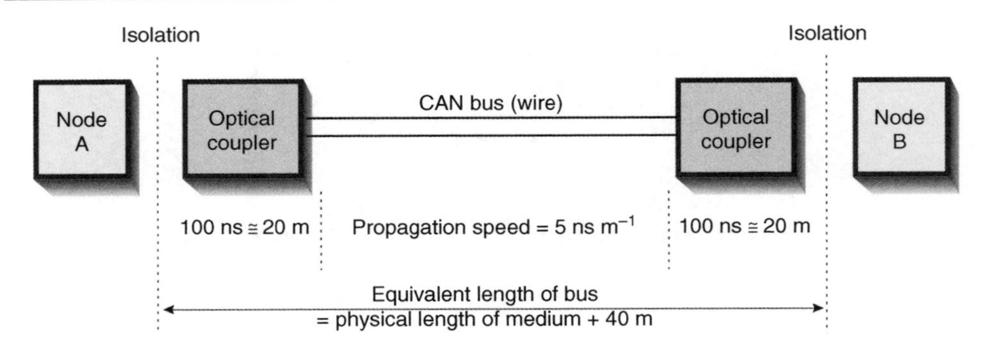


**Figure 4.36**

'HF carrier' around which the real transmission spectrum is developed. To provide a rather abbreviated but not too misleading example, let us say that a CAN frame with a bit rate of 1 Mbit $s^{-1}$ will have its spectrum located around a central (fundamental) frequency of the order of 200 kHz (the more precise calculation of the real spectrum of the carried wave is a matter outside the scope of this book).

These latter remarks imply that isolation can be provided with small HF transformers and that it will be easy to provide, at good prices, primary and secondary voltages much greater than those obtainable with optical couplers for the same cost.

The use of transformers, which are inevitably poorer at transmitting low frequencies, puts the signal retrieved at the secondary out of alignment. The signal retrieved at the secondary must generally be realigned with what are known as 'clamping' devices, using diodes and capacitors. The basis of such a circuit is shown in Figure 4.37.

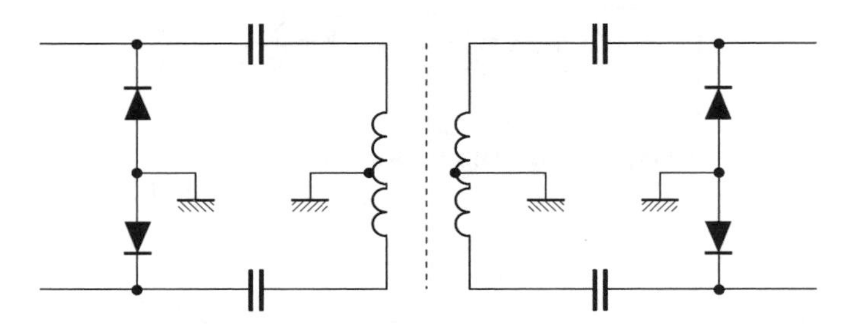

**Figure 4.37**

## *4.3.6 Wire links using CPL*

Here is another new candidate for CAN transport in the wire family: the 'CPL'. The term 'CPL' denotes a 'slow' signal, used for carrying power, with a superimposed high frequency, itself modulated by an active signal (in this case, a CAN protocol frame). A specific example is that of CAN protocol transport on a wire medium normally used for other functions, such as power transmission (with a mains voltage of 220 V AC, 50 Hz, or a continuous battery supply voltage of 12 V, 24 V, etc.), using a superimposed carrier frequency. Consequently, we can establish communications between different elements of a system while providing their remote power supply via the same wires, thus reducing the costs relating to the number of wires and pins, the contact quality (given the problems of low current flowing in CAN wire links), the prices of connectors, and the size and flexibility of the cables used.

Also, in these systems, the CAN bus seems to travel as if by magic, as there is no material indication of its presence; in some cases this can be very useful, to avoid attracting attention to additional wiring. Does this seem unimportant? If so, look a few paragraphs ahead, particularly at the section on CAN transport over 220 V AC power lines.

### Carriers, speeds and types of modulation

In general, a carrier frequency of several hundred kilohertz (around 130 kHz for the mains – see the EN 50065 standard – and from about 500 kHz to several MHz for low continuous voltages), modulated by the signal representing the protocol, is superimposed on the carrier signal (220 V, 50 Hz, or continuous voltage). The CAN bit rate which is generally used is of the order of several kilobits to several tens of kilobits per second (16–50 kbit $s^{-1}$). This speed, which some may consider slow, is generally adequate for numerous operations (command orders, etc.) at the human scale, where systems are not required to operate in real time in all circumstances.

Much has been written about CPL, the quality of networks and the quality of data transport and security, but in our case, if any parasitic elements interfere with the signal, the intrinsic robustness of the CAN protocol and the processing of transmission errors in the integrated circuits ensure the quality of the received data. Theoretically, in case of error, CAN automatically retransmits the signal until the message is carried correctly (see Chapter 2). Furthermore, the present cost of CAN components makes these solutions very attractive.

The only real problem with this kind of link is that, at any point in the network (and preferably at each node!), whenever a CAN frame is present, a minimum quantity of HF signal must be present on the node in question in order to be detected. This means that no wave originating from another node can cancel (in quantity and phase) the signal sent by another node, at any instant and at any point, particularly during the arbitration phase. There are many long-established devices (some patented) which, for example, can randomly modulate the phase of the transmitted carrier signal, so that there is an extremely low probability that there will be no signal at a given point of the network at a given moment.

Now let us examine the types of modulation which can be used to modulate the carrier wave to be superimposed on the 'carrier'.

*ASK modulation*

This is the simplest and most widely used modulation procedure, which easily supports the CSMA method (easy carrier detection and arbitration, the presence of the carrier representing the 'dominant' state of the CAN protocol). Figure 4.38 shows the operating principle.

*FSK modulation*

In this case, in order to resolve cases of arbitration and bit conflict, the values of the carriers must be carefully chosen with respect to the zero of the discriminator of the FSK demodulator. The principle of such a system is shown in Figure 4.39.

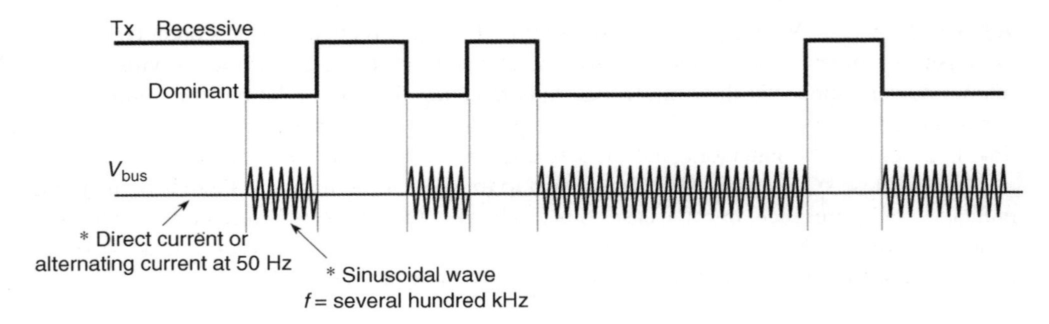


**Figure 4.38**

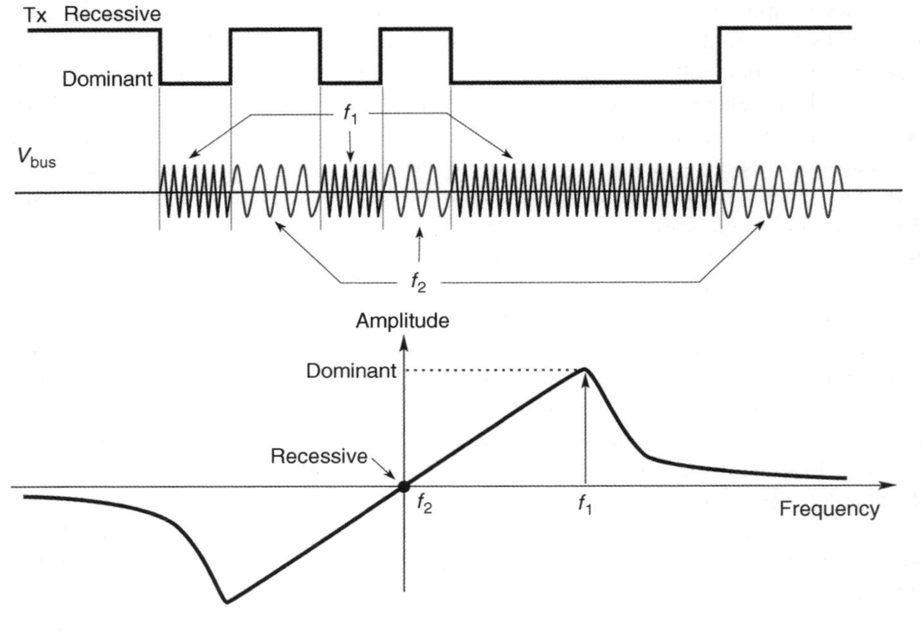


**Figure 4.39**

### On specialized lines

For their own particular reasons, some companies occasionally have to install specialized lines (alarms, security, etc.). They use these to carry data under the CAN protocol.

### On power supply lines

*Alternating type (220 V)*

An example is that of amusement arcades (Figure 4.40) (slot machines, pinball, video games, etc.) which are managed, monitored and controlled by a CAN network implemented in CPL

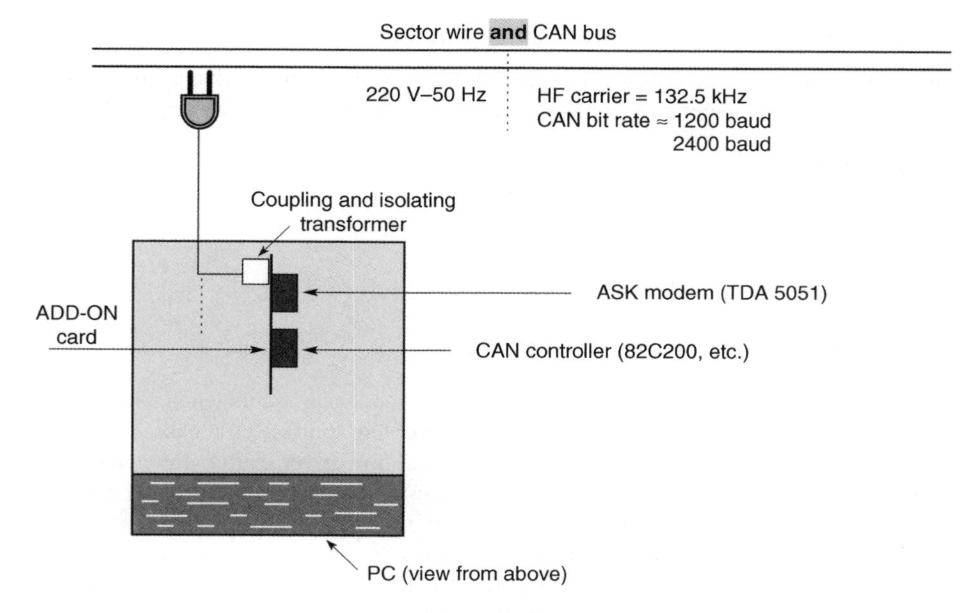


**Figure 4.40**

mode on the mains (220 V/50 Hz) by a PC which is also 'CAN-ized' with a CAN add-on card and a modem (the ASK type in this case). No new wires to install, no holes to make in the walls . . . nothing to be seen or recognized: nothing resembles an ordinary power socket as much as an ordinary power socket!

*Continuous type (for example 12 V, 24 V or other battery)*
This is a good opportunity to provide an example of links in vehicles not pre-equipped with dedicated CAN links (for the interior fittings of caravans, for example). An example is shown in Figure 4.41. Before leaving this topic of the wire aspect of the medium, here is a little light relief concerning the problems related to those mentioned above. These are the problems of the repeaters (and gateways – see Chapter 8) between one CAN network (no. 1) and another (no. 2), or between one medium and another.

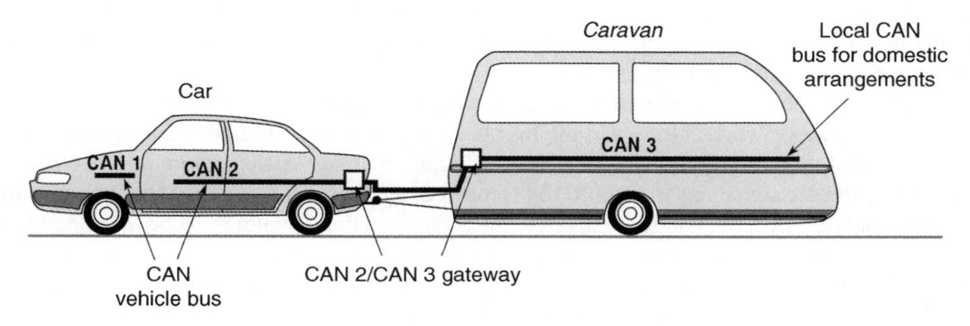


**Figure 4.41**

### *4.3.7 Repeaters*

For many reasons, we may encounter problems of repeaters in networks. To make matters clearer, here are some specific examples of situations where repeaters are required:

- the signal is weakened;
- the bus has to be split into a number of sections;
- the number of nodes is virtually unlimited;
- a requirement for very long distances;
- greater flexibility in the network topology;
- different media in different branches;
- identical bit structure throughout the network, etc.

Let us examine one of these examples in detail. A common case occurs when a large number of nodes are to be provided on a single network. The problem soon arises because of the lack of power (current) that can be delivered by a conventional line driver. For example, the 82C250 integrated circuit – although considered one of the best in the market – can (only!) drive 110 nodes simultaneously. However, according to the principle of the CAN protocol, an unlimited number of nodes can be provided on the medium, causing the designers of some locker-type applications (school lockers, railway left luggage, swimming pools, etc.) to extend the number of stations to a thousand or so, making it necessary to use CAN network repeaters.

The main task of these devices is to repeat the signal without degrading (or restoring) it, so that network portions or segments can be created and/or so that segments can be separated from each other.

On paper, the solution to the problem appears very simple, but sadly, as we often find, the reality is quite different. This is because the 'repeater' elements must provide the permanent bidirectionality required for the signal carried on the bus (for reasons of bit arbitration again). This means that the repeater must be truly bidirectional at every instant; but if this is so, its output will instantly loop back to its input and you will have invented the oscillator! Congratulations, but this was not what we wanted! This kind of situation must be avoided in all repeater designs to be recommended for use with the CAN protocol. Consequently, a deliberate delay, which is kept as short as possible in order not to damage the network performance, but which is 'known', is generally introduced into the signal processing sequence, thus avoiding the immediate feedback of the output to the input. This delay is generally of the order of 200–300 ns for a change from 0 to 1 and as short as possible for a change from 1 to 0. Now that the vicious circle has been broken, everything will work ... if allowance is made for the additional delay!

Figure 4.42 shows examples of CAN systems using repeater installations. As shown in these figures, the network is long and the bit rate is therefore not too high. This is frequently the case with applications requiring repeaters, as there are many stations more or less close to each other, and the signal has to be buffered (repeated) again. On the contrary, the inclusion of an additional delay (of tens or hundreds of nanoseconds) in the repeater must be interpreted as an apparent increase in the physical length of the network. In fact, if we consider that the electrical signal is propagated at approximately 5 ns $m^{-1}$ on a standard wire medium (differential pair, for example) and that a delay of approximately 200–250 ns has to be introduced (to ensure the fast but safe return of the correct operation of the bidirectionality of the repeater),

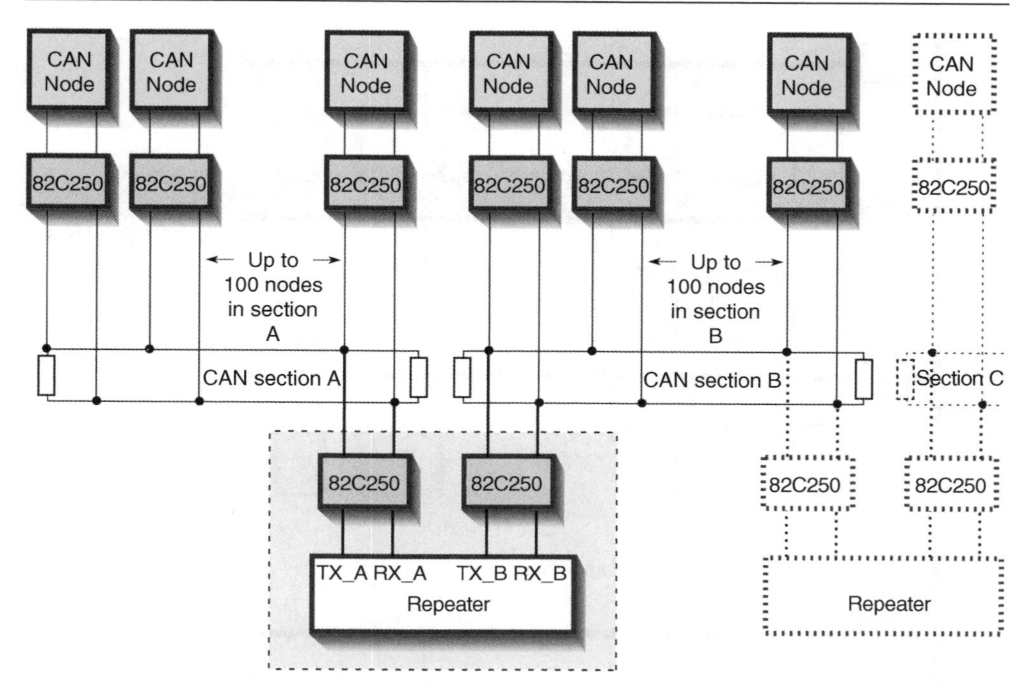


**Figure 4.42**

this means that the introduction of the repeater will be equivalent to an increase in length of the order of 40–50 m of additional wire in the (sub)network, which must be taken into account.

This distance, which is sometimes considered to be significant, can indeed become relatively significant in the case of a network which is long (from 1 to several kilometres) and which therefore operates at rather low speeds. A specific example of such a network is shown in Figure 4.43: It is interesting to note the multiple segmentation into several portions, making it possible to apparently reduce the effect of a serial accumulation of delays (i.e. of accumulated additional lengths of networks), contrary to their parallel topological configurations.

### *4.3.8 Medium-to-medium gateway*

Note in passing the benefits offered by microcontrollers with on-board CAN controllers which support and/or enable movement from one CAN network to another, using their two outputs Tx0/Tx1 and their two inputs Rx0/Rx1, and on two different media simultaneously without the addition of another component, for example using a high-speed 'CAN no. 1' network on a differential pair and another 'CAN no. 2' network electrically isolated by optical coupling, or on optical fibre. These industrial applications are much more common than it appears at first sight (Figure 4.44).

So we have reached the end of this long description of the wire medium for supporting CAN. Now let us make use of the gateways described above to move on to other types of media.

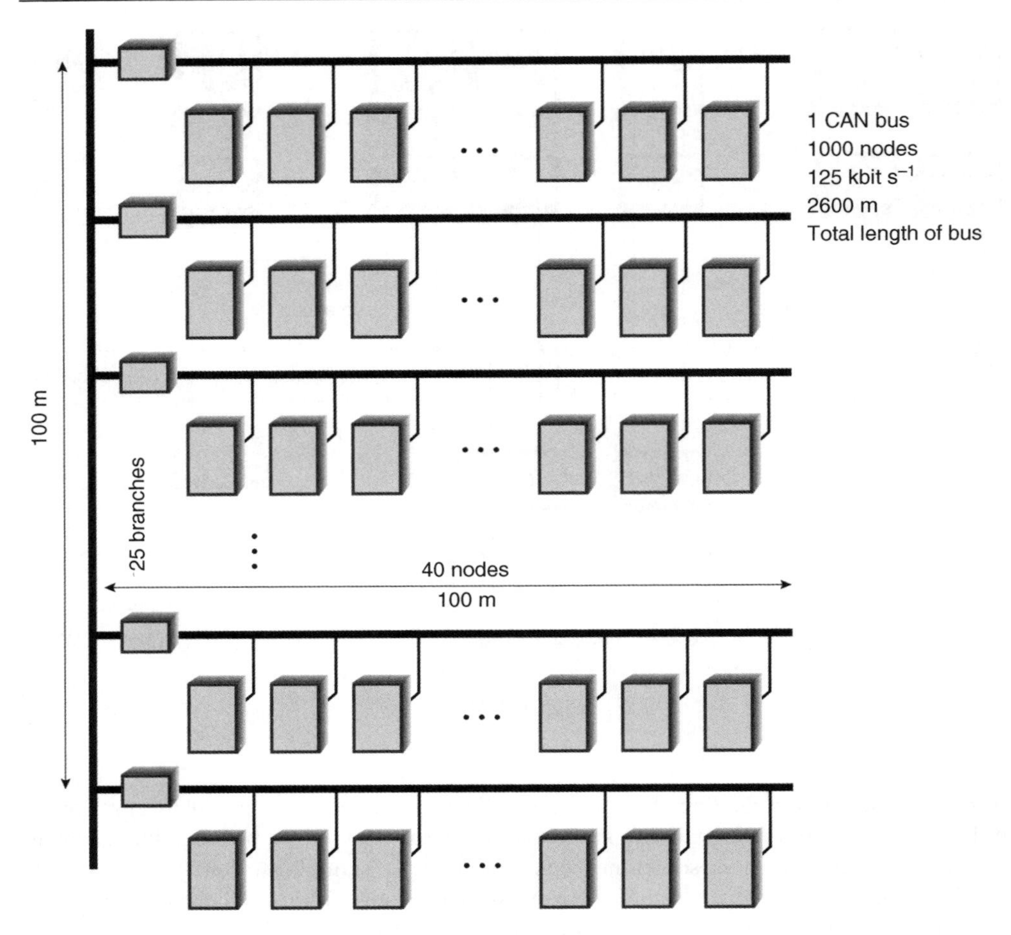


**Figure 4.43**

## 4.4 Optical Media

Optical fibres can also support the CAN bus. In this case, the dominant bit is characterized by the presence of light emission (visible or not) and the recessive state is characterized by its absence.

As regards the propagation time of an optical fibre, this is similar to that of an electrical cable, and the calculation process for speeds, distances, propagation time, nominal bit time, etc. is identical to that described above.

The main advantage of optical fibres is their structural ability to resist electromagnetic parasitic signals.

In the CAN context, the problems arising with optical fibres are mainly due to their insertion losses as a function of distance (making it necessary to use repeaters in some cases) and their poor capacity to support the 'bus' topological configuration. The last point is mainly due to the

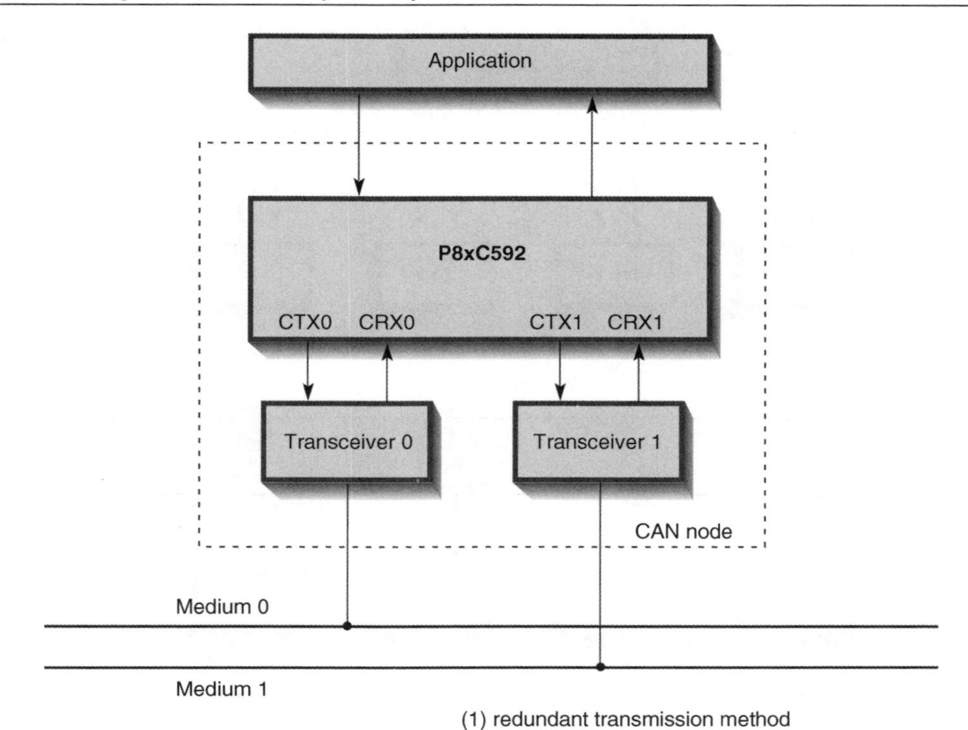


**Figure 4.44**

fact that the quantity of light transmitted is divided at each branch where a station is located and that the quantity of light remaining on the bus is smaller. Of course, some solutions do exist, and optical T devices are available in the market to form branches. In fact, this kind of link is often used to make 'point-to-point' connections, providing a CAN optical fibre gateway between two CAN wire systems (a typical example is that of a 'tractor and trailer' unit shown in Figure 4.45).

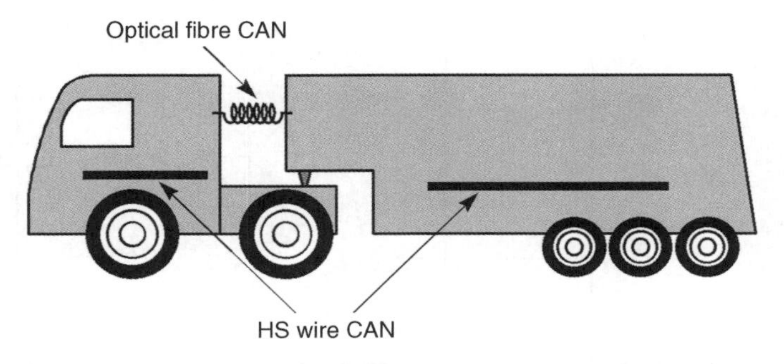


**Figure 4.45**

Figure 4.46 shows an example of the implementation of a CAN network on optical fibres.

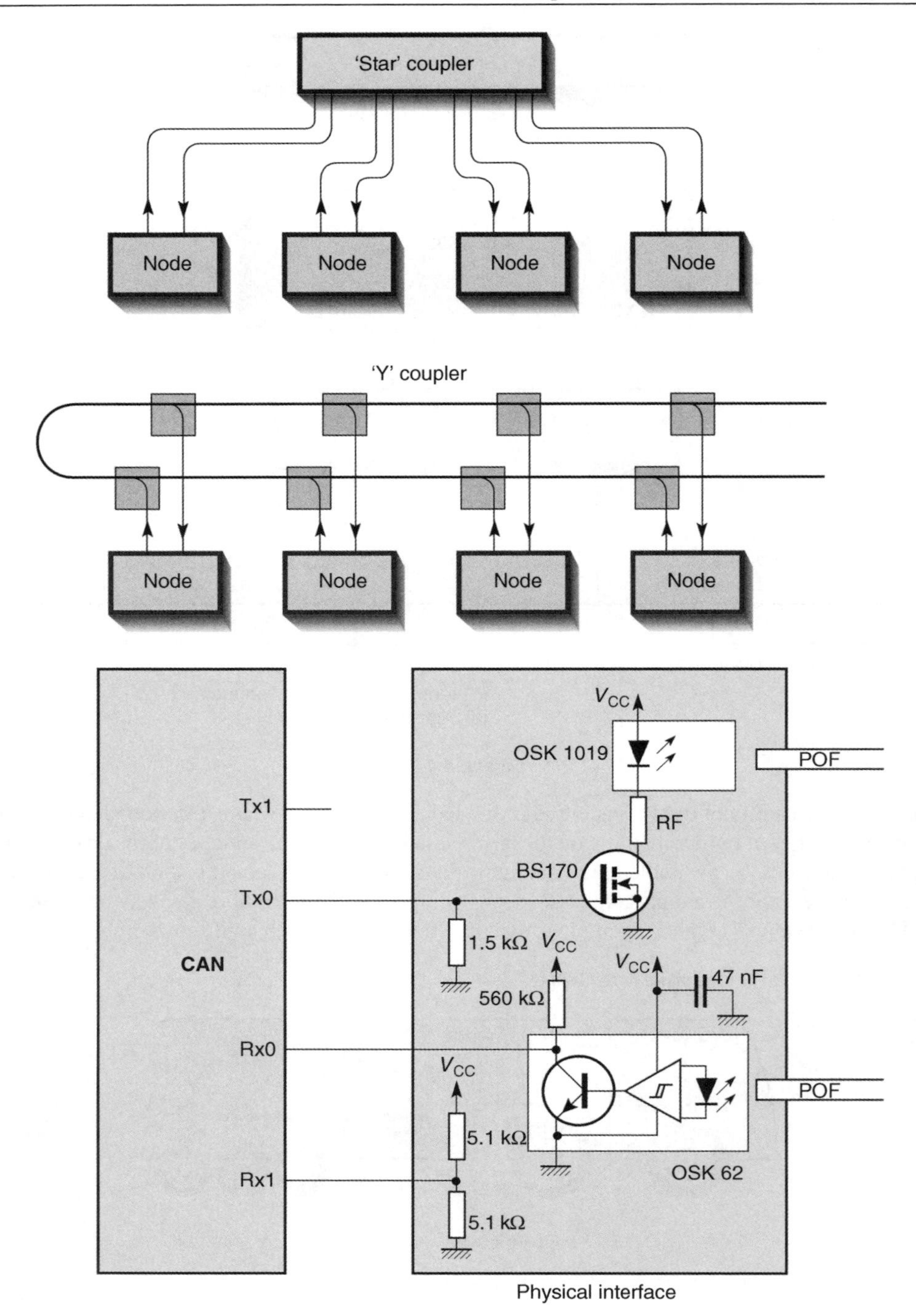

'Star' coupler
Node
Node
Node
Node
'Y' coupler
Node
Node
Node
Node
$V_{CC}$
OSK 1019
POF
Tx1
RF
BS170
Tx0
1.5 kΩ
$V_{CC}$
CAN
$V_{CC}$
47 nF
560 kΩ
Rx0
$V_{CC}$
POF
5.1 kΩ
Rx1
OSK 62
5.1 kΩ
Physical interface


**Figure 4.46**

## 4.5 Electromagnetic Media

### *4.5.1 Radio-frequency waves*

It is possible to transport CAN frames by a radio-frequency wave medium, in a similar way to that described above for 'CPL' applications. In this case, as previously, we can consider using

- either amplitude modulation with (virtually) total carrier suppression (ASK), where the dominant level is considered to be the physical state of the presence of the HF carrier and the recessive level is considered to be its absence (or the modulation residue – which is sometimes very useful for the demodulation of the incoming signal),
- or frequency modulation with frequency hopping (at two values F1 and F2 as for an FSK modem), in such a way that, for example, F1 is considered recessive and detected as such and F2 is called dominant at the point of its demodulation and detection.

The conventional problems of this form of transmission are well known. The system constructed in this way must be both a transmitter and a receiver, and the delay times used to control reception, demodulation and detection (for the receiver) and modulation and transmission (for the transmitter) must be compatible with the bit time and its different time segments, in order to eliminate problems caused by the arbitration sequence during the bit time to resolve problems of bit contention and bit-wise arbitration. Otherwise there are no major problems other than that of the automatic gain control (AGC) of the reception circuit.

There are examples of application which operate perfectly, and I regret that I cannot publish them in this book, as all those that I know of are 'proprietary' circuits.

### *4.5.2 Infrared waves*

The same principles and remarks apply to the infrared medium.

## 4.6 Pollution and EMC Conformity

This concludes, in theory, the long section relating to the choice of media, their performance and their functionality. When the choices have been made and the lines installed, everything is powered up, at which point there may or may not be a disaster in the form of pollution caused by the flow of CAN frames on the bus.

Being good engineers and citizens, we must always obey the law. In the case of Europe, the law has required compliance with the current EEC regulations since 1 January 1996 in respect of all matters concerning EMC.

In this part of the chapter, I will discuss the EMC problems caused by wire links, in the form of differential pairs (parallel or twisted) dedicated to CAN applications. The problems of ensuring EMC mainly arise when high-speed CAN is used (in the case of ISO 11898-2 and sometimes ISO 11898-3), and the 'immunity' and 'radiation' aspects of the line must then be considered.

For this purpose, we can reduce the basic diagram of the standard CAN application to that shown in Figure 4.47. This shows, in simplified form, two CAN nodes with their line drivers,

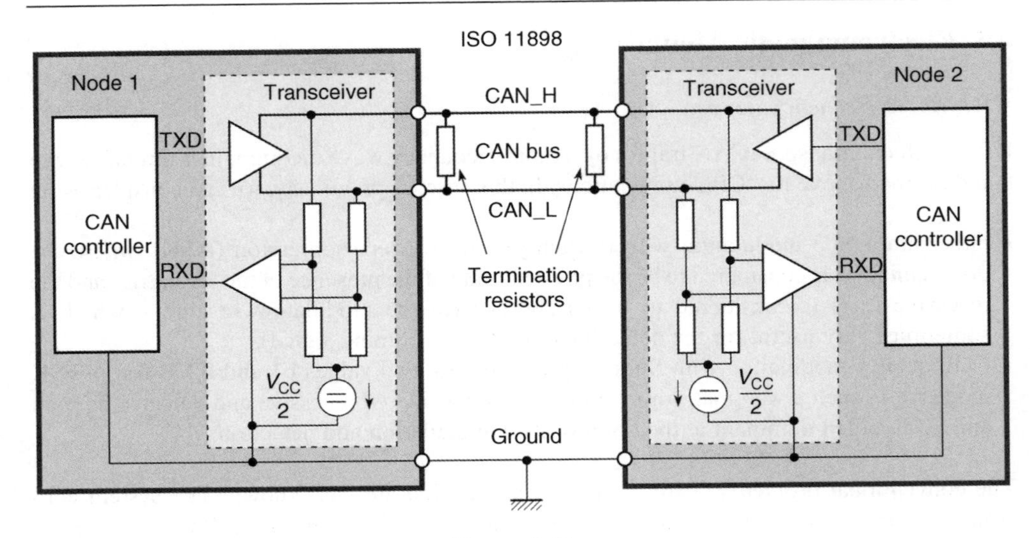


**Figure 4.47**

together with line termination resistors (120 Ω) at each end of an unscreened line, not forgetting the essential ground return, without which nothing would work between the two nodes. As the figure also shows, the bus lines are brought to the idle state (recessive level) at a potential of $V_{\alpha}/2$.

To simplify matters (but this is practically always the case), we will assume that the physical dimension of the node is small with respect to the length of the network and that consequently the 'cable' part is considered to be the main antenna of the system where problems of emission and susceptibility are concerned.

### *4.6.1 The standards*

The standards are complex (IEC 806-6) but, in a simple summary, the requirements are

- for transmission: the system must not radiate, to prevent interference with radio or mobile telephone reception, etc.;
- for susceptibility: the system must resist disturbance by the proximity of electrical fields of the order of 100–200 V m$^{-1}$ in a frequency band ranging from continuous current to several gigahertz.

#### **The theory**

The choice of a differential pair which is symmetrical and also twisted provides important structural advantages:

- It provides freedom from numerous constraints concerning the presence of parasitic or transient signals occurring on adjacent power supply line; as these affect both wires of the bus simultaneously, they will not degrade the data signal.

- The active signal (the transmitted data) develops on either side of the constant mean level, with equal and opposite values at all times.

These are the main advantages of differential systems. Thus, the immunity of this transport medium is essentially determined by what is known as its 'common-mode' performance. Now we must leave this ideal universe, where we can still dream, and enter the real world.

### The reality

In fact, signals have rise times which are never, at any instant, exactly equal and complementary to the signal developed on the other wire. Consequently, there is a difference signal (however small) which may produce electromagnetic pollution which must be allowed for and/or cancelled if possible.

As shown in the above figure, each of the two wires forms a loop comprising the bus wires, the nodes and the ground return. If they are implemented in the form of pairs, the loops have identical shapes and areas, and the whole can then be represented, for the purposes of EMC, by a diagram equivalent to that shown in Figure 4.48.

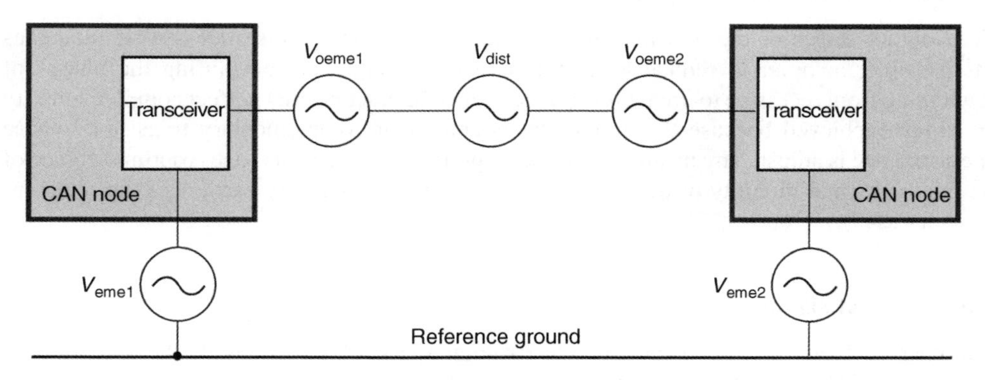


**Figure 4.48**

### Description of the equivalent diagram

Each node contributes to the radiation with two emission sources. The diagram shows

- Veme1, which is the voltage source equivalent to the pollution contributions due to the additional components of this node (such as the effect of the presence of the clock residue of the microcontroller). This source is included in the ground connection of the node in question;
- Vcem1, which represents the voltage equivalent to the emission which is produced by poor compensation between the voltages developed on the CAN_L and CAN_H wires, with respect to the reference potential (ground) of the data (in the example of voltages due to the non-symmetrical transitions of the output stage);
- Vdist, which represents the voltage equivalent to that due to the effect of the power of the electromagnetic interference captured by the CAN_L and CAN_H wires.

**Additional note on bit coding.** It is interesting to note that the choice of the type of bit coding also has an effect on EMC. This is due to the fact that the electromagnetic emission depends on the 'slew rates' (rise times) of the signal edges on the bus. Among the bit coding types, the NRZ chosen for the CAN protocol is the one which enables the slowest slew rates to be tolerated per data bit, and therefore to provide the best contribution to the improvement of the EMC performance of the network.

### *4.6.2 Measurements and results of measurements*

#### In Common mode

The common mode signal rejection is by far the most important criterion. I will give the manufacturer's guaranteed results for the 82C250 component by way of example.

This circuit, normally supplied at 5 V, has a common mode rejection voltage range from $-7$ to $+12$ V, corresponding to the equivalent injection of a sinusoidal signal of 19 peak to peak, i.e. a sinusoidal voltage of approximately 7 V effective on the bus!

#### Symmetry of the output signals

The output stages of the transmission interface must be as symmetrical and simultaneous as possible, in order to reduce to a minimum the voltages present during the phases of switching from recessive to dominant and vice versa. In practice, this perfect complementarity is almost achieved because the output stages consist of complementary transistors whose performance is adjusted by means of carefully applied physical differences (optimal choice of conductivity and mobility of the type of free minority and majority carriers of the semiconductors used).

#### Rise times of signals

Another problem related to the excessively high speed of the edges of the signals used (presence of VHF and UHF signal radiation) can be resolved by optimizing the slew rate as a function of the loop and the chosen speeds. This can be done by means of ad hoc circuitry, as shown for example in the synoptic diagram of the 82C250 (Figure 4.49) in which provision is made for the continuous adjustment (calculation below) of the signal slope (by increasing the rise time), while maintaining the speed, the precision of the sampling point, the length and the type of cable used.

**Example of calculation of the slew rate (d*V*/d*t*).** Pin 8 of the integrated circuit is used to determine the different operating modes of the circuit. The internal architecture of the circuit is such that the rise time is substantially proportional to the output current on pin 8. The latter can easily be determined with the aid of the voltage present on pin 8 (internal voltage generator equal to $V_{\alpha}/2$) and the value of the resistor $R_s$ to be connected externally between this pin and the ground. The relation between these parameters is then

$$R_s = \frac{V_{cc}/2}{k\,S_r}$$

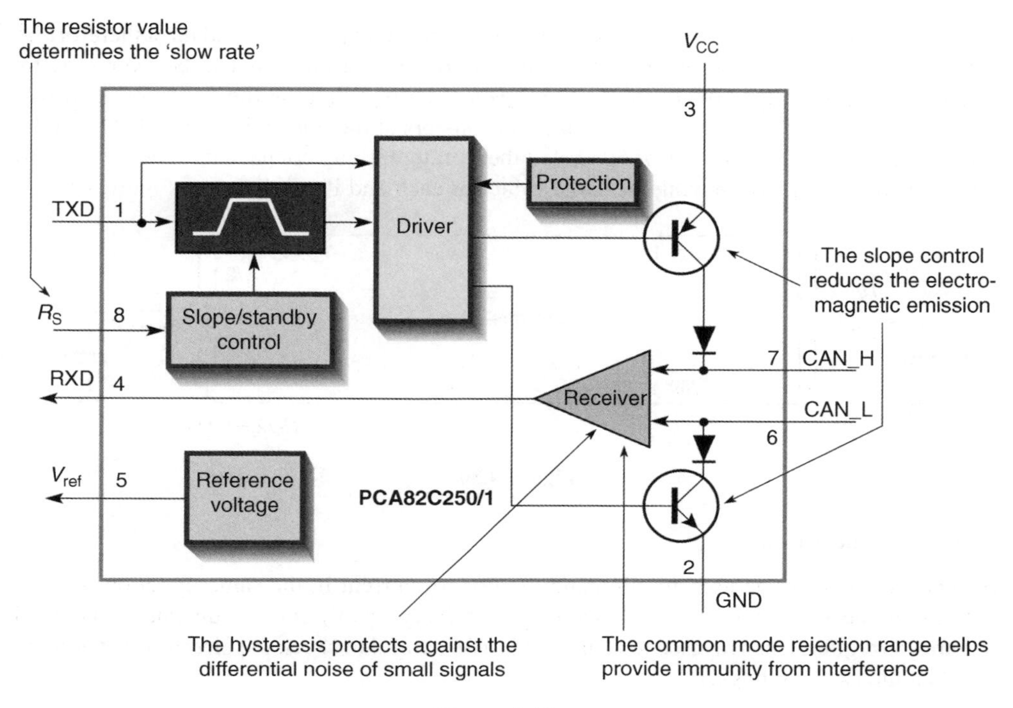


**Figure 4.49**

where $S_r$ is the slew rate to be obtained. As the manufacturer states that a slew rate of 14 μs $V^{-1}$ is obtained for a resistance $R_s$ of 47 kΩ, the internal constant $k$ of the integrated circuit can easily be found. After calculation, the value of this constant $k$ is $3.8 \times 10^{-3}$ μs $k\Omega^{-1}$.

## In differential mode

The differential noise must be low because the input stages of this kind of architecture are of the 'differential amplifier' type.

### *4.6.3 Numerous consequences and conclusions: if problems arise*

In general, all these recommendations enable us to use parallel pairs with a ground return forming part of the connecting cable. For specific problems, in order to reduce even further the effect of undesired signals in common mode, it may be necessary to use twisted pairs plus a separate ground wire or to twist all three together.

Numerous improvement methods are available in the market.

## Making the output signals symmetrical

To improve the common mode performance, the pair can be made more symmetrical with respect to ground. Normally, as the useful signal is developed differentially on the bus, the mean voltage between the two wires must remain at a constant potential (and must therefore be apparently electrically at ground in AC). Any doubts concerning this can be overcome by connecting a balancing network for the CAN_H and CAN_L lines at the output of the driver stages. For this purpose, each of the adaptation resistors at the line ends (both 120 Ω) can be divided into two series resistors of 60 Ω, and their midpoints are connected to the AC ground by means of a capacitor (having the same value for each end if possible, see Figure 4.50).

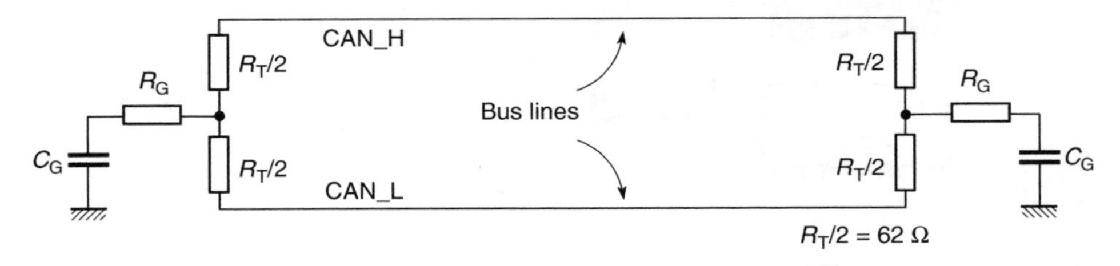


**Figure 4.50**

## Smoothing inductance

Another method can be used to eliminate the flow of current in the same direction on both outputs of the bus. Figure 4.51 shows how two (very small) surge inductances, coupled together as required, can be used to achieve this. (Be careful to check the direction of coupling between the two windings!)

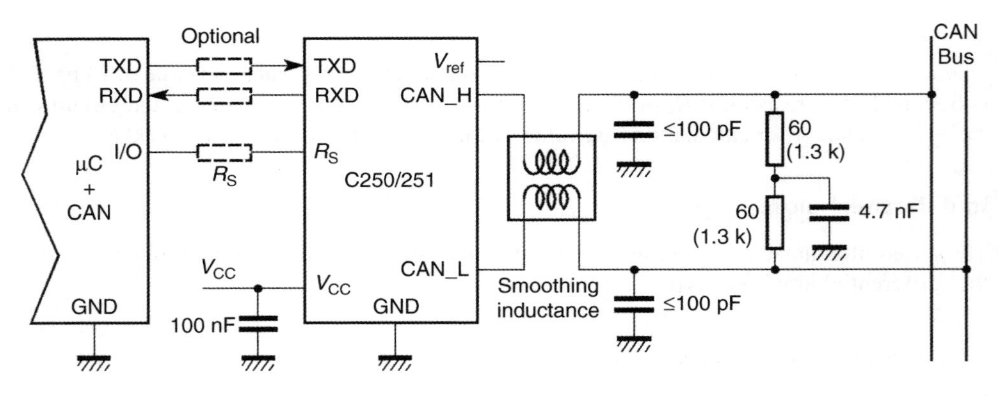


**Figure 4.51**

This method is frequently used, for the same reasons, at the inputs and outputs of 220 V mains wires.

For obvious reasons, it would be ridiculous to short circuit these coils in HF (otherwise why have them at all?), so please avoid placing a ground plane under the coils, or you will have connected these smoothing inductances for nothing, because of the parasitic capacitances present at the inputs and outputs of the coils.

For information, the use of these devices enables EMC to be achieved without problems (IEC 801-6), using an unshielded twisted pair (UTP), up to speeds of 250 kbit $s^{-1}$.

Finally, Figure 4.52 summarizes in a single diagram the main solutions used conventionally to resolve most cases.

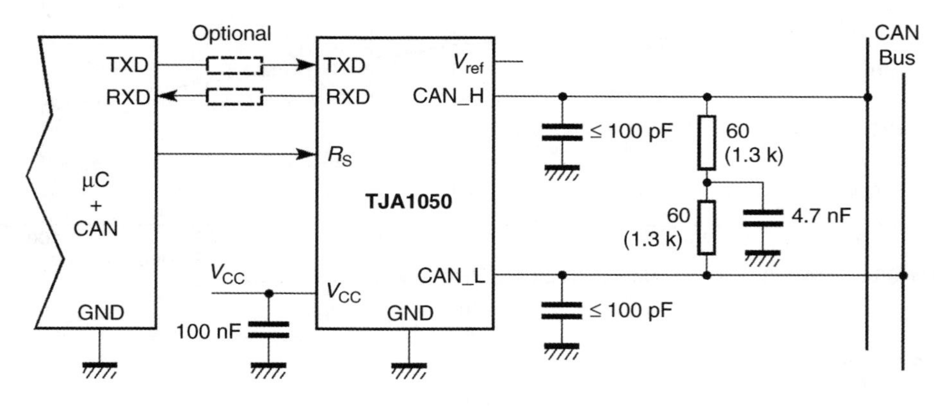


Figure 4.52

### *4.6.4 Results*

**For immunity**

The graph in Figure 4.53 shows how this network withstands an injection of parasitic signals on the bus at levels up to 20 V effective.

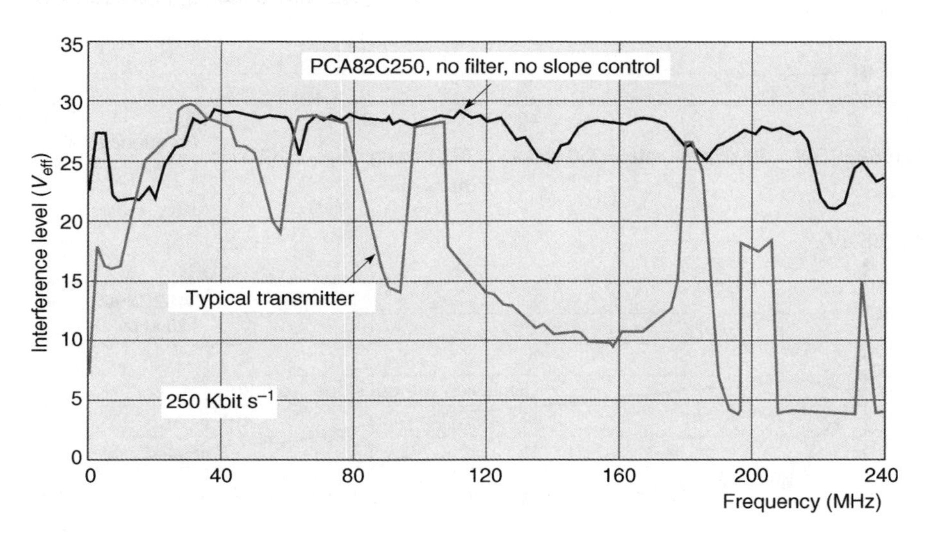


Figure 4.53

**For radiation**

The three graphs in Figure 4.54 show the effect of the different components on the non-radiating quality of the network:

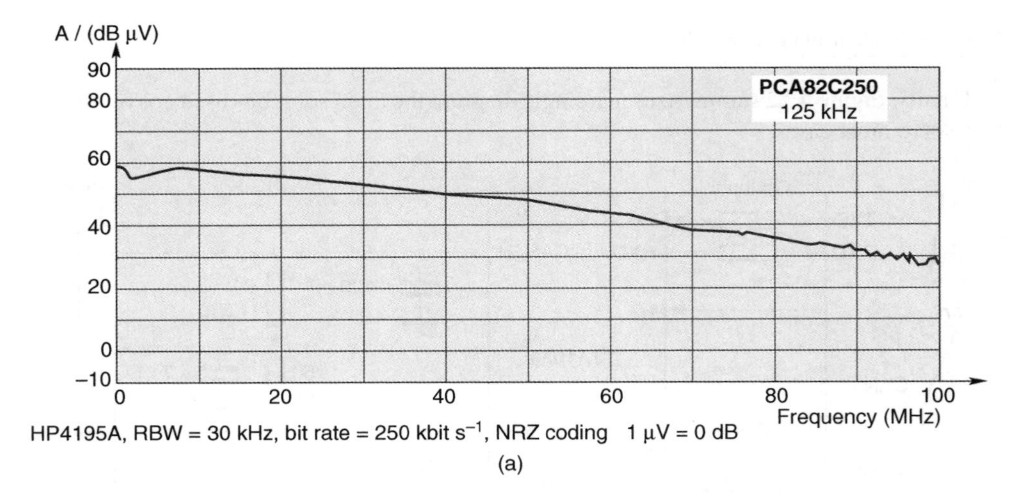

A / (dB μV)
90
80
60
40
20
0
−10
0
20
40
60
80
100
Frequency (MHz)
PCA82C250
125 kHz
HP4195A, RBW = 30 kHz, bit rate = 250 kbit s−1, NRZ coding 1 μV = 0 dB

(a)

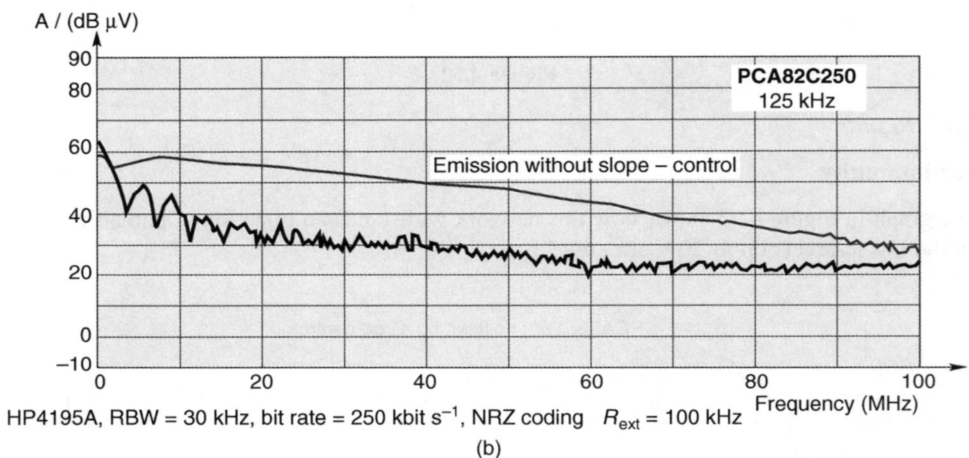

A / (dB μV)
90
80
60
40
20
0
−10
0
20
40
60
80
100
Frequency (MHz)
PCA82C250
125 kHz
Emission without slope – control
HP4195A, RBW = 30 kHz, bit rate = 250 kbit s−1, NRZ coding R ext = 100 kHz

(b)

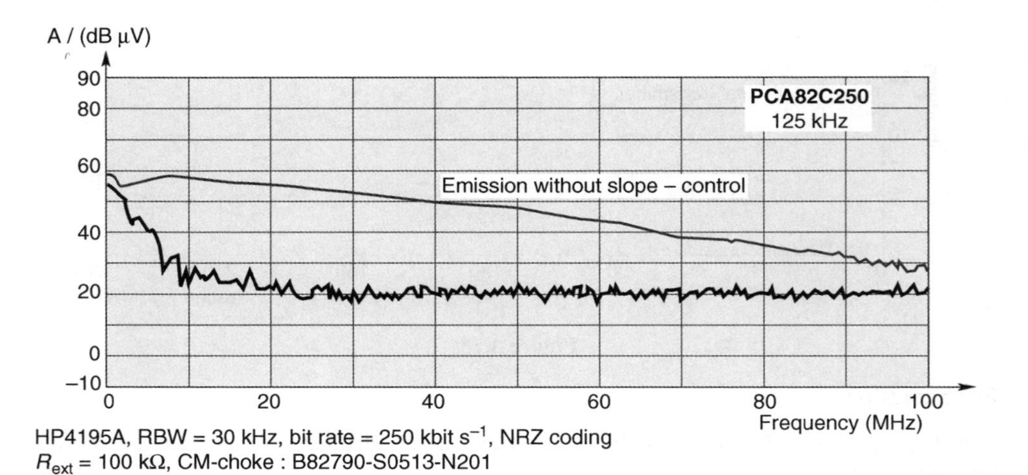

A / (dB μV)
90
80
60
40
20
0
−10
0
20
40
60
80
100
Frequency (MHz)
PCA82C250
125 kHz
Emission without slope – control
HP4195A, RBW = 30 kHz, bit rate = 250 kbit s−1, NRZ coding
R ext = 100 kΩ, CM-choke : B82790-S0513-N201

**Figure 4.54**

- without slope control, without coil;
- with slope control:
  - without coil;
  - with coil (Figure 4.55).

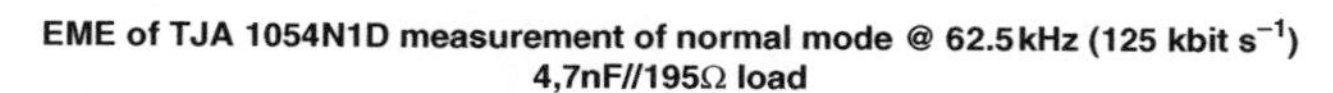


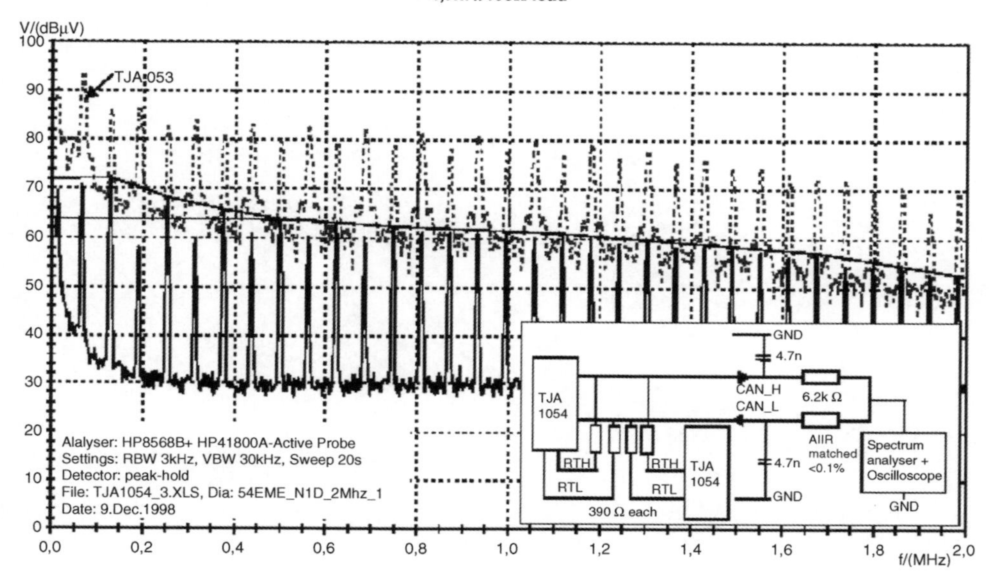


**Figure 4.55**

One last comment on high speeds, the slopes used to reduce d$V$/d$t$, d$I$/d$t$, etc. of the signal and to reduce the EMC levels. Users often protest angrily about having to add smoothing capacitors or inductances (because of their cost and area) in order to comply with specific radio interference standards and then, without saying anything, they add them to comply with the 2, 4 or 8 kV ESD (electrostatic discharge) resistance standards required for modules connected to a communication bus, which amounts to the same thing ... but this is another story.

### *4.6.5 Conclusion*

So you have now had a fairly complete survey of what is expected of a normally constructed CAN line driver, in terms of managing line short circuits, bus line faults, partial networks, low and very low power consumption, local and remote wake-up, recognition of wake-up modes, management of diagnostics, good EMC performance (in terms of immunity and susceptibility) and ESD resistance, compatibility with all the 3.3 V, 5 V and other microcontrollers in the market and resistance to heat stress.

# 5

# Components, Applications and Tools for CAN

In the preceding chapters, I have outlined the principles of CAN (controller area network) and described the details of its protocol and its numerous physical implementations.

If a protocol is to survive, it must be considered 'useful', in spite of the costs incurred by the development of special components. It must therefore be effectively supported by integrated circuits or families of integrated circuits, and there must be tools to facilitate the development of the applications.

The aim of this chapter is to describe the usual range of components supporting CAN, to provide some specific examples of electrical circuits and applications and finally to mention the development tools for assistance with the design of the applications.

## 5.1 CAN Components

Before starting this description, I would like to make an important point concerning the style of approach and intellectual integrity in relation to the following section.

It is very difficult to describe all existing components in detail; that would turn this book into a product catalogue. For practical reasons, therefore, I have chosen the following presentation:

- I shall start by describing the architectural philosophy of the existing families for components conforming to the CAN protocol;
- and I shall finish with a rather more detailed study of some of these, according to their principal functions, to give you a clearer understanding of the implementation of certain functions.

The first task is to decide on the components to be described in detail, while avoiding any hint of advertising. We therefore have a problem of choice, as, regardless of the components

*Multiplexed Networks for Embedded Systems: CAN, LIN, Flexray, Safe-by-Wire...* D. Paret

described, there will always be some degree of dissatisfaction, jealousy or discontent! Clearly, I could simply place all the component references in a hat and pull some out at random ... but this is not a very dignified solution. Consequently, having examined the currently available components, I have opted (you might have chosen differently, for many other reasons) for a solution in which, at this time, there is a complete set of components, including stand-alone protocol controllers, microcontrollers, line drivers, etc. I might replace this section with another more suitable one in a future edition, but we have to start somewhere, even if our choice can never satisfy everybody. I would therefore ask you to keep an open mind while reading the following lines, so as to obtain the maximum of information for the operation of a system equipped with the CAN bus. After all, this is the essential aim of the book.

### *5.1.1 General architecture and functional division of CAN components*

The CAN protocol is what it is. By definition, a 'CAN compliant' component must conform to it. However, this definition is not enough to provide a full description of the protocol implementation, the internal make-up and the manufacture of a component.

This is because the ISO reference document describing layers 1 and 2 leaves a certain degree of freedom to component manufacturers (for example, in relation to the depth and control of the message input and output buffers – see the specific sections on the implementation of the protocol in this chapter) so that they can provide specific, innovative and different solutions for their different markets, customer bases and applications. So we need to examine the problem rather more closely.

#### Functional division of the CAN components in a system

There are several possible functional divisions according to the level of integration of the nodes required by the network designer.

The first solution is generally one in which ISO layers 1 and 2 are controlled separately by a single component, called a 'stand-alone protocol handler'. Similarly, the part providing the interface with the medium is a separate module, thus leaving the user a large amount of freedom concerning the media used. This solution, constructed independently of the functional part of the node, gives the designer considerable freedom in the choice of the local (micro)processor/controller of the station (choice of the core, the computing power, its intrinsic resources, etc.).

Note that, in general, for the everlasting dismal economic reasons, users ultimately want a medium interface to be integrated in the same chip as the protocol handler (mainly for applications operating with differential pairs). From a technological viewpoint, this is not so simple!

In fuller integration (where this is possible), the architecture becomes reduced in the sense that users, for reasons of price and/or functionality and performance, look for components which integrate the protocol handler and the digital processor on a single chip. In these cases, the resulting solutions are called 'microcontrollers with on-board CAN handlers'. Many components of this type are currently available in the market. Figure 5.1 gives a summary of the various possibilities for dividing up the functions.

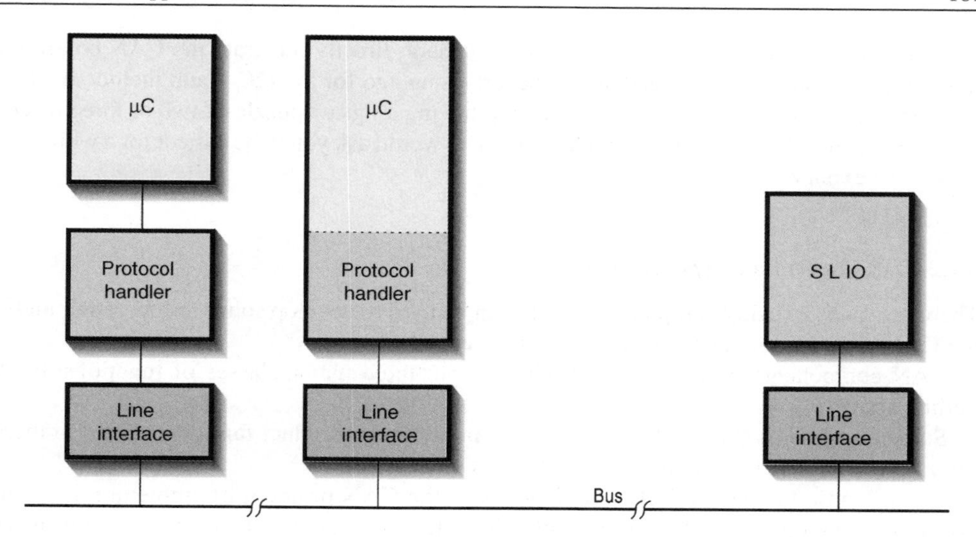


**Figure 5.1**

These general and architectural considerations of the division of a concept have given rise to different classes of components, which can be summarized under the following names:

- stand-alone protocol handlers,
- microcontrollers with on-board CAN handlers,
- line interfaces ('transceivers', or 'drivers').

## Protocol handlers

These have the task of generating and decoding the protocol. These circuits are also called protocol controllers. Of the stand-alone type, they cannot operate on their own, but must be controlled by a microcontroller (8/16/32/ ... bits). They therefore have parallel communication interfaces (address and data bus) or serial interfaces (for example SPI), suitable for one or more microprocessor families (which already complicates matters!). So far we have mentioned matters which are only virtually standard, even if complicated.

## Microcontrollers (monochips) with integrated CAN

To reduce the cost of the set of two components mentioned above, while decreasing the printed circuit area and avoiding problems of speed, CPU load, EMC, etc., every major component manufacturer must include one or more microcontrollers designed by him with on-board CAN handlers, if he is not to be excluded from this market.

## Line drivers

Clearly, the purpose of CAN is to convey information, in other words to use a medium to carry messages. There must therefore be one or more line driver circuits capable of controlling the medium.

Some of you will immediately ask why we cannot directly integrate the CAN controller into the microcontroller, as was done so cleverly long ago for the I2C – and include the line interface as well while we are at it – instead of having a jigsaw puzzle of two or three pieces (microcontroller, CAN controller and line driver). I would ask you to be patient for a while and all will be explained.

### *5.1.2 List of existing component types*

There are many existing components which can be used to develop solutions operating under the CAN protocol for carrying frames on different media.

CAN components may be divided initially into three major classes of functionality as outline above.

Some component manufacturers specialize in particular product families, whereas others offer wider ranges of components – each to his own.

It is possible to achieve full compliance with the CAN protocol by implementing it in different ways on chips. This is because the reference document makes no mention or suggestion of the way in which messages should be processed beyond ISO layer 2. This characteristic enables component manufacturers to propose, for example, various handling models for assisting and handling messages to be sent or received. Further hardware is commonly added in the form of registers to speed up the CPU of the main microcontroller, or to free it from tedious, time-consuming tasks. External communication is (fortunately!) always the same on the medium, for all possible implementations. In fact, these implementations only differ in relation to the way they handle messages within the microcontroller.

Conventionally, all CAN controllers have a common structure, consisting mainly of a protocol controller and a buffer memory unit:

- The CAN protocol controller is responsible for handling all the messages to be transferred to the medium. This includes tasks such as synchronization mechanisms (see Chapter 2 again) for error processing, arbitration, and parallel/serial and serial/parallel conversion. There is no difference in implementation between the different protocol handlers, as they are all designed for the same version of the same base protocol.
- The buffer memory unit is located between the protocol handler and the microcontroller (external or internal). It contains two entities, one dedicated to 'commands' and the other to 'messages':
  - the exchange of statuses, commands and control of signals between the CPU and the CAN protocol controller takes place in the 'command' entity;
  - the 'messages' entity is used to store messages for transmission and reception. It can also include a programmable 'acceptance filter' function.

The essential differences between these different implementations lie in the fact that some controllers have separate transmit and receive buffers and are provided with simple acceptance filter facilities, whereas others use two-port access RAM as the 'messages' entity and have more advanced acceptance filter facilities.

## CAN controllers with intermediate buffers

These controllers, often called 'BasicCAN', conform fully with the protocol. They generally have an intermediate buffer consisting of two receive buffers and one transmit buffer, together with one or more mask registers for hardware acceptance filtering limited to a width of 8 bits (actually for the 8 most significant bits of the identifier see the detailed example in Figure 5.2).

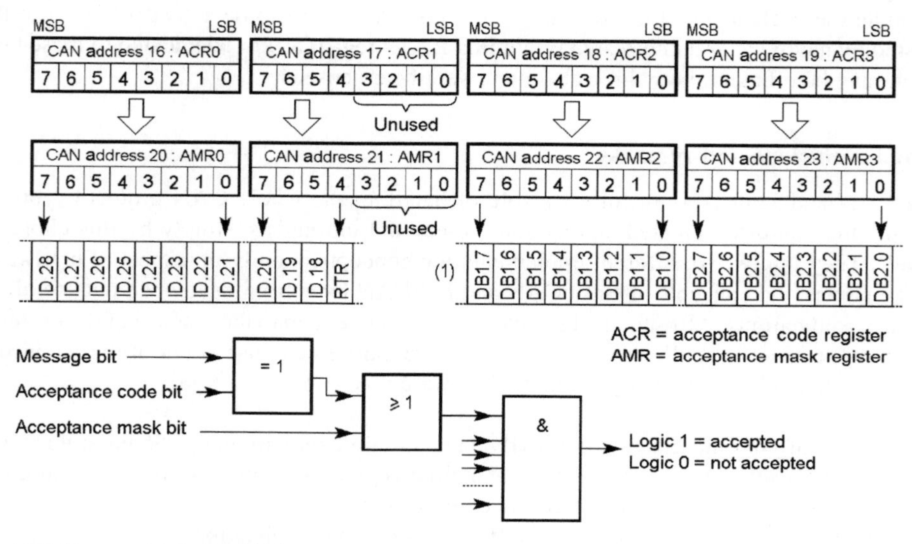


**Figure 5.2**

The choice and appropriate use of these registers (working with 8 bits) makes it easy to create groups of 256 identifiers and/or boundaries for identifiers to be used for particular time segments. (Note that, as far as I am aware, the use of 256 different identifiers covers a very large proportion of standard industrial applications in a CAN network). However, if it is necessary, for the purposes of any application at a given time, to discriminate identifiers whose values exceed 8 significant bits, in other words all the 11 bits of the protocol, the local CPU must use its software to execute the necessary operations for processing the remaining 3 bits in order to process all the possible CAN objects. Theoretically, this operation will make a small, but real addition to the CPU load. This solution is inherently suitable for the production of integrated components with a small chip area and low cost.

## CAN controllers with 'object'-oriented storage

'CAN objects' primarily consist of three parameters, namely

- the value of the identifier,
- the number of bytes contained in the transmitted message (data length code or DLC),
- the content of the transmitted data.

Controllers using 'object' storage, generally known as FullCAN, operate in the same way as those described in the preceding section, but are also capable of handling certain 'objects' to a certain extent. For example, they can be capable of determining which object is to be transmitted first, in the case of more than one simultaneous request, or to deal alone with all the problems of acceptance filtering (in a known network, therefore, and generally one having only a few participants). In this kind of application, the CPU does not have to do much to handle the exchange (some 'transmission request' bits have to be activated) because it is assisted by having a larger chip area (including RAM, registers, etc.) dedicated to these tasks which may be efficient, but which also incur significant costs.

### Some conclusions and remarks

BasicCAN and FullCAN are fully capable of handling the whole CAN protocol (you can imagine the confusion created among some users and journalists, simply by this choice of name!). They imply two different implementation concepts, mainly in respect of hardware.

The names and concepts of BasicCAN and FullCAN have been developed outside all the official specifications of Bosch, the ISO, etc., and are not related to the essence of the protocol, but simply to specific application concepts. You are quite free to use either of these concepts for your own applications, but you should be aware that

- if the field of your identifiers is primarily expressed in a maximum of 256 possible values, and exceptionally in 11 bits for particular time segments, there is only one economical choice, namely the so-called BasicCAN components;
- BasicCAN is generally satisfactory for 90% of industrial applications;
- FullCAN may be the right solution where processing speed is a consideration.

### A further subdivision

In the classification of CAN components, a further subdivision is generally made in relation to the versions of the protocol (2.0A/2.0B) that they support. If that was all, it would be too simple! In fact, for the purposes of applications, we often need to have 2.0A and 2.0B frames travelling over the same network for particular time segments, and of course we must ensure that the transmission of one of them does not trigger an error frame when the other one is transmitted (see at the end of Chapter 2 if necessary). To put it briefly, we talk about 'active' and/or 'passive' compatibility with certain types of frames (for example, a component can be both 2.0A active and 2.0B passive).

So think about the kind of system you want to set up, or the commercial and technical requests submitted to you, and the components to be associated with it, before you open your wallet! At present, practically all circuits available in the market can handle both versions of the protocol without any problem.

### Yet another subdivision

The classification of CAN components is often broken down into two further classes, relating to the maximum speeds that the network can support: low speed (0–125 kbit $s^{-1}$) and high speed (125 kbit $s^{-1}$ to 1 Mbit $s^{-1}$). See Chapter 4 for full explanations of these two operating modes.

### *5.1.3 A study of some specific components – the SJA 1000 protocol handler*

As mentioned previously, the choice of the components whose internal functions are to be detailed was guided by their performance, their presence on the professional and public markets and the degree to which they represent a full and uniform range of functions which can be implemented in a CAN network project.

The Philips-NXP semiconductors SJA 1000 is the standard example of an interface circuit between a microcontroller (Intel or Motorola/ Freescale 8/16/or 32 bit architecture) and the CAN bus (except for the line driver). Its function is to handle the CAN protocol fully, acting as a stand-alone controller. As mentioned previously, in this case a microcontroller must be connected externally to provide the 'application' layer of the system.

#### Internal structure of the circuit

Before describing this component in detail, it will be useful to outline the design philosophy of this kind of component. This integrated circuit is to be considered as a 'coprocessor' peripheral (or, more precisely, a memory-mapped input/output component) of the main CPU formed by the microcontroller of the module (Figure 5.3).

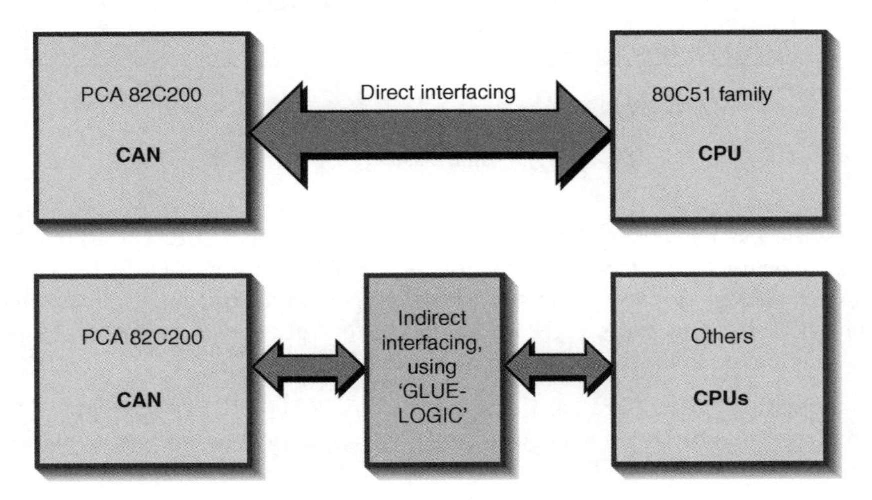


**Figure 5.3**

Its internal structure consists of numerous write- and read-addressable registers, designed to start up the different functions provided by this circuit and ensure interdependence with the microcontroller. The block diagram of this component in Figure 5.4 shows a number of clearly distinct parts:

- an interface management logic (IML), which is a logical interface providing a link between different types of microcontrollers that can be used and the integrated circuit, via an

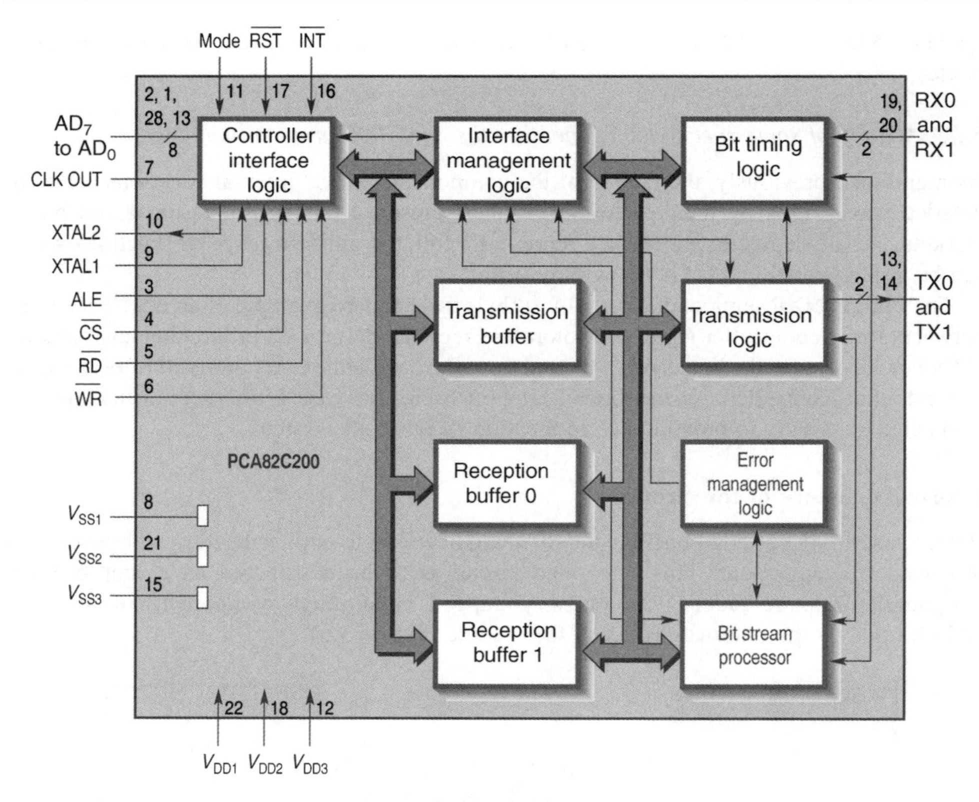


**Figure 5.4**

address/data bus comprising the conventional service signals (ALE, CS, RD, WR, etc.) which can be used with many microcontrollers available on the market (8x C xxxx, 68HC xx, 68xxx, ARM ...). This system also has the purpose of interpreting the microcontroller commands, allocating buffer registers, producing the necessary interrupts and supplying status information to the microcontroller;

- the transmission buffers (TBF) and reception buffers (RBF0, RBF1), see Figure 5.5:
  - a buffer stage, TBF (RAM with a depth of 10 bytes, representing the maximum content of an exchange on the bus – 11 identifier bits + 1 RTR bit + 4 DLC bits + 8 bytes of data), to be used for messages which are about to be transmitted along the bus, and provided with its appropriate processing logic,
  - two buffer stages, RBF0 and RBF1 (2 RAM, each with a depth of 10 bytes, for the same reasons as before), used alternately in the reception phase, thus enabling two successive information elements to arrive without disturbing the processing already in progress, and without ever creating an overload frame;
- a bit stream processor (BSP) for sequencing, commanding, etc., the data stream between the buffers TBP and RBF(x) operating with parallel access and the CAN bus operating with serial access;

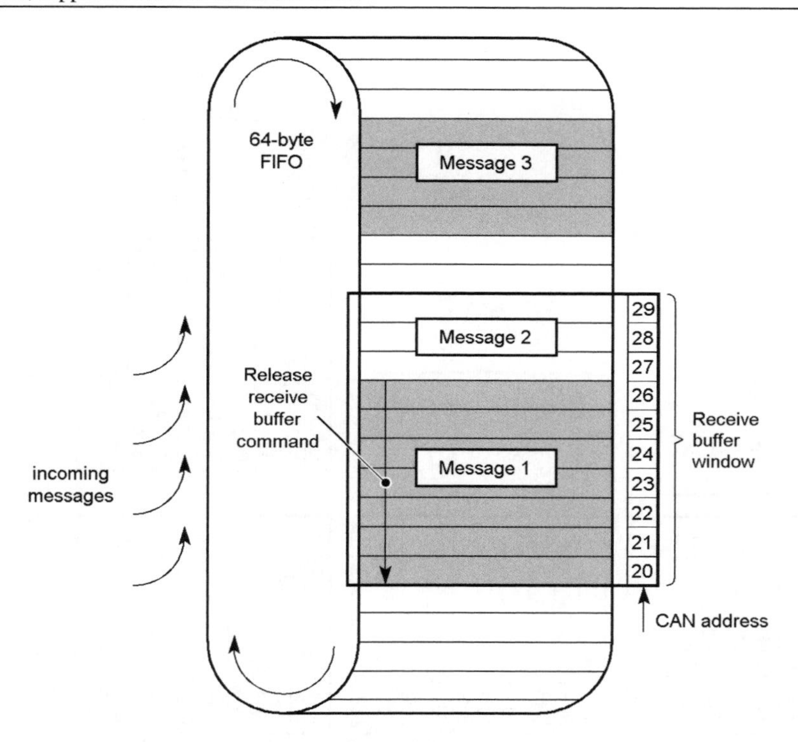


**Figure 5.5**

- a bit timing logic (BTL) whose function is to synchronize the SJA 1000 with the bit stream present on the CAN bus;
- a transmitter control logic (TCL) for controlling the output stage of the circuit;
- an error management logic (EML) for carrying out all the functions mentioned in Chapter 2 (transmission and reception error counters, error processing logic, etc.), to free the user from a lengthy, difficult and tedious processes of software writing;
- and finally, various logical 'glue' gadgets, essential for the correct operation of the whole system!

## Operation of the circuit

The circuit operates by means of registers whose allocation fields are shown in Figure 5.6. This simply shows their addresses, whereas Figure 5.7 shows the content (name, bit names, meanings, etc.) of each of these registers. The meanings of the different registers and their constituent bits provide an explanation of the hardware implementation of the CAN protocol handling.

Clearly, a complete description of the operation of this circuit would require a bit-by-bit analysis. This would take too long and would certainly bore many of you, so I have selected certain elements which I consider to be particularly interesting.

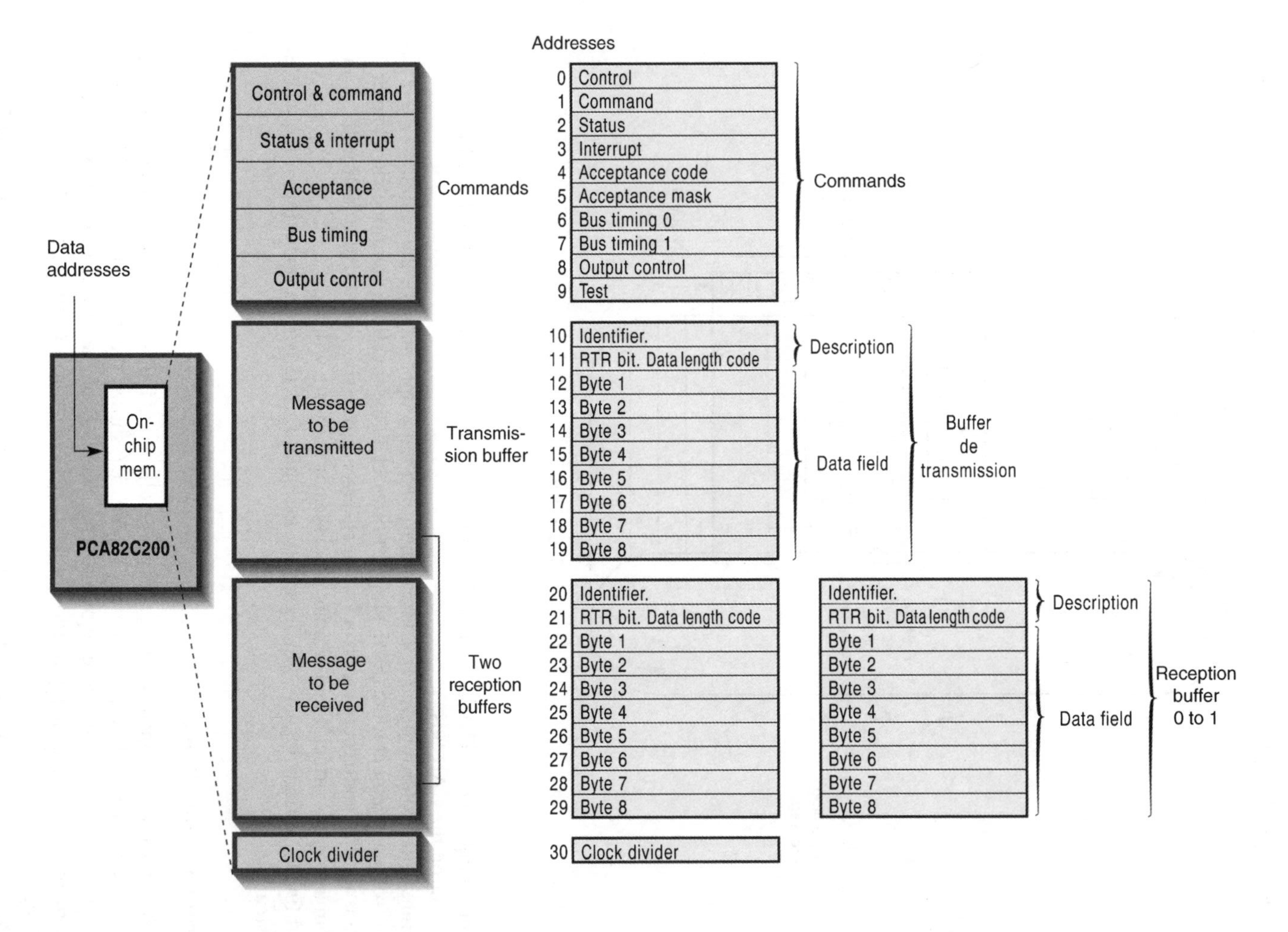

Data addresses
On-chip mem.
PCA82C200
Control & command
Status & interrupt
Acceptance
Bus timing
Output control
Commands
Message to be transmitted
Transmis-sion buffer
Message to be received
Two reception buffers
Clock divider
Addresses
0 Control
1 Command
2 Status
3 Interrupt
4 Acceptance code
5 Acceptance mask
6 Bus timing 0
7 Bus timing 1
8 Output control
9 Test
Commands
10 Identifier.
11 RTR bit. Data length code
12 Byte 1
13 Byte 2
14 Byte 3
15 Byte 4
16 Byte 5
17 Byte 6
18 Byte 7
19 Byte 8
Description
Data field
Buffer de transmission
20 Identifier.
21 RTR bit. Data length code
22 Byte 1
23 Byte 2
24 Byte 3
25 Byte 4
26 Byte 5
27 Byte 6
28 Byte 7
29 Byte 8
Identifier.
RTR bit. Data length code
Byte 1
Byte 2
Byte 3
Byte 4
Byte 5
Byte 6
Byte 7
Byte 8
Description
Data field
Reception buffer 0 to 1
30 Clock divider


**Figure 5.6**

| Title | Address | 7 | 6 | 5 | 4 | 3 | 2 | 1 | 0 |
|---|---|---|---|---|---|---|---|---|---|
| **Control segment** | | | | | | | | | |
| Control register | 0 | Test mode | Sync | Reserved | Overrun interrupt enable | Error interrupt enable | Transmit interrupt enable | Receive interrupt enable | Reset request |
| Command register | 1 | Reserved | Reserved | Reserved | Go to sleep | Clean overrun status | Release receive buffer | Abort trans-mission | Trans-mission request |
| Status register | 2 | Bus status | Error status | Transmit status | Receive status | Transmission complete status | Transmit buffer access | Data overrun | Data buffer status |
| Interrupt register | 3 | Reserved | Reserved | Reserved | Wake-up interrupt | Overrun interrupt | Error interrupt | Transmit interrupt | Reset request |
| Acceptance code register | 4 | AC.7 | AC.6 | AC.5 | AC.4 | AC.3 | AC.2 | AC.1 | AC.0 |
| Acceptance mask register | 5 | AM.7 | AM.6 | AM.5 | AM.4 | AM.3 | AM.2 | AM.1 | AM.0 |
| Bus timing register 0 | 6 | SJW.1 | SJW.0 | BRP.5 | BRP.4 | BRP.3 | BRP.2 | BRP.1 | BRP.0 |
| Bus timing register 1 | 7 | SAM | TSEG2.2 | TSEG2.1 | TSEG2.0 | TSEG1.3 | TSEG1.2 | TSEG1.1 | TSEG1.0 |
| Output control register | 8 | OCTP1 | OCTN1 | OCPOL1 | OCTP0 | OCTN0 | OCPOL0 | OCMODE1 | OCMODE0 |
| Test register (note 1) | 9 | Reserved | Reserved | Map internal register | Connect RX buffer CPU | Connect TX buffer CPU | Access internal bus | Normal RAM connect | Float output driver |
| **Transmit buffer** | | | | | | | | | |
| Identifier | 10 | ID.10 | ID.9 | ID.8 | ID.7 | ID.6 | ID.5 | ID.4 | ID.3 |
| RTR, data length code | 11 | ID.2 | ID.1 | ID.0 | RTR | DLC.3 | DLC.2 | DLC.1 | DLC.0 |
| Bytes 1–8 | 12–19 | Data | Data | Data | Data | Data | Data | Data | Data |

**Figure 5.7**

*For the 'transmit and receive' registers*
The content of the identifier (or the 'sense of the content' of the message) with a length of 11 bits has been split into 2 bytes. DSCR1 contains the 8 most significant bits and DSCR2 contains the remaining 3 bits, the RTR bit and the 4 bits giving the message length (DLC) of the data sent (Figure 5.8).

| Titre | Address | 7 | 6 | 5 | 4 | 3 | 2 | 1 | 0 |
|---|---|---|---|---|---|---|---|---|---|
| **Receive buffer 0/1** | | | | | | | | | |
| Identifier | 20 | ID.10 | ID.9 | ID.8 | ID.7 | ID.6 | ID.5 | ID.4 | ID.3 |
| RTR, data length code | 21 | ID.2 | ID.1 | ID.0 | RTR | DLC.3 | DLC.2 | DLC.1 | DLC.0 |
| Bytes 1–8 | 22–29 | Data | Data | Data | Data | Data | Data | Data | Data |
| Clock divider | 31 | Reserved | Reserved | Reserved | Reserved | Reserved | CD.2 | CD.1 | CD.0 |

**Figure 5.8**

*For the 'COMMAND' register*
*The CMR4 bit*
This bit has the name GTS (go to sleep), and tells the circuit to wake-up – or at least to open one eye, when it is in the low state. This is because, although a dominant level (i.e. activity on the bus) on the input pins RX0-RX1 will wake-up the circuit, it needs a certain time to 'stretch and yawn' before it is ready!

In fact, the local oscillator restarts, and a 'wake-up' interrupt is generated. The SJA 1000, which was asleep and is now waking up, will not be capable of receiving a message until it has detected a 'bus-free' signal (a sequence of 11 recessive bits in succession) which, if the bus speed is high, can allow a frame to pass. You must therefore be vigilant between time taken by the circuit to wake-up, as determined by the components you have added to the oscillator, and the length of the nominal bit time!

**Note on the power consumption of the circuit.** This mode can be very useful for battery-powered on-board systems. This is because the circuit draws approximately 15 mA in active mode and only 40 μA in sleep mode.

*The CMR1 bit*
This bit is named 'abort' and enables the microcontroller to suspend the next scheduled transmission in favour of another one considered to be more urgent (without interrupting the ransmission in progress, of course).

*For the 'STATUS' register*
*The SR7 bit*
The BS (bus status) bit indicates whether the component (the node in question) is in 'bus off' or 'active' mode. If the component has just switched to bus off (remember that the circuit can be 'sent off' for repeated serious misbehaviour!), it must wait for the minimum time specified by the protocol (128 occurrences of the 'bus-free' sequence) before becoming 'active' again and resetting its internal error counters to zero.

*The SR6 bit*
Called ES (error status), this bit provides information on the state of the error counters, showing that the warning phase has been reached.

### Message filtering

As mentioned previously, message filtering makes it possible to create families or classes of identifiers, opening the way to numerous possible applications (see the earlier remarks on FullCAN and BasicCAN). In the present case, I shall deal with a BasicCAN controller only. In many functional conditions (reception register empty, reset request bit set, etc.), the acceptance filter for messages carried on the bus is provided in the form of two registers (Figure 5.9):

- The acceptance code register (ACR). Bits AC.7 to AC.0 of the acceptance code and the 8 most significant bits of the message identifier must be equal to the bits shown as significant by the bits of the acceptance mask.
- The acceptance mask register (ACM), which determines which bits of the acceptance code will be accepted as 'significant' or 'insignificant' for filtering. If the following condition is found, global acceptance is granted:

| ACR | Addresses 4 | | | | | | |
|---|---|---|---|---|---|---|---|
| 7 | 6 | 5 | 4 | 3 | 2 | 1 | 0 |
| AC.7 | AC.6 | AC.5 | AC.4 | AC.3 | AC.2 | AC.1 | AC.0 |

| AMR | Addresses 5 | | | | | | |
|---|---|---|---|---|---|---|---|
| 7 | 6 | 5 | 4 | 3 | 2 | 1 | 0 |
| AM.7 | AM.6 | AM.5 | AM.4 | AM.3 | AM.2 | AM.1 | AM.0 |

| Acceptance mask bit | Value | Comments |
|---|---|---|
| AM.7 to AM.0 | High (don't care)<br>Low (relevant) | This is position is 'don't care' for the acceptance of a message.<br>This bit position is 'relevant' for the acceptance filtering. |

**Figure 5.9**

[(ID.10 to ID.3) = (AC.7 to AC.0)] or [(AM.7 to AM.0)] = 1111 1111

When a message is passed by the acceptance filter, it is transmitted and written to the reception buffer register RBFx.

## The other registers

*The bus timing register*

This register can be used to define all the parameters mentioned in Chapter 3 concerning the time values of the bit, in other words the durations of various phase segments, the position in time of the bit sampling instant and the resynchronization jump width; see Figure 5.10.

**Bus timing register 0 bits**

| BTR0 | Addresses 6 | | | | | | |
|---|---|---|---|---|---|---|---|
| 7 | 6 | 5 | 4 | 3 | 2 | 1 | 0 |
| SJW.1 | SJW.0 | BRP.5 | BRP.4 | BRP.3 | BRP.2 | BRP.1 | BRP.0 |

Baud rate prescaler (BRP)

**Bus timing register 1 bits**

| BTR1 | Addresses 5 | | | | | | |
|---|---|---|---|---|---|---|---|
| 7 | 6 | 5 | 4 | 3 | 2 | 1 | 0 |
| SAM | TSEG2.2 | TSEG2.1 | TSEG2.0 | TSEG1.3 | TSEG1.2 | TSEG1.1 | TSEG1.0 |

**Selection of sampling**

| Bit | Value | Comments |
|---|---|---|
| Sampling (SAM) | High (3 samples)<br>Low (1 sample) | Three samples are taken.<br>The bus is sampled once. |

**Figure 5.10**

*The output stage command register*

To provide for different configurations of the output stages and connection of different physical layers, a register is provided to control the configuration of the output stage of the integrated circuit, both during normal operation of the bus and while it is asleep (Figure 5.11):

**Output control register bits**

| OCR | Addresses 8 | | | | | | |
|---|---|---|---|---|---|---|---|
| 7 | 6 | 5 | 4 | 3 | 2 | 1 | 0 |
| OCTP1 | OCTN1 | OCPOL1 | OCTP0 | OCTN0 | OCPOL0 | OCMODE1 | OCMODE0 |

**Figure 5.11**

- at reset, the bus is floating;
- when asleep, the bus is recessive;
- when active, the output stage is either push-pull, pull-down or pull-up (Figure 5.12).

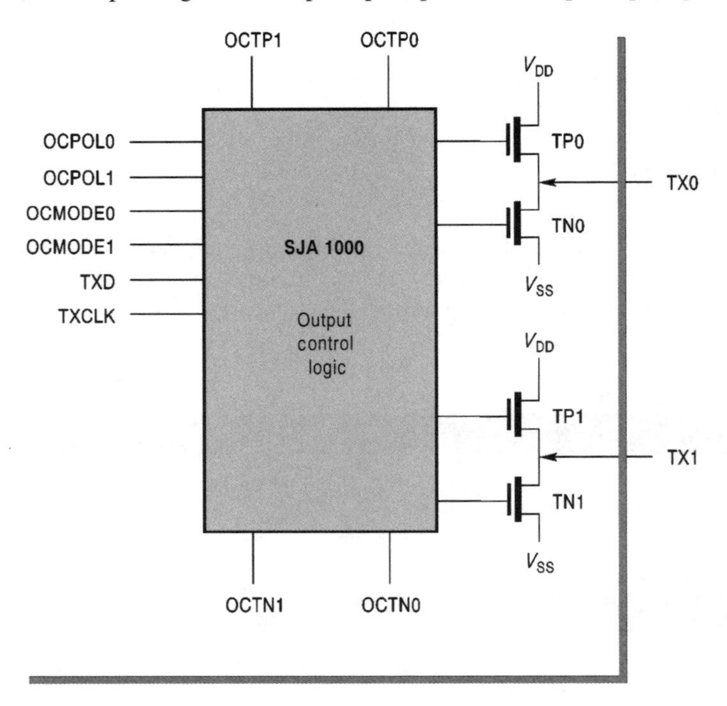


**Figure 5.12**

The signal polarity can be chosen to match the ISO-recommended configurations for the high-speed or low-speed bus modes.

**Note:** Outside the CAN protocol in its strict sense, this circuit can output a bit stream which is bi-phase coded, instead of being NRZ coded according to the rules. This bit-coding mode enables the user

- to have a transition inside each transmitted bit and thus facilitate the synchronization of the incoming bit stream if this is necessary for a particular application;
- and/or to couple the bus controller to the medium, if desired, via a transformer, to provide galvanic isolation. This known bit-coding mode has the effect of suppressing the continuous component found in NRZ coding in the frequency spectrum of the transmitted signal, which makes it incompatible with a link created by means of a transformer.

If you are still curious, you will have to examine the meaning of each bit in order to gain some idea of the subtleties of the circuit (see the example in Figure 5.13).

**Description of the output mode bits**

| OCMODE1 | OCMODE0 | Description |
|---|---|---|
| 1 | 0 | Normal output mode; TX0, TX1 : bit sequence (TXD; note 1). |
| 1 | 1 | Clock output mode; TX0 : bit sequence, TX1 bus clock (TXCLK). |
| 0 | 0 | Bi-phase output mode. |
| 0 | 1 | Test output mode; TX0 : bit sequence, TX1 : COMPT OUT. |

**Figure 5.13**

Having swallowed all these bits one by one (a large and indigestible meal, and we regret that no bicarbonate of soda is provided!), we must activate the registers and bits in a certain order which is different for each application.

By way of example, Figure 5.14 shows some characteristic algorithms, and the 'circuit + associated software' assembly for compliance with all the features of the CAN bus protocol, version 2.0A, particularly as regards the total number of identifiers available (2,032), for providing a maximum latency for priority messages, the multicast and broadcast aspect of the bus, etc.

Now let us look at some of its big brothers.

### *5.1.4 Microcontrollers with integrated CAN handlers: the 8x C 592*

This microcontroller is one of the first 'CAN-ized' derivatives of the 8x C552, a member of the large 80C 51 family produced by Philips-NXP semiconductors. All the standard software routines, UART, A/D conversion, PWM, etc., can be recovered, transported and linked. It is only necessary to spend any 'real' time on developing the new CAN part alone. This can save us a remarkable amount of money sometimes . . . but after this burst of enthusiasm and this honeymoon period, let us take a closer look at the block diagram in Figure 5.15.

At first sight, there is nothing very special here, except for the fact that the SJA 1000 protocol manager described above is integrated with the CPU (this is why I chose this system for description), and this requires some important remarks:

- In addition to the conventional internal 256-byte RAM (80C 52 family), a second 256-byte RAM, officially called the 'auxiliary RAM', is provided on the chip;

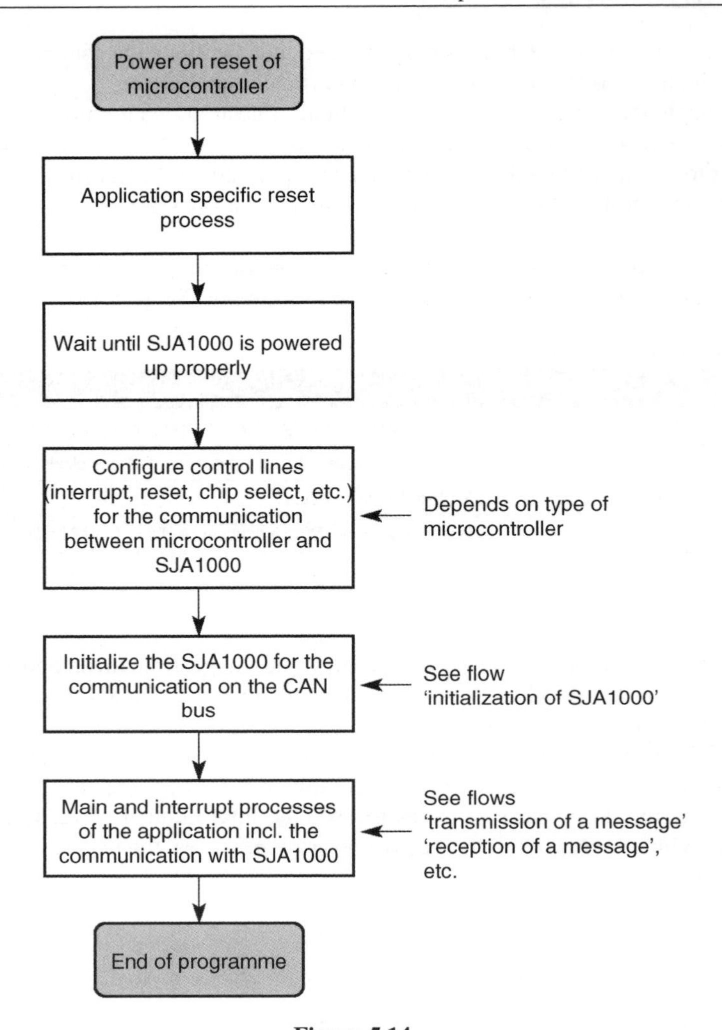


**Figure 5.14**

- In addition to this feature, if you examine the diagram carefully you will see an additional internal bus, the DMA (direct memory access) bus, between the CAN section and the ordinary RAM. This architecture accelerates the transfer of exchanges between the entry point of data from the CAN bus and the CPU of the microcontroller (see below).

### The CAN handler

This has a family resemblance to the handler of the SJA 1000 – not surprisingly, because it is just the same. Almost!

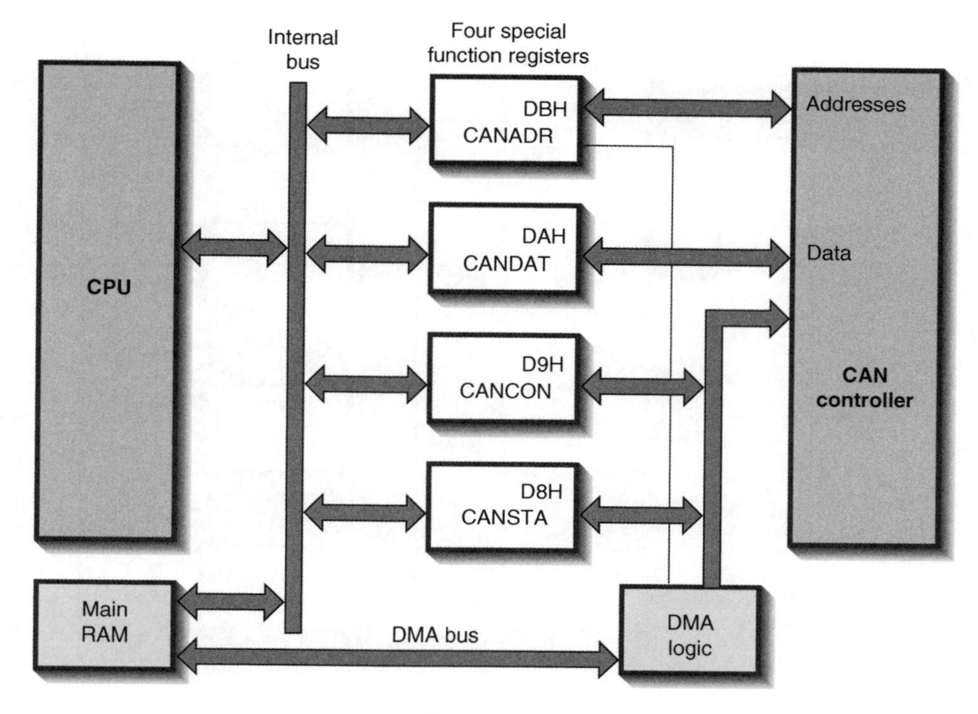


**Figure 5.15**

The major features of the SJA 1000 reappear in the CAN section of this IC, and a few modifications have been made to the control section connected to a microcontroller to optimize and improve it.

As you will see on the left of the CAN section of Figure 5.16, the protocol handler and the CPU of the microcontroller communicate via two (internal) buses:

- the address bus, which has the function of indicating the locations where the commands will be placed for execution by the CAN interface;
- the data bus, which positions the data in the command registers of the CAN interface, and also transports CAN data between the CPU (actually its RAM) and the protocol handler, for both the transmission and the reception of a CAN message. A DMA architecture has been implemented on the chip to provide 'quasi'-instantaneous and transparent data transfer between the main internal RAM and the transmission and reception buffers in the CAN section of the integrated circuit;
- the DMA. In fact, strictly speaking, the time taken for a data block to travel via this DMA is equal to not more than two instruction cycles, after the DMA operating command bit has been set. This is all done as a 'background task' and does not affect the normal running of the program.

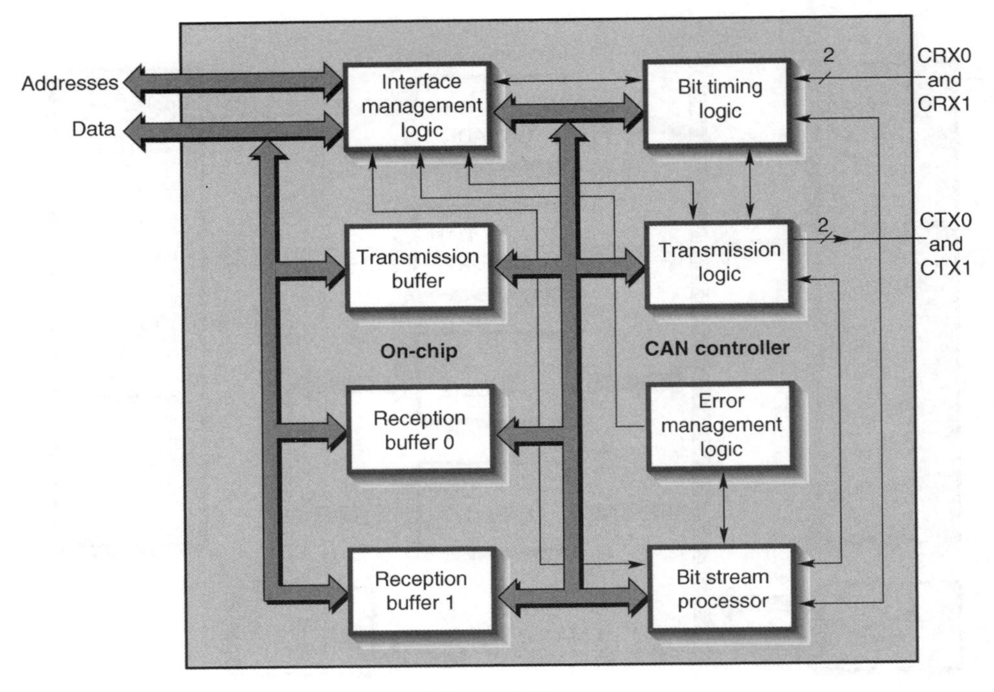


Figure 5.16

## Relations between the CPU and the CAN sections

Relations between the CPU and the CAN interfaces are provided with the aid of four SFRs (special function registers) with the following names:

- *CANADR CAN address,* located at the DBh addresses,
- *CANDAT CAN data,* located at the DAh addresses,
- *CANCON CAN control,* located at the D9h addresses,
- *CANSTA CAN status,* located at the D8h addresses.

Figure 5.17 gives a general idea of the functionality controlled by the various bits of these registers.Now let us look at the line interfaces.

## The CAN line interfaces

The internal circuitry of the interface with the CAN physical layer of these integrated circuits is made up of the following elements (Figure 5.18).

*Transmission unit*
The transmission unit (Tx) allows a direct differential output (CTX0 and CTX1 for CAN Tx0 and CAN Tx1). The configuration of these outputs can be modified by software (capable of supplying ±10 mA) in different ways:

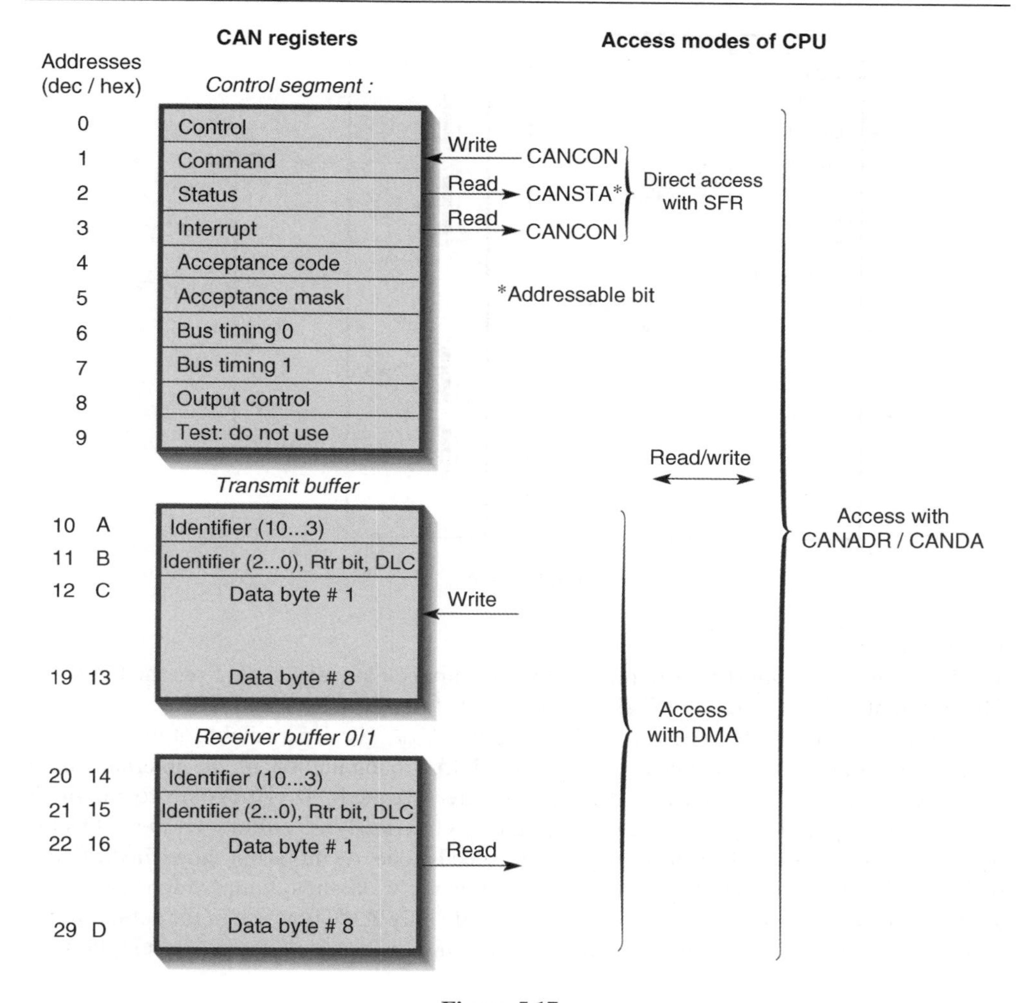


**Figure 5.17**

- in open drain mode (either to ground or to the positive of the power supply),
- in a push–pull circuit,
- in asymmetric mode with respect to ground to provide a DC drive to the external bus drivers,
- or, lastly (outside the CAN standard in respect of the bit specification), in bi-phase mode.

The last mode provides a useful extension of the protocol, as it enables an edge to be present in all the bits conveyed, so that other stations can be synchronized with the aid of this process, and also enables the bus drivers or coupling transformers to be driven in AC for the creation of galvanic isolation.

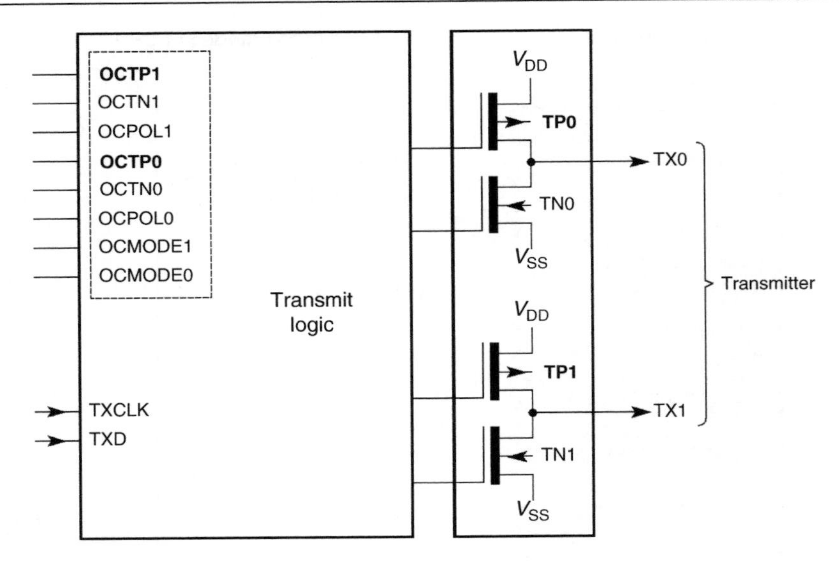


**Figure 5.18**

*Reception unit*
The function of this unit (Rx) is to process the incoming electrical signals from the CAN bus. The differential inputs from the CAN bus are applied to pins CRX0 and CRX1.

You will see that a reference voltage $V_{ref}$ (equal to $V_{dd}/2$) is also available at the output for all possible applications of the CAN bus. By skilful manipulation of the internal switch provided for this purpose (which can be software-activated), it is possible to internally connect, or output, or input a reference voltage to electrically determine the mean level of the line voltage, which will be necessary in high-speed mode. As this is not shown in the figure, it should also be noted that this circuit has an automatic wake-up system (which can also be software-activated), so that the controller is woken up only at the moment of the actual start of communication on the transmission line, thus avoiding unnecessary power consumption.

**Note:** Two drivers on a single microcontroller! Why have two outputs to the CAN bus, as this allows different sub-assemblies to be combined? A good question. This has been done so that two completely separate CAN networks can be created simultaneously, using a single component (refer back to the section on 'gateways').

### *5.1.6 CAN line drivers*

Whatever types of line driver are used, they must all have the following properties:

- in the transmitter:
  - correct matching of the differential command,
  - protection of the outputs from line surges,
  - the most symmetrical configuration possible, to minimize radiation;

- in the receiver:
  - correct matching for differential reception,
  - protection of the inputs from surges,
  - no adverse effect on the common mode rejection qualities of the line.

In the preceding chapter (Chapter 4), I detailed all the CAN line interfaces: high speed, fault tolerant low speed, one wire, etc. To avoid unnecessary repetition, I shall leave you to review this chapter for yourself.

To conclude this section, Figure 5.19 shows, for guidance, how CAN frames can also be transmitted on differential lines whose electrical levels have been made compatible with the RS 485 standard.

This completes, for now, our detailed survey of the different types of components that can be used for CAN systems.

## 5.2 Applications

In the previous chapters, I have described several aspects of CAN, particularly its protocol, the line aspect, the existing components, etc., and I have also outlined some applications. The aim of this section is to present different examples and possibilities for CAN applications and systems based on 8-, 16- and 32-bit microcontrollers which may or may not have dedicated on-board CAN hardware interfaces. The examples are not intended to cover the range of applications that can be implemented with CAN, but should enable most of you to gain an understanding of this concept. So you will be able to use this information to make your electronic dreams a reality *can* you do it? Yes, you *can*!

Clearly, the areas of application of CAN are determined by the characteristics which I have already described in detail. They are primarily concerned with short messages (commands, data, etc.), but may also deal with fragmented longer files (see the book cited in the references) which have to be transported rapidly over short to long distances, at speeds used in highly polluted environments. A high level of data transfer security must be provided. An example is one of the well-known applications of use with the multiplexed wiring of conventional cars and industrial vehicles. This example springs to mind because this protocol was developed by Bosch which is well known in this market, but we should not forget the thousands of other applications in industry, building automation and domestic automation which have already been implemented by numerous manufacturers using the CAN bus as a high-performance economical system based on a wide range of competitive components.

For your information, CAN has been used for a long time in the following fields, in addition to all the motor industry applications (light vehicles, heavy goods, utility, agricultural vehicles, etc.):

- the monitoring and control modules of high-speed trains,
- a large number of machine tool control systems,
- a large number of sensors (temperature, moisture, position, etc.),

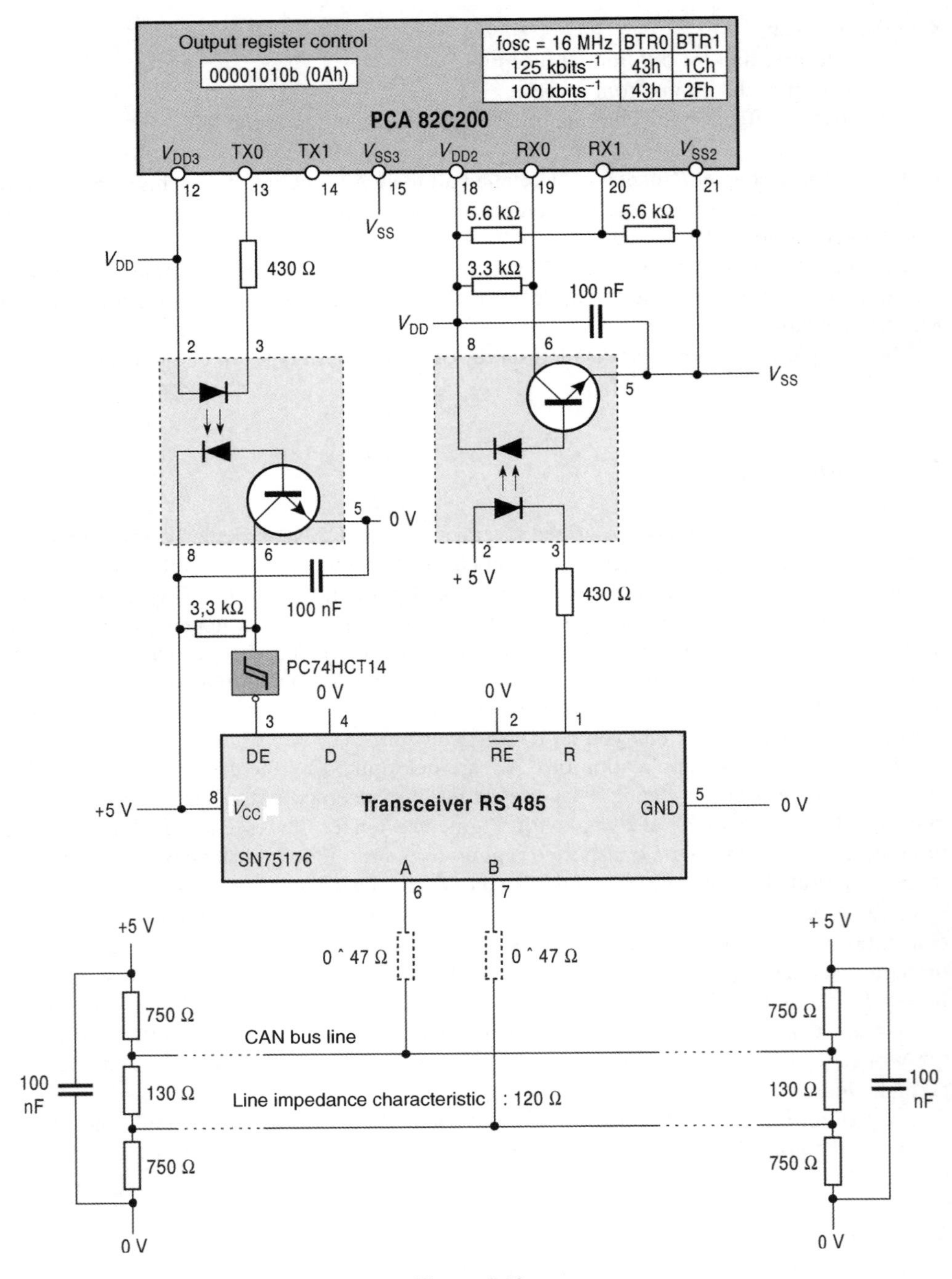

Output register control
00001010b (0Ah)
fosc = 16 MHz | BTR0 | BTR1
125 kbits−1 | 43h | 1Ch
100 kbits−1 | 43h | 2Fh
PCA 82C200
VDD3 TX0 TX1 VSS3 VDD2 RX0 RX1 VSS2
12 13 14 15 18 19 20 21
VSS
VDD
430 Ω
5.6 kΩ
5.6 kΩ
3.3 kΩ
100 nF
VDD
VSS
0 V
3,3 kΩ
100 nF
+ 5 V
430 Ω
PC74HCT14
0 V
0 V
DE D RE R
Transceiver RS 485
+5 V
VCC
GND
0 V
SN75176
A B
+5 V
0 ˆ 47 Ω
0 ˆ 47 Ω
+ 5 V
750 Ω
CAN bus line
130 Ω
Line impedance characteristic : 120 Ω
100 nF
750 Ω
0 V
0 V


**Figure 5.19**

- monitoring and control systems for modern lifts,
- oil well drilling probes,
- crane control and monitoring,
- domestic and building automations (heating and air conditioning and management of buildings, hotels, etc.),
- agricultural and food applications (animal feeders, etc.),
- glasshouse controllers and private and industrial watering systems,
- medical instrumentation (radiology, MRI, medical equipment for invalids, etc.),
- and finally, in the new Airbus A380 (containing more than twenty different CAN networks!).

### *5.2.1 Physical and functional divisions of a CAN-based system*

The division of a CAN-based system must be considered from different viewpoints:

- the physical and functional divisions of the functions and stations to be provided,
- the choice of the medium and operation of the physical layer,
- and then, at software level:
  - the protocol handling,
  - and finally the development of the messaging system and associated software of the application layer.

An application is a physical, concrete reality, and we need components to build it and make it work – it would be a miracle otherwise! A conventional network generally comprises a number of protocol handlers (at least two to form a network), given that each node is expected to operate independently of the others, because of the shared memory principle supported by the CAN protocol.

An example of a network (with a bus topology) is shown in Figure 5.20.

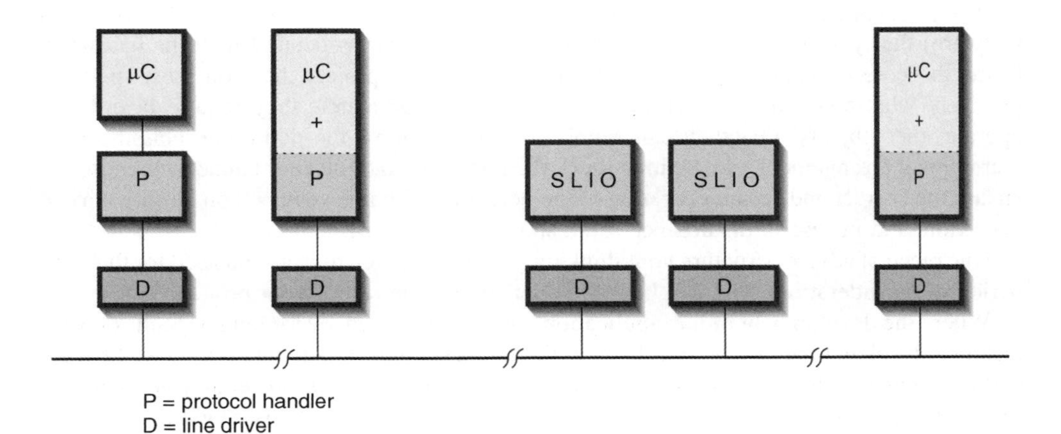


**Figure 5.20**

In this network, the different stations cohabit, but they must all operate at a single speed. As mentioned in Chapter 4, this speed can be set at any level from several kilobits per second to $1\ \mathrm{Mbit\ s^{-1}}$, generally in one of the two sub-classes officially called 'slow' (from 0 to $125\ \mathrm{kbit\ s^{-1}}$) or 'fast' (from $125\ \mathrm{kbit\ s^{-1}}$ to $1\ \mathrm{Mbit\ s^{-1}}$).

The network can then be designed, with its stations, participants, links, distances and all the rest of it. A chance for you to create grand designs, wonderful schemes ... however, you will have to come back to earth (with a more or less soft landing) when you count the cost of your project! Drastic cuts may appear in the offing as soon as you start to count the number of components. At this stage of the project, designers often have to reconsider the whole hardware and messaging architecture (the distribution of distributed intelligence), evaluating the necessary functionality in the light of budgetary constraints.

At this phase, it is common for nodes with local intelligence to be replaced with nodes having simpler functionality and a new conception and organization of the network messaging system.

### Physical layer

I will not discuss this again, as it has been amply described in the preceding chapters.

### Management of the CAN protocol

This is generally provided by dedicated microcontroller components, stand-alone protocol handlers and dedicated software routines, called 'low-level' routines, enabling the integrated hardware CAN interfaces to operate correctly.

### Application layer

This apparently insignificant generic term actually covers most of the time that you will spend on developing your application. You have been warned!

It is the software in this layer that will make your application work. So this is the layer (and software) that you will have to develop yourself. Its design, associated with the hardware, forms the core of your project. It is at this stage of your project that you must spell out precisely what tasks the stations are to carry out, the parameters they require in order to operate, the 'objects' that they can supply to other stations, the definition, classification, hierarchical arrangement, etc., of identifiers, the forms of handling and redundancy chosen to ensure the correct and secure operation of the network when it is switched on, during normal operation, and in case of disturbance ... in short, a lot of work!

One piece of advice: structure your thinking and take as much time as you need for this long period of consideration. This will be very beneficial for the rest of your project.

Where the development of this application layer is concerned, everything depends on your technical level and the size of your budget. For amateurs, you should be aware that there are certain 'gadgets' which, on a personal level, will help you to develop and debug your models at low cost (e.g. development using ROM simulators, etc.). It does not cost much ... except in terms of time!

Moving on to a higher (professional) level, you should know that, here too, some large companies have developed high-level software 'application layers' for the CAN protocol, which will facilitate your work on integrating your own application. This avoids an unnecessary expenditure of time and allows you to concentrate on the specification of your own application rather than the rest. As time is money, everything must be paid for, including advanced software packages. For guidance, I would mention the availability on the market of large standard 'software packages' of application layers which are 'encapsulated' and physically transported by CAN. These include, in particular, DeviceNet and SDS, and a more general-purpose one, widely used in numerous applications, called the CiA CAL (CAN application layer).

### *5.2.2 Examples of applications*

As I cannot cover everything, I must be selective again to avoid overloading this book. As regards the type of component for handling the protocol for a conventional version, and the 8-bit microcontroller with integrated protocol handler for a more economical version, I have chosen to use the SJA 1000 and the 80C 592 (80C 51 family) respectively, as these are widely used in the market. The same applies to the 32-bit solution, whose ARM core is supported by many companies. Solutions using other families competing with the components used for these examples of application are very similar as regards their circuits and use.

**Based on 8-bit microcontrollers**

- Discrete 80C 31 + SJA 1000 assembly. The circuit has been designed around the 80C 31.
- 80C 592 solution with integrated CAN handler.

**Based on 16-bit microcontrollers**

- A general-purpose solution: the '68000 (or 68xxx) + SJA 1000' assembly.
- Another general-purpose solution, the 'XA + SJA 1000' assembly.
- Finally, an integrated XA CAN hardware solution.

Why these last two suggestions? It is very simple: Following the previous presentation of 8-bit solutions designed around versions of the 80C 51, the XA (extended architecture) microcontroller family is the 16-bit family compatible with C51 source code with on-board CAN.

**Based on 32-bit microcontrollers**

The ARM core is well known to users in the car and other industries for 32-bit applications. Here again, many microcontroller versions with integrated CAN handlers are available in the market.

### *5.2.3 Example of applications based on 8-bit microcontrollers*

**CAN central unit with 80C31 and SJA 1000**

The circuit of this CAN CPU is shown in Figures 5.21. The architecture of this family of microcontrollers is well known, and I have described the operation of the SJA 1000 in detail at the beginning of this chapter. Clearly, the service signals required for the correct operation of the SJA 1000 are conventional (read, write, chip select, etc.), and, if you so wish, you can easily use another type of microcontroller from another 8- or 16-bit family.

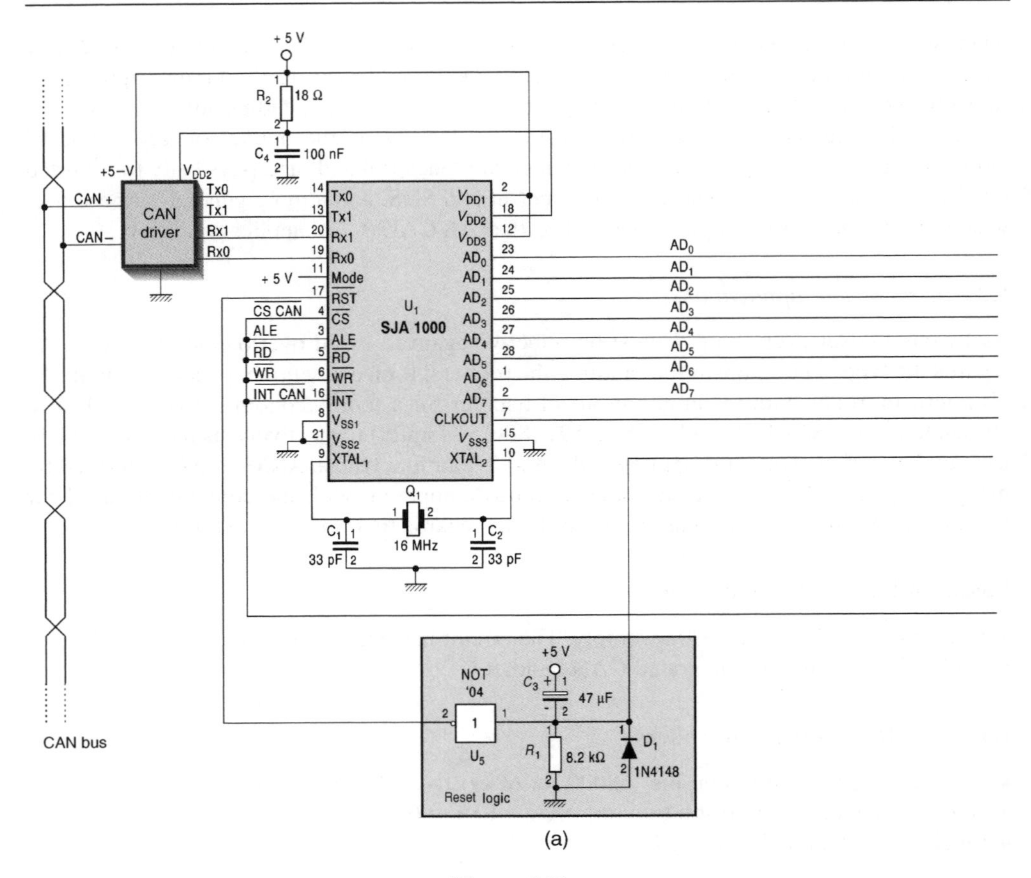

(a)

**Figure 5.21**

One reason for my choice is that there is a microcontroller (80 C 652) of the 80 C 51 family which is pin-for-pin compatible and has an integrated I2C hardware interface (output on ports P1.6 and P1.7). Furthermore, if an implementation requires the simultaneous presence of both buses, this concept offers greater flexibility of use. In the latter case, this solution becomes a 'CAN' CPU, with an external hardware CAN protocol handler, controlled by the microcontroller, and simultaneously an internal hardware I2C. This provides ample scope for the installation of many kinds of consumer equipment such as audio, radio and television, or installations on board motor vehicles (car radio, navigational aids, caravan fittings, etc.).

### CAN central unit based on 8x C 592

Before starting on the specific description and performance of this CAN CPU, I must make a few important remarks:

- As you will see in the diagrams, EPROM external memories are provided, but, if you wish, you can use OTP/Flash or ROM-ized versions of this microcontroller to operate in internal program memory mode only.

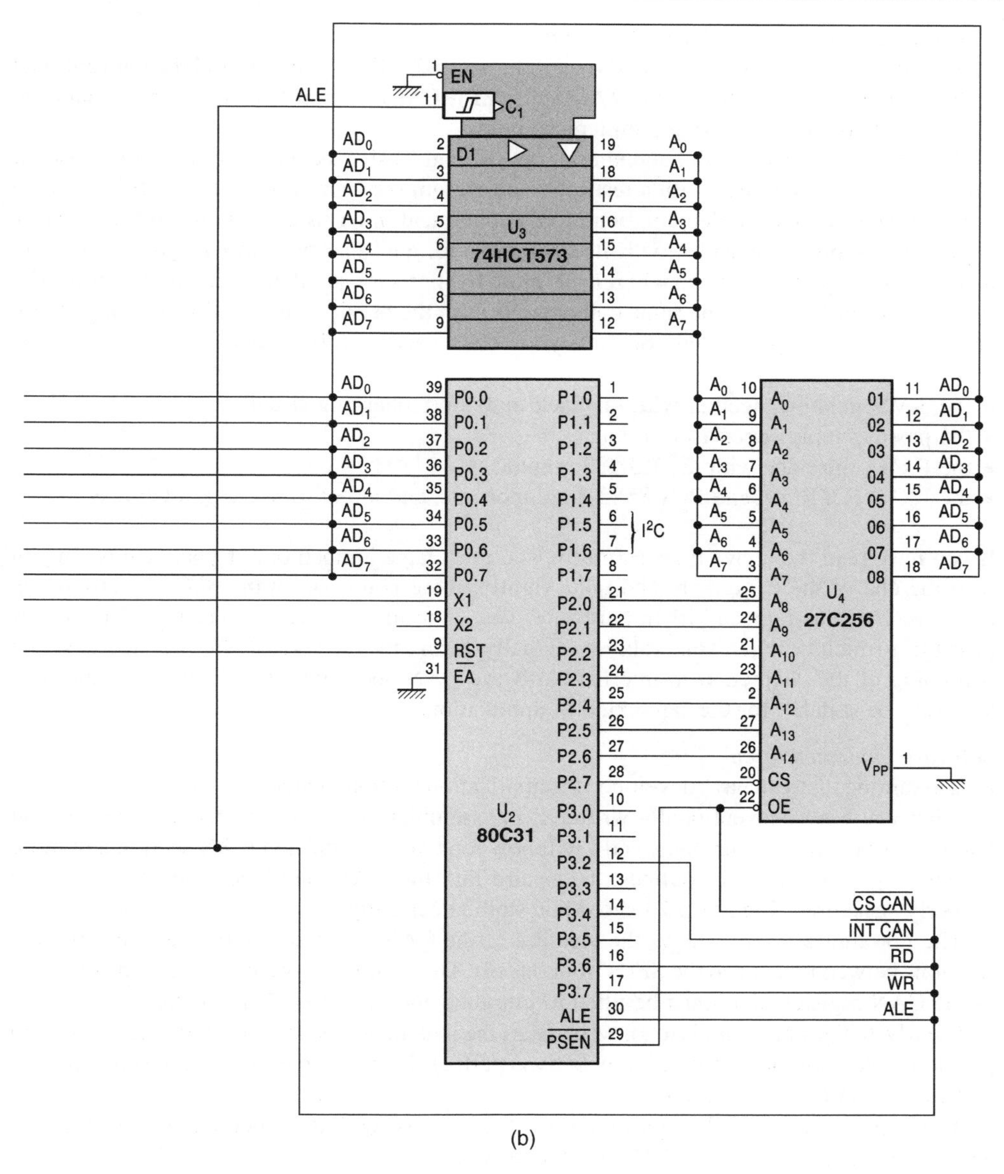


(b)

**Figure 5.21** (*Continued*)

- I have chosen the TJA 1050 integrated circuit for the control of the physical line of the bus (differential pair). As far as I know, although it exists in a DIL casing, it is mainly offered in an SO casing (surface component) ... so consider your implementations carefully.

After these remarks, let us go on to examine the subject itself.

*General structure of the CAN CPU circuit*
The complete CAN CPU circuit is shown in Figure 5.22. Here you will find the conventional address latch (74 HCT 573), the RAM and (EP)ROM (program) memory space, and the NAND gate for superimposing the memory fields.

A more economical implementation option can easily be envisaged in the case of fully masked applications (with a restrictive maximum program limit of 8 kB). It is possible to make the proposed implementation more economical in terms of cost and surface area by replacing the microcontroller with its OTP version, and also by removing the previously mentioned components, which will release more I/O ports for supplementary applications. For the rest, the card contains the same components as in the previous application, although these are shown here as optional (although they are often very useful), namely

- the 5-V regulator, together with the back-up battery charging circuit,
- the power supply supervisor circuit,
- the Rx/Tx interface with the RS 232 output,
- the I2C – E2PROM and PCx 8582-2 components, and a real-time clock, PCF 8583.

If you have read the above lines carefully, you must have jumped out of your seat when you read the end of the paragraph, and quite rightly: there is no I2C in the 8xC 592 any more, because CAN has replaced it! In hardware, yes – but in software, no, because it is always possible to include one. This solution actually forms the inverse of that described at the beginning of the chapter, with internal hardware CAN and internal software I2C, and may therefore be suitable for the same kind of applications.

*Software implementation*
I am assuming that you have developed the application section, which has to be specific to you, and that you have conventionally written it in assembler or C51 (allowing for some special features of this microcontroller), and I will now look at the dedicated CAN section and more particularly these specific routines. To ensure that the CAN interface control software is properly structured, Figures 5.23 should be studied carefully.

The first figure clearly shows the location of the CAN interface control registers and the relations between them and the SFRs (for control). The second shows the names and addresses of the CAN registers that must be filled to initialize and launch exchanges, etc.

Clearly, the ultimate aim is to know by heart the meaning of each bit of these registers, or to post it up over your desk! However, nobody is perfect. So we need only to fill in some lines of software to start up the interface.

You may be eager to ask a question now: Why persist with this microcontroller when the SJA 1000 could do everything? Here is the answer.

*Load on the CPU during CAN communications*
Now for a few words on a subject which is often skipped over during the design of this kind of system, but which is fundamental. This is the evaluation of the time load on the CPU due to communication management on the network. First of all, what does the 'CPU load' mean?

The CPU load is defined as the percentage of time taken by the CPU to 'serve' communications (CAN in this case) out of a specified reference period. In the arcane language of microcontrollers, the word 'serve' denotes that the appropriate interrupt routine is executed,

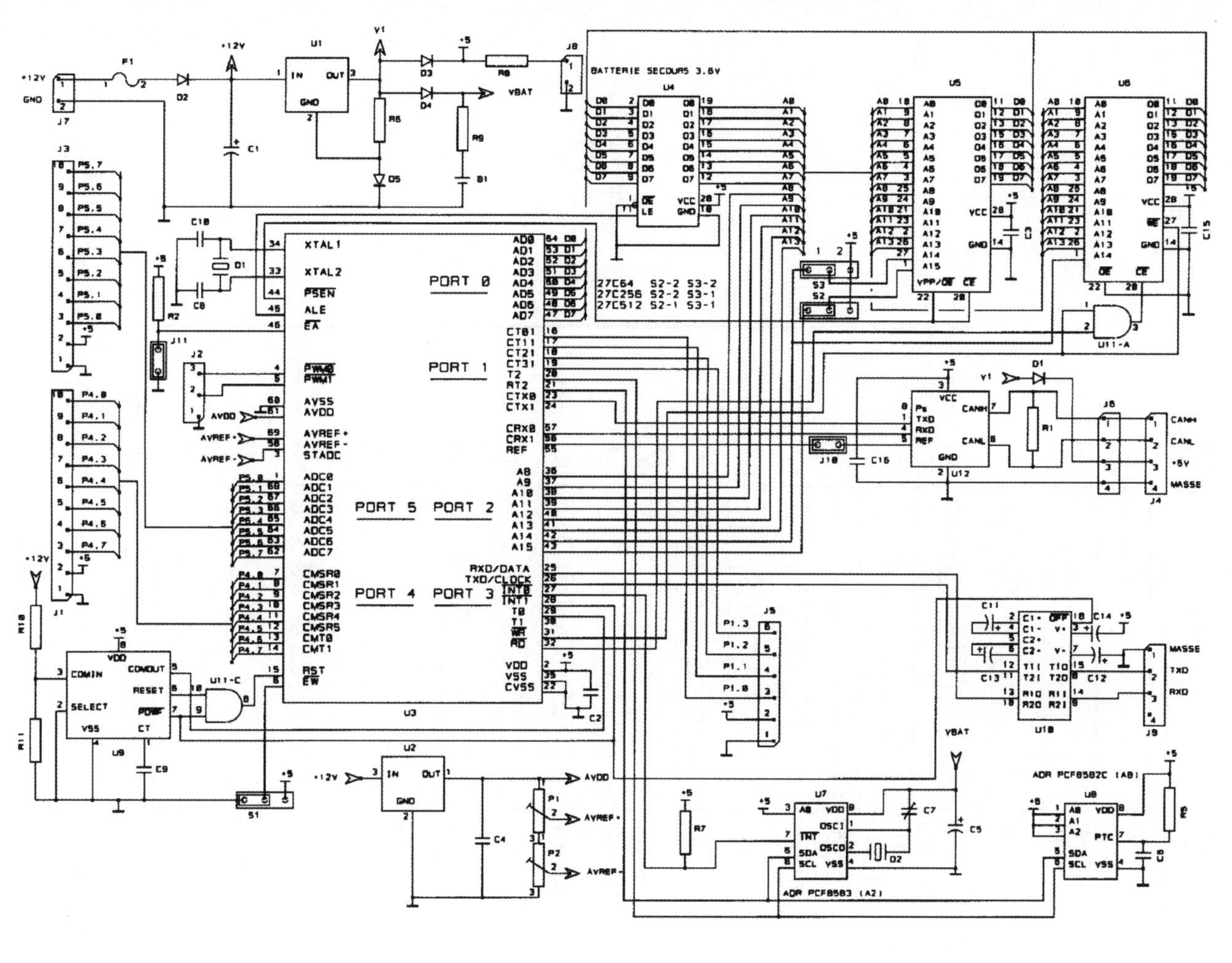
BATTERIE SECOURS 3.6V
27C64 S2-2 S3-2
27C256 S2-2 S3-1
27C512 S2-1 S3-1
PORT 0
PORT 1
PORT 5
PORT 2
PORT 4
PORT 3
ADR PCF8582C (A0)
ADR PCF8583 (A2)

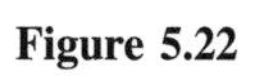

Figure 5.22

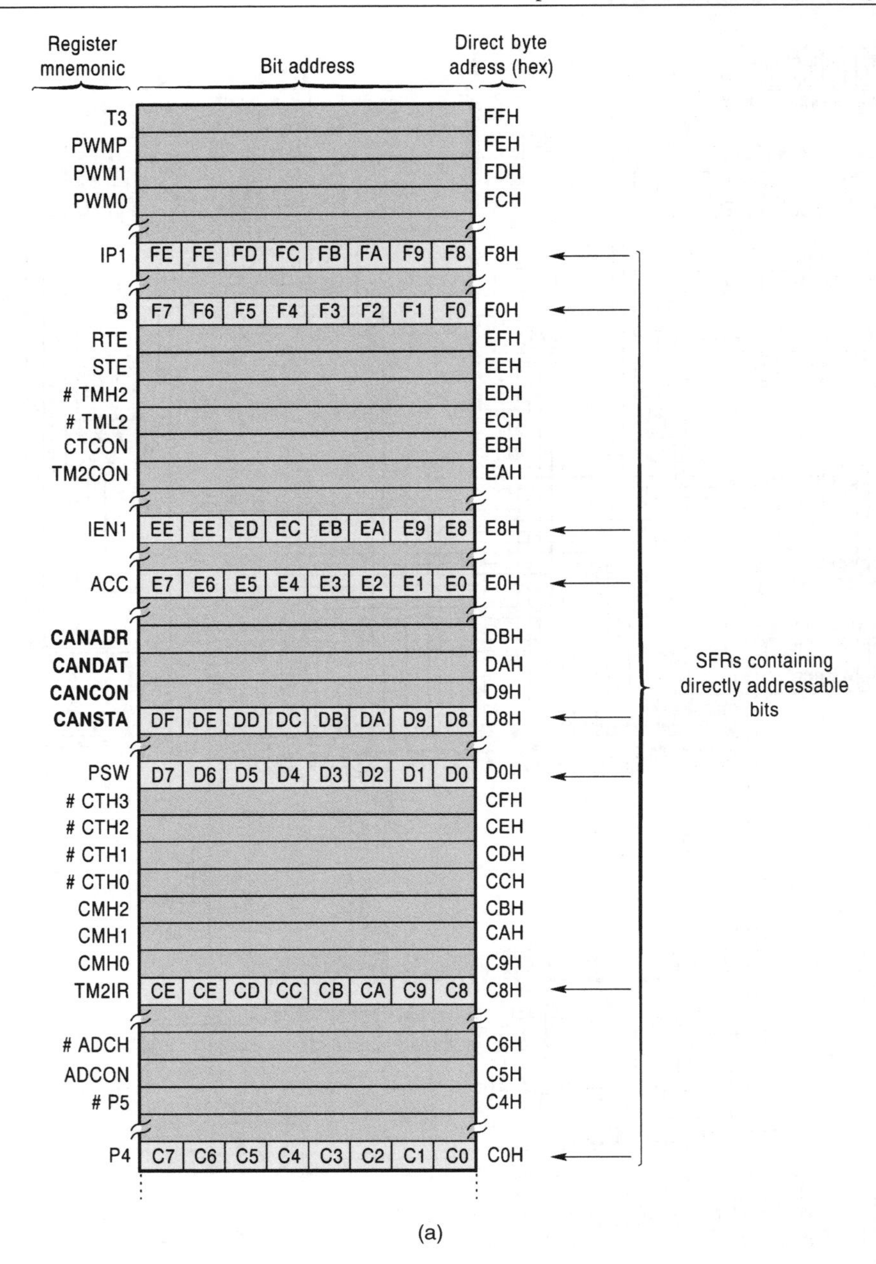
Register mnemonic
Bit address
Direct byte adress (hex)
T3
PWMP
PWM1
PWM0
IP1
B
RTE
STE
# TMH2
# TML2
CTCON
TM2CON
IEN1
ACC
CANADR
CANDAT
CANCON
CANSTA
PSW
# CTH3
# CTH2
# CTH1
# CTH0
CMH2
CMH1
CMH0
TM2IR
# ADCH
ADCON
# P5
P4
FE FE FD FC FB FA F9 F8
F7 F6 F5 F4 F3 F2 F1 F0
EE EE ED EC EB EA E9 E8
E7 E6 E5 E4 E3 E2 E1 E0
DF DE DD DC DB DA D9 D8
D7 D6 D5 D4 D3 D2 D1 D0
CE CE CD CC CB CA C9 C8
C7 C6 C5 C4 C3 C2 C1 C0
FFH
FEH
FDH
FCH
F8H
F0H
EFH
EEH
EDH
ECH
EBH
EAH
E8H
E0H
DBH
DAH
D9H
D8H
D0H
CFH
CEH
CDH
CCH
CBH
CAH
C9H
C8H
C6H
C5H
C4H
C0H
SFRs containing directly addressable bits

(a)

**Figure 5.23**

| Address | Access | Bit 7 | 6 | 5 | 4 | 3 | 2 | 1 | 0 |
|---|---|---|---|---|---|---|---|---|---|
| CANADR | | | | | | | | | |
| DBH | R/W | DMA | Reserved | Autolnc | CANA4 | CANA3 | CANA2 | CANA1 | CANA0 |
| CANDAT | | | | | | | | | |
| DAH | R/W | CAND7 | CAND6 | CAND5 | CAND4 | CAND3 | CAND2 | CAND1 | CAND0 |
| CANCON | | | | | | | | | |
| D9H | R | Reserved | Reserved | Reserved | WUI | OI | EI | TI | RI |
| | W | RX0A | RX1A | WUM | SLP | COS | RRB | AT | TR |
| CANSTA; bit addresses of CANSTA (7 to 0) are DFH to DH8; do not use a RMW instruction | | | | | | | | | |
| DFH to D8H | R | BS | ES | TS | RS | TCS | TBS | DO | RBS |
| | W | RAMA7 | RAMA6 | RAMA5 | RAMA4 | RAMA3 | RAMA2 | RAMA1 | RAMA0 |

(b)

| SFR | ADR | ACS | MSB 7 | 6 | 5 | 4 | 3 | 2 | 1 | LSB 0 |
|---|---|---|---|---|---|---|---|---|---|---|
| CANADR | D8H | R/W | DMA | Reserved | Autoinc | CANA4 | CANA3 | CANA2 | CANA1 | CANA0 |
| CANADT | DAH | R/W | CAND7 | CAND6 | CAND5 | CAND4 | CAND3 | CAND2 | CAND1 | CAND0 |
| CANCON | D9H | R | Reserved | Reserved | Reserved | Wake-up int | Overrun int | Error int | Transmit int | Receive int |
| | | W | RX0 active | RX1 active | Wake-up mode | Sleep | Clear overrun | Release Rx buffer | Abort tansmit | Tansmit request |
| CANSTA | D8H | R | Bus status | Error status | Tansmit status | Receive status | TX compt status | TX buffer status | Data overrun | RX buffer status |
| | | W | RAMA7 | RAMA6 | RAMA5 | RAMA4 | RAMA3 | RAMA2 | RAMA1 | RAMA0 |

(c)

**Figure 5.23** (*Continued*)

and the normal running of the main program is then resumed (thus it implies a knowledge of a reference 'interrupt routine' which is not described here). You should consult Figures 5.24 for a fuller understanding of the architectural design and flowchart of the interrupt routine.

As you know, a CAN node is constantly monitoring the identifiers of messages travelling in the network. At this point in the process, it decides, on the basis of their specific values, either to ignore the message totally by 'rejecting' it, or to 'copy' its content (data) to make use of it.

In both of the above cases, the CPU is active and thus uses machine time via the interrupt routine. Figure 5.25 shows an example of the number of cycles occupied in the two cases (fortunately, the time taken to reject an 'irrelevant' message is shorter than that required to record a 'relevant' message).

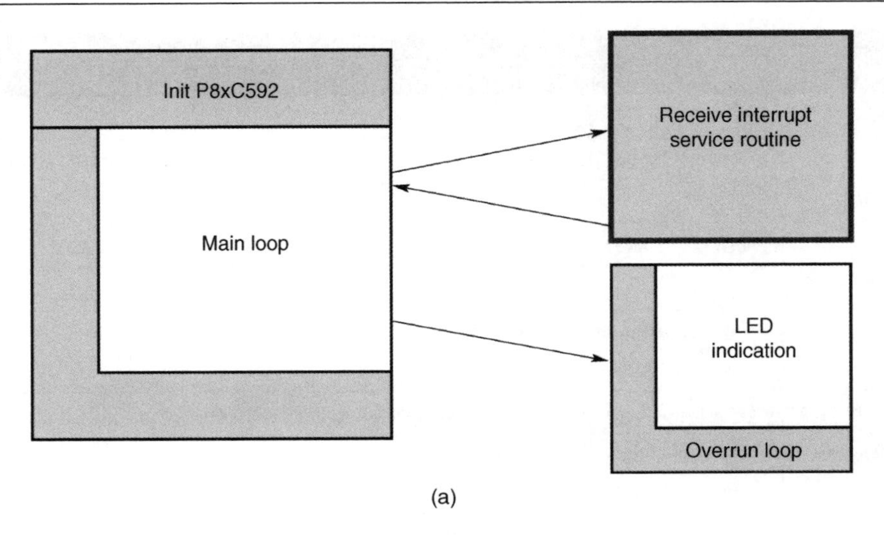


(a)

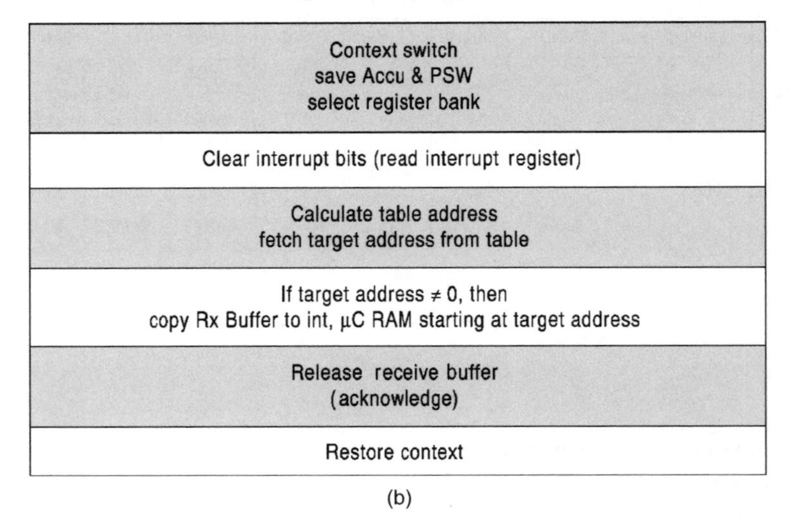


(b)

**Figure 5.24**

Clearly, the length of the message sent along the bus (from 0 to 8 bytes transmitted per data or request frame) has an effect on this CPU load, as does the frequency of the microcontroller clock (and therefore the gross bit rate of the bus). An example of the time required is shown in Figure 5.26 for a 16 MHz clock.

| Macro name | Machine cycles | Copy | Reject | Rx buff release |
|---|---|---|---|---|
| Max. interrupt response time | 5 | x | x | x |
| Long jump to interrupt service routine | 2 | x | x | x |
| Context switch | 6 | x | x | x |
| Clear int. bits | 1 | x | x | x |
| Calculate target address | 15 | x | x | x |
| Copy Rx buffer | 8 | x | Ð | x |
| Acknowledge | 2 | x | x | x |
| Restore context | 6 | x | x | Ð |
| Sum of cycles | | 45 | 37 | 39 |

**Figure 5.25**

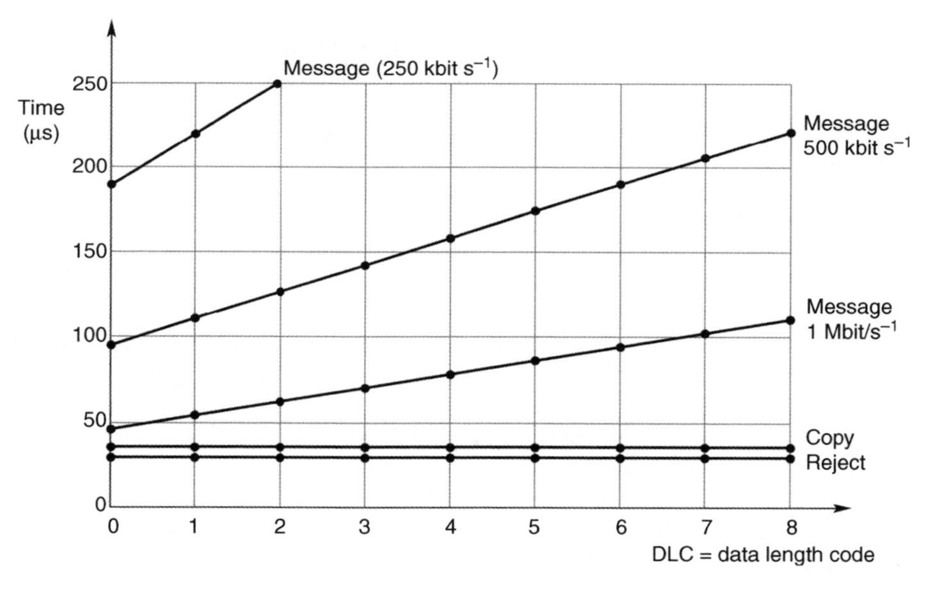


**Figure 5.26**

Having clarified this point, we will move on to the calculation of the CPU load. As mentioned above, it depends on three main parameters:

- the execution time of the INTerrupt routine,
- the bus speed,

- the bus load (the percentage of time when the bus is active for message transfer), which itself depends on
  - the number of messages transferred per time interval,
  - the length of each message, and therefore the number of bits transmitted and the duration of each bit.

This can be expressed in the following equations:

$$\text{Bus load} = \frac{\text{Time for which the bus is active}}{\text{Reference time interval}}$$

$$\text{Bus load} = \frac{\text{Number of messages} \times \text{Message length} \times \text{length} \times \text{Bit time}}{\text{Reference time interval}}$$

and therefore the number of messages is

$$\text{Number of messages} = \frac{\text{Bus load} \times \text{Reference time interval}}{\text{Message length} \times \text{Bit time}} \qquad [5.1]$$

and therefore, finally, the CPU load is

$$\text{CPU load} = \frac{\text{Number of messages} \times \text{dINT routine execution time}}{\text{Reference time interval}} \qquad [5.2]$$

Assuming that the bit rate is equal to 1/bit time, and incorporating equation 5.1 into equation 5.2, we find

$$\text{CPU load} = \frac{\text{Bus load} \times \text{Bit rate} \times \text{dINT routine execution time}}{\text{Message length}}$$

$$\text{CPU load} = \text{Bus load} \times \text{Bit rate} \times \tau$$

The value of $\tau$ depends on the message length. Its variations are shown in Figure 5.27 and can be used to calculate the CPU load for each type of message (or ‘average’ message) transmitted.

All this lengthy section is summed up in Figure 5.28.

**Note:** Every designer of a CAN system should theoretically carry out the same type of calculation, according to the components he chooses. Sometimes there may be pleasant or unpleasant surprises!

To provide some context, the first solution described, including the standard microprocessor of the same 80 C 31 family and provided with an SJA 1000 external protocol handler, has a CPU load three to five times that of the dedicated 8x C 592 microcontroller.

### CAN and 16-bit microcontrollers

Here are two applications based on 16-bit microcontrollers.

The first is built around the well-known Freescale/MOTOROLA 68 000 or its derivatives.

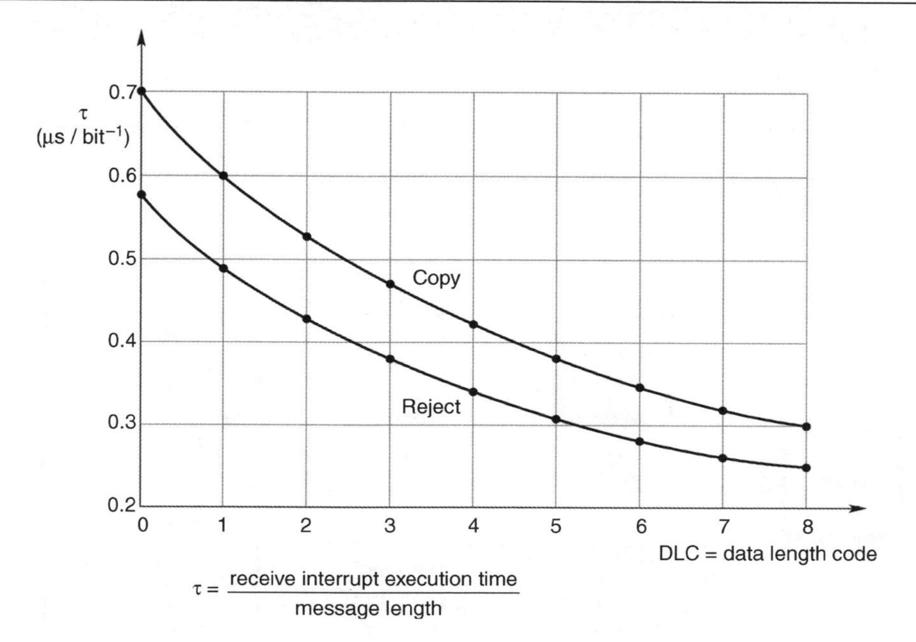


**Figure 5.27**

| Bit rate | 10% Bus load | 20% Bus load | 50% Bus load |
|---|---|---|---|
| 100 kbit $s^{-1}$ | 0.4% | 0.8% | 2.1% |
| 250 kbit $s^{-1}$ | 1.0% | 2.1% | 5.2% |
| 500 kbit $s^{-1}$ | 2.1% | 4.2% | 10% |
| 1 Mbit $s^{-1}$ | 4.2% | 8.4% | 21% |

**Figure 5.28**

The second is built around a CPU called 80C51 eXtended Architecture, or alternatively 80C51 XA. This 16-bit family is 'source compatible' with the former 80C51 8-bit devices, thus allowing all fans of the 80C51 to enhance the power without the need for tens or hundreds of tedious hours of software development work.

*68 000 (or 68 xxx) and SJA 1000*

Figure 5.29 shows a block diagram of this implementation. Numerous applications operate in the field around the 68 xxx family. Note also that there are numerous derivatives of this family with integrated CAN hardware.

*The XA and SJA 1000 family*

A diagram of this application is shown in Figure 5.30. Except for the CPU with a much higher performance, this environment resembles the '51' type.

| | Asynchoronous bus interface 68000 | Parallel bus interface 82C200 |
|---|---|---|
| Timing | Asynchronous | Synchronous |
| Data/address bud | 24/16 bits, non-multiplexed | 5/8 bits, non-multiplexed |
| Control signals | AS, R/WRN, DTACKN, CSION, UDS, LDS | RDN, WRN, CSN, ALE |

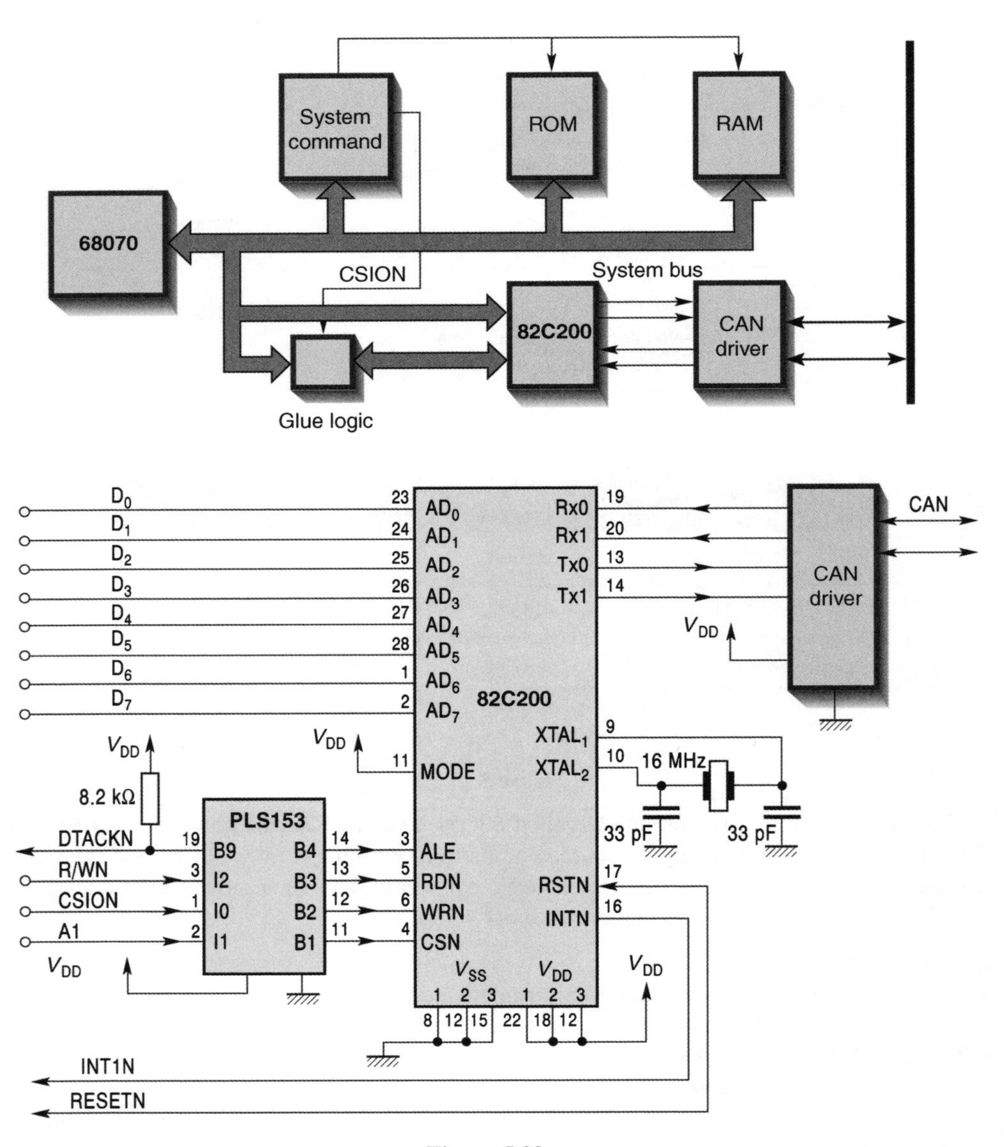


Figure 5.29

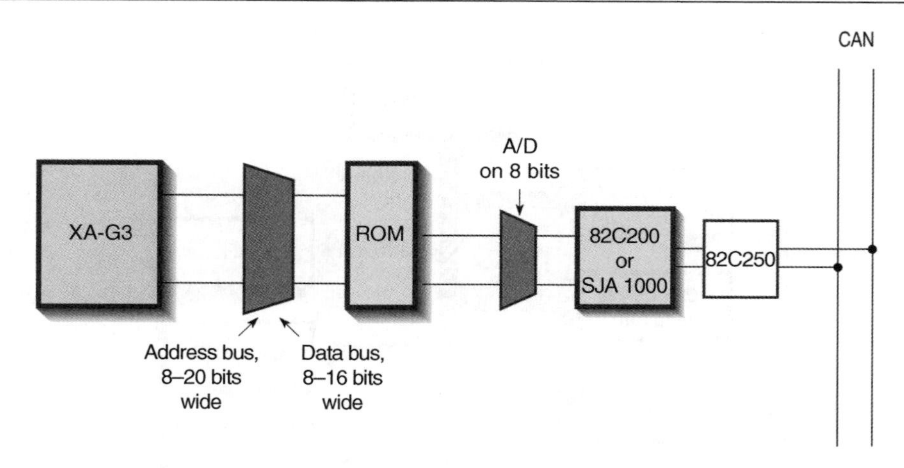


**Figure 5.30**

*The CAN version of XA*

The XA family has a CAN version, the XA-C3, with an on-board CAN 2.0B protocol handler (in hardware, of course). Figure 5.31 shows the block diagram. This component is perfectly suitable for DeviceNet applications, because it also includes part of the disencapsulation of the DeviceNet application layer, particularly at the level of fragmentation and defragmentation of CAN messages carrying more than 8 bytes.

**CAN and 32-bit microcontrollers, ARM core**

To complete this long list, I must also point out that many 32-bit microcontrollers supporting CAN are available in the market. One of many possible examples is shown in Figure 5.32 which is a block diagram of one of the microcontrollers of the LPC 21xx family with an ARM7 core, comprising several I2C interfaces, SPIs and CANs simultaneously. These microcontrollers are generally intended for industrial use, because most of them cannot fully meet the requirements of operation in motor vehicle temperature ranges (from -40–125°C or even 150°C).

This concludes our description on examples of CAN CPU implementation. You are now free to choose your favourite applications and set up your own CAN local networks, for operation with numerical control and analogue inputs/outputs of all kinds.

## 5.3 Application Layers and Development Tools for CAN

This brings us to the end of this long section on CAN, and up to this point I have restricted myself to a description of the fundamentals of the protocols and the implementation of components capable of handling communications and controlling stations, etc. All of this plays its part in the correct operation of a network, but what is missing is the real application part specific to your network, constructed on the communication protocol (in the sense of ISO

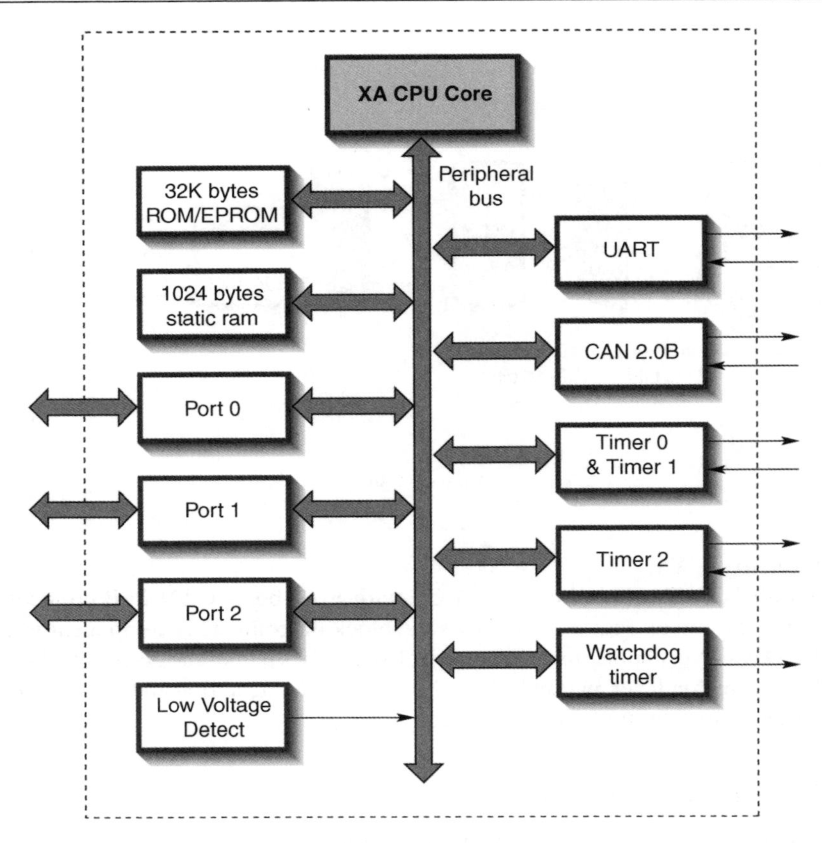


**Figure 5.31**

layer 2). In fact, I have mentioned the application layers corresponding to ISO layer 7 without really discussing them.

However, this part is vitally important for the functional design of the network, as it brings together practically all the organization and operation of the system that you wish to implement. The description of all these elements is a huge topic which deserves at least half a book to itself. Why not a whole book, you may ask. The reason is simple. When you have completed this part relating to the design and implementation of the application layers, you will have only one more thing to do: test it . . . and test its consequences on the system! It may well appear elementary and childish to you, but you will have to go through this stage to ensure the correct operation and reliability of your design.

To be consistent, therefore, the other part of this second book would have to deal with the network test tools and methods (for debugging, analysis, emulation, simulation, etc., and for validating the operation of the network). You should know that there is at least one detailed work on these CAN-related fields available for you to consult (see Appendix C).

The following text is intended to provide some general information on the problems raised by this and to give some outlines of responses to these problems.

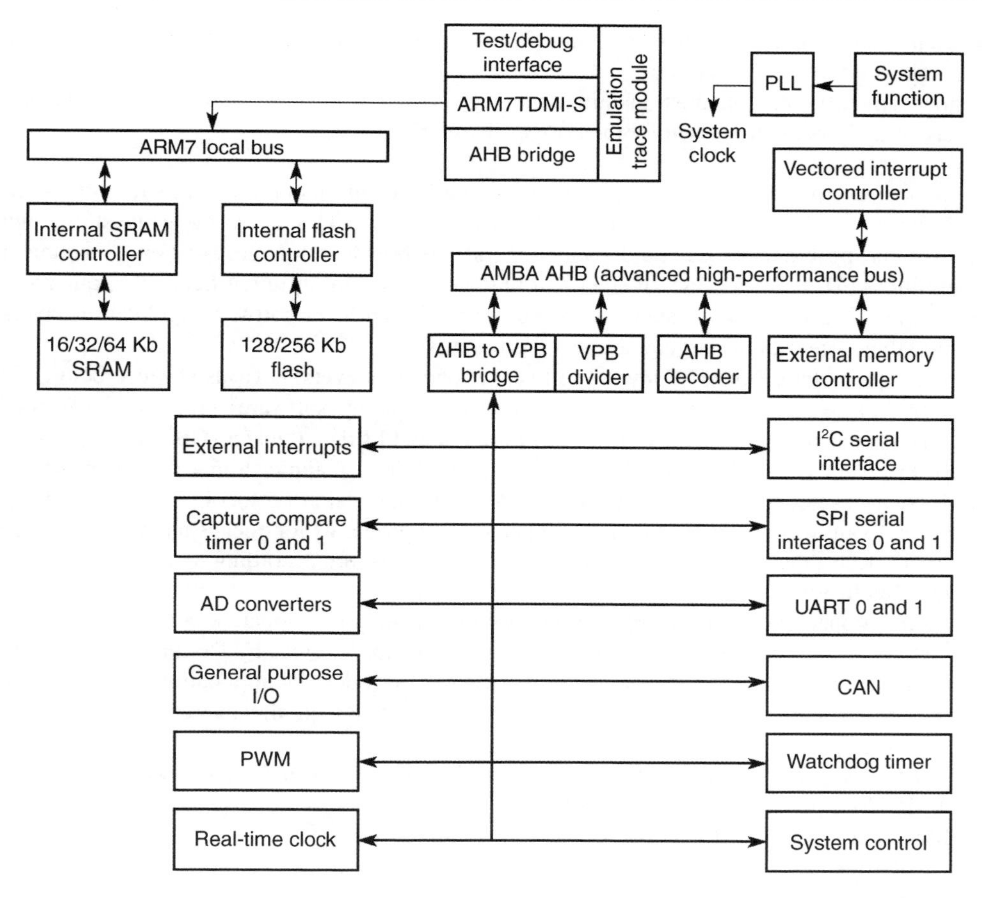


**Figure 5.32**

### *5.3.1 CAN application layers*

At present, there are few official regulations covering the application layers (ISO level 7), which are left to the discretion of each user. Theoretically, therefore, everyone can choose what he wants and do as he likes. The reality is very different: let us start from the beginning.

For example, the CAN communication protocol

- permits the identification of the content of the transmitted message,
- permits the transfer of a maximum of 8 bytes per frame,
- provides a 'correct transmission' acknowledgement,
- offers a high level of security for data transfer.

If we consider just the first two points, many questions immediately arise:

- what should we do if we need to transmit a message containing more than 8 bytes of data (such as a file, for example)?
- how can we be sure that the transmitted message is used in any particular way, or even that it has actually been processed by the correct recipient? – etc.

CAN does not provide any answer in these two cases, even though they are straightforward and common; this is because its rationale lies elsewhere. To answer these questions and resolve the problems raised (and many others), ad hoc working groups (WG), technical committees (TC), forums, etc., made up of many experts from different fields of application, have attempted to draw up answers in the form of high-level software application layers (which therefore require large amounts of software).

These procedures (ISO/OSI layer 7 software application layers) have also been structured in the form of straightforward 'protocols'. These new (purely software) protocols have been designed for transfer by means of the CAN protocol and its data transfer (OSI 2) and physical (OSI 1) layers. Thus, in plain English, 'encapsulations' of application software protocols have been developed and provided in the CAN communication protocol. Since everyone is thinking about his own system, there have been many proposals for encapsulation of software protocols according to taste, or according to colour . . . in short, according to each user and his specific applications.

As mentioned above, these application layers are worthy of a separate book describing all the details of their operation (see the book *Bus CAN – Applications* by Dominique Paret). To summarize, I would point out that, as far as their use for industrial applications is concerned, there are four main ones for CAN, with the following names (in alphabetical order):

- CAL/CANopen CAN applications layers, produced by CAN in Automation (CiA);
- Device Net, produced by Allen Bradley;
- SDS (smart distributed system), produced by Honeywell;
- OSEK (open systems and interfaces for distributed electronics in cars), produced by the European car industries users group;
- plus others, more dedicated to specific applications: CANKingdom (produced by Kvaser), J1939 (produced by SAE, the Society of Automotive Engineers in the USA), MMS, etc.

**Notes:** CAL/CANopen were produced by a large group of industrial companies (CiA, i.e. 'CAN in Automation' – see the appendix to this book) and were designed to cover a wide range of non-proprietary applications, whereas SDS and DeviceNet are initially proprietary developments.

All these layers have a family resemblance, because of their ultimate encapsulation in the CAN transfer frame. In fact, it will nearly always be necessary to sacrifice one or two bytes of the data field to provide other supplementary information in the content of the transferred message (such as the status, the message fragmentation, etc.). This means that the useful data speed is slightly reduced; but it has these advantages:

- it offers new possibilities,
- it decreases the potential high risk of low-quality system software,
- it reduces the high degree of specificity of each development,
- it increases the prospects of re-use and interoperability,

- it creates 'information objects',
- it promotes the development of shared-memory distributed systems,
- it establishes open systems.

### CAL and CANopen

When the CiA group was set up, its members resolved to create an open system based on the CAN communication protocol. This was the origin of the CAN application layers (CAL), and subsequently led to CANopen. Depending on their implementation options, these application layers, written in C language, require 4–30 kB of ROM, a certain amount of RAM, and a certain amount of CPU execution time. The aim of the architecture of these layers is to conceal the details of the basic CAN communication handling from programmers, and to provide them with a first step in implementing shared-variable systems. Figure 5.33 shows their positions in the ISO/OSI reference model; you will notice that layers 3 to 6 are vacant.

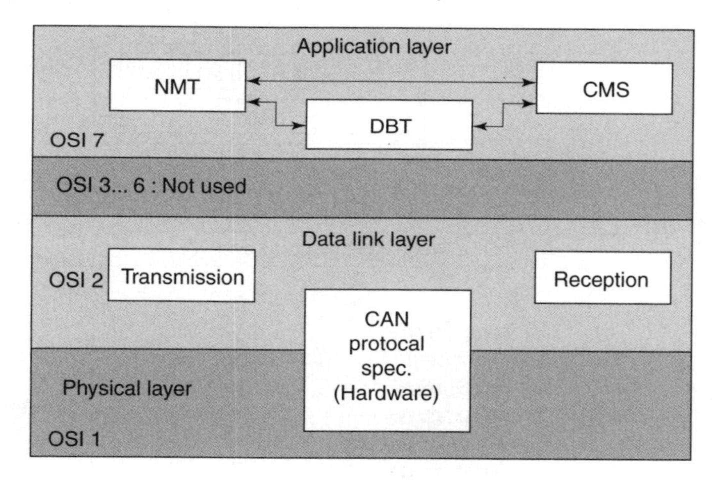


**Figure 5.33**

CAL and CANopen are interfaced upwards with the 'user' applications and downwards with the hardware, in other words the integrated circuits handling the CAN protocol.

Figure 5.34 shows all the 'sub-layers' of CAL and CANopen, listed briefly below.

The lower part of the figure, called the hardware level, contains

- the HSL (hardware specific level), which is the interface available for many types of integrated circuits handling the CAN protocol, including those of the PC platform, all with real-time handling options;
- the CCM (communication chip management), having the function of concealing the communication details of the specific component.

The core of a CAL contains the CMS (CAN-based message specification), the NMT (network management), the DBT (COB-ID distributor) and the LMT (layer management).

CMS is a language describing how the functionality of a module can be provided via its CAN interface. The CMS defines the communication services between the different CAL

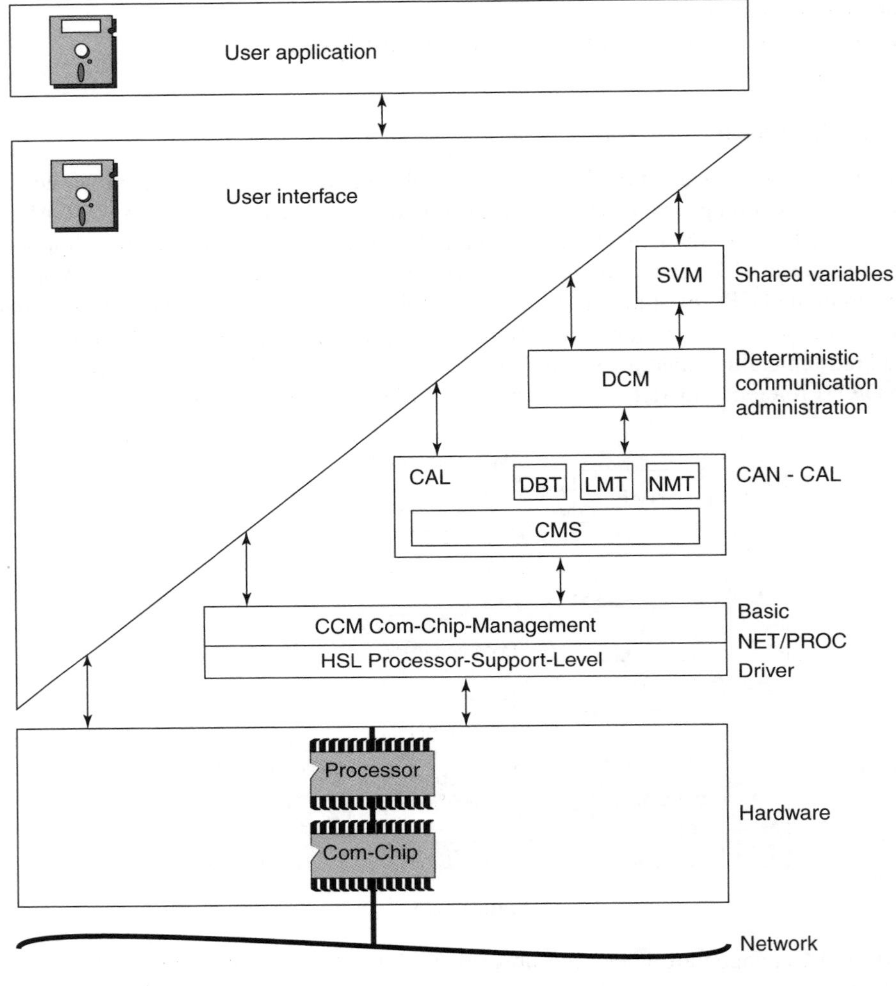


**Figure 5.34**

modules. The elementary data types (bit, integer, group, structure, etc.) of CAN messages are defined in the encoding rules. The CMS object variables can be basic variables of the following types:

- *read only* (used by a client to collect data by means of a request frame);
- *write only* (used by a client to request one or more servers to execute a command);
- *read write* (to collect data or command execution and receive an execution report);
- *domains* (to transmit messages with more than 8 bytes of data);
- *events* (spontaneous events, asynchronous data).

Note that these CMS objects are accompanied by many attributes (not shown here).

The NMT specifies a global service for the control and management of the network. It is based on a master/slave relationship with one NMT master and 255 NMT slaves. It configures and initializes the CAN network and is responsible for the network error management. The NMT services are divided into three groups:

- the module control and monitoring services,
- the error control and monitoring services,
- the configuration control and monitoring services.

The DBT enables identifiers to be distributed dynamically during operation if required. It is based on a relationship with an optional DBT master. It is responsible for the distribution of the COB-IDs (communication object identifiers: the COB-ID determines the priority of the COB) of the COBs used by the CMS service elements.

The LMT defines the services for configuring certain parameters of the CAL layer, such as the individual numbers of node identifiers and the CAN time parameters, over the whole network. It is based on a master/slave relationship with an optional LMT master. It is used to configure the parameters of each layer in the CAN reference model, via the CAN network.

To facilitate the implementation of CAL/CANopen in a system, the software package includes a design aid procedure, the system builder, and a system tester. Using this 'toolbox', we can develop an application quickly without concerning ourselves about the physical part of the network.

The details of these functions are described in CiA documents CiA/DS 201 to 205, and CiA DS 310. These specifications are freely available to any company (no licence required). A large number of companies are now using CALs in all kinds of applications.

### DeviceNet

The DeviceNet application layer was designed and developed by Rockwell's subsidiary Allen Bradley around the CAN protocol, mainly because of its performance in an industrial environment and its low cost. These application layers, with the associated connection systems and physical media, were created to fill the former gap between sensor or actuator buses and general-purpose networks (Figure 5.35).

Here again, these specifications are available to any company on a completely free basis (no licence required). Anyone can obtain them and develop his own application software; this is an open system. Physically, DeviceNet takes the form of a double differential pair: one for data communication under CAN, the other for the power supply to remote equipment (Figure 5.36).

### The DeviceNet application layer

The messaging system is based on the CAN 2.0A frame with its 11-bit identifier field. The DeviceNet protocol is carried in the CAN data field (destination addresses, services to be provided, addresses of data for processing, etc.). The messages may be fragmented if they are too long for a single CAN frame. There is no limit on the message length, and each segment is subject to arbitration. The data exchange can be launched in different ways, by

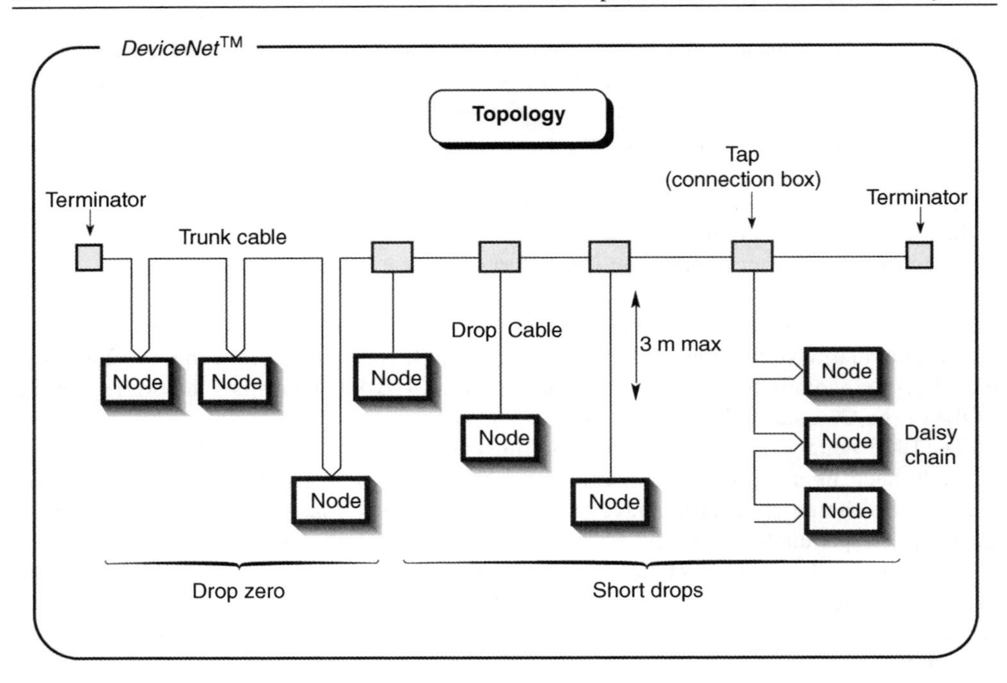


Figure 5.35

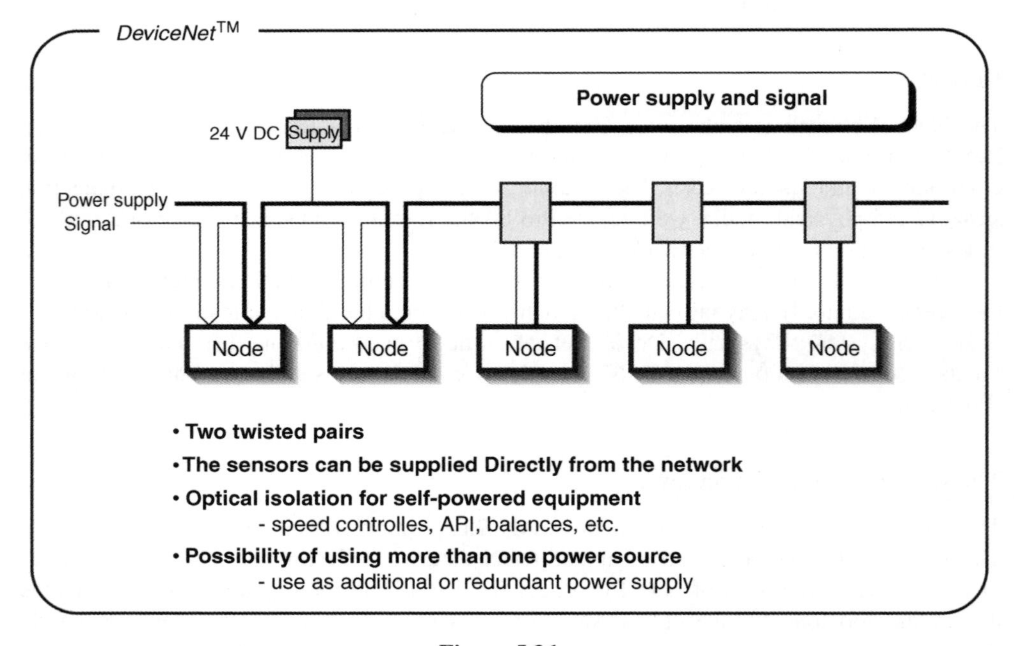


Figure 5.36

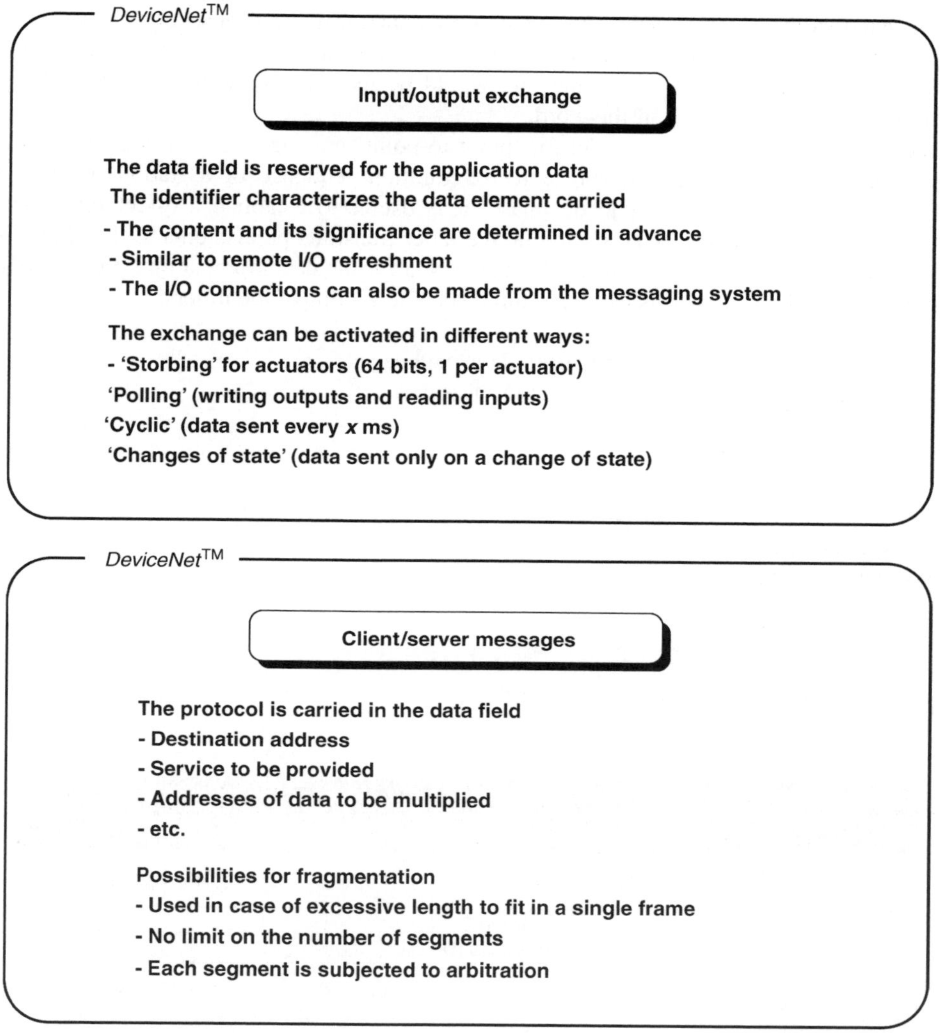

DeviceNet™
Input/output exchange
The data field is reserved for the application data
The identifier characterizes the data element carried
- The content and its significance are determined in advance
- Similar to remote I/O refreshment
- The I/O connections can also be made from the messaging system
The exchange can be activated in different ways:
- 'Storbing' for actuators (64 bits, 1 per actuator)
'Polling' (writing outputs and reading inputs)
'Cyclic' (data sent every x ms)
'Changes of state' (data sent only on a change of state)

DeviceNet™
Client/server messages
The protocol is carried in the data field
- Destination address
- Service to be provided
- Addresses of data to be multiplied
- etc.
Possibilities for fragmentation
- Used in case of excessive length to fit in a single frame
- No limit on the number of segments
- Each segment is subjected to arbitration

**Figure 5.37**

- strobing, for actuators (64 data bits, 1 for each actuator);
- polling, for writing outputs and reading inputs;
- cyclic operation, with data sent every $x$ ms;
- a change of state.

Different operating modes can be used for data distribution (Figure 5.37)

- according to a producer/consumer model for input/output data. In normal operation (high-priority traffic), the data are handled in this mode to optimize the bandwidth of the network (the bandwidth is defined as the number of useful data transmitted per second). A 'producer' produces data and, by using CAN broadcast distribution, a number of 'consumers' can use these data without the loss of time that would be incurred by addressing the different stations in the strict sense of the word.

- according to a client/server model for 'point-to-point' messages. Less frequently (with a low priority), for example during a reconfiguration sequence of a station or piece of equipment, or during a diagnostic phase, it is useful to establish a direct relationship between a 'client' and a 'server', so that the latter transmits precise data to the client that are particular relevant to it. This is also the reason for the development of a point-to-point relationship. This is established using three specific groups of identifiers:
  - group 1: high priority,
  - group 2: medium priority for master/slave exchanges,
  - group 3: low priority, used for sending messages.

The protocol also defines the following:

device address (often called media access control identifier, MAC ID),
class identifier,
instance identifier,
attribute identifier,
service code.

Figure 5.38 defines the fields of the identifiers.

| Identifire bits | | | | | | | | | | | Hex range | Identity usage |
|---|---|---|---|---|---|---|---|---|---|---|---|---|
| 10 | 9 | 8 | 7 | 6 | 5 | 4 | 3 | 2 | 1 | 0 | | |
| 0 | Group 1 message ID | | | | Source MAC ID | | | | | | 000 – 3ff | Message group 1 |
| 1 | 0 | MAC ID | | | | | | Group 2 message ID | | | 400 –5ff | Message group 2 |
| 1 | 1 | Group 3 message ID | | | Source MAC ID | | | | | | 600 –7bf | Message group 3 |
| 1 | 1 | 1 | 1 | 1 | Group 4 message ID (0-2f) | | | | | | 7c0 –7ef | Message group 4 |
| 1 | 1 | 1 | 1 | 1 | 1 | 1 | X | X | X | X | 7f0 –7ff | Invaled CAN identifier |
| 10 | 9 | 8 | 7 | 6 | 5 | 4 | 3 | 2 | 1 | 0 | | |

**Figure 5.38**

*Physical layer*

Because certain wire diameters are used, implying certain line capacities and specific characteristic impedances of the cables, the DeviceNet specification sets out requirements for the speeds, distances and topology of the networks which can carry the data (Figure 5.39).

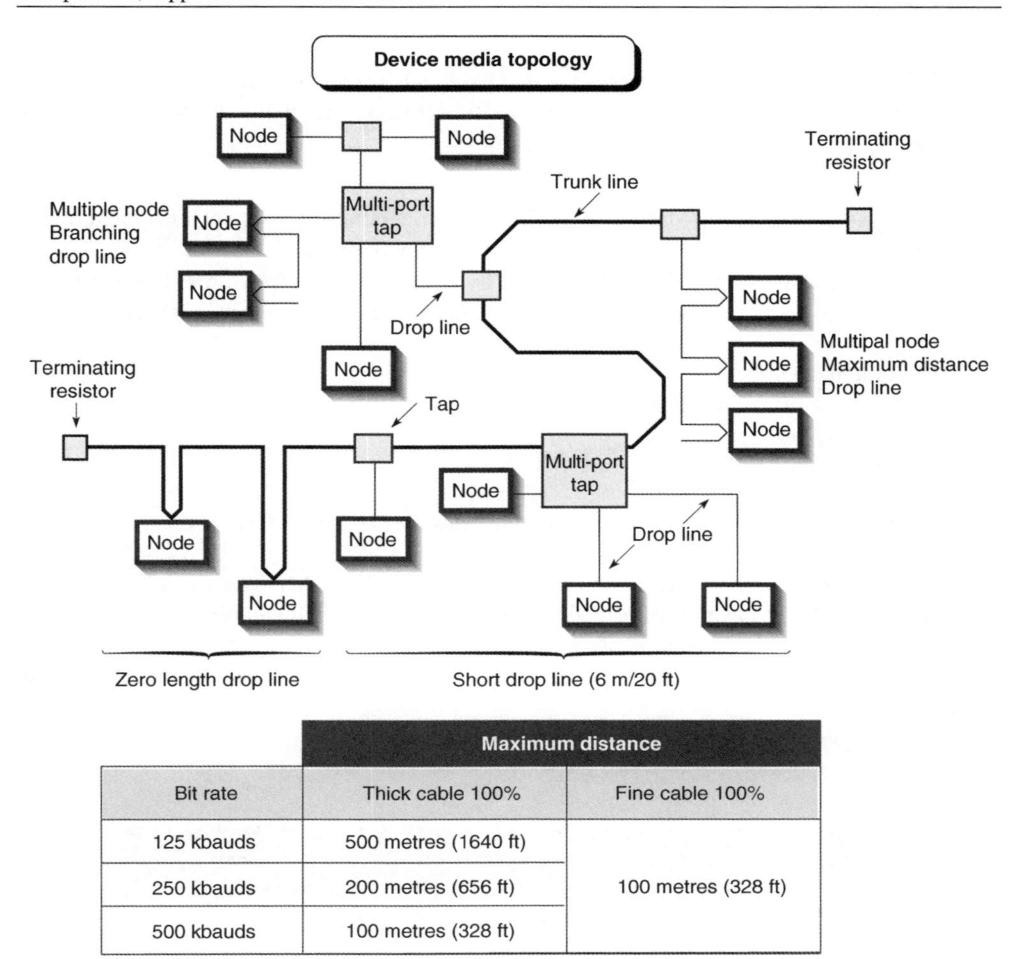


| | Maximum distance | |
|---|---|---|
| Bit rate | Thick cable 100% | Fine cable 100% |
| 125 kbauds | 500 metres (1640 ft) | 100 metres (328 ft) |
| 250 kbauds | 200 metres (656 ft) | |
| 500 kbauds | 100 metres (328 ft) | |

**Figure 5.39**

Many companies working in these fields of application have now joined the group of DeviceNet users, and have established a society (ODVA, Open DeviceNet Vendor Association) for the promotion and technical support of DeviceNet and for the development of its specifications.

## SDS

Proposed by Honeywell, the smart distributed system (SDS) has many similarities in respect of design (because of the use of CAN) with CAL and DeviceNet which have been described above.

*Physical layer*

The SDS connector has two differential twisted pairs: one for CAN communication, the other for the remote supply to the different stations in the network, and finally a fifth wire to fix the potential of the screen provided in the connector.

*Application layer*

The SDS layer, supporting 64 stations, is built around four generic service classes: *read*, *write*, *action* and *event*, with functions similar to those described above.

Like its predecessors, SDS is based on the 10 bytes of the CAN 2.0A frame. Yes, I did write 10, not 8! This is because a 'CAN object' is generally written as follows in terms of bytes:

- CAN frame header:
  - identifier, 11 bits = 8 + 3 bits,
  - RTR bit = 1 bit,
  - data length code = 4 bits;
- data field:
  - data transferred = 64 bits, or, expressed in bytes, 1 + 1 + 8 = 10 bytes.

Described in this way, the encapsulation of the SDS protocol in CAN takes the following form (see Figure 5.40 for the examples of APDU, application protocol data unit, and, for information, Figure 5.41 which gives an idea of the attributes field).

Here again, you should consult the SDS specifications cited in the references for further information.

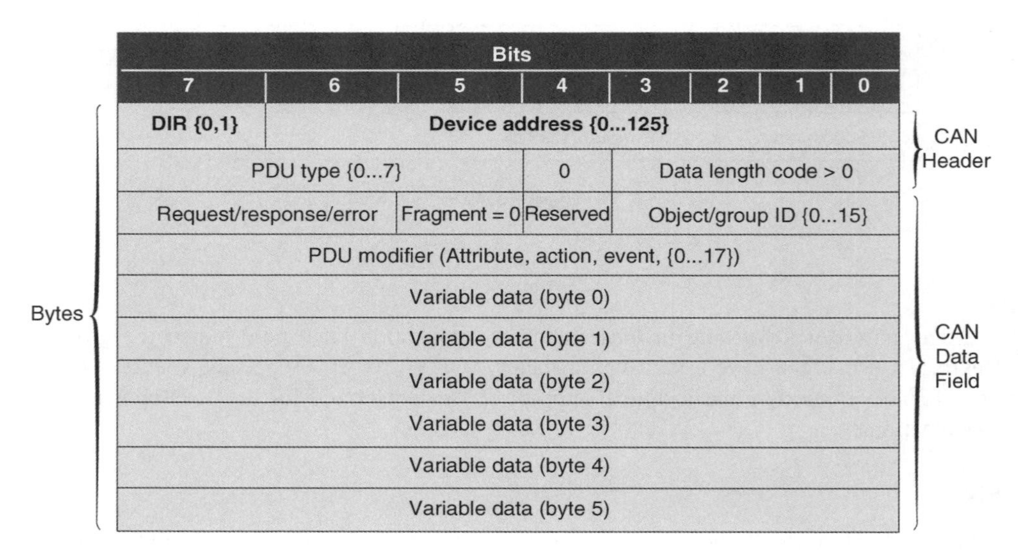


**Figure 5.40**

<table>
<tr><th colspan="11">11-bit identifiers</th><th rowspan="2">Range of values (hex)</th><th rowspan="2"></th></tr>
<tr><th>10</th><th>9</th><th>8</th><th>7</th><th>6</th><th>5</th><th>4</th><th>3</th><th>2</th><th>1</th><th>0</th></tr>
<tr><td>0</td><td colspan="4">Groupe 1 message ID</td><td colspan="6">Source MAC ID</td><td>000–3FF</td><td>Message Groupe 1</td></tr>
<tr><td>1</td><td>0</td><td colspan="6">MAC ID</td><td colspan="3">Groupe 2 message ID</td><td>400–5FF</td><td>Message Groupe 2</td></tr>
<tr><td>1</td><td>1</td><td colspan="3">Groupe 3 message ID</td><td colspan="6">Source MAC ID</td><td>600–7BF</td><td>Message Groupe 3</td></tr>
<tr><td>1</td><td>1</td><td>1</td><td>1</td><td>1</td><td colspan="6">Groupe 4 message ID</td><td>7C0–7EF</td><td>Message Groupe 4</td></tr>
<tr><td>1</td><td>1</td><td>1</td><td>1</td><td>1</td><td>1</td><td>1</td><td>X</td><td>X</td><td>X</td><td>X</td><td>7F0–7FF</td><td>Invalid CAN identifiers</td></tr>
</table>

Increasing priority

**Figure 5.41**

### *5.3.2 Development tools*

In this very short section, I will not deal with all the ramifications of the test procedures, but will quickly run through the details of the development tools for the local networks described in the preceding chapters.

During the development of a protocol-based design, it is sooner or later necessary to test and validate the concept that is to be implemented. This requires the use of high-performance tools, at the following four levels at least:

- The first is an introductory level (optional, depending on the technical level of the end user). It generally consists of demo boards and evaluation or takeover boards (starter kit).
- The second level, designed for the actual development itself, and the test for the codes to be supplied to the microcontroller handling the exchange to ensure the correct operation of the main program and the on-board CAN routines, is intended for professionals. It includes cross-assemblers, linkers, compilers and logic simulators for the software part, and real-time emulators (for microcontrollers) and programmers (for flash microcontrollers) for the hardware part;
- The third level applies to the development and testing of the network itself. In order to distinguish between the terms used below and those described above, we must now know the intrinsic qualities and performance of the network, for example by knowing the number of frames carried during a given time interval, the number of error frames generated, the bus occupation rate, the number of messages having the same identifier, statistics, residual errors, etc. Here we are again concerned with network simulators, network emulators, network analysers and network monitors, but with the previous meanings (hardware and software).
- Finally, a complementary level which can be either a development assistance tool or an end in itself. The typical embodiment of this level is the add-on card for a PC, which enables it to act as a CAN bus generator (or hardware simulator), or to form an integral part of the developed system.

### Demo boards

Every component manufacturer offers demo boards and/or starter kits for his products in his catalogue. These boards generally provide a quick assessment of the main performance levels that can be expected from the components used. They are generally driven via serial or USB connectors from a PC, and have an on-board 'monitor' program which is often rudimentary, but adequate for low-level applications.

These systems are useful for a quick simulation of the operation of layers 1 and 2 of the protocol. Because of their low cost and high performance, they often form useful laboratory resources for technical colleges and universities. For the sake of completeness, I should also point out that many industrial companies, as well as component manufacturers, can provide demo boards and starter kits supporting practically the whole range of existing components.

### Industrial tools

Turning to professional tools, here again we find that many companies have developed implementation and testing resources for all the buses described above. Clearly, their performance level is much higher (as are their prices!), but their functionality and the desired reliability of the networks often make it necessary to use them.

These tools are essential for a knowledge of the performance of the system being developed. For this purpose, the simulators and emulators are used initially to validate the concept in terms of its structure, system, communication, messaging, real-time operation, etc. The network analysers are then used to determine the real extent of problems which may occur during normal operation, with particular regard to the statistics on the frequency of messages exchanged, the bus loads, the error rates, etc., and finally all the types of malfunction which can occur; the aim is to achieve maximum reliability of the network.

These highly specialized tools thus enable a whole application to be scrutinized in great detail, and allow the performance of a specific network to be validated. Theoretically, no application should be able to escape these 'torture instruments' before being released to the market.

### Bus card controllers for PCs

CAN add-on cards for PCs are very useful tools which are needed at different stages of prototype design and development. They are multi-purpose systems. They can be used for

- a quick start of the CAN protocol using a protocol handler circuit;
- familiarization with the CAN protocol, using special-purpose software;
- controlling a network with a PC if you so wish;
- helping you to develop your own autonomous networks subsequently.

These systems generally comprise the PC interface card, CAN PC software, a user's manual and an application manual, with specific DLLs being supplied if necessary.

These boards comply fully with the protocols concerned, and can communicate at the maximum speeds supported by the specific protocol, at 1 Mbit $s^{-1}$, and, depending on the media and speeds used, can operate in bus mode over distances of several kilometres (10 km at 5 bits $s^{-1}$). They may therefore be very useful for the design of your own local networks and/or site buses.

This concludes our quick survey of the start-up and development assistance tools for networks using CAN. Without embarking on an exhaustive list, I will just mention the names of the best-known suppliers in this market (in alphabetical order).

- CiA
- KVASER
- STZP
- VECTOR
- etc.

# 6

# Time-Triggered Protocols – FlexRay

## 6.1 Some General Remarks

By way of introduction to this chapter, I shall discuss a few matters which are common to all network architectures. You should know that this book is now about to change direction by assessing the properties of CAN (controller area network), describing its main limitations and devising new solutions to give new prospects for the decades ahead. First, CAN, created more than 20 years ago, is perfect for today's applications, but inevitably the passage of time reveals some of its limitations. In fact, as I will demonstrate shortly, CAN is very much event triggered (i.e. communications are initiated by events), and it lacks a 'real-time' orientation, or in other words, a 'time-triggered' philosophy. To overcome this, the first response was to create an upper layer called 'TTCAN', triggered by temporal events, to help revive CAN (see below).

The gross bit rate of CAN is limited to 1 Mbit $s^{-1}$ and future applications will be more orientated towards gross bit rates of 5–10 Mbit $s^{-1}$. So everything must be redesigned. The word 'everything' may surprise you, but it is true! Everything will have to be rethought and redesigned because the maximum CAN bit rate of 1 Mbit $s^{-1}$ practically corresponds to the limits of a technical philosophy in which we can still avoid too much mention of line propagation phenomena, reflection coefficients, stubs, Smith charts, etc., but above this rate we cannot consider designing physical layers and protocols without taking these very physical parameters into account.

Finally, it is rather difficult, if not impossible, to create a network architecture/topology on CAN principles which provides redundancy of communications in the physical layer and thus offers the prospect of designing systems controlled entirely by links operating according to the well-known X-by-Wire ('all-by-wire') concept.

Note that the last two problems have been resolved by the advent of the FlexRay concept which I will describe later.

*Multiplexed Networks for Embedded Systems: CAN, LIN, Flexray, Safe-by-Wire...* D. Paret

## 6.2 Event-triggered and Time-triggered Aspects

### *6.2.1 The intrinsically probabilistic aspect of CAN*

By its very nature, the CAN protocol encourages the transmission of communication frames when events occur in a node: it is called an event-triggered system. This is because a participant often sends a message following an action or a request for information, as required by the application.

As explained above, CAN messages are ordered hierarchically by the system designer, using values that he assigns to the identifier values. Theoretically, no node will be aware at any given instant whether its message has been transmitted immediately, because of the conflict management and arbitration based on the values of the identifiers en route. This means that CAN messaging is 'probabilistic' because it is subject to the arbitration procedure, which itself is dependent on the corresponding values of the identifiers of the messages competing to take over the bus. In terms of timing, this makes the transmission and its associated latency time very dependent on the probability of the appearance of the respective values of the identifiers. The only truly 'deterministic' message is the message whose identifier is 0000 hex. On the contrary, the latency for this identifier only is known and is equal to 'one CAN frame, less one bit, plus the interframe time (3 bits)', as this message (of higher priority) lost the priority for accessing the bus in the previous round.

For the other messages (with identifiers other than 0000 (hex)), it is a matter of the probability applied to the respective values of the network activity model and the respective values of the competing message identifiers. Moreover, the chance that this phase of arbitration will take place is exceedingly high because when the bus is occupied – as is often the case – all the nodes that have been unable to access the bus wait for the favourable moment to attempt to take over the bus, and as they all start at the same moment immediately after the interframe phase, they are all immediately subjected to the arbitration procedure.

A problem arises, therefore, if we want to be sure of communicating (transmitting or receiving) at a given instant – in other words, to be temporally deterministic. The principle of CAN means that this cannot be done. It is therefore necessary to create systems in which certain actions are triggered deliberately at precise instants; these are commonly called time-triggered systems, including in our case the three concepts TTCAN, TTP/C and FlexRay, which are described in detail below.

### *6.2.2 The deterministic aspect of applications*

In a large number of applications, it is necessary, or becomes necessary, for some actions to be triggered deliberately at precise instants: these are time-triggered (TT) systems, also known as 'real-time' systems. When systems have to operate in 'real time' (which is theoretically non-existent and is a misnomer – see the note), the major problem arises when we wish to be sure of communicating (transmitting or receiving) at a given instant or in a specific time slot . . . thus making the communication 'deterministic' in nature.

As mentioned previously, the CAN principle is such that this cannot be achieved. In these cases, we must therefore establish, in the upper layers of the OSI (Open Systems Interconnect) model (layer 5, the 'session layer', or layer 7, the 'application' layer), a real-time (time-triggered) mini-operating system, or else integrate this type of operation into the definition of the protocol in order to resolve all or part of this problem.

To do this, we usually establish special, closely defined, time windows for data to be circulated in the network. Theoretically, there are no limitations on the methods of establishing these time windows. The only difficulty is that of ensuring perfect synchronization between all participants, so that each can talk or respond in turn, without encroaching on the time windows of others. This generally makes it necessary to transmit a 'reference clock' cyclically over the whole network, so as to rest the clocks of all participants.

**Note:** To avoid any misunderstanding, the term 'real time' is customarily used to denote known latencies; in other words, we can be sure that something supposed to be done is in fact done at a precise instant. However, the term 'real time' and the concepts of 'deterministic' systems are frequently conflated or confused. The following text will explain how, under certain deterministic conditions, it is possible to make the system operation resemble 'quasi-real-time operation', in other words, with deterministic access to the communication medium and known latencies.

## 6.3 TTCAN – Time-Triggered Communication on CAN

Because of the dominant position of CAN in the market and the increasing complexity of on-board systems, a demand soon arose for protocols providing 'real-time' responses, deterministic solutions and high security. Consequently, systems using global time devices were designed. The first of these, TTCAN (time-triggered CAN), was proposed by the CAN in Automation (CiA) group and the Bosch company in the ISO TC 22/SC 3/WG 1/TF 6[1] at the beginning of 2002 and was to become the ISO 11898-4 standard.

### *6.3.1 TTCAN – ISO 11898-4*

TTCAN is a protocol layer at a higher level than CAN itself, without any modification of the data link layers (DLL) and physical layers (PL) of the latter. TTCAN is located primarily in the session layer of the OSI/ISO (International Standardization Organization) model; in other words, TTCAN is 'encapsulated' in the CAN transfer protocol. The aim of TTCAN is to keep the latency of each message to a specified value, independently of the load on the CAN bus itself. This protocol can be implemented on two levels:

- level 1 is limited to the transfer of cyclic messages;
- level 2 supports what is known as a 'global time' system.

My purpose in this book is not to describe this (very comprehensive) standard in detail – you should consult it for further information – but to outline its content in a few paragraphs.

Note that TTCAN is inserted between CAN and FlexRay, and its use can enable some existing CAN networks to be released (temporarily) or controlled. Its specification corresponds to a session layer (5) of the OSI model (located between the data link layer 2 and the

[1]For those unfamiliar with the language of the ISO, this stands for *Technical Committee 22 (road vehicles), Subcommittee 3 (electrical and electronic equipment), Working Group 1, Task Force 6.*

application layer 7), and it is fitted into the original compressed model of CAN. It will be easier to understand the TTCAN concept if we outline the essentials of the session layer of the OSI model.

### Session layer

The ISO definition says 'the purpose of the session layer is to provide the means necessary for cooperating presentation entities to organize and to synchronize their dialogue and to manage their data exchange. To do this, the session layer provides services to establish a session connection between two presentation entities and to support orderly data exchange interactions.' This layer provides the functions for supporting the dialogue between processes, such as initialization, synchronization and termination of the dialogue, and also makes the constraints and characteristics of the implementations in lower layers transparent to the user.

Because of this, the references to the different systems consist of symbolic names instead of network addresses. It also includes elementary synchronization and recovery services for exchanges.

### Operating principle of TTCAN

TTCAN is based on a deterministic temporal exchange, based on a time window of a predetermined operating cycle, whose global operation can be represented in the matrix of rows and columns in Figure 6.1, which summarizes the general operating principle of this protocol.

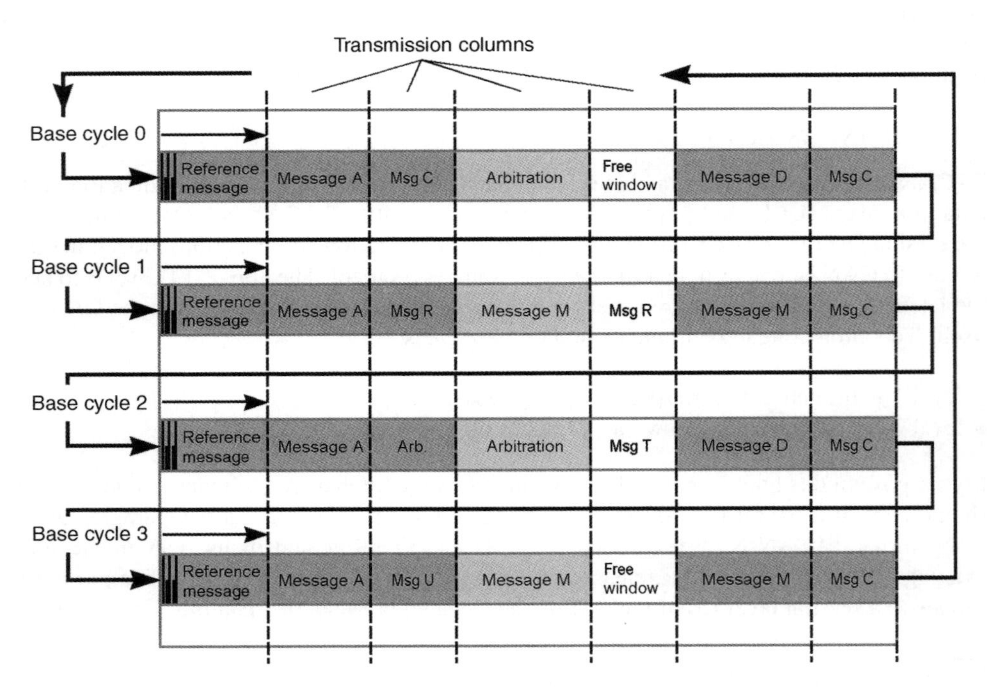


**Figure 6.1**

All the messages travelling in the network between the CAN nodes are organized as elements of an $X \times Y$ matrix. This matrix time system consists of time windows organized in 'basic cycles', identical time values (represented as the totality of each row of the matrix) and numerous time intervals (windows) during which transmissions are authorized (represented by the columns of the matrix). Thus, it defines the relationship between the time windows and the presence of messages in the network.

The TTCAN operating principle is based on the fact that one of the nodes of the network is responsible for organizing the time division and allocation involved. This is because, when the system starts up, one of the nodes allocates the reserved time phase(s) to each of the others. The system thus becomes deterministic, as each node has the right to transmit at a precise moment known to it, for a closely specified period. Clearly, this does not constitute a real-time system at all, but if the complete cycle is executed quickly enough, there is a rapid return to the same node, and to all the participants this appears to be 'quasi-real-time' network access.

To make matters clearer, here is a very exaggerated, but representative, example of the principle followed.

Being the controller of the system, I call myself node 1, and I decide unilaterally to allocate the following time phases to the other four participants in the network. To do this, I start by sending them a minimum information required for the correct operation of the system, by means of a special generic message called a 'reference message', using a special identifier, indicating the following.

The time interval of the basic cycle will be 1 h from now on and until further instructions. This is the time window sequence for each of you:

- node 2, you do not have much to say, so you can talk from 00 to 05;
- I, node 1, am known to be very talkative, so I shall speak from 05 to 20;
- node 3, you will talk from 20 to 25;
- all of you can talk from 25 to 30, subject to CAN arbitration;
- node 2, you do not have much to say, so you can talk again from 30 to 35;
- node 4, you will talk from 35 to 45;
- node 2, you can talk again from 45 to 50;
- node 5, you can talk from 50 to 55;
- from 55 to 00 it is free for all, everyone is welcome to talk – subject to CAN arbitration.

So that you can synchronize your watches, I shall now tell you that it is 10.54 precisely.

I hope that this colourful example would have helped you to understand the basic mechanism of TTCAN. You would certainly have noticed in passing that each basic cycle starts with a reference message and that we have given node 2 several chances to talk because, though it does not say much, it has to report information frequently; theoretically, the choice of time distribution is absolutely free and left to the discretion of the cycle manager. Note that, for obvious reasons of synchronization and possible drift, the reference message must be sent periodically by the 'time master'.

In a more structured way, TTCAN specifies that (Figure 6.2)

- periodic messages are included in exclusive time windows,
- occasional messages are included in arbitrating time windows,
- and free time windows are reserved as free spaces for all traffic.

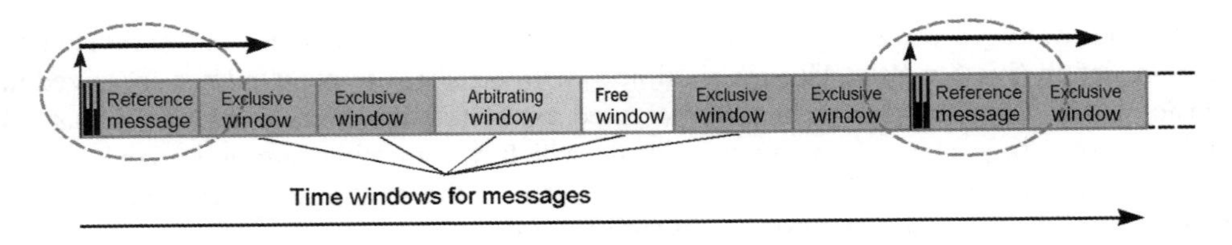


**Figure 6.2**

Of course, the daily reality of applications is quite different, thanks to a multitude of constraints due to the systems used; the time sequence often has to be reconfigured according to external events, whether expected or unexpected.

Here is the simplest possible summary of the principle of the TTCAN concept, in a few paragraphs. For further details on this protocol, here are two useful addresses: see the ISO 11989-4 standard for the factual information or contact CAN in Automation directly (see Appendix D for the address), and they will provide you with a basic application support for using TTCAN.

## 6.4 Towards High-Speed, X-by-Wire and Redundant Systems

What are these strange and outlandish terms? High speed, no problem there. Redundant and redundancy imply a degree of doubling or tripling of functions, message transmissions, physical media, etc. to provide the desired security of operation. But what about 'X-by-Wire'? Let us summarize all this.

### *6.4.1 High speed*

To meet increasing data processing requirements, future systems and concepts must be able to support high communication bit rates, with all that this implies in terms of physical layer performance, transmission quality, internode synchronization, etc.

### *6.4.2 X-by-Wire*

The generic term X-by-Wire conceals all applications using 'systems controlled by wire links', and therefore not having 'any control provided via a mechanical interface'. In fact, this is not new at all. Numerous on-board systems used in the aeronautical industry have been operating according to '...-by-Wire' models for several decades. It is a long time since aircraft flaps and rudders were controlled mechanically with rods, hydraulic actuators and other mechanical systems; these have been replaced by electric motors controlled by wire networks, interconnected and arranged in bus and other topologies ... What is more, it all works – otherwise we would soon know! Surprising though it may seem, a high level of security can be provided for these systems by taking a large number of structural precautions, without any software and/or hardware redundancy.

### *6.4.3 Redundancy*

Clearly, an even higher security of operation of all these systems requires multiple levels of additional redundancy, in terms of communication, the physical layers and combinations of the two, etc. However, you should know that the car industry and other industrial applications are becoming very interested in these technologies – with the start of serial production planned for the years 2008–2010 – for substantially the same reasons as those which have guided their predecessors, with mechanical controls being progressively replaced with control 'by wire'. In the first place, then, we can dispense with steering columns that can impale the driver in case of impact, steering racks (and all their associated fittings), brake master cylinders subject to leakage, suspension springs that can become fatigued (even when stationary!), roll bars to assist in roadholding and accelerator cables that can be trapped or breaked. The benefits are a reduction in weight of on-board equipment, resulting in lower fuel consumption, reduced pollution and improved passive safety provision in case of impact. Moreover, this technology will allow greater flexibility in the overall mechanical design and external and internal styling of vehicles (Figure 6.3), or even, in a single model,

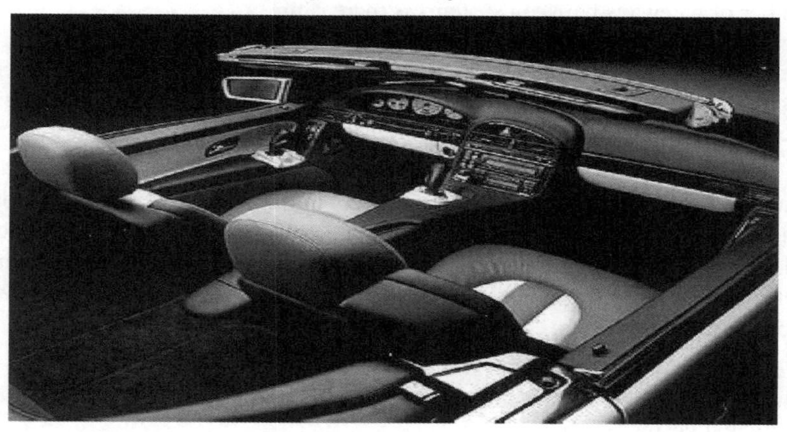

**Figure 6.3**

the option of providing minor adaptations to simplify the provision of right- or left-hand steering, innovative dashboard design, the possibility of eliminating the brake and clutch pedals, etc., not to mention the constant question of reducing costs. In short – this is the future!

Following these amazing predictions of my favourite crystal ball (technologies which will go into volume production in 6–10 years' time), we must come down to earth and consider the numerous problems to be resolved. We will take a look at these now.

### *6.4.4 High-level application requirements*

Let us start from the beginning of the story, examining our future on-board system in terms of the trends in the architecture of the vehicles and on-board systems of tomorrow. Globally, the number of communication systems is increasing. In case you did not know, a top-range car now (in 2005) has 60–70 CPUs (!) and also has 5 or 6 CAN networks (high speed + fault tolerant, low speed) and 6 or 7 LINs (local interconnect networks).
Therefore

- the number of gateways between systems is increasing;
- topologies are becoming increasingly complex;
- there are more and more very high-level interactions between the different systems;
- the electronic architecture must be common to a number of vehicle platforms, so as to provide a high level of synergy and rapid transfer of innovation.

As each manufacturer or equipment maker is becoming increasingly specialized in its own specific fields of competence, the electronic and electrical architecture of a planned system must be designed on a modular basis, and in such a way that it can evolve with a variable scale or geometry (the well-known concept of 'scalability'). This structural elasticity has several implications at different levels:

- different models and brands of modules must operate on different platforms (effect on scalability and costs);
- the interfaces that are created must be completely open (increase in the number of applications);
- the application agreement becomes essential throughout the product design process;
- the complexity of systems is reduced by more carefully specified interactions between applications;
- the architecture of the communication network and the nodes must permit
  - the manufacture of economy vehicles and top-range models on the same platform;
  - the support of communications by simple, double or mixed/combined physical communication channels;
  - clear visibility of the network for each field of application (chassis, security, power systems, environmental detection, etc.);
  - the use of inexpensive components (for example, piezoelectric resonators instead of quartz ones).

Let us briefly review the functional requirements of these new technical and industrial strategies.

### High efficiency

From the outset, and depending on the planned applications, the devices must be optimized in every possible way, up to the physical limits of the principles followed and the components of the physical layer.

### Speed of communication

Because of the larger amount of data to be conveyed, and the increase in the content and quality of the data, the bit rate of the high-speed CAN network (1 Mbit $s^{-1}$) used hitherto is no longer sufficient. The gross bit rate required for these new systems must be of the order of 10 Mbit $s^{-1}$ on a 'single-channel' medium (net bit rate approximately 5 Mbit $s^{-1}$) or on a 'two-channel' structure at a higher rate with the possibility of redundancy.

### Physical layer

The physical medium used for the communication must be capable of being supported by at least two different technologies, for example wire (differential pairs, for example) and optoelectronic (plastic optical fibres, for example), and must allow the network nodes to be put into sleep mode and woken up by means of the medium in question. Additionally, signals travelling along the physical layer must not pollute the radio-frequency band (low emission of radio interference) and must have a high immunity to external signals. Furthermore, containment errors must be managed with the aid of an independent bus monitoring element, called the 'bus guardian'.

### Medium access and control

Experience has shown that the temporal access to the medium is always an important and complicated matter in this kind of concept. In order to meet all the functional and security requirements of the network

- the transmission of 'static' or 'real-time' data must be deterministic, using for example the principle of time slots (in the same way as for those used in TTCAN);
- the transmission of event-triggered 'dynamic' data must also be provided for and ensured;
- there must not be any interference between the aforementioned 'static' and 'dynamic' transmission modes in any circumstances;
- the transmission principle must be totally free of any arbitration system;
- the bandwidth (the bit rate) of the network must be variable, and it must be capable of being allocated dynamically;
- it must be possible to send different and complementary ('differential') data during the same time slot over two physically different communication channels;
- different nodes can use the same time slot in different channels.

## Synchronization method

A reliable synchronization method must be established to ensure perfect synchronization between the operations of the different elements in the network. This requires the provision of a number of elements, particularly

- what is known as a 'global time' device, using distributed synchronization and/or triggered by a time reference ('time triggered');
- synchronization provided by a master ('master synchronization').

The system must also support momentary disappearance of nodes from the network and their reconnection, together with cold and hot starting of the network.

## Network topologies

As explained in more detail below, if we want to improve communication bit rate and reliability, the topological aspect of the network becomes increasingly important. We therefore have to consider new topologies, in addition to the everlasting 'bus' configuration on which thousands of users have depended so far. We will therefore consider the following topologies (Figure 6.4):

- passive bus ... as before!
- passive star networks,
- active star networks and their possible serial connection,
- and finally, a nice mixture of active stars and passive buses.

A fine menu, isn't it?

*System-level requirements*

On the basis of these different topologies and the varying performance levels required from the physical layer as mentioned above, it is also necessary to devise a system which is fault tolerant, with a double transmission channel, capable of detecting its own errors and sending diagnostic messages. We must also seriously consider the establishment of redundancy between the CPUs in each node, to ensure reliability of operation based on a combination of multiple physical redundancy and transmission redundancy. Furthermore, the systems must be standardized (according to ISO if possible), interoperable, re-usable (a fashionable current requirement), open to all without exclusion clauses or payment of royalties, and qualified and certified by recognized test procedures; they must offer a wide range of development tools during the design phase (emulators, simulators, etc.) and during the systems integration phase (monitoring, fault injection, etc.), and of course the components must be available from numerous suppliers, in other words, the usual list of requirements – not to forget the classic refrain of 'more for less'.

This summarizes the functional context of these new networks which, as you will see, are very different from CAN but complementary to it. We shall now look at some proposed systems for meeting all these requirements.

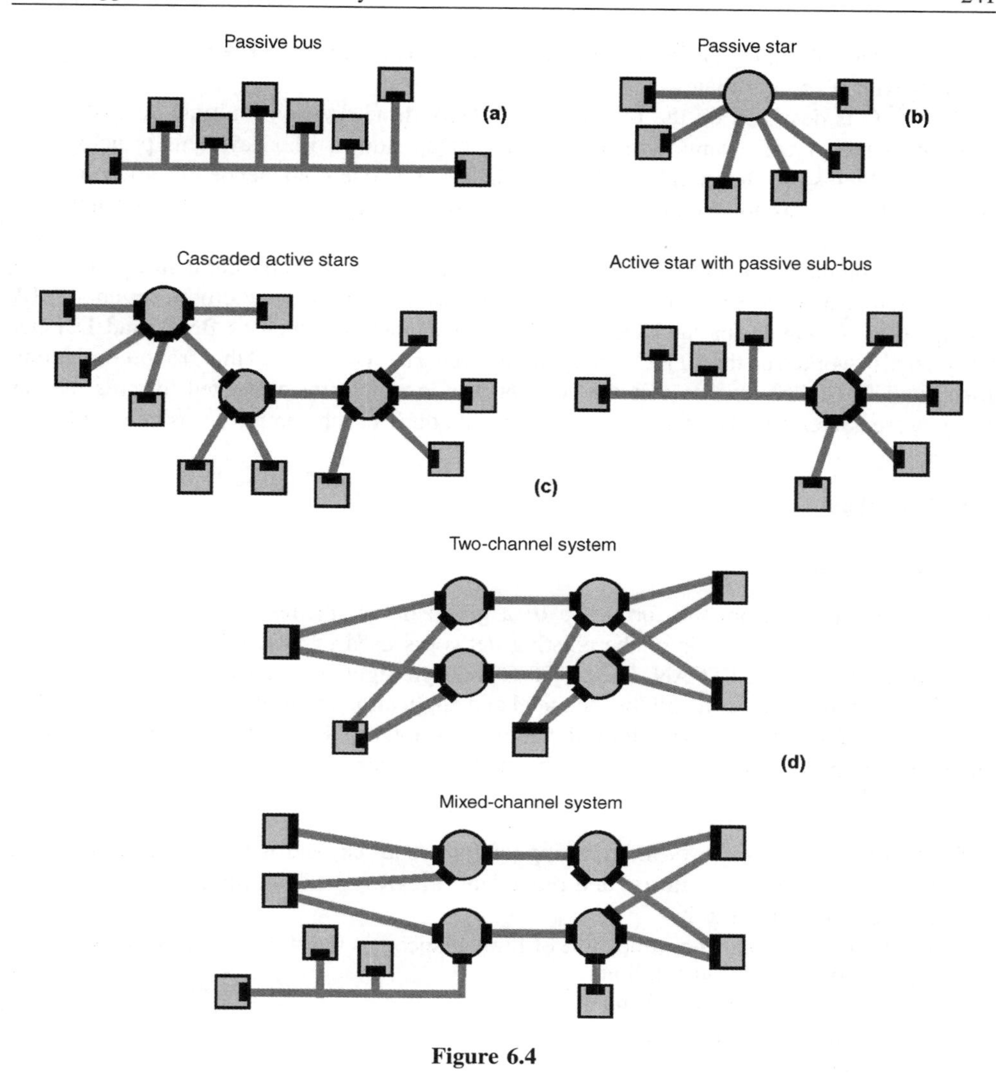


**Figure 6.4**

### *6.4.5 TTP/C – time-triggered protocol*

By way of a reminder, I will mention this system which has been proposed for several years now as the solution to the problem outlined above.

The TTP/C system, a member of the great family of time-triggered protocols (the '/C' indicates that it meets the criteria of class C of the SAE – Society of Automotive Engineers – for the real-time and fault-tolerant aspects of communication in the car industry), was developed by Professor Hermann Kopetz of the Vienna University of Technology,

Austria. It was subsequently adopted by the TTTech[1] – Time-Triggered Technology company, together with several affiliates.

TTP$^{TM}$/C is designed on the principle of TDMA (time division multiple access) to the medium, which I will examine in detail in my discussion of FlexRay. This principle can resolve problems of interoperability between CPUs developed independently of each other. Note that TTP/C was not originally intended for motor vehicle applications, but for industrial applications in general.

At a certain date, around 1998, following several presentations to the car industry, some car manufacturers such as Audi (and Volkswagen who are part of the same group) set up a 'TTA Group' (the 'A' stands for 'architecture'). Other manufacturers such as BMW and Daimler Chrysler also worked on the TTP/C concept for a number of years, but left this group, considering for specific technical reasons that the concept was insufficiently orientated towards the car industry (see Section 6.5) and that collaboration with the TTTech company was rather difficult.

## 6.5 FlexRay

### *6.5.1 The origins*

FlexRay originated from the formation of a group of companies wishing to conduct an exhaustive technical analysis of the existing networks used or available for use in the car industry, namely CAN, TTCAN, TCN, TTP/C and Byteflight, to discover whether any one of them was capable of meeting all the technical and application requirements stated above. The conclusions of this study clearly showed that this was not the case, leading to the development of a new proposal, called 'FlexRay'. The findings concerning the existing solutions are summarized below:

- CAN is not fast enough for the new applications required, and it is difficult to make the transmission truly deterministic and redundant. FlexRay will not replace CAN, but will operate as an addition to it.
- TTCAN inevitably has the same fault of limited speed as CAN. It also fails to provide
  - support for optical transmission;
  - a redundant transmission channel;
  - a fault-tolerant global time;
  - a bus guardian.
- For TTP/C, the content of the data carried by the frame is considered too small. Furthermore, in spite of the use of TDMA for the bus access, there are not many properties in common with FlexRay. Also, TTP/C provides no flexibility, by comparison with FlexRay, regarding
  - the combination of synchronous and asynchronous transmission sections;
  - the multiple transmission slots for a single node in the synchronous section;
  - nodes acting on single, double or mixed channels;

[1]TTTech (Time Triggered Technology Computertechnik GmbH) is a 'house system', developed for the work carried out at the Vienna University of Technology on the TTP/C and trade protocol; it provides a relationship between users and this technology for different applications.

  - a ‘never give up’ strategy in relation to the control of the communication system with respect to the application;
  - the problems of membership status in FlexRay and licence and service rights.
- Byteflight does not offer enough functionality. If FlexRay is used in purely asynchronous mode, it can be configured to be functionally compatible with Byteflight.

### *6.5.2 The FlexRay Consortium*

In 2000, a consortium agreement was signed between the following six companies (‘always the same ones’, somebody is bound point out!):

- the car manufacturers Daimler Chrysler AG, BMW AG and General Motors Corporation,
- the equipment manufacturer, Robert Bosch GmbH,
- the chip makers Motorola GmbH (now Freescale) for the communication controller and Philips GmbH for the components of the physical layer (also involved in the specification of the protocol),
- and, somewhat later, Volkswagen AG, completing the roll call of core partners of the FlexRay Consortium, each being expected to bring its own specific skills to the project. The two classes of Premium Members and Associated Members were defined at the same time.

Figure 6.5 shows the various participants and their levels of operation at a given date. Note that all the major car manufacturers and related companies were involved in the FlexRay Consortium by the end of 2004, including

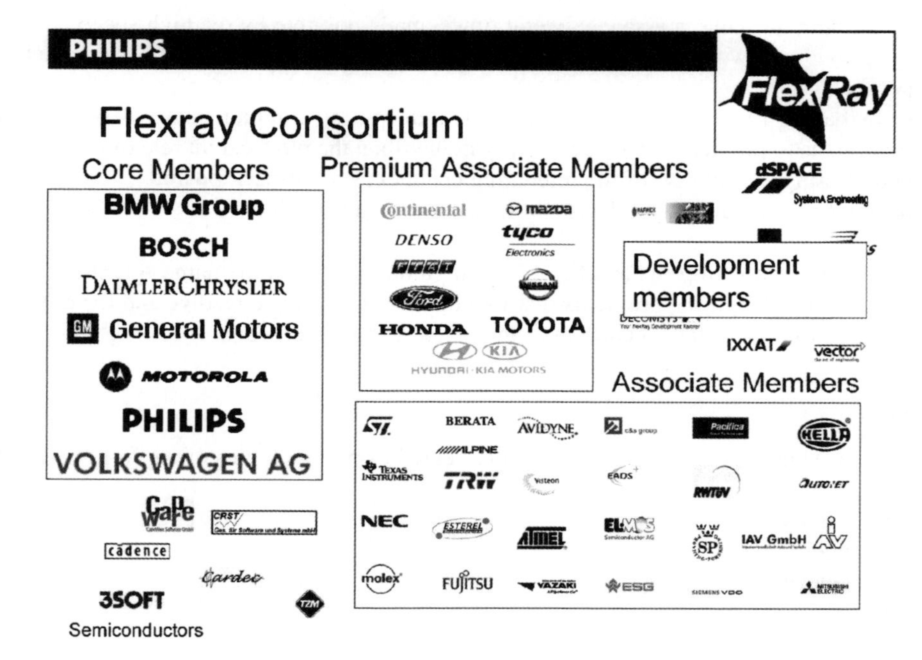


**Figure 6.5**

- the aforementioned core members, BMW, Daimler Chrysler, General Motors and Volkswagen AG;
- the Premium Associate Members (in alphabetical order), Fiat, Ford, Hyundai – Kia Motors, Mazda, Nissan, PSA-Peugeot Citroën, Renault, Toyota, etc. and the equipment makers Continental, Delphi, Denso, Tyco, plus 50 or so Associate Members.

This obviously gives a degree of importance to the chosen system!

Note that the development tools company DeComSys (most of whose staff came from the team that developed TTP/C) is the official administrator of the consortium, which unveiled the FlexRay concept at a public conference at Munich in April 2002 and at the first FlexRay Product Day in September 2004 at Böblingen. About 250 people attended each event.

### *6.5.3 The objective of FlexRay*

Since its origin (so it says), the objective of the FlexRay Consortium has been to create a communication system for controlling and monitoring applications at three different levels:

- at high bit rates, so as to enhance, complement and supplement the applications limited by the bit rate of CAN;
- capable of implementing X-by-Wire solutions;
- developing solutions offering redundancy:
  - by sending the same messages several times, made possible by the high speed,
  - by providing two separate communication channels transmitting the same data in parallel,
  - or by having two communication channels transmitting complementary data in normal time, so as to provide a speed apparently higher than the physical bit rate of the protocol and offering the full use of the remaining channel as a fall-back position;
- capable of serving all future electronic functions in motor vehicles.

Put simply, the aim is to provide a flexible configuration – hence the name 'FlexRay'.

Here is a brief list of the problems that FlexRay must be able to resolve and the contributions it can make, classified according to the major technical aspects.

#### Topology

FlexRay must support communications implemented using topologies of the following type:

- single-channel,
- two-channel,
- bus, and bus with stubs,
- passive and active stars,
- multiple stars with optional sub-buses,
- and all combinations of the above.

### Communications

FlexRay must

- have a large digital data transmission bandwidth (10 Mbit $s^{-1}$);
- transmit data in synchronous and asynchronous modes (scaleable);
- carry deterministic data transmissions in the network, with known and guaranteed message latency and jitter;
- have different simultaneous speeds, using simple bandwidth allocation for the nodes;
- be able to carry out static and dynamic segmentation of data transmission
  - for distribution of requirements;
  - for distribution of areas of functionality;
  - for supporting a configurable number of transmission time slots for each node and for each operating cycle;
- support variable geometry hardware and software redundancy (single-channel, two-channel and combined system);
- support the sleep mode and wake-up of participants via the network;
- support power consumption management for the participants in the network;
- detect and signal errors very quickly;
- withstand synchronization errors of the global time base;
- provide fault-tolerant and time-triggered services by means of hardware devices;
- manage error containment in the physical layer by means of an independent bus guardian;
- be capable of supporting the presence of a decentralized bus guardian in the physical layer (all topologies combined);
- be such that the protocol is independent of the use of a central bus guardian (optional bus guardian, no interference);
- permit the addition of new nodes to an existing system without the need to reconfigure the existing nodes. The configuration data must comply with the need-to-know principle (via a special intrinsic 'knowledge search' device).

### Security requirements

For X-by-Wire systems, FlexRay must

- be able to manage redundant communications;
- provide deterministic network access (synchronous redundancy);
- avoid collisions for bus access;
- provide an option for variable geometry redundancy (single-channel, two-channel and combined system);
- operate by a 'never give up' strategy, thus ensuring that any unavailable communication system automatically disables distributed backup mechanisms;
- be such that the restarting of a node in an operating system implies much more than the simple restarting of its communication;
- maintain communication as long as the communication between other nodes is not compromised.

Maximum security can only be achieved by a combination of the hardware (two-channel architecture, independent bus guardian, fault-tolerant CPUs), the mechanisms incorporated in the protocol and the application support. It is also necessary to ensure

- the robustness of the system against transient faults and external radiation;
- a minimum of radiation external to the systems, EMC (electromagnetic compatibility) protection, etc.

### Application requirements

FlexRay must ensure that the application

- is always fully responsible for any decision in terms of network security or availability;
- always takes the final decision;
- is always in control of the communication system – not the other way round!
- maintains reception as long as possible because a break in communication is a critical decision that must also be taken at the application level;
- is capable of providing different operating modes in respect of communication:
  - normal or continuous operation;
  - operation in dedicated degraded mode:
    - warning: continuous operation, but with notification to the host node;
    - reduced operation with errors: stoppage of transmission and notification;
    - fatal error: operation stopped. All the pins and the bus guardian are switched to fail-safe status;
- is such that the nodes can be configured to survive several cycles without receiving communication frames.

These general parameters and these requests will enable future requirements to be met for three classes of application not yet covered by CAN or by other existing protocols. These are

- class 1 – communication with high bandwidth;
- class 2 – deterministic communication with high bandwidth;
- class 3 – deterministic and redundant communication with high bandwidth.

On a time scale of about 10 years, these ideas will be manifested industrially in new hierarchies of networks used in on-board applications of the FlexRay type in the three classes of applications described above, the use of CAN as a sub-bus of FlexRay and the use of LIN as a sub-bus of CAN.

### *6.5.4 The FlexRay protocol*

On 30 June 2004, three reference documents were posted on the FlexRay web site. These were version 2.0 of the protocol, the physical layer and the bus guardian. The final version, 2.1 (May 2005), includes some additional refinements and minor modifications. One day, I may go so far as to dissect this protocol for your information and examine its multiple applications in detail. But have patience – it is rather too soon at present. However, if you so wish, you can download these documents, but you are warned that they are rather hard to follow – especially if you were not a party to all the minor mysteries discussed during the development of the protocol. Meanwhile, in lieu of a detailed exposition, here is a summary of the main points of these documents.

**Protocol handling**

The original document describing the FlexRay 2.0 protocol contains 224 pages; we would need at least another 450 to detail the way in which this protocol operates. The aim of this brief section is to give an outline of this concept, rather than an exhaustive study. Real fans will have to visit the FlexRay web site!

Before moving on to the content of the FlexRay frame, let me start by stating its operating principle, which is radically different from that of CAN and other protocols. This will take several paragraphs, but is necessary for a full understanding of the subtleties contained in this protocol.

First, in terms of structure, FlexRay was designed to provide a communication system in which collisions for access to the medium are impossible; in other words, the nodes are not subject to arbitration on the transmission channel, and collisions should not occur in normal operation. However, collisions may arise during the starting phase of the protocol on the transmission medium. The physical layer does not provide any means of resolving these collisions, and therefore the application layer must take over to handle such problems.

In order to provide the system with the greatest flexibility of application, it is necessary

(a) to allow operation in 'real time', in other words, to communicate at a precise, known instant for a known maximum time, with the assurance of being the only station present at that moment on the physical communication medium, thus making collisions impossible;
(b) to allow communication at a variable bit rate if required, and thus to require an unspecified communication time.

These two points are totally contradictory. So the circle has to be squared ... and that is just what has been done, without the use of '$\pi$'!

For this purpose, FlexRay follows TTCAN in proposing communication based on communication 'loops', called 'communication cycles', in which access to communication is provided synchronously in time slots clearly specified by the designer (i.e. you, me, etc.) of the system.

The time taken for this cycle is divided into equal time slots, which belong exclusively to the CPUs which enable them to transmit their data. The exclusive allocations (to the CPUs) of these slots are carried out offline, in other words, during the system design phase, thus inherently eliminating all competition for access to the network and other side effects during the normal operating phase which is called 'online'.

Of course, the network designer or manager is responsible for the correct choice of these values. This principle of access to the medium by compliance with predetermined time slots (TDMA, see below) structurally eliminates any possibility of message collision ('collision avoidance').

Once this has been specified, FlexRay combines two paradigms,[1] the first being time controlled and the other controlled by external events. For this purpose, we must create two completely distinct areas inside a 'communication cycle': one is called the 'static segment' and the other is called the 'dynamic segment', these being dedicated strictly to (a) and (b) above, respectively . . . and the job is as good as done (Figure 6.6).

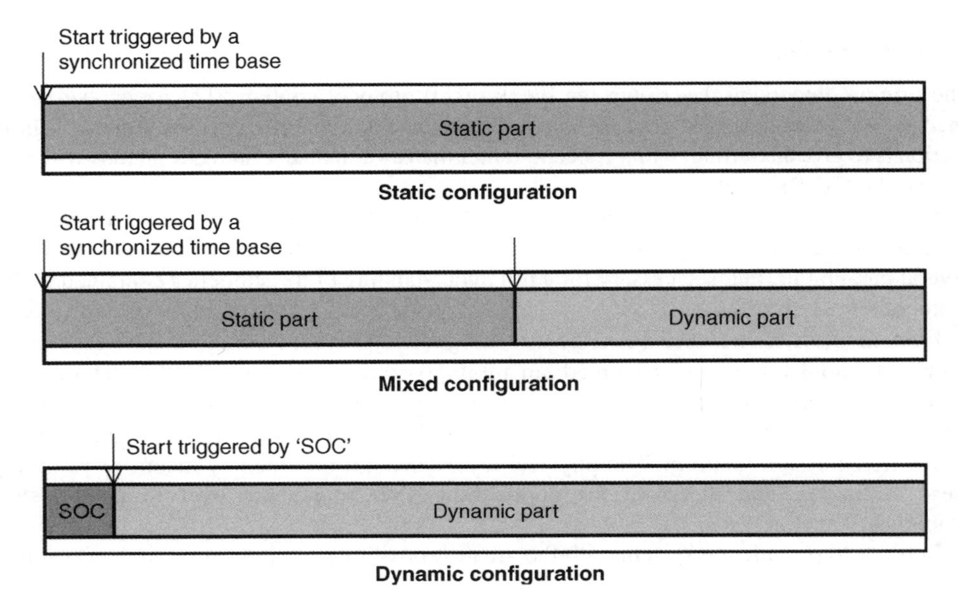


**Figure 6.6**

### Communication cycle

Let us briefly examine the general structure of the FlexRay communication cycle.

It is illustrated in Figure 6.7. Communication in FlexRay takes place with the aid of recurring communication cycles, each comprising a 'static segment', a 'dynamic segment', an 'optional symbol window' and finally a phase in which the network is in idle mode, called the network idle time (NIT). This cycle is initialized by the network manager node.

I will describe all these segments in a few lines.

### Communication frame

Let us now examine the detailed content of the FlexRay communication frame. It is illustrated in Figure 6.8.

[1]Paradigm: a grammatical model for conjugation. Example: the verb 'to be' is conjugated 'I am, you are, he/she is', etc. The term 'paradigm' is often used about FlexRay, because of the number of possible variations of basic principles.

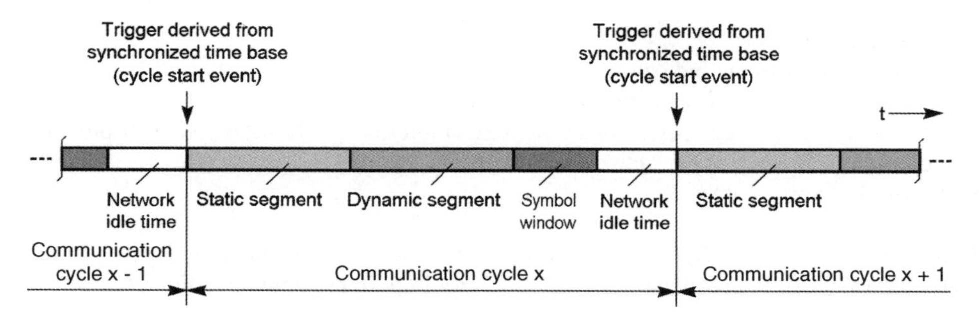


**Figure 6.7**

It comprises three very clearly separated data fields:

- the header;
- the payload, i.e. the field containing the useful load (data);
- the trailer, i.e. the end of frame field.

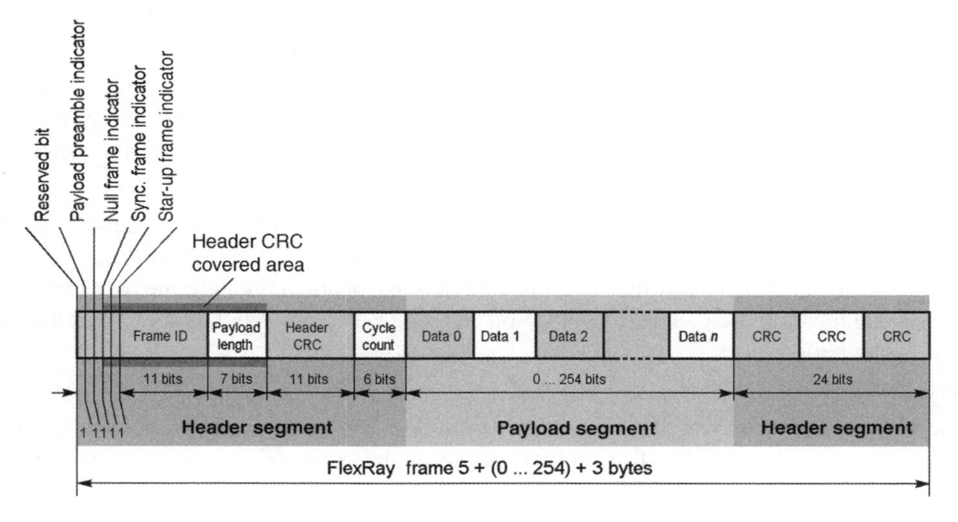


**Figure 6.8**

*Header*

Having started with a sequence called the frame start sequence (FSS), consisting of 8 bits with a value of '0' without a start or stop bit, the header is encoded in 40 bits (5 + 11 + 7 + 11 + 6), making a total of 5 bytes. The meanings of the first 5 bits (in order of appearance) are as follows:

- reserved = 1;
- payload preamble indicator;

- null frame indicator;
- synchronization frame indicator:
  - '0' indicates the transmission of a normal frame;
  - '1' indicates the transmission of a synchronization frame to be used for synchronizing the clocks;
- start-up frame indicator.

*Frame identifier (frame ID).*The value of the frame identifier, encoded in 11 bits (from 1 to 2047, the value '0' being illegal), defines two very important communication parameters:

- first, the position of the time slot in the static segment;
- second, the priority level in the dynamic segment. A low value indicates a high priority.

Additionally, two controllers are prohibited from transmitting frames having the same identifier in the same transmission channel.

*Payload length.* The 7 bits (i.e. a maximum of 128 values) of this field indicate the value (divided by 2 because the word length is 16 bits) of the number of words transmitted in the payload field (maximum 254), i.e. from 0 to 123. Values exceeding 123 are treated as errors.

Note that the payload can include message identifiers (in 0 or 16 bits).

*Header CRC.* The CRC of the header field is encoded in 11 bits to protect the length and the synchronization bit with a Hamming distance of 6. This CRC is configured by the host controller responsible for the communication and of course is checked by the controller receiving it.

*Cycle counter.* This field encoded in 6 bits (giving 64 values) indicates the number of the current communication cycle and acts as a 'continuity index' for the communication. It is automatically incremented by the controller and must be identical for all frames transmitted in one communication cycle. Because of its 6-bit coding, its incrementation is not infinite and has a recurrent periodicity.

*Payload field*

The payload field, carrying the useful data (from 0 to 254 bytes), consists of two principal elements essential for the FlexRay protocol; these are the static and dynamic segments which we will now examine (Figure 6.9).

*Static segment.* As indicated by the left-hand part of this diagram, the portion of the communication cycle during which medium access is controlled and monitored by a static time division multiple access (TDMA) principle is called the 'static segment' in the FlexRay protocol.

The purpose of this segment is to provide deterministic communication, to define the semantics (the sense or meaning) of status messages and to manage the distributed systems and control closed loops. As I will demonstrate later on, this offers major benefits for the design and simulation of distributed functions.

The payload field is initially divided into equal time slots, including some reserved spaces (silences in this case), which are well known and well structured, and therefore start and end at precise instants defined mainly by application groups (clusters). Each time slot is identified by

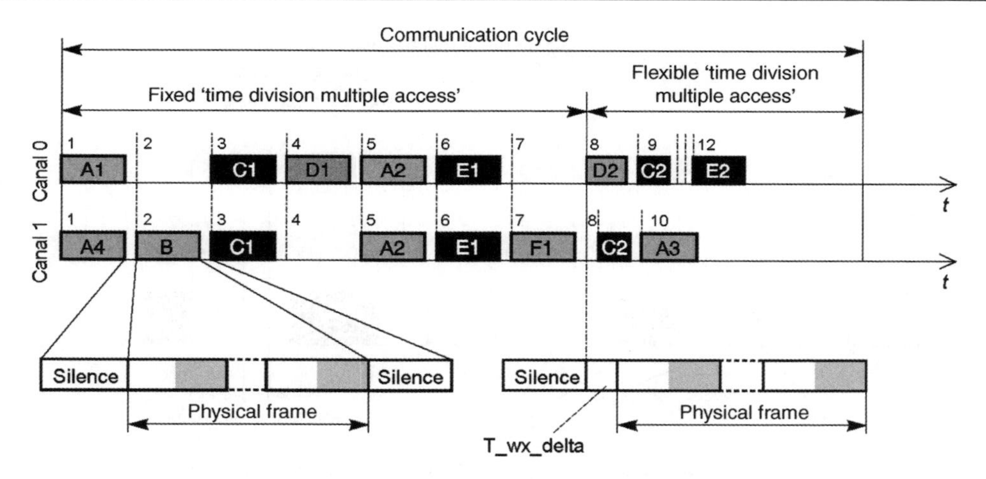


**Figure 6.9**

a unique identifier (ID). All we have to do now is to allocate these time slots by name to certain nodes or tasks, and this structure prevents any encroachment on communications between nodes. Thus, we have

- provided time-distributed medium access (TDMA);
- avoided the creation of collisions;
- designed a real-time system because the latencies are known;
- designed a system with known bandwidth for a given bit time.

*Dynamic segment.* The right-hand part of the same diagram shows the portion of the communication cycle during which medium access is controlled and monitored by 'mini-slotting', also known as flexible time division multiple access (FTDMA). In the FlexRay protocol, this is called the 'dynamic segment'. In this segment, medium access is allowed dynamically according to the priority level assigned to the nodes having data to transmit, via the value of the frame ID contained in the message header. The purpose of this segment is to provide communication within limits of time and bandwidth.

This segment will also be initially divided – presumably by the system designer – into equal mini time slots, which start and end at precise instants (see Figure 6.10 for the details of this mini-slotting). Each of the nodes is then (subject to the control of the system designer) free to vary the duration of the slots according to the quantity of data that it wishes to transmit, to suit the requirements to be met, at a certain level with arbitration of access to the medium for

- providing easy access to the time-distributed medium (flexible time-distributed medium access, FTDMA) and a variable-bandwidth system,
- enabling arbitration to be carried out in the network,
- transferring spontaneous messages,
- providing burst transmission,
- easily managing diagnostic data,
- and, in general, transferring all kinds of messages in an ad hoc way.

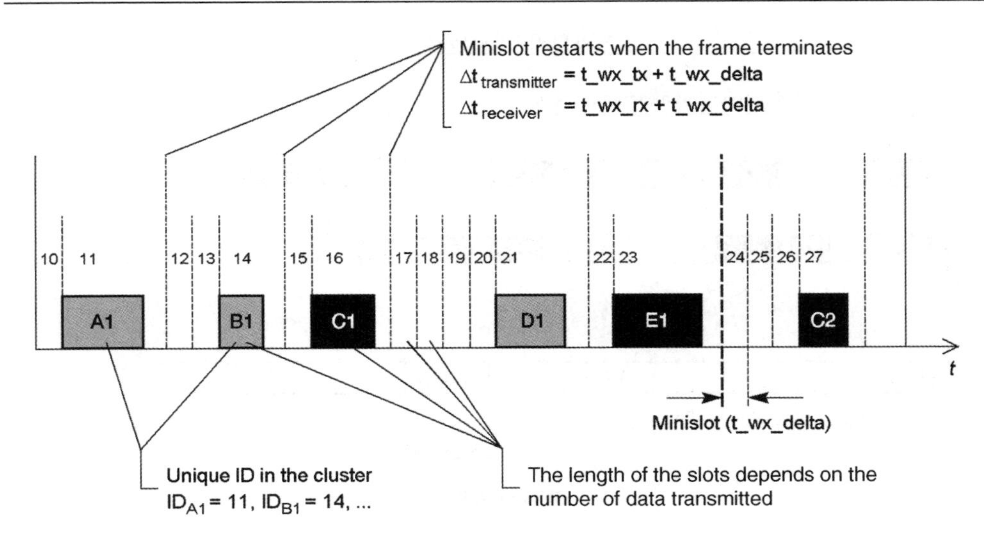


**Figure 6.10**

*End of frame field*
Finally, the frame concludes with an end of frame field containing a single 24-bit CRC field. This protects the integrity of the frame with a Hamming distance of 6.

### Access to the medium

For a better understanding of all the technical, industrial and economic features of the possible applications of FlexRay, we must return to the methods used to access the medium, which form one of the key points of this protocol.

To begin with, each designer has to define the tasks of each node in the network (point-to-point links, centralized tasks, etc. and, most commonly, distributed tasks) and must carefully define the tasks which are to access the network in a deterministic or other way. Thus, the designer can use, in conceptual terms:

- on the one hand, two possibilities depending on the form of network access:
  - either via the static segment, for 'quasi-real-time' tasks enabling multiple time slots to be used for each node if necessary during the same communication cycle,
  - or via the dynamic segment, for asynchronous event-triggered ('spontaneous') tasks, subject to arbitration and with a bandwidth that can be adapted dynamically during operation according to the operating requirements of the system;
- on the other hand, in a communication cycle, the relative division of one segment with respect to another, bearing in mind that, theoretically, there will be no encroachment or interference between the above two segments, that totally different data can be sent in the two channels during the same time slot and that different nodes can use the same time slots in different channels.

Figure 6.11 will help us to explain the operation and possibilities of this very special configuration.

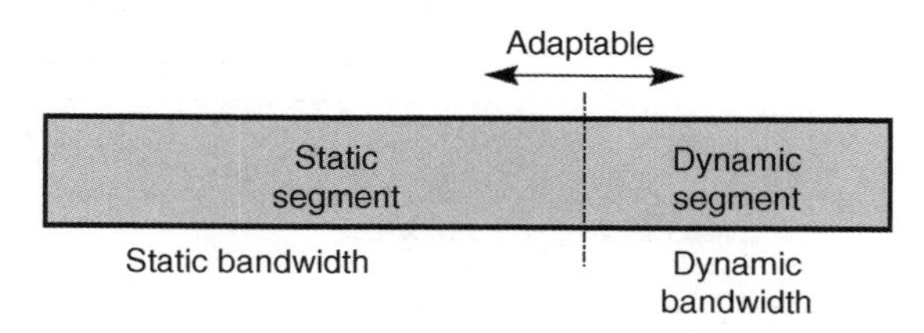


**Figure 6.11**

This diagram shows the set of two segments, static and dynamic, of the communication frame, with a separation chosen by the designer.

This choice is based on the desired allocation of bandwidth (in terms of Mbit $s^{-1}$), according to the operating requirements of the planned system, the 246 data bytes of the payload being divided into two segments of adequate size.

When this has been done, each participant in the network requiring 'quasi-real-time' access must be assigned its own time slot in the static segment. This is shown in Figure 6.12, in which we can see, for example, that application A (front right-hand brake) can communicate during the first time slot, application B (suspension) can communicate during the second time slot, application E (clutch control) can communicate during the fifth time slot and so on.

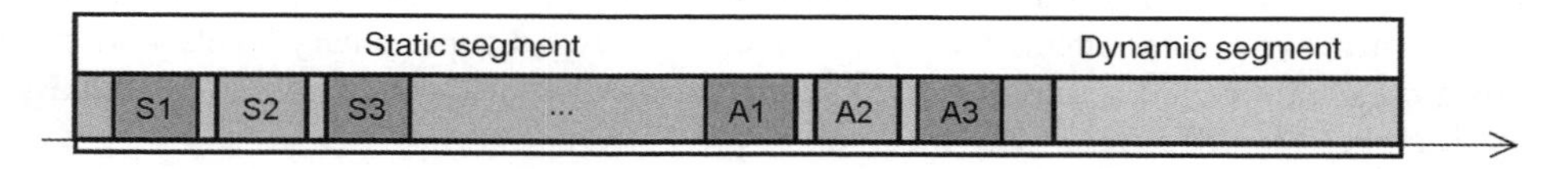


**Figure 6.12**

All quite simple up to this point! But there is a lot hidden behind the technical description and interpretation of this simple little diagram. Let us take a closer look.

In fact, because of the principle of allocation of time slots, and as we can be sure that there will be no overlapping of time slots, the system integrator (a car manufacturer or industrial consortium, for example) is quite free – for reasons of skill, costs, testability, etc. – to choose, for example, the same equipment maker/supplier 'X' for applications A, E, H, etc. and a second company 'Y' for the other applications, as indicated by the expanded view in Figure 6.13.

Leaving aside the notion of 'divide and rule' which some malicious observers may mention at this point, this enables the system manager to integrate all the modules much more quickly than with CAN, as all the data for the applications are strictly separated, without any concept of arbitration or random latency of tasks in respect of time. Testing time and costs are thus greatly reduced, and the flexibility of development is considerably increased, as it is easy to produce variants; the competitive choice of industrial partners for developing the system is also increased. This gives FlexRay an undeniable advantage over other protocols.

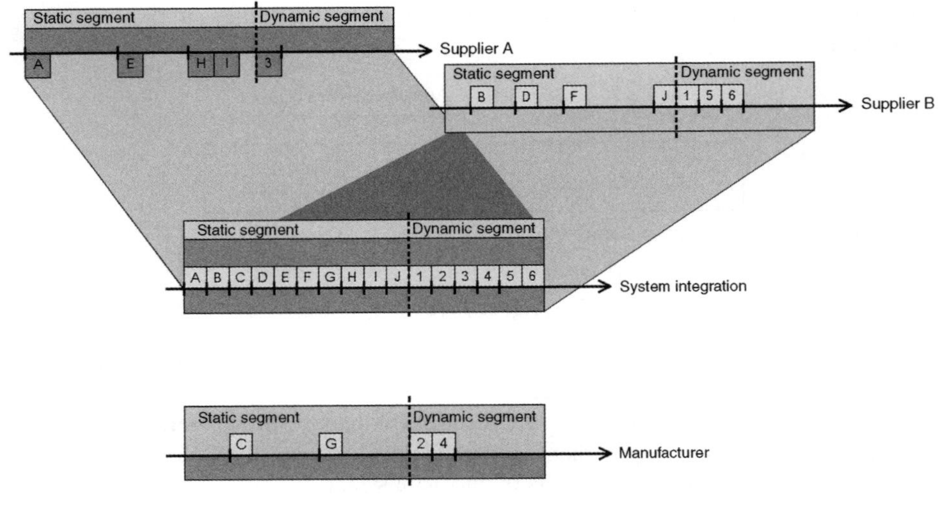


**Figure 6.13**

## The FlexRay physical layer

Now let us examine the physical layer.

Rather as in the case of CAN, the physical layer medium described in the official FlexRay specifications is not explicitly defined, so there are various possibilities, including differential pair wire media or optical fibres, if we limit ourselves to two distinct binary levels whose use forms a bit stream called a communication element. In view of this, version 2.0 of FlexRay fully describes the differential wire medium.

A very distinctive feature of FlexRay with respect to CAN is that a node must be capable of supporting up to two completely independent channels of the physical layer, defined as 'channel A' and 'channel B'. This is specified for two purposes: first, as will be demonstrated below, so that communication can take place at higher speeds while the network is operating well, and, second, so that functional redundancy of data transmission can be provided in the case of an incident occurring on one of the two transmission channels, thus enhancing the fault-tolerant aspect of the system.

## Bit coding

The bit coding used is 'non-return to zero' (NRZ 8N1); in other words, the value of the physical signal does not change during the whole of the bit time. The transmission of each byte (of 8 bits) is framed by a start bit and a stop bit.

## Physical representation of the bits

The FlexRay 2.0 specification declares that the electrical values of the binary elements 1 and 0 are both represented by dominant states. Curious! In fact, the electrical signal is transmitted in 'differential mode' as in CAN, but, as shown by Figure 6.14, it is more like the low-speed

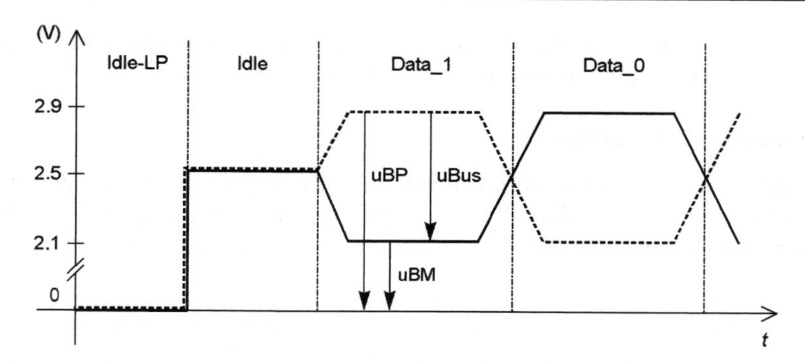


**Figure 6.14**

CAN signal than the high-speed CAN version, because the differential electrical levels alternate (change their sign) to represent 1 and 0. Note that the recessive level is reserved exclusively for the idle mode of the bus and that there is also a fourth level (two dominant, one recessive, plus this additional level) for the 'low power down' mode.

The differential voltage is of the order of 700 mV, and thus the signals present in the network only make a very small contribution to the electromagnetic radiation.

### Bit rate

FlexRay specifies that the nominal value of the bit time is exclusively 100 ns, implying a gross bit rate of 10 Mbit $s^{-1}$, and therefore the delay times, the signal propagation times and the network topology will all have major effects on the quality and integrity of the bits carried.

### Propagation time

Regardless of the topology used to create the network (direct line, active stars, repeaters, etc.), the maximum propagation time must not exceed 2500 ns.

As these times are highly dependent on the media used, the FlexRay specification clearly states that the use of 'cable' wire links (twisted pairs, with or without electrical screens) depends on their performance and sets characteristic impedances in the range from 80 to 110 Ω, for which the maximum propagation time is 10 ns $m^{-1}$ and the maximum attenuation is 82 dB $km^{-1}$. Furthermore, if two transmission channels (A and B) are used, it is strongly recommended that the differences should be minimized and that the delays due to signal propagation in the two channels should be balanced as far as possible.

### Notes

1. Do not be too hasty to conclude from the above lines that the point-to-point length of a network could be 2500 ns/10 ns $m^{-1}$ = 250 m, but bear in mind that the total propagation time must not exceed 2500 ns.
2. The use of active stars (see the discussion later in this chapter) can be considered as equal in terms of time to the use of a repeater, which obviously has a time requirement that is added to the pure signal propagation time. This time is generally estimated to be approximately 200 ns.

In fact, the topologies used often include two active stars per network, which immediately removes 400–500 ns from the total propagation time available for the FlexRay protocol.

## Single-channel system topology

The very high speed means that the possible topology or topologies of the network must be very carefully considered. Indeed, the propagation time, the rise and decay times, the radiation, the data redundancy, etc. are the main factors to be borne in mind.

*Point-to-point link*

Let us start with the simplest link, which is the 'point-to-point' link between two nodes.

Given the parameters mentioned above, the FlexRay specification indicates that the maximum length is 24 m. In the case of a wire link created with a differential pair, the 'point-to-point' link can be designed with bidirectional line drivers as shown in Figure 6.15, in which a termination resistor (line impedance matching, at the cost of standing waves and degradation of the signal's shape and received power) must be provided, in the same way as for CAN.

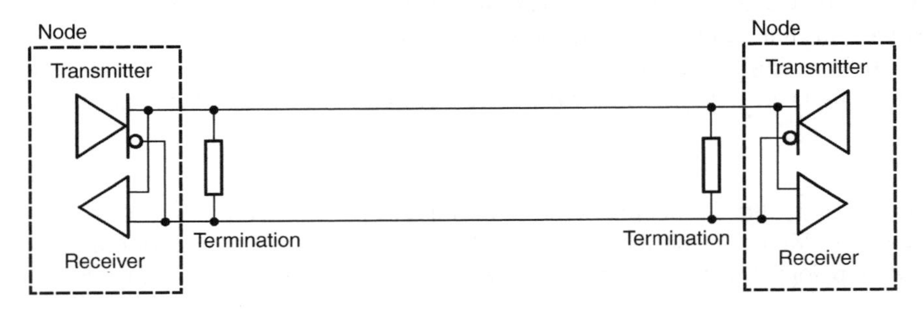


**Figure 6.15**

*Link using a passive linear bus*

This is the link topology which immediately comes to mind when we consider how to connect different nodes. Unfortunately, in terms of distance, the nodes are not all directly connected to the bus, but are often linked to it by means of longer or shorter 'feed lines', as shown in Figure 6.16. Because of the high frequency associated with the speed of FlexRay,

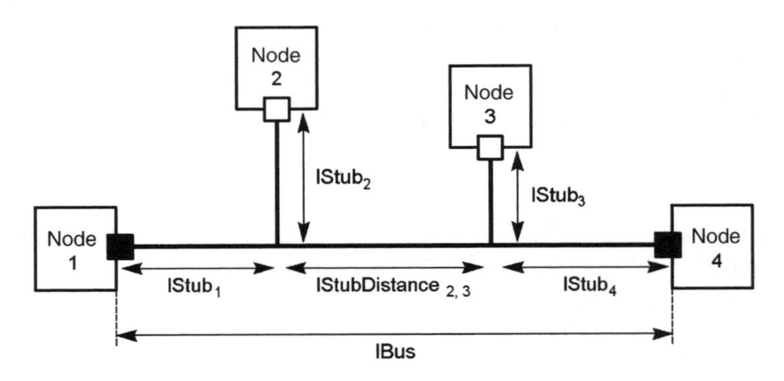


**Figure 6.16**

these feed lines act as branches of a communication line and form what are known as 'stubs', according to their relative lengths, in the same way as those found in line designs for UHF and SHF. This may cause the appearance of antinodes and wave clusters on the bus, which locally cancel out or amplify the voltage present at the line feed connection points.

In order to overcome these problems, and in view of its own specifications, FlexRay sets certain precise values which must not be exceeded:

- maximum distance between two nodes in a system = 24 m;
- distance between two splices in the network = 150 mm;
- maximum number of nodes with stubs = 22.

*A star is born! The star link*

*Passive star.* To avoid the problems mentioned in the previous section concerning stubs, their length and their relative positions in the network, we can simply establish what is known as a passive star topology, in other words, a star structure in which the common centre point can be compared with a 'large blob of solder' (Figure 6.17).

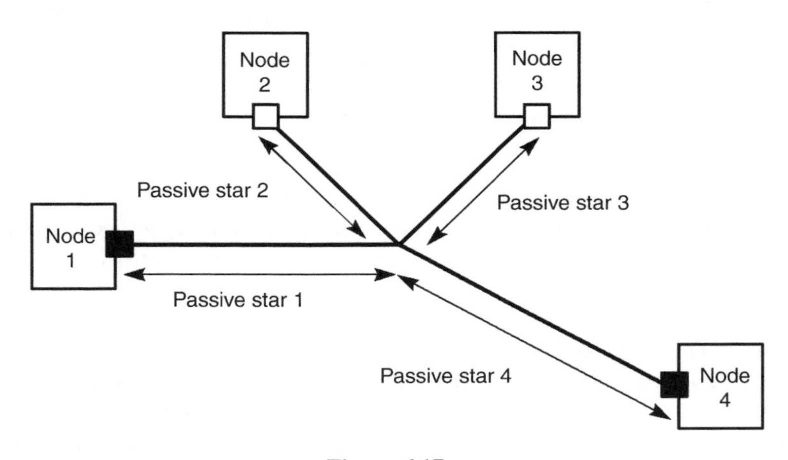


**Figure 6.17**

In this case, all the nodes have practically the same function, the only difference being found in the distance (their distance from this common point). Theoretically, if all the nodes are at the same distance from the centre point, the propagation times will be identical from node to node, but in view of the fact that this is never the case in applications, due to a host of practical and mechanical factors, certain values are also specified:

- maximum distance between two nodes in a system = 24 m;
- maximum number of nodes possible in a star system = 22.

Note that only the two nodes located at the farthest ends of the network have line terminations in order to secure the impedances.

*Active star.* In the previous section, the star was passive because it was totally incapable of interpreting any information. In the present case (Figure 6.18), the star includes or can include on-board 'intelligent' electronic systems, which can be used for many functions. Examples are the switching of a message towards the correct node or the disconnection of one or more of these branches if a link malfunctions, etc. Clearly, it can also strengthen a weakened signal (by acting as a repeater). As this star is electronically active, it must also be considered as a true line termination, and therefore has line matching at each of its ports.

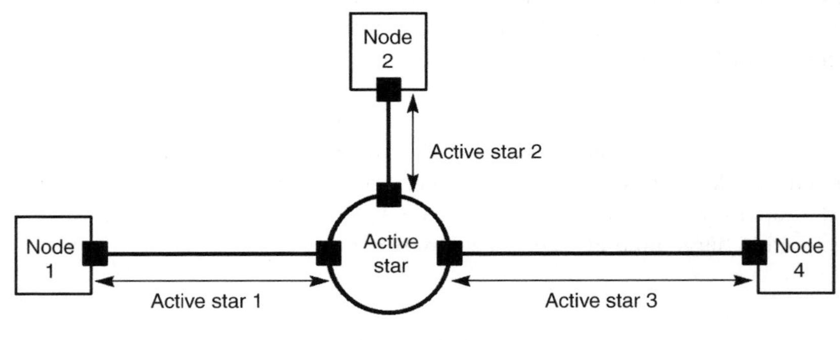


**Figure 6.18**

This method also has its limits in the range of applications concerned. Certain mandatory values are also specified:

- length of a branch from a node to an active star = 24 m;
- number of branches of a star = 2 at least (the maximum is not specified), thus permitting many different network architectures.

*Cascaded active stars*

We are now entering an area of topological complexity, but in fact these are the most commonly encountered types. This is because, in a motor vehicle, it is never easy to position CPUs just where we would like, and the elements to be commanded and controlled are often placed in strictly determined locations. So, to avoid the use of stubs, while having clear signals and known propagation times, etc., network architectures are often built around solutions based on cascades (or even torrents!) of active stars. This produces a real fireworks display! As a picture says more than a thousand words, this topology is outlined in Figure 6.19. Each entity in the network is linked by a single bus between active stars.

This method also has its limits. Again, there are some very closely specified values:

- maximum number of active stars on the signal path between any node and any other node of the network = 2;
- (electrical) distance between two active stars = 24 m.

This opens up new prospects for many possible network architectures.

*Hybrid solution*

The cherry on the cake! And what is the recipe for the cake? Take all the systems described above, mix them carefully over a low light for a few minutes, and you will have the system

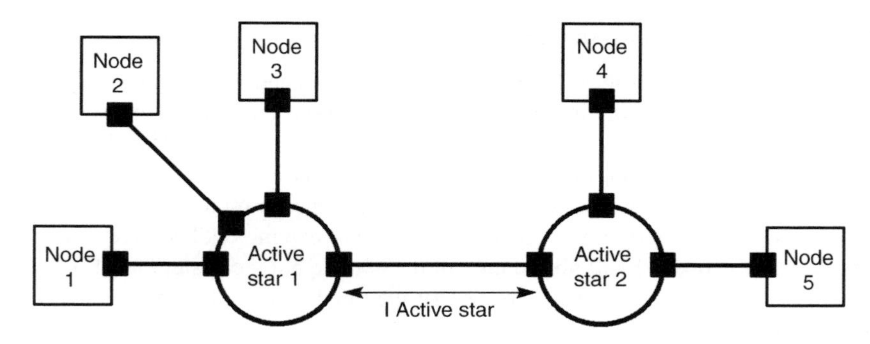


**Figure 6.19**

shown in Figure 6.20, as long as you have taken care to comply with all the constraints stated in the previous paragraphs.

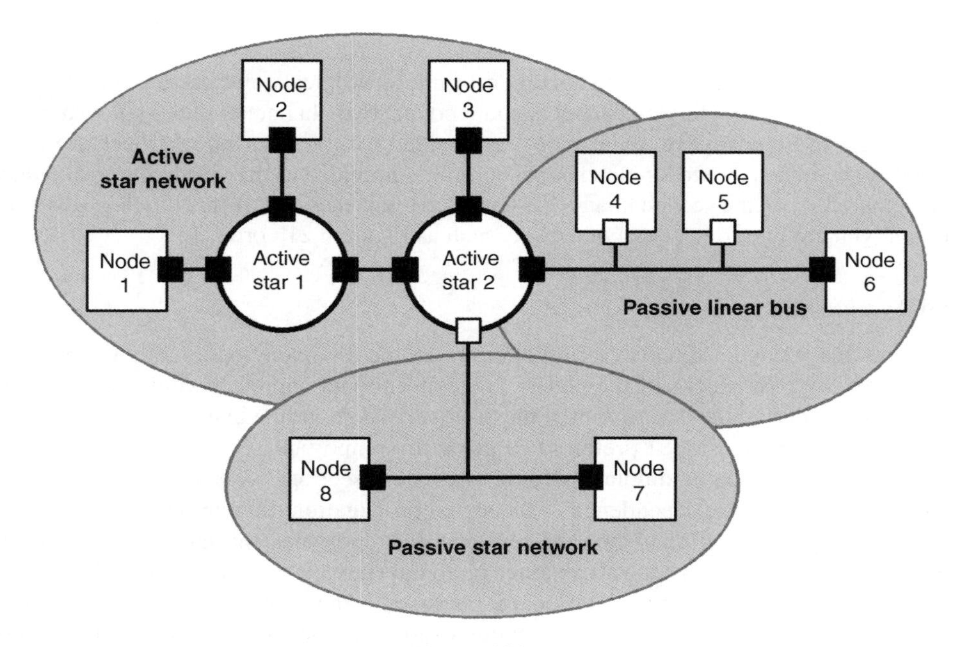


**Figure 6.20**

I have summed up in these few lines the complexity of high-speed networks, of which FlexRay is a worthy representative, particularly for motor vehicle applications.

It is true that the fundamental problems to be resolved – speeds of 10 $\mathrm{Mbit\,s^{-1}}$, a vast number of possible topologies, the management of worst possible cases – will make us think twice (or even three or four times) before finalizing the network architecture of a new vehicle or on-board system. In view of the safety aspects involved, we must not create any dead ends! My professional colleagues, who design all kinds of multiplexed networks, should know that I have great sympathy with their present and future troubles.

### Topology of a two-channel redundant system

After this introduction dealing with non-redundant high-speed architectures using a single communication channel, let me interpose a few words about redundancy in systems.

Wire control of brakes (brake by wire) is a typical example of a system in which a degree of operating redundancy is preferable. It dispenses with master cylinders subject to leakage, all kinds of pipes and special fluids, etc. Instead, we have electric motors (and their associated electronics) mounted directly on the brake callipers and driving endless screws to press the brake pads on to the discs. I will not go into all the details, but the system must operate in all circumstances. As it is better to be safe than sorry, we can double the communication networks with the aim of

- enabling two non-redundant channels to be used in normal operation to transmit complementary data at an even higher rate; for example, if there is a net bit rate of 7.5 Mbit $s^{-1}$, the global bit rate will be 15 Mbit $s^{-1}$;
- supporting concurrent transmission of redundant data, in case one of the two transmission channels fails.

Some conventional examples of combination of the topological solutions described above, used in a single system (single-channel at some points, two-channel at others), are shown in Figure 6.21. In Figure 6.21a, redundancy is provided for the front brake wheels by using two networks, and standard single-channel control is provided at the rear in order to reduce wiring and costs, as the second channel is considered unnecessary. If this topology does not suit you, you are free to devise another one, such as Figure 6.21b or 6.21c!

## *6.5.6 Synchronization is all!*

We cannot leave this technical description of the protocol without mentioning the problem of synchronizing the clocks of the nodes. The conventional approach would have been to raise this point during the description of the protocol and the frame management, but, in the interests of comprehension, I preferred to place this important section at the end of the description. The reason is simple. After the preceding sections, you would have become aware of the very strong dependence of the signal propagation time on the composition of the physical layer, in relation to the numerous possible topologies, the variable delays due to supplementary filtering introduced to reduce radio interference, the variety of nodes present in the network, each having an on-board microcontroller dedicated to its application and therefore its own clock, etc. These matters should remind you (for other reasons) of the lengthy chapters dealing with the problems of synchronization and resynchronization in CAN. In this case too, for similar reasons, synchronization is absolutely essential to ensure that TDMA accesses to the medium via the time slots do not elbow each other out of the way...

### Synchronization of the FlexRay TDMA system

The synchronization of the CPUs of a distributed intelligence system operating with medium access of the TDMA type is crucial. As I began to explain in the preceding

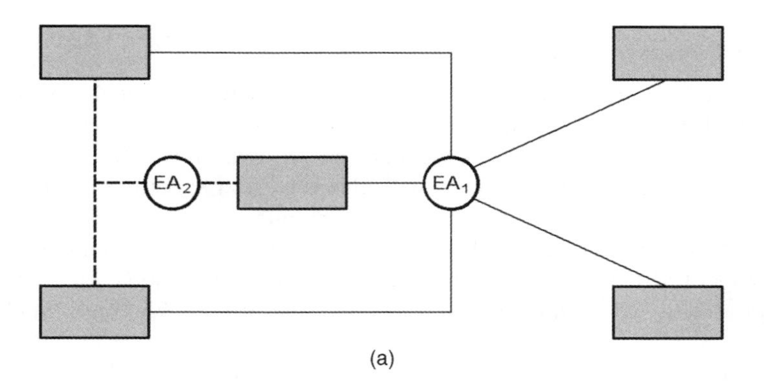
EA2
EA1
(a)

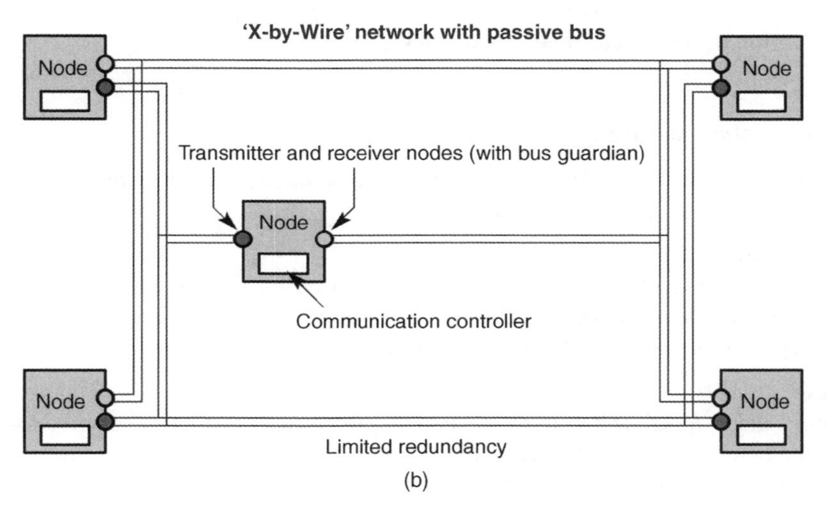
'X-by-Wire' network with passive bus
Node
Node
Transmitter and receiver nodes (with bus guardian)
Node
Communication controller
Node
Node
Limited redundancy
(b)

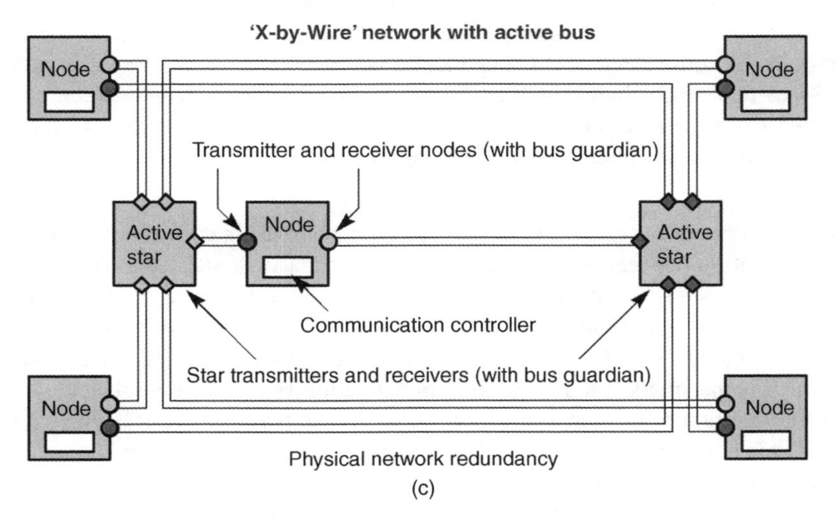
'X-by-Wire' network with active bus
Node
Node
Transmitter and receiver nodes (with bus guardian)
Active star
Node
Active star
Communication controller
Star transmitters and receivers (with bus guardian)
Node
Node
Physical network redundancy
(c)

**Figure 6.21**

section, each node in the network has a specific microcontroller which operates with the aid of the clock (at $x$ MHz) which is most suitable for its task (ARM, MIPS, etc.) and which has a functional sequencing dependent on its own clock. The network development manager or architecture designer will also choose a specific bit rate for his own reasons and will specify his messaging system (the number of time slots and their duration) to meet his own global application requirements. The problem is therefore to achieve compatibility between all these time requirements, some of which may be directly opposed to each other. This is the purpose of the FlexRay synchronization mechanism.

This synchronization is carried out in two stages or at two levels called 'macroticks' and 'microticks'. Figure 6.22 shows an overview of the operating principle of FlexRay message synchronization using 'macro' and 'micro' ticks.

### The FlexRay time hierarchy

Figure 6.23 shows the tree diagram of FlexRay time hierarchy, including the communication cycle, the macroticks and the microticks.

*Communication cycle*
The communication cycle consists of a whole number of macroticks. The number of macroticks per cycle is identical for all nodes of a single group (cluster) and remains the same from one cycle to the next. All the nodes must have the same cycle number at every instant.

*Microticks*
At the other end of this time hierarchy, we find the 'microticks', which are created locally in a node by the node itself, generally after the division of its local clock (quartz oscillator). By definition, therefore, they are integer multiples of the clock period of the local microcontroller of the CPU. Each node thus creates its own microticks whose structure is such that they cannot be affected by any synchronization mechanism. Speaking metaphorically, we could say that the microtick is constructed by a tiny electronic 'mechanic'. Thus, all the nodes have microticks of different values which represent the specific granularity of each node. They will be used to create macroticks and, in conjunction with these, to carry out the phase of synchronization between microticks and macroticks.

*Macroticks*
Let us start with an outline of their function. The purpose of the macroticks is to provide a first pre-synchronization between the physical signal in the network and the microticks.

The macrotick is a time interval relating to a specific set of network participants (cluster-wide), calculated by an algorithmic synchronization routine. So it is not created by a simple electronic 'mechanic', like the microtick, but is the outcome of a skilled calculation. In fact, the macrotick is the smallest unit of global time granularity of the network.

The local duration of each macrotick of a node inevitably consists of a whole number of microticks, this number being calculated or adjusted by the clock synchronization algorithm. As the value of the microticks is specific to each node, as mentioned above, being dependent on the frequency of its local oscillator and the setting of its pre-divider, the values of the macroticks will also differ from one node to the next.

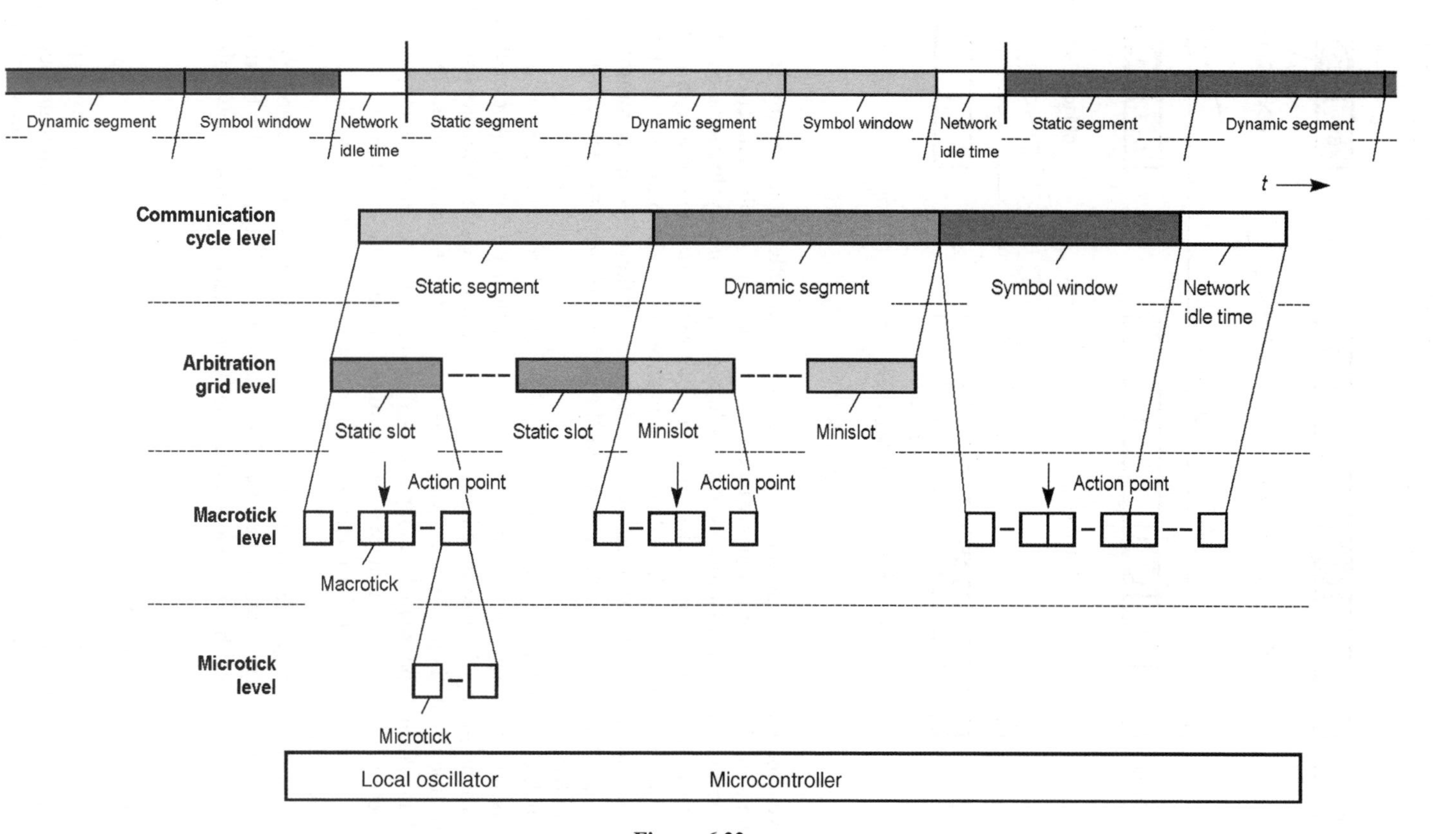
Dynamic segment
Symbol window
Network
idle time
Static segment
Dynamic segment
Symbol window
Network
idle time
Static segment
Dynamic segment
t
Communication
cycle level
Static segment
Dynamic segment
Symbol window
Network
idle time
Arbitration
grid level
Static slot
Static slot
Minislot
Minislot
Macrotick
level
Action point
Action point
Action point
Macrotick
Microtick
level
Microtick
Local oscillator
Microcontroller

**Figure 6.22**

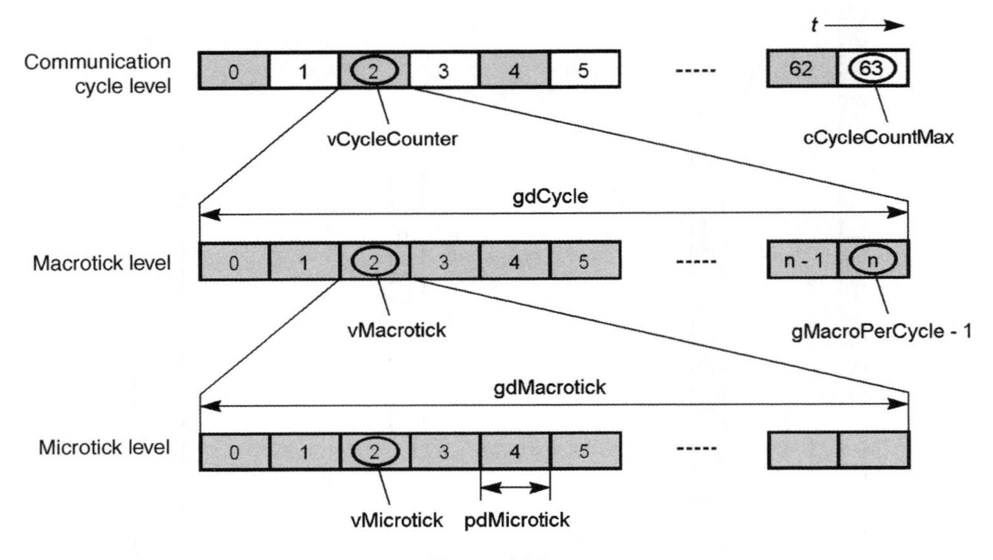


**Figure 6.23**

Moreover, the number of microticks per macrotick can differ from one macrotick to the next, within a single node. Although any macrotick consists of a whole number of microticks, the mean duration of all the macroticks of a whole communication cycle can be a non-integer value, in other words, a whole number of microticks plus a fraction of a microtick. These calculated time adjustments of the value of the macroticks – which themselves are directly related to the microticks determined by the frequency of the CPU microcontrollers – are responsible for the synchronization between the signals present in the network and the microcontrollers.

This synchronization principle is well known to television enthusiasts, as (for similar reasons, based on two stages and double release) it is practically a copy of that used for many years for synchronizing the line time bases of the familiar television receivers using double synchronization. First, there is a synchronization between the incoming video signal and the local oscillator, using a phase locked loop (PLL), followed by a second phase loop providing a strict positioning of the image on the screen, regardless of the variations (tolerances, time drift, etc.) of the storage times of the transistors and other components of the (horizontal) line scanning power circuit.

Figure 6.24 illustrates all these concepts, showing the instants in which the phase errors between the network signals and the macro/microticks are calculated, and also the moments in which the time values of the time slots are digitally corrected, to make the whole network temporally coherent.

### *6.5.7 Architecture of a FlexRay node*

Figure 6.25 clearly shows how the architecture of a FlexRay node conforms to the division of the ISO/OSI model. This conformity enables anyone (designer, equipment manufacturer, etc.) to develop each of the entities shown in Figure 6.25 independently and without difficulty.

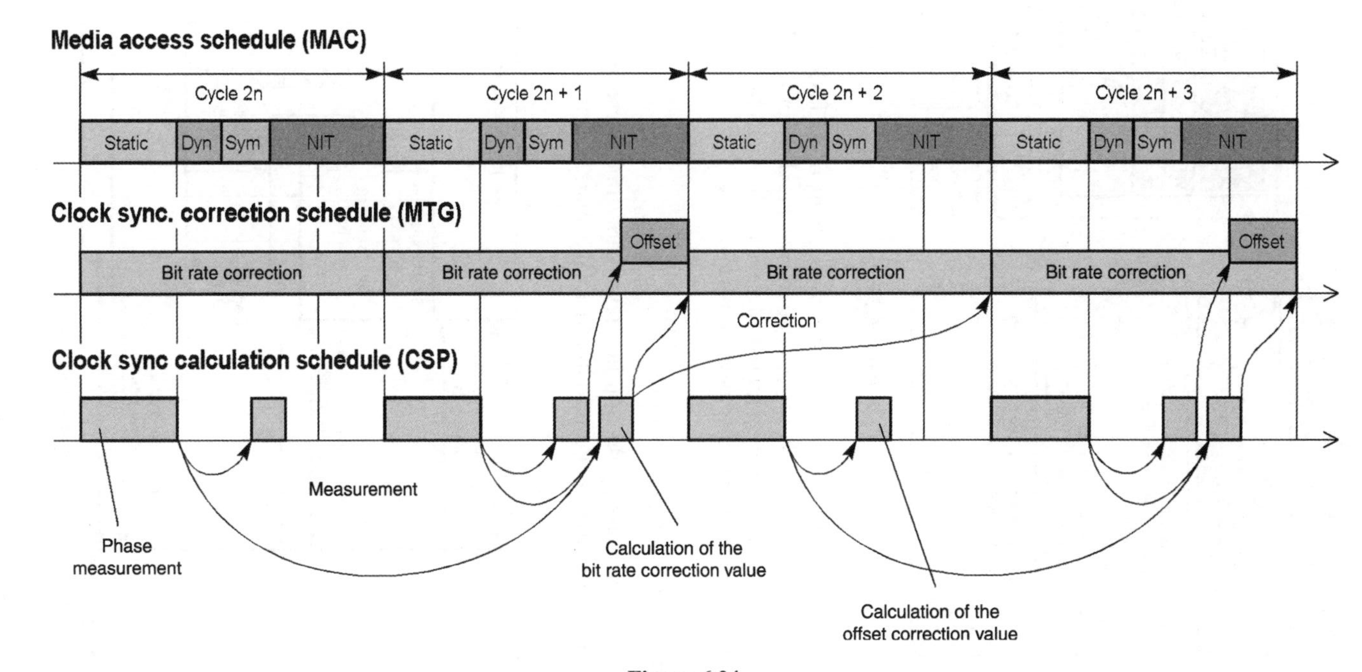
Media access schedule (MAC)
Cycle 2n
Cycle 2n + 1
Cycle 2n + 2
Cycle 2n + 3
Static
Dyn
Sym
NIT
Clock sync. correction schedule (MTG)
Offset
Bit rate correction
Correction
Clock sync calculation schedule (CSP)
Measurement
Phase measurement
Calculation of the bit rate correction value
Calculation of the offset correction value

Figure 6.24

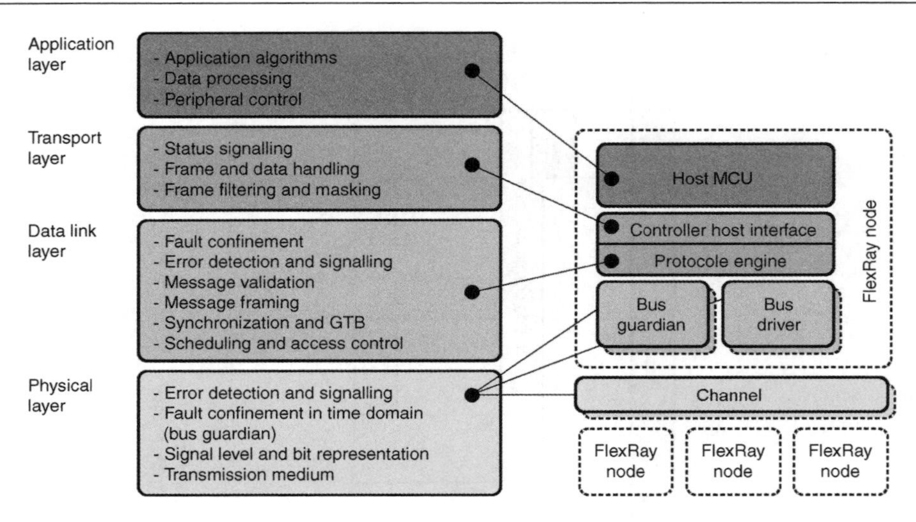


**Figure 6.25**

Figure 6.26 shows the details of the internal connections of a FlexRay node between the host controller in charge of the whole application, the FlexRay protocol handler, the line driver(s) (singular or plural, depending on whether the applications have one or two communication channels) and finally the bus guardian, if necessary.

### 6.5.8 Electronic components for FlexRay

Before they are put into mass production in a few years' time, these advanced systems and technical novelties will have to be introduced to the market via the appearance of new

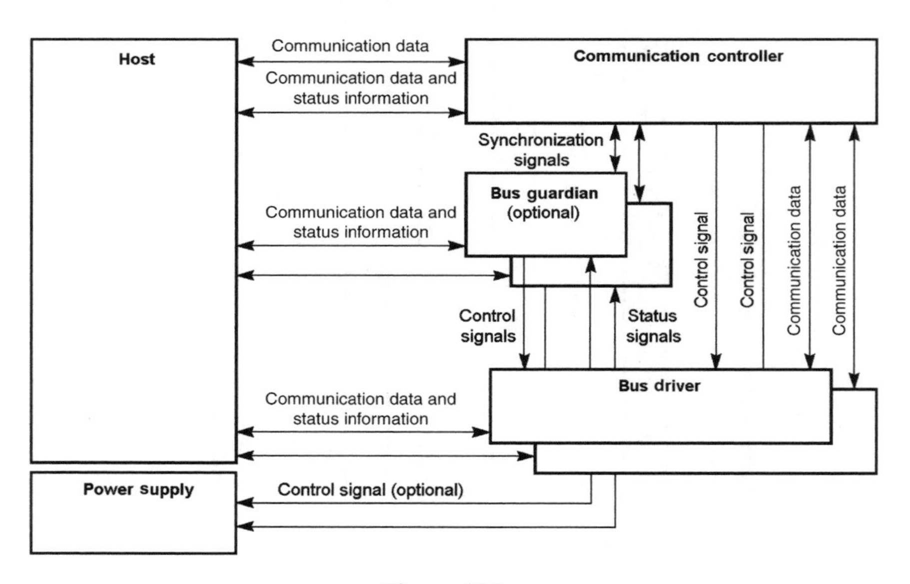


**Figure 6.26**

generations of top-range, and therefore expensive, motor vehicles – as happened in the early days of CAN. There are not too many of these models worldwide. Moreover, anyone in the industry will tell you that you need about 30 years to develop a new aircraft design, 20 years for a high-speed train and about 7 years for a car, starting from nothing. Knowing this rule, it is easy to guess when a new vehicle will next make its appearance. Taking all the predictions together, it has been known for several years that the most likely period for the industrial launch of this kind of project runs from the end of 2006 to the beginning of 2007. If you count backwards, you will of course find that everyone has been making preparations since 1999–2000 and that most electronic components were already in the pre-approval/final certification phase 3 years before the launch (i.e. in 2003). Does this come as a surprise? In fact, the line drivers (made by Philips/NXP Semiconductors) and communication controllers (made by Freescale, formerly Motorola Semiconductors) have already made their appearance, and the procurement forecasts from the car manufacturers are currently being refined.

To provide the historical background, Figure 6.27 shows the initial development plan of the FlexRay project, which at the start of 2005 is by no means inaccurate. For once, this is worth noting.

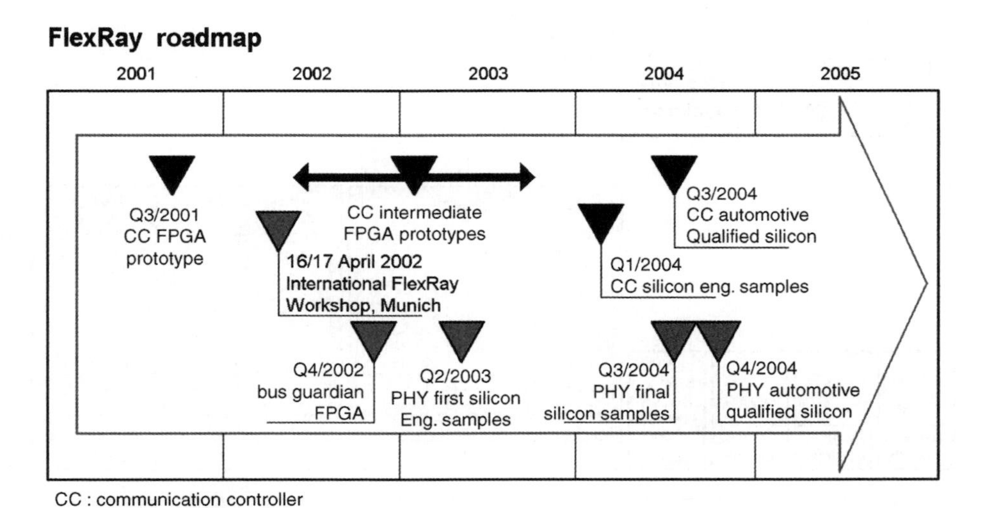


**Figure 6.27**

To return to specifics, the set of FlexRay components is generally similar to the CAN set, consisting of a protocol handler, one or two line drivers (depending on whether the applications are redundant or not) and finally a special element in the form of a bus guardian. The first two companies to take the plunge are Freescale and Philips Semiconductors, to be followed by many other semiconductor manufacturers (ST, Infinéon, Fujitsu, Renesa, Samsung, etc.), all eager to supply this new market.

Here is a list of this growing family of components.

**FlexRay protocol handler**

Following the development of FlexRay, the first protocol handlers required by the consortium were produced by Freescale in separate boxes, in the form of FPGA programmable circuits. Clearly, the FlexRay protocol handler will very soon be integrated with the (micro)controller – the ARM9 or ARM11, for example – which also controls the application. An example of a block diagram of a microcontroller with integrated FlexRay handler, which also provides a link to the CAN protocol, is shown in Figure 6.28.

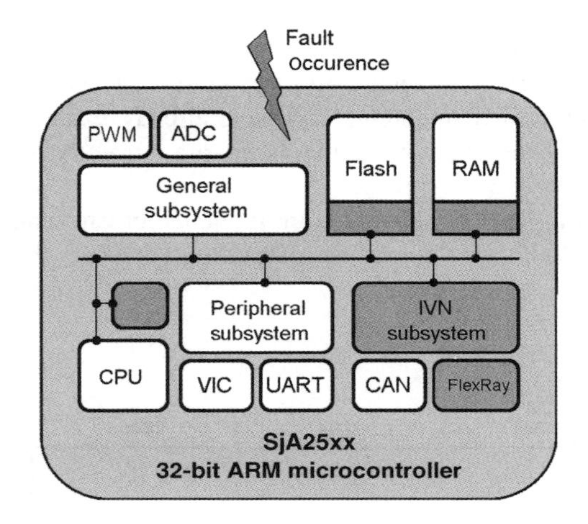


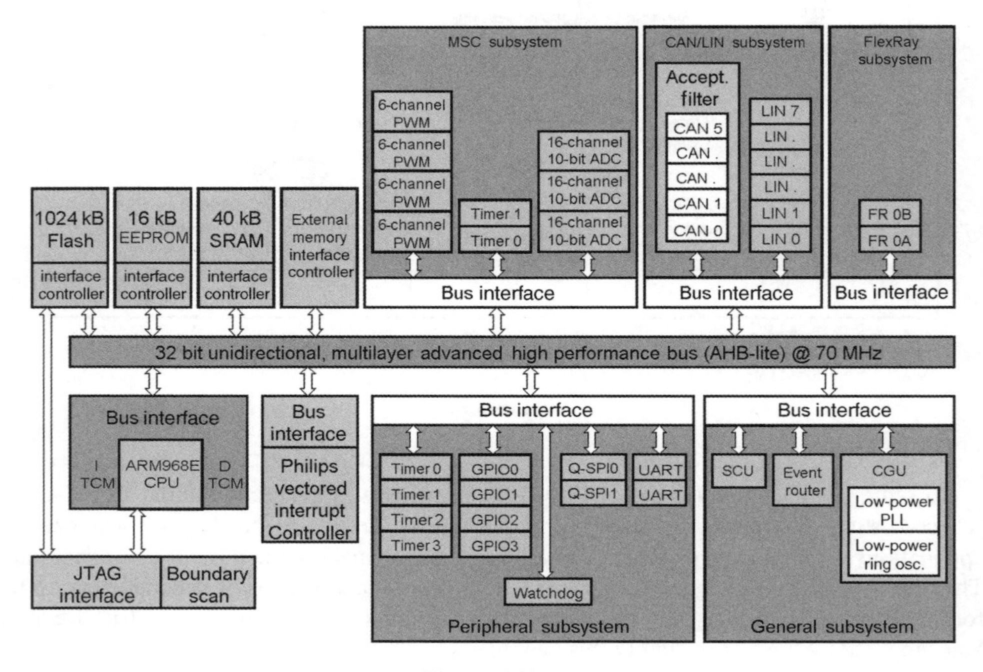


**Figure 6.28**

## Line driver

The first line driver available in the market (the Philips Semiconductors TJA 1080) clearly meets the 2.0 specifications of the FlexRay Consortium. Its block diagram is shown in Figure 6.29. Nothing special to note here, with the exception of two special interfaces for the use of the bus guardian and the host, which are not present in the CAN version.

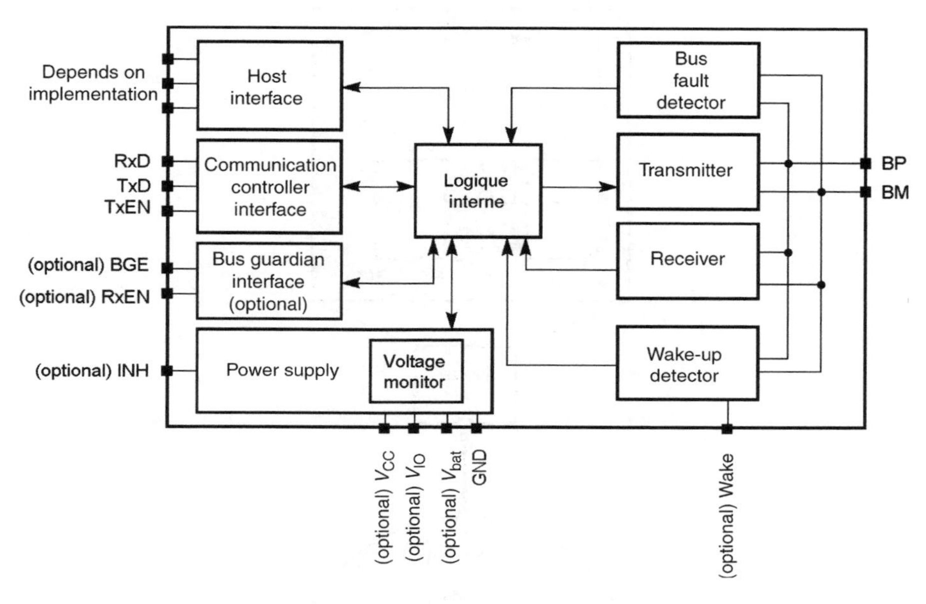


**Figure 6.29**

As regards the resolution of EMC and radio-frequency pollution problems, the same problems require the same remedies (Figure 6.30).

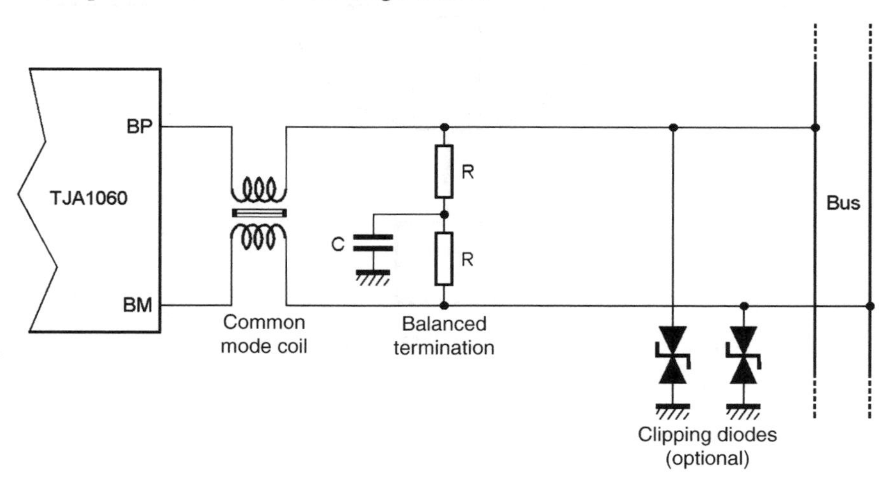


**Figure 6.30**

Note also that this circuit was designed to enable active stars to be implemented easily without additional components. Figure 6.31 is a schematic diagram of this implementation and Figure 6.32 shows an actual example, using the TJA 1080.

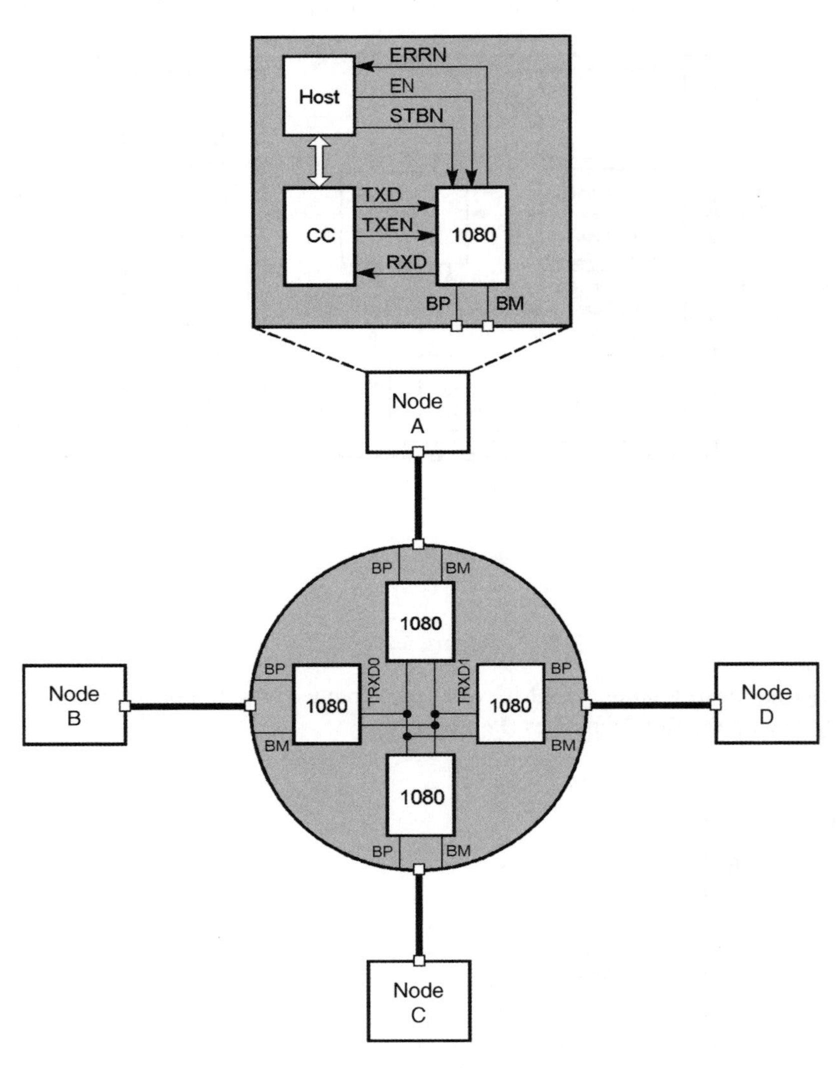


**Figure 6.31**

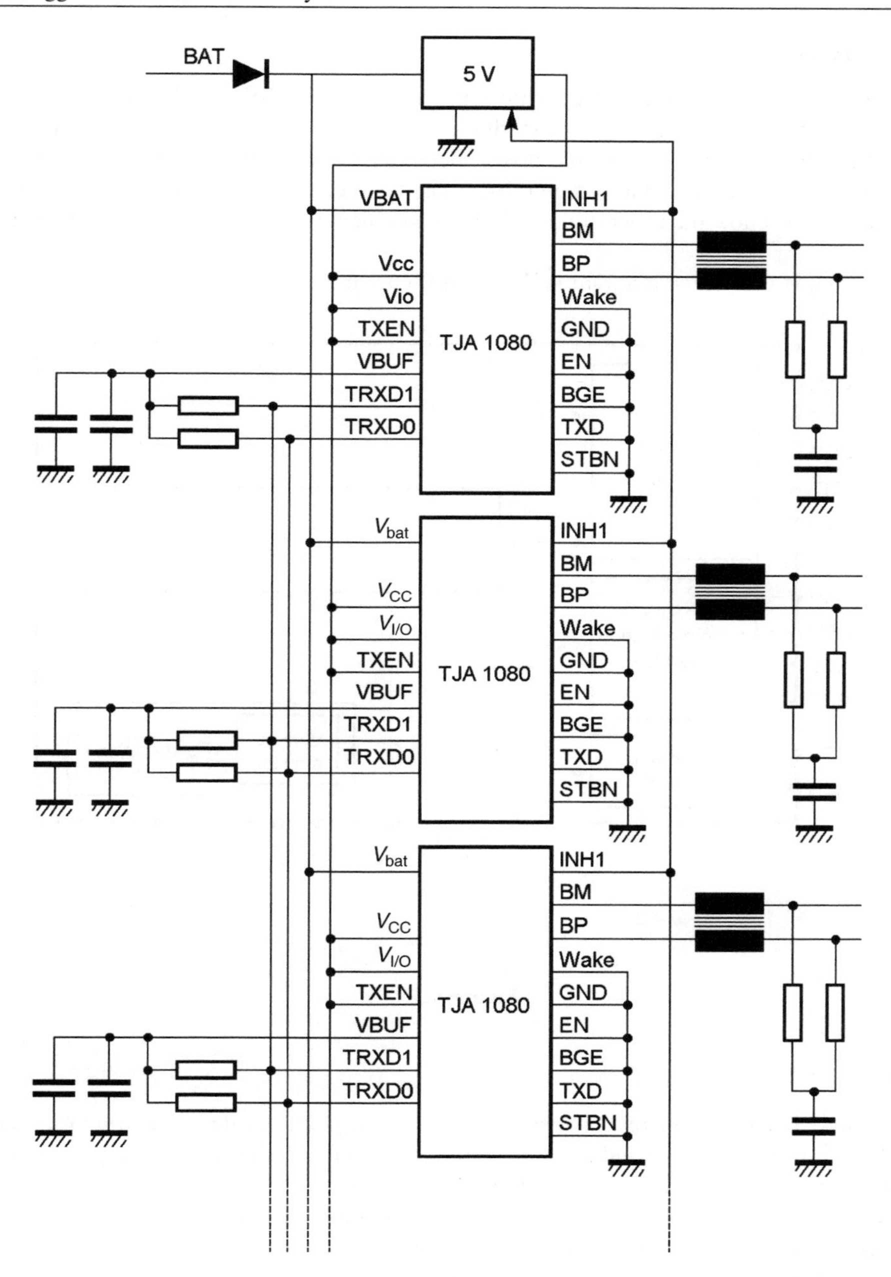
BAT
5 V
VBAT
Vcc
Vio
TXEN
VBUF
TRXD1
TRXD0
TJA 1080
INH1
BM
BP
Wake
GND
EN
BGE
TXD
STBN
$V_{bat}$
$V_{CC}$
$V_{I/O}$
TXEN
VBUF
TRXD1
TRXD0
TJA 1080
INH1
BM
BP
Wake
GND
EN
BGE
TXD
STBN
$V_{bat}$
$V_{CC}$
$V_{I/O}$
TXEN
VBUF
TRXD1
TRXD0
TJA 1080
INH1
BM
BP
Wake
GND
EN
BGE
TXD
STBN

**Figure 6.32**

### Bus guardian

The main functions of the bus guardian are to prevent what are known as 'babbling idiot' faults in the bus and to prohibit access to the medium in a bad time slot. It has the task of enabling or preventing the operation of the line driver, detecting errors and supervising access to the medium. For this purpose, the bus guardian must be synchronized with the communication controller, must know the exact timing of the communications and must have an independent clock.

An example of a block diagram is shown in Figure 6.33.

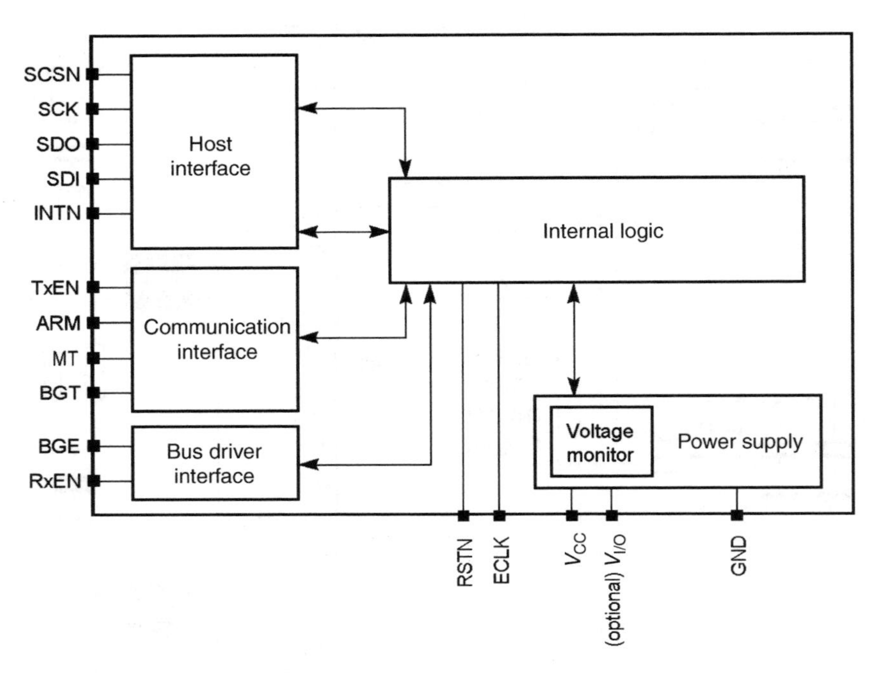


**Figure 6.33**

## 6.5.9 Conclusion

To conclude our summary of this new concept, I must point out the great similarity and/or parallelism of the technical and strategic approaches to the design and development of both FlexRay and CAN. As you would certainly have noticed, the design and production of FlexRay and CAN have been masterminded by practically the same organizations, with the same proactive characteristics and remarkable determination. Resemblance can be a good thing sometimes.

I hope that this chapter would at least have given you some new ideas for your future systems, and I look forward to meeting you again in 2010–2012 to share our first reactions, after taking the necessary time out to consider the introductory phase of this system. As far as this topic goes, it's goodbye for now.

# Part B

# New Multiplexed Bus Concepts: LIN, FlexRay, Fail-safe SBC, Safe-by-Wire

It may be strange to use the word 'parallel' to describe serial links, but it is true that many other serial communication protocols have been developed in 'parallel' with CAN (controller area network).

Following my presentation of the HS (high-speed), LS (low-speed), LS FT (fault-tolerant) and one-wire CAN buses, I shall now devote some fairly brief chapters to a description of other major protocol families and allies of CAN. These are used in the world of on-board multiplexed networks, found throughout the motor vehicle, aeronautical and industrial markets, and all related more or less closely to CAN, this system often acting as the 'backbone' of a set of networks.

These other serial links are dedicated to data transfer applications which may operate at lower speeds than CAN, or more commonly at higher speeds. They include, for example, digital data transfer, high-speed communications, audio, video, navigational aid and safety links, etc.

To make matters clearer, I have broken down these newcomers into two major groups: those in which the links consist of wire elements ('wired') and those in which the links consist of radio waves, known as 'wireless' systems. The most fashionable systems include

- in 'wired' technology:
  - LIN,
  - TTCAN,
  - X-by-Wire: TTP/C and FlexRay,
  - Safe-by-Wire,
  - I2C/D2B,
  - MOST,
  - IEEE 1394,
  - CPL;

- in 'wireless' technology:
  - Bluetooth,
  - Zigbee,
  - IEEE 802.11,
  - NFC,
  - RKE, PKE, passive go,
  - TPMS,
  - TiD,
  - etc.

Naturally, each of these has its own specific characteristics related to its speed, operating distance, possibilities, performance, cost, etc. It will therefore be more or less suitable for any particular application. It would be excellent if we could draw up a table showing their principal characteristics without any preconceptions, enabling each designer to find the right solution for his own requirements. However, such a table would not be easy to produce, as it would have to include a large number of different headings, to prevent the correct solution from being missed for lack of information on one or other criterion. To help you to draw up this table according to your own criteria, I shall start with a brief description, but with as much detail as possible, of the intrinsic properties of each solution.

As you will have noticed from the beginning of this book, I have based my descriptions on examples of motor vehicle applications, for obvious reasons of simplicity and comprehension. I will continue to use examples from this sector to support my descriptions, but will systematically extend their applications to other fields.

Starting from this point, and bearing in mind that the 'e-vehicle' (electronic vehicle) of tomorrow is likely to be designed not only as a means of safe comfortable travel, but also as an extension of the home and the office, let us plunge into the heart of the topic of multiplexed systems, their operation and their multiple applications. First, we will envisage the scope of the applications of these concepts by imagining the futuristic, but still very realistic, prospects for the motorcar.

Theoretically, the main function of a car is to provide safe and comfortable travel. This includes

- the power train,
- the brakes,
- the suspension,
- control of the passenger compartment: secure access (doors, boot, etc.),
- comfort: air conditioning unit, seat adjustment,
- active and passive protection: airbags, etc.,
- data communications,
- navigational aids, etc.

Note that each of these systems generally requires a dedicated multiplexed network

- for the engine and gearbox: high-speed CAN;
- for the passenger compartment and comfort: low-speed fault-tolerant CAN;
- for protection: Safe-by-Wire.

The car is increasingly becoming an extension of the home, with its highly sophisticated 'domestic' equipment (on wheels, of course):

- audio,
- video,
- video games,
- telephone, etc.

It is also increasingly becoming an extension of the office:

- on-board printer,
- the communicating car:
  - GSM,
  - Bluetooth (at different speeds),
  - Zigbee,
  - NFC (near-field communication).

This requires superimposed layers of specific, dedicated networks and of course numerous gateways between these networks. You will find ample detail in the following chapters.

Some troublemakers may claim that cars now run entirely on cables, and they will not be entirely wrong: those who announced, 15 years ago, that the introduction of buses would decrease wiring on vehicles will have to eat their words because the cable manufacturers are still very much in business! Clearly, these buses and this wiring are needed for numerous functions which did not exist, or could not even be imagined, in those earlier days.

The aim of this section of the book is therefore to guide you through the complexities of these networks and their interconnections. Now that you have been warned, let us go to work!

## The vehicle infrastructure

A modern vehicle includes a large number of intersecting and interlacing layers of links and communications. The same applies to an aircraft or any other high-level on-board system. Admittedly, some of the buses used for the latter applications are different, but the problems are substantially the same. To give you a more accurate idea of the situation, the diagram below shows a view, taken from a very high level, of the conventional architecture of a vehicle in the near future (i.e. about 2010).

Note that some motor manufacturers have already assigned different colour codes to different networks, with the aim of facilitating the reading of the electrical diagrams of their vehicles. By way of example, Daimler Chrysler often uses the following colours:

- LIN $\leq 20\,\text{kbit s}^{-1}$: green;
- CAN (fault-tolerant) $\leq 125\,\text{kbit s}^{-1}$: red;
- CAN (HS CAN) $\leq 500\,\text{kbit s}^{-1}$: blue;
- CAN (HS CAN) $\leq 1000\,\text{kbit s}^{-1}$: purple;
- Telematics (MOST) $> 1000\,\text{kbit s}^{-1}$: orange.

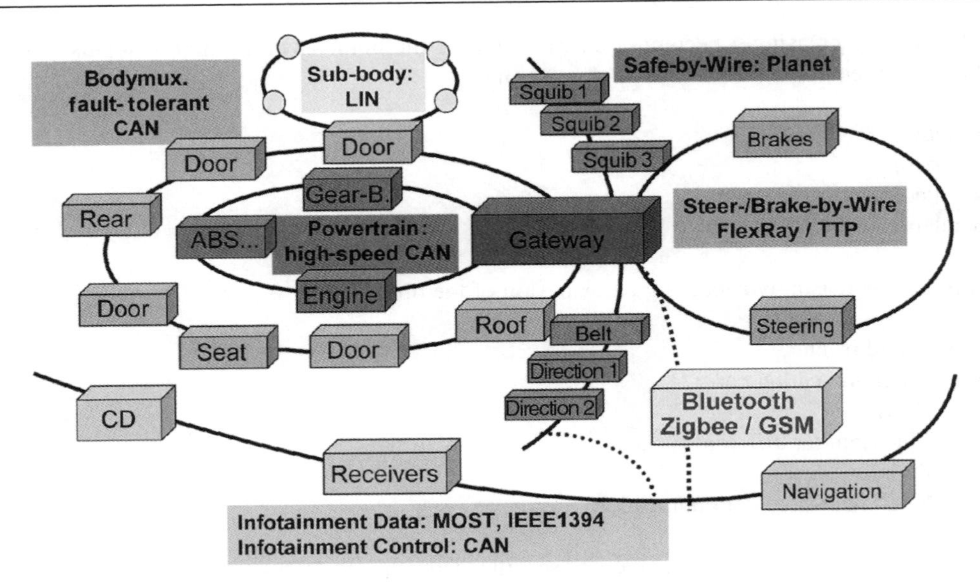


**Figure Part B.1**

## 'Internal' and 'external'

Before going further, I must make a small digression on the subject of the internal and external applications of a system in a car or in the industrial world. This is because, with today's increasing development of radio-frequency links, our list must also include the new communication protocols for establishing wireless (radio-frequency) networks, enabling us to extend the communication range and distances. Networks are beginning to show the influence of a degree of freedom regarding distance (of the order of 10–50 m), without the constraint of any 'umbilical cord', and wired/wireless gateways are appearing in the market or are already present in large numbers (for example, CAN/Bluetooth or CAN/IEEE 802.11b gateways).

*Simple solutions*

The simplest protocols and buses were designed for reasons of cost and functionality. Of course, the performance in terms of bandwidth, speed, safety and functionality is more limited. The oldest and best known of these are the I2C, D2B and SPI buses. A newcomer has recently arrived on the scene, in the form of the LIN bus, described in Chapter 7.

*The complex solutions*

These are often high-performance and highly specialized protocols and buses orientated towards specific applications.

For wired bus applications, these are protocols conforming to the TTCAN, X-by-Wire, FlexRay, Safe-by-Wire *Plus*, IEEE 1394 and other standards. There are fast, high-performance, real-time application orientated buses, which may be self-powered, security protected, redundant, etc., together with protocols for radio-frequency applications such as DECT, Bluetooth, Zigbee, IEEE 802.11x, Wi-Fi, etc.

## For internal use within a vehicle

Numerous protocols are already found in top-range vehicles or concept cars, in the form of 'wired multiplexed' systems. Each has its own characteristics relating to speed, possibilities, performance and cost. The most popular names are HS, LS, LS FT, one-wire CAN, TTCAN, LIN, I2C, D2B, MOST, IEEE 1394, CPL, X-by-Wire, FlexRay, Safe-by-Wire, etc.

*Wired communications*

Let us start with the major nerve centre of the system: high-speed CAN.

• *High-speed CAN*

As indicated in the above diagram, the links in the power section, including the engine control, gearbox and brake, are provided by a high-speed CAN link, because of the numerous exchanges that must take place at high speed (250 kbit $s^{-1}$ or more generally 500 kbit $s^{-1}$ and 1 Mbit $s^{-1}$). This area is highly technical and very critical. It is the core of the vehicle's operation and is designed by a process of very close collaboration between the vehicle manufacturer and its regular equipment suppliers. The development of the hardware and software layers is based on well-established specialist expertise, a strongly proactive approach and large research and development budgets, with everybody involved working incessantly on new vehicle projects including new principles, new messaging systems, etc.

• *Low-speed CAN*

This link layer is a peripheral layer of the one described above. In fact, it forms an integral part of the design of the vehicle, although to a (supposedly) slightly lesser degree. These links primarily serve the nodes located in the passenger compartment (body part) areas, which are therefore remote from the power system described previously. They appear to operate at a rate that is more understandable to humans, in other words, with slower reaction times . . . which is why low-speed CAN is used. The conventional speeds used (62.5 and often 125 kbit $s^{-1}$) cover links between nodes such as the 'openings' (doors, opening roofs, boots) and also seat adjustments, links with the dashboard incorporating the car radio, the conventional car radio display, the internal and external temperatures and the on-board computer (these applications are often on the borderline between high- and low-speed CAN).

As users may want to modify their cars, there must be a capacity to handle incidents due to short circuits and power cuts to elements of the links which do not endanger the overall operation of the network in use. This network is basically fault tolerant, unlike the one mentioned in the preceding section which, because of its functions, is implemented with the maximum provision for preventing any possibility of faults of this kind.

Note that some vehicles have up to five or six CAN networks (high and low speed) which are physically separate from each other (and may or may not run at the same speeds), each having a limited but highly dedicated number of participants, with the aim of providing very specific functionality or sharing the risks of malfunction.

- *LIN*

Considered by its designers as a sub-bus of CAN, LIN is mainly used for low-speed links (maximum 20 kbit $s^{-1}$), to reduce the cost of nodes (the performance required for these nodes is low but non-strategic). To make matters clearer, I will describe the example of electrically controlled adjustment, heating and rear view mirror folding, within the 'door' function. Of course, when everything is electrically operated, that is just perfect . . . but in case of failure or vandalism we need to be able to perform these operations manually. The same applies to the adjustment or pre-adjustment of the seats. We should note that LIN has appeared, next to FT LS CAN, at the 'top of the steering column', in the area often containing many control systems (horn, direction indicators, headlight controls, remote control reception, tyre pressure monitoring system (TPMS) reception, etc.). We should also note that some vehicles have up to five or six LIN buses which are physically separate from each other (and may or may not run at the same speed), each having a limited number of highly dedicated participants for providing very specific functionality. Clearly, this requires gateways between all the buses mentioned above. For example, before the driver leaves the vehicle, it is useful to signal to him, on the display element of the dashboard (high-speed or low-speed CAN), via the low-speed CAN controller located in the door, that the rear-view mirror (LIN) is not folded back. Similarly, it must be possible for the vehicle to start even if the mirror has been broken off or if the door has been dented by an impact.

- *CPL – current power line*

For the same applications as above, CPL links can send data and control commands to systems (low speed, of the order of 20–50 kbit $s^{-1}$ maximum), using an HF carrier with ASK or FSK modulation, this carrier being superimposed on the '$+V_{bat}$' wire carrying the power to all the network modules (see the solutions described in Chapter 4). Here again, the main purpose of these applications is to reduce costs.

- *X-by-Wire and FlexRay*

As you will see in Chapter 6 (on the operating principles of TTCAN, FlexRay and X-by-Wire applications), communications based on these protocols are very fast and can be considered 'real-time'. At present, the devices controlled according to these protocols are the ones whose weight and volume are to be reduced and the ones which are to be made more intelligent, such as steering column systems, brake, clutch, suspension, roadholding systems, etc. This form of communication requires high speeds (of the order of 7–10 Mbit $s^{-1}$) and a degree of hardware and software redundancy, depending on the planned applications. The first production vehicles fitted with these systems are expected around the beginning of 2007, and of course the top-range vehicles, with 6–15 nodes each, will be the first to benefit from the new systems (as was the case with CAN's first appearance). As everybody knows, anyone will be able to have the optional extra for 1 euro just a few years later! Here again, it is essential to have gateways between the CAN and FlexRay networks.

- *Safe-by-Wire*

Point-to-point links for safety systems, such as the airbag triggering systems and safety belt pre-tensioning devices should disappear eventually and be replaced by a bus topology forming part of what is known as the Safe-by-Wire Plus ASRB-2 solution (see Chapter 9). These links, directly related to the physical safety of the vehicle occupants, must be able to operate as soon as there is an impact on the vehicle, and they are naturally based on principles of fast communication between the detected data (impact detector, accelerometer, inertial controller, etc.) and the squib actuators, seat belt pre-tensioning and other devices. Once again, gateways are essential.

• *Audio and video systems*
Audio and video distribution is becoming common in motor vehicles. As the technologies evolve, we have seen the establishment of communications networks with data transfer via the I2C and then the D2B bus, followed by MOST at the present time, and IEEE 1394 in the near future. In anticipation of Chapter 10 which describes these communication systems, here is a brief introduction to the audio and video applications supported by them.

*I2C bus.* I2C (inter-integrated circuit), the ancestor of these systems, is still the bus most commonly used for mass production vehicles, mainly for providing links for control commands (stop, play, change tracks, etc.) between the car radio and the CD changer located in the boot.

*D2B bus.* For a long time, D2B (domestic digital bus) was the main bus used for transferring digital audio data to car radios from conventional CD audio players which are often located in the boots of vehicles.

*MOST.* MOST (media-oriented systems transport) is a bus having the primary function of audio distribution in the vehicle. It is also used for video applications in a vehicle, when the bit rates can be adapted to the on-board applications (for example, MPEG1 or MPEG2 in SIF or quarter SIF format).

*IEEE 1394.* We have now seen the arrival of the IEEE 1394 bus (and its derivative IDB) which enable even higher speeds to be achieved. Clearly, it can be used for the same applications as MOST and can multiplex different audio/video sources in a single wired medium, for example

- data from the family of audio and video CDs and CD ROMs of roadmaps for navigation support;
- all types of video data intended for screen display at some point, including data for video display, communications and office technology, video games, DVDs, GPS location and navigational aids either in their basic form or complemented with dynamic data supplied via a GSM link.

*Radio-frequency communications*
Still in the field of internal applications, there are numerous applications using a radio-frequency medium for digital communications.

• *Radio-frequency receivers for all applications*
*Radio.* Let us start with the most conventional kind of receiver, namely the AM/FM/digital radio receiver, which may or may not have antennae incorporated in the windscreen or in a rear screen de-icing system. These systems are entirely conventional, except in

the case of voice commands (voice synthesis and/or recognition) for tuning and volume control, etc.

*Immobilizer or start prevention.* The immobilizer function (vehicle start-prevention device) using RF transponders (operating on a 125 kHz carrier) has been well known for some years. A contactless element (transponder) located in the head of the contact key (or in a badge) is made to communicate at a few kilobits per second, with encryption, to allow the operation of the ignition/injection part of the vehicle in which the decryption system is located (as a buried layer) and to enable the engine to be started.

• *Bluetooth*
The Bluetooth system is now well known. Where local regulations vary from one country to another, making it impossible to use the same bandwidth at 2.45 GHz (and therefore the same number of transmission channels) and the same radiated power (EIRP), it is easy to use Bluetooth for internal purposes in a vehicle. The first, very common, example is that of the use of earpieces which enable a driver to use a mobile telephone while driving, when he must legally keep both hands on the steering wheel. Other examples can be found in the construction of 'pico' networks in vehicles, linking all the on-board data processing and office automation modules (e-mail printer, etc.), and also at the boundary between internal and external systems:

- applications in the reception departments of garages, branches, dealers, etc.;
- applications in passive entry systems;
- applications in anti-theft or anti-intruder systems, based on principles of relay attacks, etc.

Let us now look at the radio-frequency applications for linking the interior of a vehicle with the outside world.

*Radio-frequency communications*
• *Remote control*
Having used infrared links for many years, remote control systems now provide greater user-friendliness, mainly in terms of reduced directionality and longer operating distances, by using UHF radio-frequency links (mainly in the 433 MHz band, less often at 866 MHz in Europe). The carrier wave is modulated either by ASK (OOK, on–off keying) or FSK. The communication bit rates are of the order of several (tens of) kilobauds and the protocols used are generally proprietary ones. The messages transmitted by these remote control systems, which are also encrypted, usually with the encryption used in the immobilizer system, are subsequently transcoded into CAN frame on reception, while awaiting the decoders. The same electronic system decrypts the signals arriving from the immobilizer function and those of the remote controller located in the key or badge.

• *PKE (passive keyless entry)–passive go*
PKE–passive go covers the functionality of 'hands-free entry' and 'push-button starting' (without the use of any visible key). This RF link is bidirectional, with the uplink from the

vehicle to the key, operating on a 125 kHz carrier with a speed of several kilobits per second. In the downlink direction, from the key/badge towards the vehicle, it operates with an RF carrier generally located in the 433 MHz band. The communication protocol (generally proprietary) must be able to control and process message and/or carrier conflicts because all operations are hands-free, and more than one passenger with a badge (e.g. a husband and wife, etc.) may approach the vehicle simultaneously.

• *TPMS*
The TPMS is designed to measure the individual tyre pressures, correct them according to the tyre temperature and transmit them to the driver; it sends these data or alerts at radio frequency (433, 868 or 915 MHz, depending on the country) every 1 or 2 min when everything is running smoothly, or much more frequently if there is a problem.

• *TiD – tyre identification*
*Warning*: Never confuse ‘tyre identification’ with the TPMS system above! What we are concerned with here is everything relating to RF identification, at 900 MHz,[1] of the tyres (type, model, place of manufacture, whether new or retread, etc.), but not the measurement of their pressure. What is the purpose of such a system, apart from ensuring traceability of tyres in the course of manufacture and delivery? You will be surprised! These matters will be examined in detail in Chapter 11.

*Wireless networks*
I shall conclude this overview with a brief description of wireless networks operating at radio frequency.

• *GSM*
Apart from providing the well-known simple mobile telephone link, this kind of RF link also opens the way to many new services which will be detailed in Chapter 11.

• *GSM and Bluetooth*
Depending on the local regulations in force in different countries, it is possible to transmit at different EIRP power levels (classes 1, 2 and 3), and consequently over different communication distances (from several metres to several hundreds of metres). This is sufficient for developing internal applications. See Chapter 11 for a description of these new applications.

• *IEEE 802.11x*
The IEEE 802.11x concept, also known by the name of Wi-Fi, is not aimed at the same applications as Bluetooth, but it would take a cleverer person than me to predict the future areas of application of these two concepts with any certainty! I shall attempt to indicate a few applications of this system in Chapter 11.

• *NFC*
The newcomer in the radio-frequency family, NFC, permits communications over a distance of approximately 10 cm. Can this concept be of any use in the external environment, when this tiny communication distance appears to place it firmly in the ‘internal’ category? For the answer, see Chapter 11.

[1] RFID: See numerous books by the present author, published by Dunod.

## Grading in terms of performance versus costs

Before concluding this brief introduction to these new protocols, I shall show their grading, now and in the near future, in terms of the performance/cost ratio, the relative importance and uses of each with respect to the others, and the annual and cumulated consumption. The diagram below shows the speed of these concepts as a function of cost. If you want quality, you have to pay for it: The faster the system, the higher the price. There is nothing new about this! Of course, the relative costs shown on the horizontal axis are given for guidance only, with the usual reservations. They will have to be revised on a day-to-day basis, according on your precise application requirements.

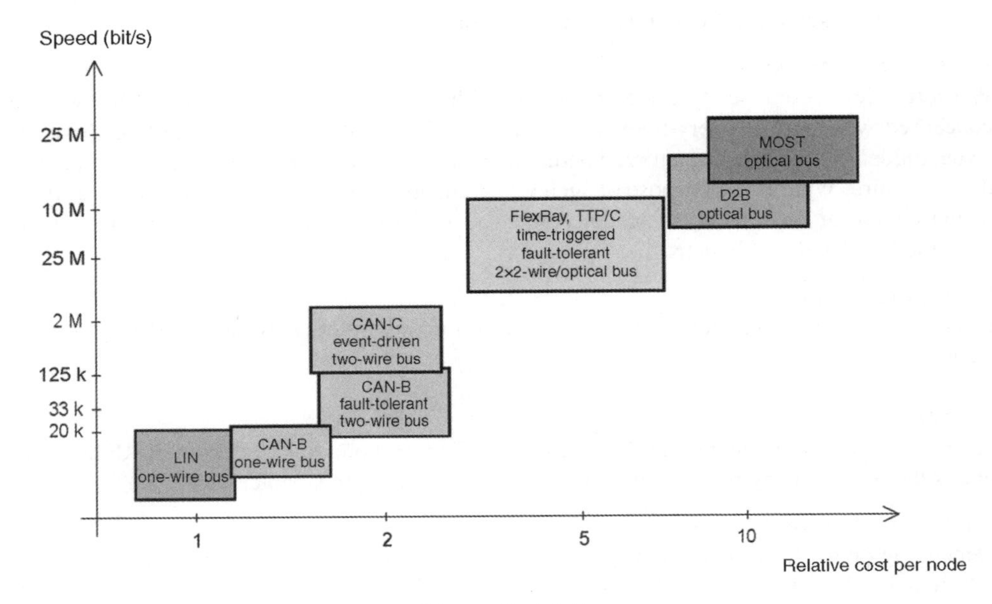


**Figure Part B.2**

## Grading in terms of fields of application

To complete this introduction with some specific details, the following diagram shows in terms of the total number of nodes (numbers cumulated from their origins) the distribution of conventional wired buses, all forms of CAN, LIN and others (with the same reservations as before), and the distribution of these 'buses' in motor industry applications.

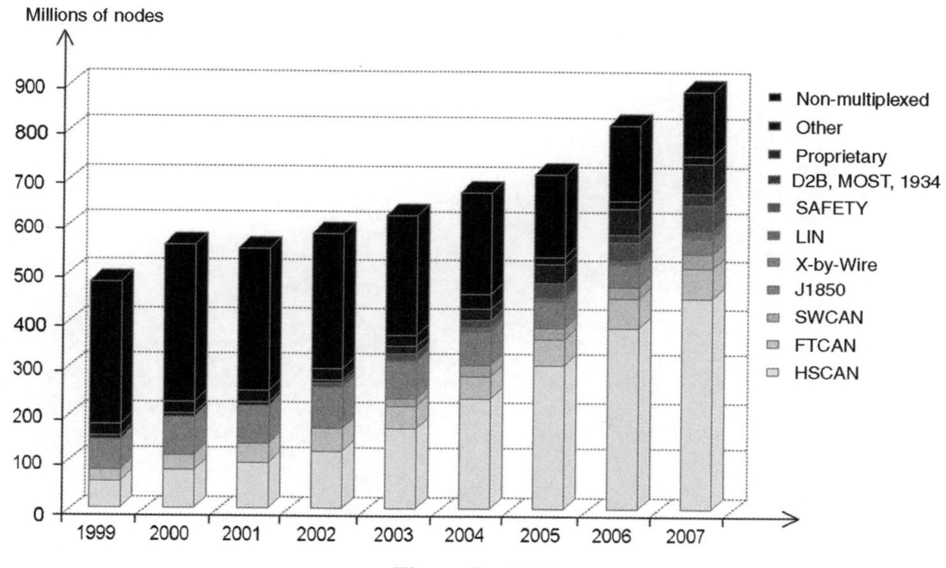


**Figure Part B.3**

# 7

# LIN – Local Interconnect Network

Rome was not built in a day. Neither was the local interconnect network – better known as the LIN bus.

The major developer of the LIN concept was the Motorola company (now Freescale) and the first specification, 'LIN rev. 0', appeared in July 1999 (33 pages). A consortium was soon created, in March 2000, including the car manufacturers Audi, BMW, Daimler Chrysler, Volkswagen and Volvo Car Corporation, as well as Motorola Inc. and Volcano Communication Technologies AB. Note that Motorola was in charge of the LIN consortium from 2000 onwards.

After several intermediate versions, particularly LIN 1.2 (November 2000) and LIN 1.3 (December 2002), allowing for various adjustments of the time aspects and electrical signal tolerances of the protocol and software resources, and negotiations with our American colleagues of SAE J2602, the final specification, 'LIN rev. 2.0' (125 pages long!), was issued in September 2003. How useful is it to have a stable protocol to enable reliable systems to be developed and perfected!

To make matters clearer for you: 'LIN 2.0 is a superset of LIN 1.3, and is the version recommended for all new developments', says the Consortium. So it is goodbye to the many (too many) earlier versions. Note also that this protocol belongs to the LIN Consortium and is not at present standardized by the ISO, and is supplied 'as is' (according to the Consortium).

## 7.1 Introduction

The LIN protocol is mainly intended to support the control of 'mechatronic' elements found in distributed systems for motor vehicle applications, but of course it can be applied in many other fields.

The LIN protocol concept is a multiplexed communication system whose level and associated performances are clearly positioned below what we can expect from CAN

*Multiplexed Networks for Embedded Systems: CAN, LIN, Flexray, Safe-by-Wire...* D. Paret

(controller area network). This is because it is very different from, and much simpler than, CAN; it is even simpler than the I2C bus, being based on the concept of a (sub)network containing only one master with a finite set of slave nodes. As the network is controlled by the master alone, the communication system is deterministic, being totally dependent on a time sequencing fixed by the task control provided by the master. The primary and original purpose of LIN is therefore to provide a 'sub-bus' for CAN, with reduced functionality and lower costs, in other words to provide an economical solution when the requisite performance level is not high. Thus it can be used where the bit rate and bandwidth of the network are low and the reliability and robustness found in CAN are not required. Note that there are no problems of conflict and arbitration, etc., for the simple reason that the system operates in a single master and multiple slave modes.

In a motor vehicle, for example, there are numerous network nodes/participants that operate satisfactorily at this performance level (Figures 7.1 and 7.2):

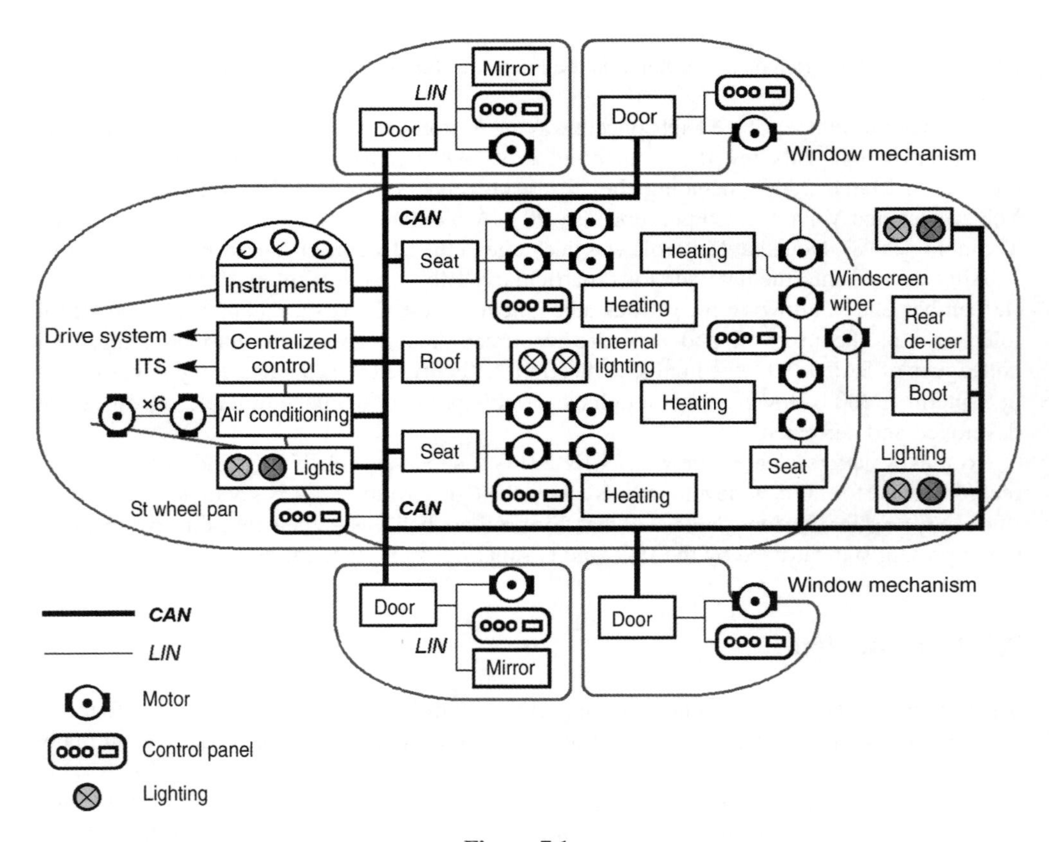


**Figure 7.1**

- sun roof (open, close, inclination, etc.);
- rain detector, automatic headlight switch-on;

- seats (all seat adjustments and functions);
- top of steering column, column-mounted controls, etc.;
- doors (window winders, etc.), wing mirrors (position, de-icing, etc.);
- windscreen wiper control;
- interior light control, etc.

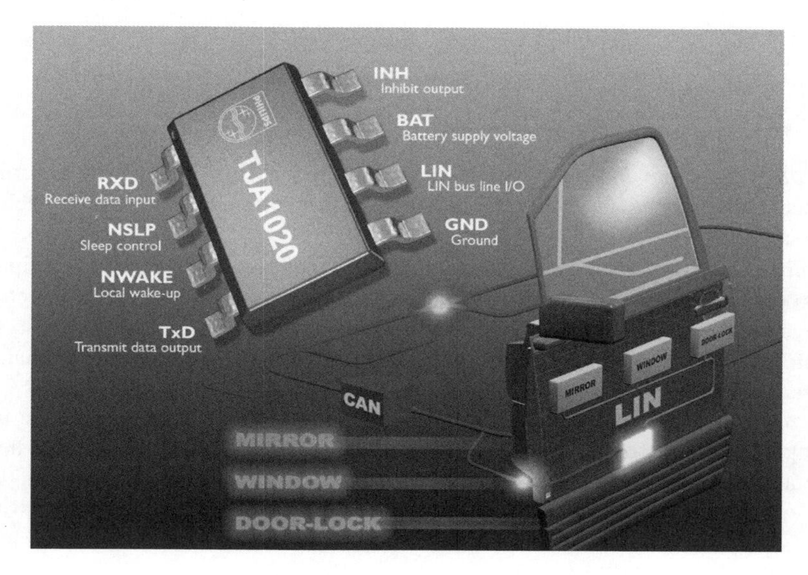


**Figure 7.2**

## 7.2 Basic Concept of the LIN 2.0 Protocol

Now let us examine the main features of the LIN 2.0 ‘single master and multiple slave’ concept, where no bus access arbitration is needed. The LIN 2.0 specification describes a serial communication protocol, constructed on the basis of the division of the two lower layers: layer 1 (physical layer) and layer 2 (data link layer), and the well-known ISO/OSI reference model (Figure 7.3), and provides some guidelines for the intermediate layers.

The authorized text of the ‘LIN specification package – revision 2.0, September 2003’ is divided into several parts. The aim of this chapter is not to transcribe all the contents of this document, but to summarize the key points, as well as raising some important detailed matters which are unmentioned in the specification. For further information, you should download the whole document which is freely available on the LIN web site. It comprises

- *Protocol specification*, which describes the data link layer;
- *Physical layer specification*, which deals with the physical layer and covers the speed (bit rate), clock tolerances, etc.;
- *Diagnostic and configuration specification*, which describes the services that LIN can provide above the data link layer for sending diagnostic and node configuration messages;

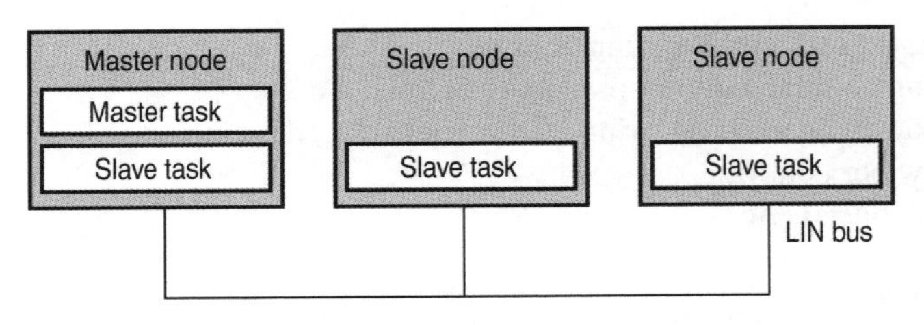


**Figure 7.3**

- *Application program interface specification*, which has the purpose of describing the interface between the network and the application program and which includes the diagnostic module;
- *Capability language specification*, which describes the format of the 'LIN description' file, used to configure the whole network, and also acting as a common interface between car manufacturers and suppliers of the various network nodes. This specification also serves as a development assistance tool, and for the design of on-board production and maintenance analysis tools;
- *Node capability language specification*, which has the purpose of describing the format of the language to be used for designing rack-mounted LIN modules which can be used in plug-and-play mode.

### *7.2.1 Operating principle of LIN*

Let us start with some general information on the operating principle of LIN and its physical layer.

The LIN link is based on asynchronous communication (thus without supplementary clock data transmission). The physical implementation of the internode link is therefore based on 'one' wire (plus ground, of course!), and, compared with CAN, reduces the wiring and connector costs. Subject to certain limits of operation, performance and communication reliabilities, the LIN concept offers a multicast reception solution with synchronization and self-synchronization devices which can be constructed without the use of quartz oscillators or ceramic resonators in the slave nodes. LIN also provides

- a guaranteed maximum latency of the transmitted signal;
- a flexibility of physical or daisy-chain configuration.

It offers

- security of transmitted data, based on CRC and error detection;
- the detection of faulty nodes in the network.

To minimize the size (i.e. surface area) and cost of the integrated circuits, the hardware implementation of the LIN protocol can be based either on a (virtually) conventional UART/

SCI hardware interface or on a software equivalent, or again on a pure state machine that can be implemented in wired or programmable logic.

### *7.2.2 The data link layer (DLL)*

Current discussions of LIN frequently focus on the 'LIN cluster'. A LIN cluster (i.e. a group, or a network/group) consists of a 'master task' and a number of 'slave tasks'. A master contains a master task, but can simultaneously execute a slave task. The other nodes contain only, or execute only, slave tasks.

#### **Messaging system**

The traffic on the bus is initialized and controlled solely by the 'master task' of the network, which is the only entity that decides which frames are to be transferred on the bus, and when this is to take place (i.e. it determines the start of the frame on the bus). The data are then supplied to the bus by any member of the network, which may be the master itself or the slaves. The slaves cannot supply data unless they have been invited on the basis of a scheduling table established by the master. This rule of operation is very important, because it enables us to predict the moment when a message will be supplied on the bus. Thus it provides the protocol with a degree of determinism, as the most unfavourable case of latency is known, and this is essential for the development of reliable networks.

The data are transmitted by means of communication frames (see below) with a fixed format but with an adaptable length. All the master and slave tasks form an integral part of the frame handler.

#### **Arbitration**

As we have seen, the properties of a single-master system are such that the LIN protocol requires no arbitration procedure, and therefore this protocol is deterministic and has known latencies. Therefore, only the slave task matching the transmitted identifier is authorized to respond. On the contrary, the other nodes can use data carried on the bus if necessary, if they are relevant (see the description of the 'multicast' aspect below).

#### **Communication frames**

A communication frame consists of a header provided by the master task and a response provided by a slave task. The header (see below for the details) consists of a break, a synchronization sequence and an identifier. Only the slave task authorized to supply the response associated with the transmitted identifier can respond. The response consists of a data field and a checksum field. The data associated with the identifier concerned is relevant to certain slave tasks (there may be several of these) which receive the response, verify the checksum and can use the transmitted data if they wish. This means that we have the following properties:

- Flexibility of the system: nodes can be added to a LIN cluster without the need for hardware or software modifications in the other nodes already present on the network (see below for the possibilities of daisy-chain configuration);

- Message routing: the destination and content of a message are determined by the identifier. Note that this does not indicate the destination of the message, but describes the significance of the data contained in the message. Because of the number of bits available in the identifier field, the maximum number of identifiers is 64;

By its nature, a node in a LIN system does not use any information about the system configuration, with the exception of the name of the single master node.

- Multicast: a number of nodes (any number) can simultaneously receive and act on a single frame. . . . but note that only one of them is authorized to respond.

### Data transmission

Two different types of data can be transmitted in a frame, namely data/signals and diagnostic messages.

### Signals

'Signals' are scalar values or byte arrays which are packed in the data field of the frame. A 'signal' is always present in the same position of the data field for all the frames that has the same identifier. An example is shown in Figure 7.4.

### Diagnostic messages

The diagnostic messages are carried in frames with each having two identifiers specially reserved for this purpose.

### Security

In keeping with the spirit of the CAN specifications, the following sections describe what the LIN implementation (hardware or software) must be able to do.

### Error detection

The sender of the message must have a monitoring device, to compare the electrical level of 'what ought to be' with 'what actually is' on the bus at any instant. Two other parts of the frame can also be used to detect errors if necessary:

- 2 parity bits at the end of the identifier field;
- 1 CRC at the end of the transmission frame.

**Figure 7.4**

### Performance of the error detection

All the local and global protocol errors of the transmitter are detected.

### Error signalling and error recovery time

Because of the single-master design, it is not possible to have an error-signalling device.

However, the errors are detected locally and these data can be supplied on request, in the form of 'diagnostic messages'.

### Error confinement

LIN nodes are capable of distinguishing short-term malfunctions from permanent ones, and can carry out their own local diagnoses and take corrective action.

### Communication frame

Figure 7.5 shows the overall structure of the frame format. It consists of a break sequence, followed by 4 to 11 byte fields. Each message frame comprises a number of data fields, and each field consists of a 10-bit byte, encoded by the well-known UART/SCI 8N1 coding (1 start bit, always dominant, to start the exchange, then 8 data bits and finally 1 stop bit which is always recessive). Note that the transmission takes place in LSB (low significant bit) first mode.

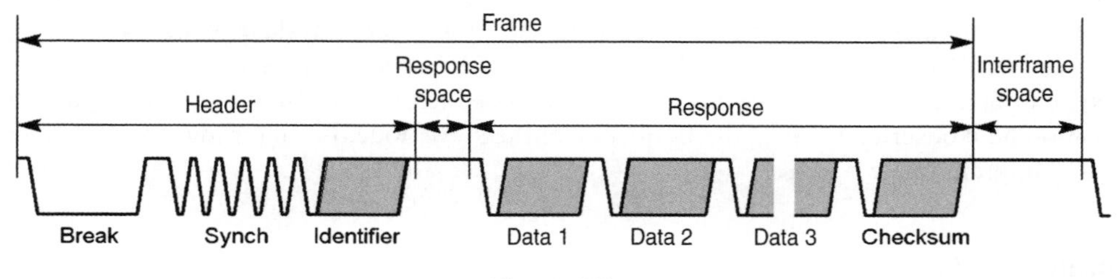


**Figure 7.5**

### Frame header

The frame starts with a header field, consisting of three main parts, all transmitted by the master. This header field consists of a synchronization break field, a break delimiter field and a synchronization field.

*Synchronization break field*
This field is formed from at least 13 dominant bits. Note that this field has no specified maximum number of bits!

*Break delimiter field*
The synchronization break field is followed by a break delimiter, whose recessive value is at least 1 bit. Its length characteristics are the same as those of the preceding field.

A brief note in passing concerning the fact that the maximum values of these two fields are unspecified: this may cause structural problems of network interoperability, due to different time-out values between nodes, where the presence or absence of short circuits on the LIN line is to be indicated.

*Synchronization field*
These are followed by a 'synchronization field' (SYNCH field), with the conventional value of 0101 0101 = hex 0×55, framed by the dominant start bit '1' and recessive stop bit '0'. We thus have a binary sequence of 10 bits (1 0101 0101 0), making numerous bit transitions available to facilitate the evaluation of the bit rate of the bus if required.

**Note on clock recovery and synchronization**. As the master starts each message frame with the synchronization break, slave nodes that have lost their clock synchronization can identify the synchronization field during which they may have a chance of becoming synchronized. This is one of the distinctive properties of an asynchronous system, compared with a synchronous system. This field is followed by the synchronization field (SYNCH field), which includes several downward transitions or fronts (i.e. transitions from 'recessive' to 'dominant' bits), at time distances which are multiples of the bit time. This distance can be measured (by a capture counter, for example) and used by the slave to calculate the current bit time. This is obviously very useful, but it increases the cost, thus detracting from the much-vaunted 'low-cost' image of LIN. However, it makes it possible to design automatic bit rate recognition systems, known as 'auto baud rate' systems. While on this subject, you should also note that LIN specifies that the master node oscillator tolerance must be less than 0.5% (allowing for all factors such as power supply, temperature variations, etc.), and that the tolerance for slave nodes not having (re)synchronization devices must be less than 1.5%. If a slave node has an on-board resynchronization device, this tolerance range can be increased to 14% and, in the synchronization field described above, the frequency of the slave node oscillator must be picked up in a maximum range of 2%.

### Identifier field

Next is a 10-bit identifier field, which uniquely defines the subject, i.e. the deep meaning of the frame being transmitted. It consists of (see figure 7.6)

- 1 dominant start bit;
- 6 identifier bits (bits 0 to 5) which can be divided into
  - 4 identifier bits,
  - 2 bits specifying the length of the data field;
- 2 parity bits (bits 6 and 7, called P0 and P1);
- 1 recessive stop bit.

The special architecture of the byte contained in the identifier field makes it possible to divide the first 6 bits of the byte, providing 64 possible identifiers (from 0 to 63) in 4 sub-

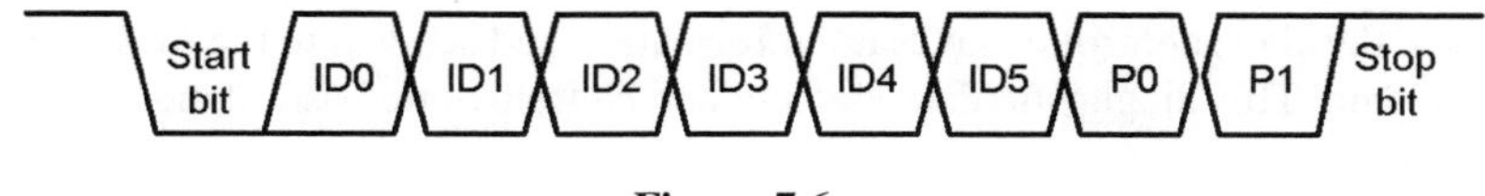


**Figure 7.6**

groups (via the 2 bits, numbered 4 and 5) of 16 identifiers (via the 4 bits numbered 0 to 3), each having data fields with a length of 2, 4 or 8 bytes. The division of the 64 LIN identifiers is as follows:

- from 0 to 59: unconditional frames, event-triggered frames or sporadic frames;
- 60 and 61: frames used for carrying diagnostic data, or for user-defined extensions;
- 63: value reserved for future enhancements of the LIN protocol.

Note that, since the data are supplied to the bus by any of the network members, which may be the master itself or any of the slaves, a message can come only from a single source (slave or master tasks), but more than one member of the network can use it; this is a distinctive property of LIN.

### Data field

The frame header is followed by the 'response' of one of the slaves (or the master task). The slave task consists of sending

- the 'data field' (LSB first);
- the 'check field', a modulo 256 inverse sum with the remainder of the MSB carried over to the LSB, calculated for all bits transmitted in the frame.

A data communication field can carry 1 to 8 data bytes plus 3 command and security data bytes. The number of bytes in each frame depends on the specific identifier of the transmitted frame and will be set by agreement between the issuer and all the subscribers of the network.

### Interbyte space

The frames are separated by a time interval called the 'interframe'.

### Acknowledgement

Unlike CAN, LIN does not specify the acknowledgement of a correctly received message. The master node verifies the consistency of the message initialized by the master task and received by one of its slaves. If there is an inconsistency (for example, no response from the slave, incorrect checksum, etc.), the master can retransmit the message. If the slave detects an inconsistency, the slave controller saves the information and sends it to the master in the form of diagnostic data. This diagnostic data element can be transmitted as a standard frame of the data frame type.

### *7.2.3 Physical layer (PL)*

I will start by providing some details of the physical layer used by LIN, which is actually an enhancement of the ISO 9141 physical layer which has long been used for the diagnostic bus of motor vehicles.

#### Single-wire bus ... or almost

As in most systems operating in asynchronous mode, the LIN bus consists of a single communication channel. A data resynchronization system can be implemented on the basis of the received signals.

The physical medium of the bus consists of a single wire (plus the usual ground wire, of course), connected to the positive pole of the battery via a potential recovery (pull-up) resistor and a diode which will be described later (Figure 7.7). The bus is activated by means of the line driver, by setting the communication line to ground potential, in other words the 'dominant' level.

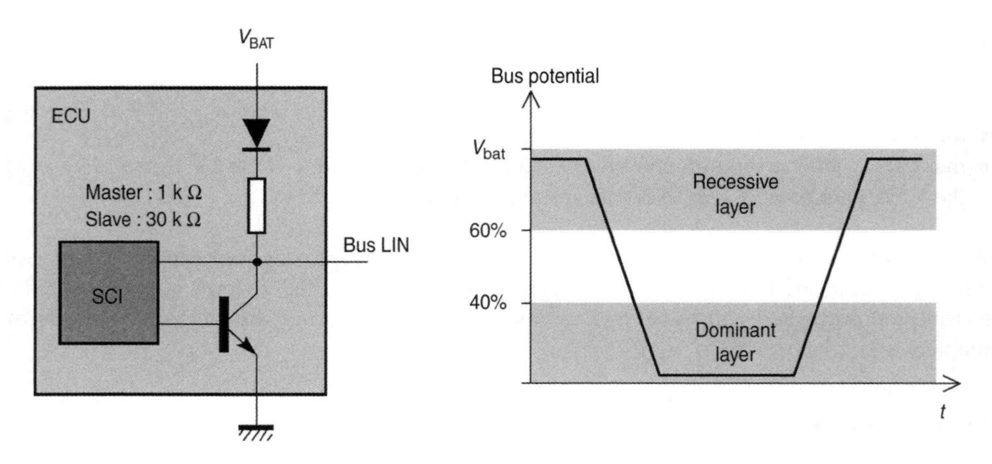


**Figure 7.7**

#### Bus bit rate

The maximum bit rate of a LIN network is 20 kbit s$^{-1}$. For historic reasons, related to the UARTs used to form RS 232 serial links, LIN bit rates are very often multiples of 300, i.e. 600, 1200, 2400, 4800, 9600 and 19200 bit s$^{-1}$. Another value is also used: this is the bit rate of 10.4 kbit s$^{-1}$, widely used for what is called the diagnostic K-line.

### Bit coding

The bit coding is NRZ (non return to zero), which, for an asynchronous link, is not ideal if a high quality is wanted for bit detection and clock resynchronization, or clock losses due to the relative slippage between the slave and master on-board oscillator frequencies. I will return to this point a little further on.

### Values of the bus

As with CAN, the LIN bit on the bus can have two complementary logical values, namely 'dominant' and 'recessive'. Unlike the CAN specification which is completely open as regards the physical transmission medium, the LIN specification leaves no doubt as to the possible type of medium, but states, in a way that cannot be questioned, the direct correspondence between the values of the bits and the voltages to be provided:

- dominant: '0', ground potential;
- recessive: '1', the + ve terminal of the battery.

### Maximum number of connection points of a LIN network

Theoretically, the maximum number of nodes that can be connected to a LIN network is 64 – but there is no chance of achieving this! For practical purposes (disregarding all the false promises of marketing departments), the total number of nodes is, as usual, limited by physical parameters such as the signal propagation time on the line (approximately 5–7 ns $m^{-1}$ for wire links), or by the electrical loads (resistive and capacitive) of the set of nodes connected to the bus, and therefore by the peaks and variations of the rise and decay times due to the number of line drivers connected to the network. In fact, if we come down to earth from the infinite theoretical plane,

- the maximum 'recommended' number of nodes in a sub-network 'should not exceed' (quote) 16 (!).... even though the protocol can support a maximum of 63;
- the length of the 'galvanic' (wire) link of a network must not exceed 40 m;
- the termination resistance is 1 kΩ for the master node, and 20–47 kΩ for the slave nodes;
- the capacitance of the master node must be greater than those of the slave nodes, so as to eliminate the major differences in value which could occur depending on the number (and variety) of the nodes connected to the network.

Figure 7.8 summarizes the principal parameters relating to the line aspect of LIN.

### Line interface

Figure 7.9 is a conventional physical representation of the stage used as a bidirectional electrical interface with the LIN bus. In spite of its simplicity, similar to that used for the I2C bus, this stage includes some hidden problems.

| | | | min | typ | max | unit |
|---|---|---|---|---|---|---|
| 3.6.1 | Total length of bus | $LEN_{BUS}$ | | | 40 | m |
| 3.6.2 | Total capacity of system | $C_{BUS}$ | 1 | 4 | 10 | nF |
| 3.6.3 | Global time constant of system | | 1 | | 5 | $\propto$s |
| 3.6.4 | Capacity of master node | $C_{MASTER}$ | | 220 | | pF |
| 3.6.5 | Capacity of slave node | $C_{SLAVE}$ | | 220 | 250 | pF |
| 3.6.6 | Capacity of line | $C_{LINE}$ | | 100 | 150 | pF/m |

**Figure 7.8**

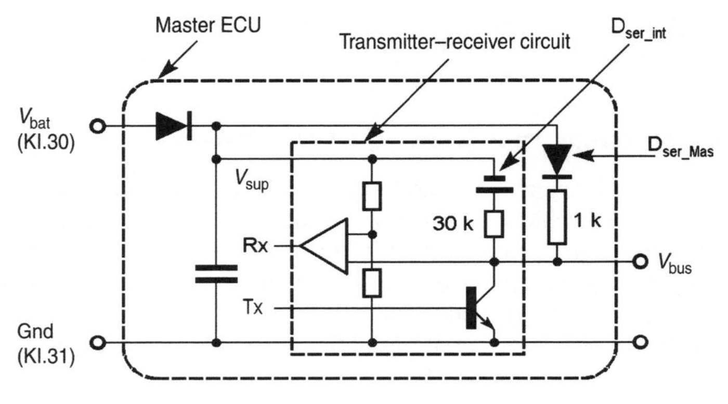


**Figure 7.9**

The presence of the diode D1 connected in series with the output transistor load resistor is necessary to prevent any possible resupply of the node in question via the bus if the node suffers a local ground loss. Note also that the electrical values shown in the LIN specification sometimes refer to the voltages present outside the node ($V_{bat}$) and sometimes to the voltage present within the node ($V_{sup}$). It is therefore important to know and allow for the exact value of the voltage drop (related to the intrinsic consumption of the node in question) found at the terminals of the second diode D2, called the anti-return diode, which protects against polarity inversion of the power supply and is connected in series with the node power supply. Figure 7.10 shows the electrical levels on the LIN line – but it does not reveal the underlying problems.

These are concealed in the latent ambiguity of LIN which has to be cheap, insensitive to fluctuations in supply voltage, ground differences, variations in network topology, the number of participants, the degree of radiation if any, etc., while being usable for industrial purposes at up to 20 kbit s$^{-1}$, to match the hopes and dreams of all designers ... otherwise all the marketing efforts are in vain!

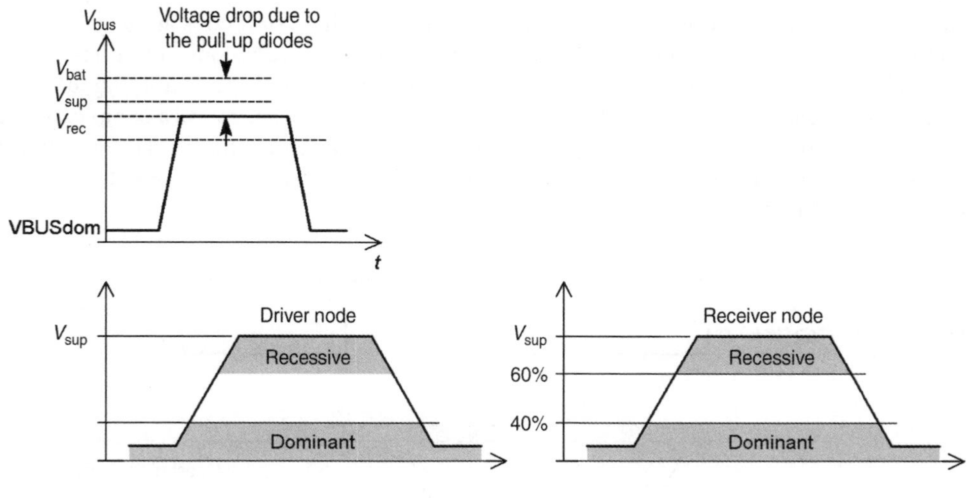


**Figure 7.10**

Clearly, these concepts are technically contradictory; otherwise we would be in trouble. In other words, there is a very good chance that we will eventually reach a very critical compromise – and this is indeed the case. I shall now attempt to explain this.

In this case, theory and reality are separated not by a gap but by a yawning chasm. We have to allow for

- the fact that each node has its own clock (based on an RC circuit if possible, to save expense), with a known initial accuracy of its nominal value and tolerances and fluctuations depending on time and temperature. We must not forget that some nodes can be very hot, inside a vehicle, whereas others are subject to freezing temperatures outside the vehicle in winter;
- propagation delays, due to
  - the length of the line formed by the bus,
  - the internal electronics of the integrated circuits themselves;
- the numerous capacitors added to the networks for the purposes of
  - limiting radio interference, EMC,
  - protection against electrostatic discharge (ESD) (4 or 8 kV);
- resistive and capacitive loads contributed by the nodes, which result in totally asymmetric rise and decay times, playing an important part in the temporal and electrical distortions of the bit according to the internal voltage thresholds (and their tolerances) of the integrated circuits;
- and also, for economic reasons, the fact that the values of the electrical signals, being directly related to the value $V_{bat}$, are directly subject to the effects of the latter as well as the repercussions of ground differences.

This all further complicates the precise definition of the 'sampling point' at which each network participant decides to sample the electrical signal representing the bit in order to detect its binary value. After long discussions and the definition of a constant-slope system (LIN rev. 1.2) and/or a constant-rise system (LIN rev. 1.3), the LIN consortium has finally specified, in LIN 2.0, that the best way of establishing, quantifying and qualifying all these problems and achieving a good quality of signals carried on the bus is to use a complex set of cyclic ratios of the signals (Figure 7.11) so that the signals representing the bit value are almost always present at an UART which is more or less normally constructed, enabling it to process these signals correctly.

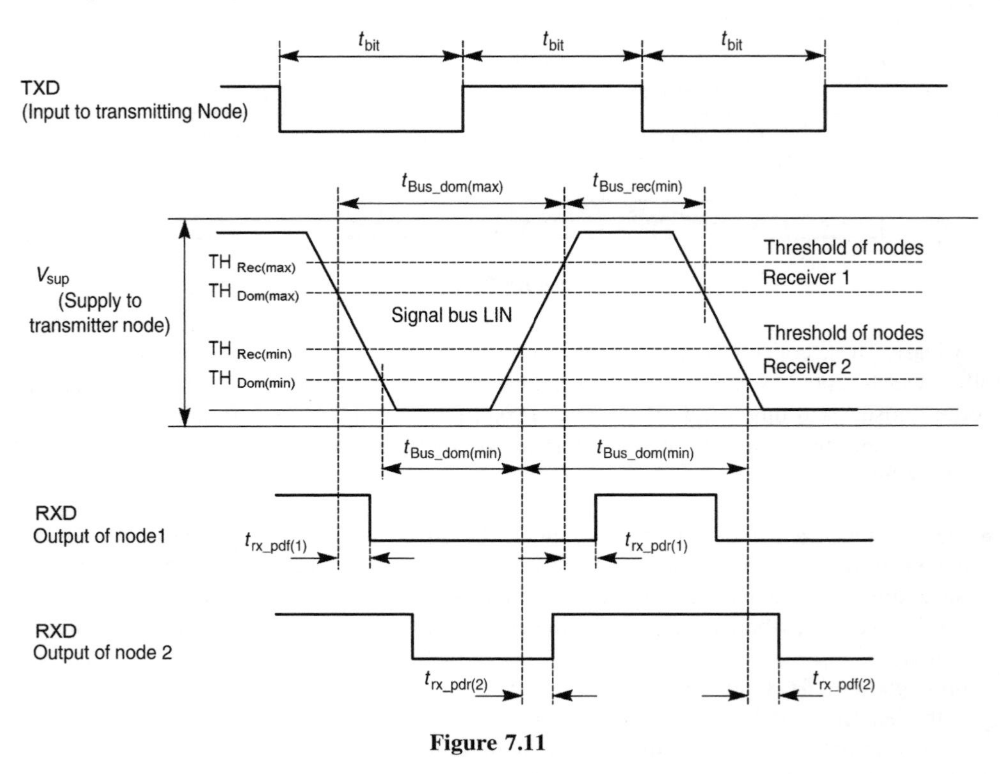


**Figure 7.11**

To conclude this section dealing specifically with the physical layer, note that the line driver and receiver specifications comply for practical purposes, electrically speaking, with ISO 9141 (called the K-line, supporting diagnostic messages for vehicle components), supplemented with a few enhancements here and there to provide better EMC performance.

### Sleep mode and wake-up

To reduce the system's power consumption, a LIN node can be put into sleep mode by a special message broadcast over the network. For this purpose, a special frame, called the 'sleep mode frame', is reserved for putting the whole network into sleep mode. Its identifier is (hex $0 \times 80$). In sleep mode, with no internal activity and a passive line driver, the level present on the bus is recessive.

### *7.2.4 Performance of LIN 2.0 compared with LIN 1.2 and 1.3*

As many circuits and applications using LIN 1.2 and 1.3 components are still in the market, this section is intended to shed some light on the basic differences between the LIN 2.0 final version and the earlier versions, LIN 1.2 and LIN 1.3.

First of all, and most importantly, it must be borne in mind that LIN 2.0 is a 'superset' of its predecessors, and that downward compatibility is maintained and assured. However, many novel features have been introduced, in both the communication protocol and the physical layer, mainly at two levels:

- on the one hand, there is the correction or addition of information on certain technical shortcomings of the earlier versions, for the example the fact that
  - larger variations of supply voltage can be withstood,
  - ground differences similar to those actually found in vehicles can be withstood,
  - larger variations in temperature, local oscillator fluctuations and tolerances, etc. can be withstood;
- on the other hand, there is the addition of new options to LIN, for example
  - the network can be self-configured by a daisy-chain procedure,
  - automatic speed detection is possible, permitting the construction of LIN plug and play modules, etc.

Rather than embark on a long and tedious discussion, I shall offer you an overview of all this in the form of two comparative tables, the first of which, Figure 7.12, shows the main differences in the protocol itself, whereas the second, Figure 7.13, is concerned only with the physical layer.

| LIN 2.0 | LIN 1.3 |
|---|---|
| Number of data byte is independent of the identifier | Number of data byte is correlated with the identifier |
| Two checksum models: classic and enhanced | One checksum model: classic |
| Node configuration via mandatory and optional commands | Node configuration not specified |
| Diagnostic via three optional methods | Diagnostic not specified |
| Product identification is mandatory | Product identification not specified |
| Network management timing is defined in seconds | Network management timing is defined in bit times |
| Status management with one mandatory response error bit | Status management with five different message error bits |

**Figure 7.12**

Note that the latest modifications of the protocol at the level of the physical layer have finally met the requirements of motor vehicles in terms of performance (see the range of performance covered in Figure 7.14) and cost.

### *7.2.5 The same old problems*

Despite this, we still have the everyday problems to solve, namely the problems of radio interference, radiation, susceptibility, electrostatic discharges, etc.

| LIN 2.0 | LIN 1.3 |
|---|---|
| Transmitter is optimized for all worst-case conditions | Transmitter is based on insufficient LIN 1.2 spec. |
| Electrical AC parameters are defined generally via duty cycle | Electrical AC parameters are defined specifically for different transceiver types |
| *resulting in*<br>Proper LIN communication within worst-case system environments:<br>– Full GND/BAT shifts (up to 10% of $V_{bat}$)<br>– Full oscillator tolerance (up to 2% between master and slaves)<br>– Full communications peed (up to 20 kbit $s^{-1}$)<br>– Full polarity protection diode drops in BAT line (0.4 V...1.0 V)<br>All parameters can have their worst-case values at the same time. | *resulting in*<br>LIN communication only within limited system environments:<br>– Reduced GND/BAT shifts and/or<br>– Reduced oscillator tolerance and/or<br>– Reduced communication speed<br>NOT all parameters can have their worst-case values at the same time. |

**Figure 7.13**

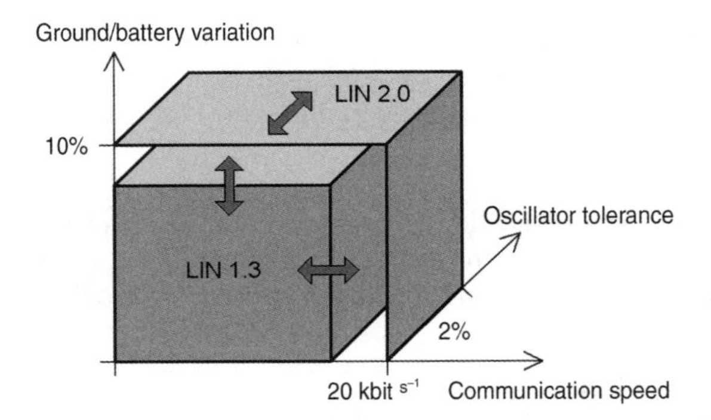


**Figure 7.14**

One of the basic problems of serial buses in asymmetrical physical implementations with respect to ground (the I2C bus, the one wire CAN bus, etc.) is the radiation and radio interference that the transmitted signal can cause during its propagation along the bus. These problems limit the maximum speed of the network for any given bit coding (NRZ coding in this case).

At present, because of local regulations (FCC in the USA, ETSI in Europe, ARIB in Japan, etc.), and the proprietary standards that may or may not be in force around the world, this limit is frequently in the vicinity of 15–20 kbit $s^{-1}$, speeds for which the level of emitted pollution is close to that emitted by a CAN differential pair operating at 500 kbit $s^{-1}$, and falls within the pollution limits to be complied with.

Clearly, we can and must provide filter elements (inductances, capacitors, etc.) or vary the signal slopes and thus have to perform a tricky juggling act to ensure the resynchronization of signals above 10 kbit $s^{-1}$. Unfortunately, in this case most or even all of the arguments concerning the low cost of this LIN link lose their validity; we will probably return to the use of low-speed CAN, in differential mode, which provides greater reliability and security of communication for the same cost.

## 7.3 Cost and Market

This factor was, and still remains, at the heart of the promotion of LIN by its designers. After the repeated hyping of this system as cost-effective (quote: 'two or three times cheaper than a CAN node'), many users are becoming rather disenchanted. In fact, if we allow for the essential security features, even at a minimal level, we find that many extras are needed in the integrated circuits (resynchronization system, automatic speed detection, a dedicated UART that is rather more intelligent than a normal UART, ground disconnection, etc.) to enable this 'simple' concept to operate correctly. Clearly, the cost of each solution is affected by this.

But let us return to this well-known phrase, 'A LIN node is $x$ times cheaper than a CAN node'. Everything depends on the '$x$'. Let us be very frank.

- The price of a LIN line driver is very close to that of a one-wire CAN driver.
- Admittedly, in absolute values, in terms of $mm^2$, there is a significant difference between the area of silicon required for a LIN protocol handler and that needed for a CAN handler, but
  - where the master task is concerned, this difference is not very significant, because a master needs a built-in microcontroller (with a surface area that is not negligible), and the difference between a microcontroller + CAN, a microcontroller + LIN and a microcontroller + UART is small with respect to the total surface area of the microcontroller. So we fall into the classic 'per cent of a per cent';
  - where the slave task is concerned, the LIN handling can be implemented in different ways (small microcontroller, wired logic, input/output functions, etc.).

But here again, the difference becomes small when an automatic baud rate detection device is implemented, to withstand large variations in the oscillator frequency, etc.

In this comparison, we must also consider the power of the CAN protocol and the absence of hardware fault management (fault tolerance), the lack of automatic frame retransmission and network information provided by error counters, etc.

I will conclude this section with a few remarks about the market. At present, an ordinary vehicle has 3–10 LIN nodes; this may increase to 35–40 nodes, sometimes arranged in six different LIN networks, for certain top-range models. The best forecasters estimate the global market for LIN nodes to be worth 1–1.5 billion per year in 2008–2010.

## 7.4 Conformity of LIN

### *7.4.1 General*

The LIN protocol belongs to the LIN consortium. At the time of writing, it has not been standardized by the ISO, and, as mentioned at the beginning of this chapter, it is supplied 'as is'. On the contrary, the concept of a standard implies new problems of conformity and tests for conformity with the proposed standard.

For your information, you should know that, whenever an ISO standard is developed, then, not surprisingly, another standard called 'conformance tests' is always developed at the same time, describing in detail the measurements to be made to establish conformance with the standard in question. For evident reasons (the desire for rapid access to a new market), the designers of LIN have put the protocol through many different versions (1.0, 1.2, 1.3, . . . up to 2.0). It is not a simple matter to develop this famous document at the same time, to the point that there can be official independent laboratories and authorities accredited to grant their recognition of any conformance.

On 15 August 2004, the LIN consortium officially granted its approval to the internal issue of a document of this kind, namely the 'LIN 2.0 conformance test plan', and its forthcoming external distribution. Now, although LIN is not considered to be a secure bus, for simple reasons of functional interoperability, car manufacturers can at least rely on certified LIN 2.0 products – without even considering the constraints of possibilities of 'double source' procurement that are so attractive to the purchasing departments of every business.

In parallel with this, as the demand for nodes is already increasing daily, many laboratories have now completed the process of bringing their test equipment into conformity with this document, to counteract all the latent problems of compatibility and interoperability which may arise between nodes in networks. In conclusion, we should also note that the conformity should obviously relate to the conformity of the physical layer as well as to the proper cohesion and handling of the protocol.

### *7.4.2 Hardware and software conformity*

To make matters entirely clear, these well-known certification tests will certainly leave a trail of 'corpses', because, for the simple reason of the low cost announced by the promoters of LIN, the users very often implement LIN protocol handling in software form, in 'miniature' microcontrollers fitted with simple UARTs, contrary to the CAN protocol, which can only be practically implemented in hardware by semiconductor producers. Consequently, certification may be refused simply on the basis of a bit error in programming. In terms of certification, this leads to a loss of time (there are long queues at the entrances to these laboratories) and money for repeated attempts (leading to the aforementioned self-promotion of certain laboratories, which see a new abundant source of potential revenue and will be remarkably thorough in determining protocol conformity, down to the last bit!). We should therefore welcome the year 2008 which will see the final conformity tests for both the physical layer and the communication protocol level .

By way of example, Figure 7.15 shows a facsimile of 'conformance test' certificates (issued by the well-known C&S company) for the Philips semiconductors LIN TJA 1020 line driver circuit, concerning the conformity of the physical part with LIN 1.3 (Figure 7.15a) and LIN 2.0 (Figure 7.15b).

## 7.5 Examples of Components for LIN 2.0

Here again, the potential size of the market for LIN applications means that there are numerous component suppliers (for example, in strictly alphabetical order: AMIs, Freescale, Infinéon, Melexis, Philips-NXP Semiconductors, ST, etc.), each of which has chosen a slightly different range of functionality according to its technological skills and specialization in relation to semiconductors and marketing.

Fachhochschule
University of Applied Science

Salzdahlumer Straße 46/48
D-38302 Wolfenbüttel

Fachhochschule Salzdahlumer Str. 46/48
C&S group – Prof. Lawrenz D-38302 Wolfenbüttel

Philips Semiconductor
International Product Marketing Automotive
Business Line
Rob Bouwer

Tel.+31 24 353 4018
Fax. +31 24 353 4100
Rob.Bouwer@philips.com

C&S communication & systems group
c/o Informatik

Prof. Dr.-Ing. W. Lawrenz
Director

Tel +49 5331 939 6600
Fax +49 5331 939 6602
Email W.Lawrenz@cs-group.de
Web www.cs-group.de

| Auftragsnummer (unbedingt mit angeben) | Ihr Zeichen, Ihre Nachricht vom | Mein Zeichen | Wolfenbüttel, den |
|---|---|---|---|
| 2004 – 182 | | WL/TR/ TJA1020_Report | 2004-10-07 |

**Philips TJA 1020**

**LIN OSI Layer 1 - Physical Layer Conformance Tests**

**Final-Report**

**Performed Tests and References**

The following tests according to the referred specification have been performed:

- **LIN OSI Layer 1 – Physical Layer**
  **For the LIN Protocol Specification Revision 1.3 (Dec. 12, 2002)**
  **Version 1.3**

**Test Equipment**

The following test equipment and test system have been used:

- ***Tester***
  - LIN Conformance Tester Version 1.3
- ***Implementation Under Test (IUT)***
  - TJA 1020 [4E5C8 n4351]

**Test Results**

- The LIN module **passed the LIN OSI Layer 1 PL CT tests successfully**

For detailed information see chapters *Problem History* and *Test List* at the following pages.

LIN Conformance Testing C&S
© 2004 C&S -1- communication & systems group
Fachhochschule Braunschweig/Wolfenbüttel

(a)

**Figure 7.15**

As LIN 2.0 is currently recommended by its consortium for all new applications, I shall briefly describe two circuits (and a half – see Chapter 7, on 'Fail safe SBC LIN') which conform strictly to LIN 2.0. These are the TJA 1020/1021 line drivers, meeting the LIN 1.3 and 2.0 specifications, and the UJA 1023 I/O slave and its associated demo board.

Fachhochschule C&S group — Salzdahlumer Str. 46/48 D-38302 Wolfenbüttel

Philips Semiconductor
International Product Marketing Automotive
Business Line
Rob Bouwer

Tel.+31 24 353 4018
Fax. +31 24 353 4100
Rob.Bouwer@philips.com

Salzdahlumer Straße 46/48
D-38302 Wolfenbüttel

C&S
communication & systems group
www.cs-group.de
c/o Informatik

Thomas Roskam
Tel +49 5331 939 6618
Fax +49 5331 939 6602
Email Th.Roskam@cs-group.de

| Auftragsnummer (unbedingt mit angeben) | Ihr Zeichen, Ihre Nachricht vom | Mein Zeichen | Wolfenbüttel, den |
|---|---|---|---|
| 2004 – 214 | | TR/AK/ TJA1020_Report | 2004-11-30 |

# Philips TJA 1020
# LIN OSI Layer 1 - Physical Layer Conformance Tests
# Final-Report

### Performed Tests and References

The following tests according to the referred specification have been performed:

- LIN Conformance Test Specification
  LIN OSI Layer 1 – Physical Layer
  For the LIN Physical Layer Specification Revision 2.0 (Sep. 23, 2003)
  Version 1.0

### Test Equipment

The following test equipment and test system have been used:

- *Tester*
  - LIN Conformance Tester Version 1.3
- *Implementation Under Test (IUT)*
  - TJA 1020 [4a4H0 nG417]

### Test Results

- The Implementation Under Test has passed the LIN OSI Layer 1 Physical Layer Conformance Tests successfully.

For detailed information see chapters *Problem History* and *Test List* at the following pages.

| | i.A. Roskam | i.A. Kurz |
|---|---|---|
| Wolfenbüttel, 2004-11-30 | Roskam, Project Manager | Kurz, Test Execution |

LIN Conformance Testing
© 2004 C&S
-1-
C&S communication & systems group
Fachhochschule Braunschweig/Wolfenbüttel

(b)

**Figure 7.15** *(Continued).*

*7.5.1 The TJA 1020/1021 line drivers*

Figure 7.16 shows the block diagram of the TJA 1020/1021 circuit. Once again, it is encapsulated in an 8-pin case – and even without considering the pins, the principle of this circuit shows a family resemblance to the associated 'one wire CAN' design. Here again, very low power consumption is essential, as is provision for earth connection breaks, freedom from

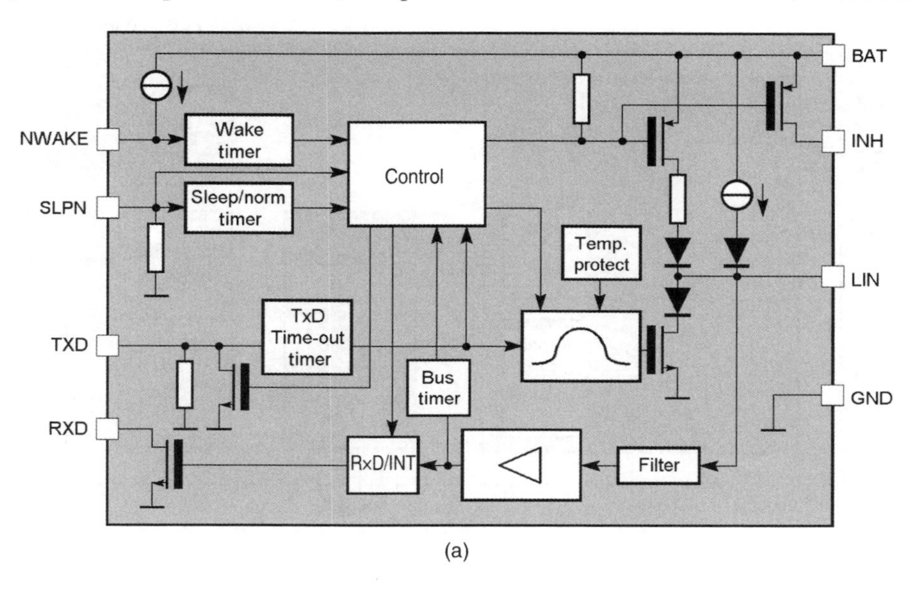


(a)

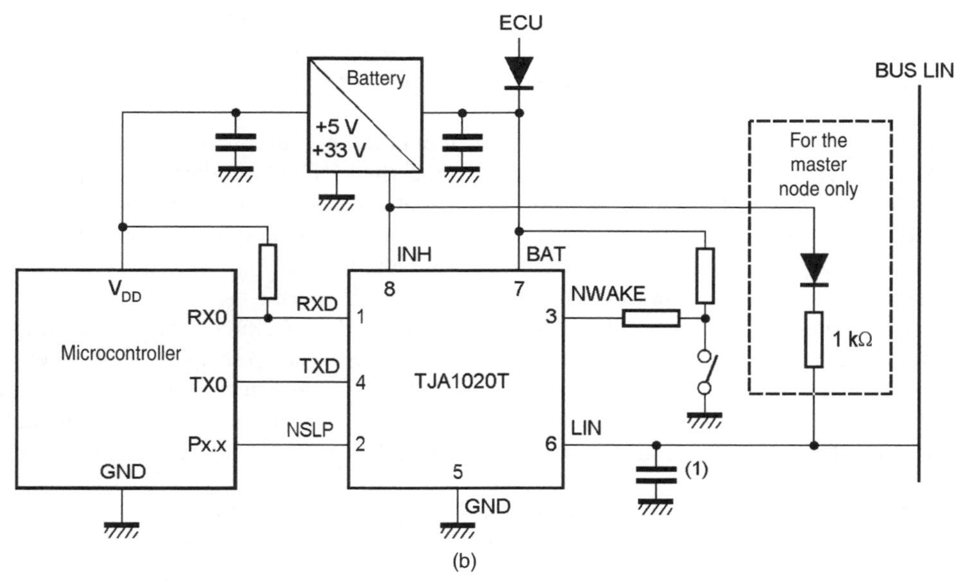


(b)

**Figure 7.16**

short circuits, heat protection and the possibility of adjusting the signal slopes to minimize EMC problems.

### *7.5.2 The UJA 1023 LIN I/O slave*

The block diagram is shown in Figure 7.17.

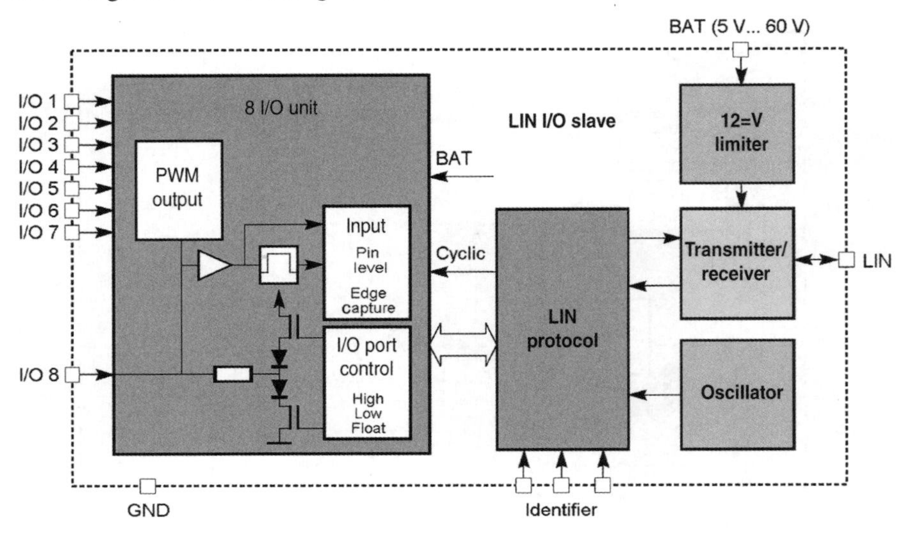


**Figure 7.17**

These circuits handle numerous analogues and digital inputs and outputs and A/D and D/A conversions, so a fail-safe security system and a fall-back position ('limp home') is very important, as it covers the entire LIN 2.0 handling, including

- automatic bit rate recognition. An oscillator is built in for this purpose, covering the various operating ranges and bit rate tolerances described in the LIN 2.0 specification. Its presence also makes it possible to handle all possible time outs, in case of communication problems and/or line faults such as permanent short circuits (a permanent dominant state on the LIN) and thus enables coherent diagnostic messages to be sent;
- diagnostic messages, sent in the form of responses to LIN messages;
- the unique number of the circuit (manufacturer and model), making it easy to develop a configuration (which may or may not be based on the daisy chain principle – see below) for mass-produced nodes or modules, even if the exact functions for which they will be used are not known. Furthermore, as the module type, model and version are known, the module control software implemented in the microcontroller cannot confuse certain specific characteristics and create 'tangles' with other functions;
- possible automatic reconfiguration on the daisy chain principle. The LIN 2.0 version describes a very specific procedure for configuring the nodes of the network by what is called a daisy chain procedure. Clearly, this works only if the component allows it, which is the case here (see below) because of a special little electronic device conforming to the LIN protocol commands;
- etc.

### Configuring the elements of a network using UJA 1023 circuits

When I/O slave circuits are used, we often find that many of them are connected to the same LIN network. They must therefore be addressed separately, if only to enable them to be configured according to their own functions. LIN 2.0 allows two configuration solutions to be used, if the integrated circuits support them, which is the case with UJA 1023 circuits.

### The hardware solution

The first solution is based on a possibility of hardware configuration which can be implemented using special pins which help to define the specific address of the component. Clearly, this solution is easy to implement, but it has certain requirements (Figure 7.18):

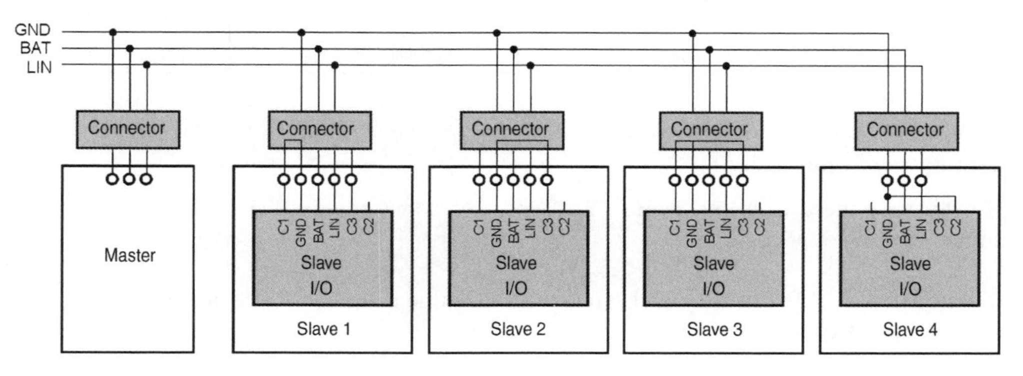


**Figure 7.18**

- mechanical devices (which may be more or less costly) in the connector of the module on which the integrated circuit is located, to enable any module to be fitted in different parts of the system, for example either on the right or on the left of a vehicle, while maintaining dedicated right-to-right and left-to-left operation;
- or the mounting of the integrated circuit on a dedicated printed circuit, one for each module, the specific configuration being formed by the tracks. In this case, a 'right-hand' module remains a 'right-hand' module and can never be used as a 'left-hand' module.

### The daisy chain solution

In this case, no special provision is required. This is because, for a given wiring system, we know the first circuit served by the strand of the cable bundle/harness, and, in this case, this is the only circuit that has a local device in its connector (Figure 7.19). Therefore, the configuration of this first circuit is carried out first of all, and then, when this is finished, its configuration is declared complete and the switch between its ground pin and pin C3 is closed so that we can move on to the configuration of the next circuit. . . and so on, as shown in the diagram, following the well-known 'daisy chain' procedure. This solution has the advantage of making it possible to manufacture and store (for after-sales purposes, for example) identical, 'universal' modules which can be fitted equally well in place of other modules.

All these possibilities make this circuit very suitable for multiple-use applications.

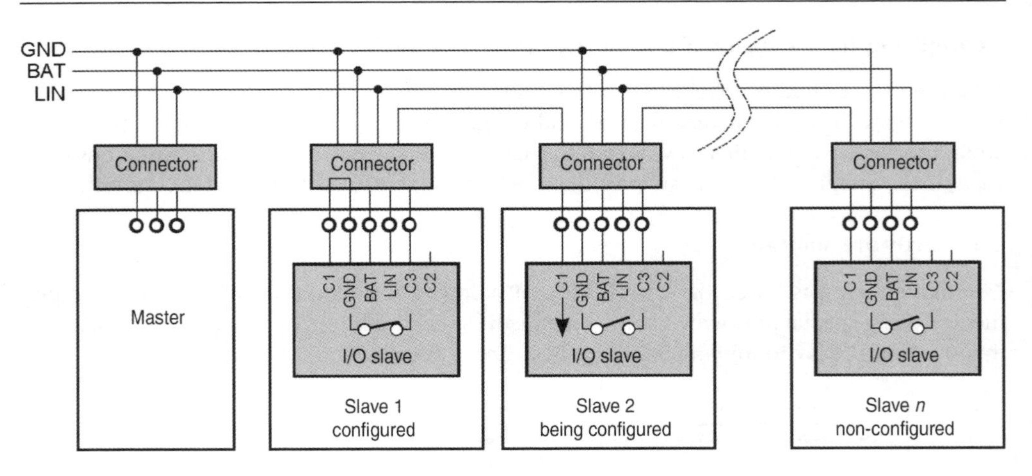


**Figure 7.19**

## Examples of applications

To illustrate these concepts, Figure 7.20 provides some non-exclusive examples of these numerous LIN applications:

- on/off inputs, LED display and luminance adjustment via the PWM acting as a digital-analogue 'pseudo-converter';
- numeric keypad matrix control;

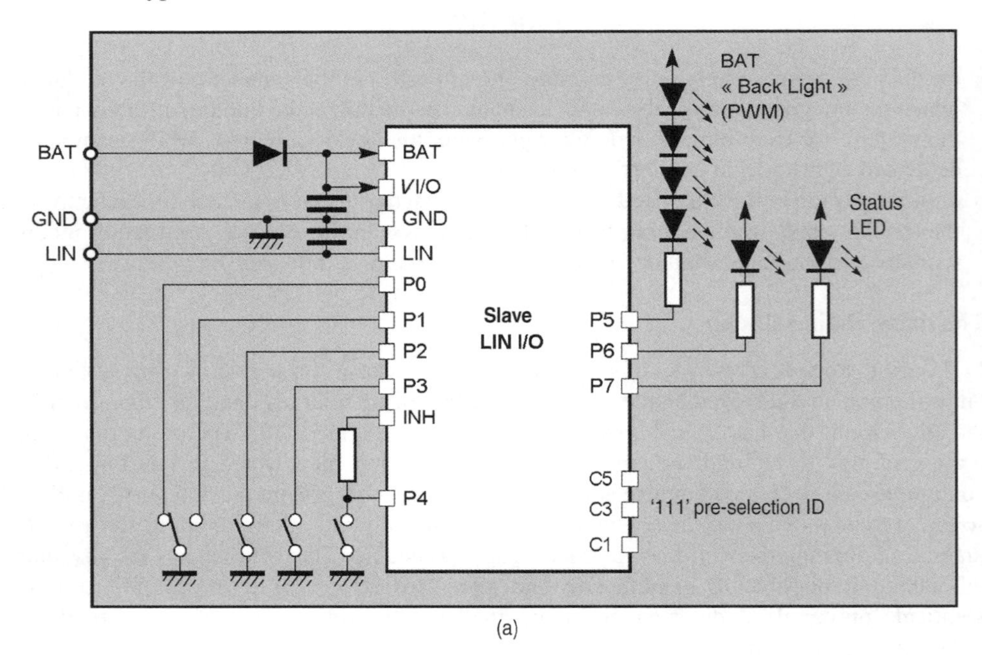


**Figure 7.20**

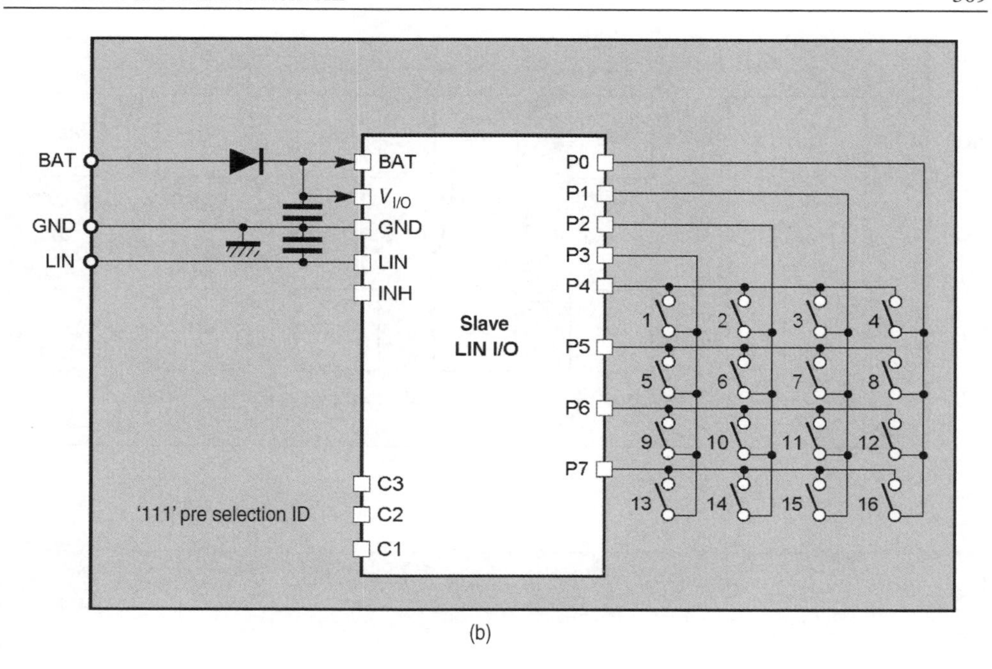


(b)

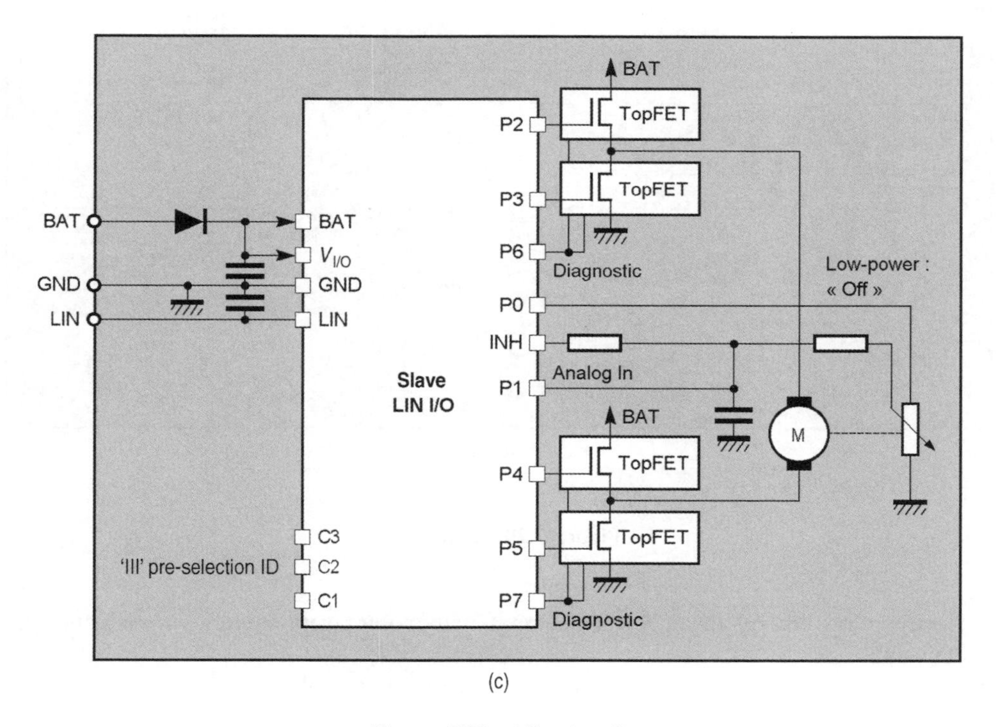


(c)

**Figure 7.20** *(Continued)*

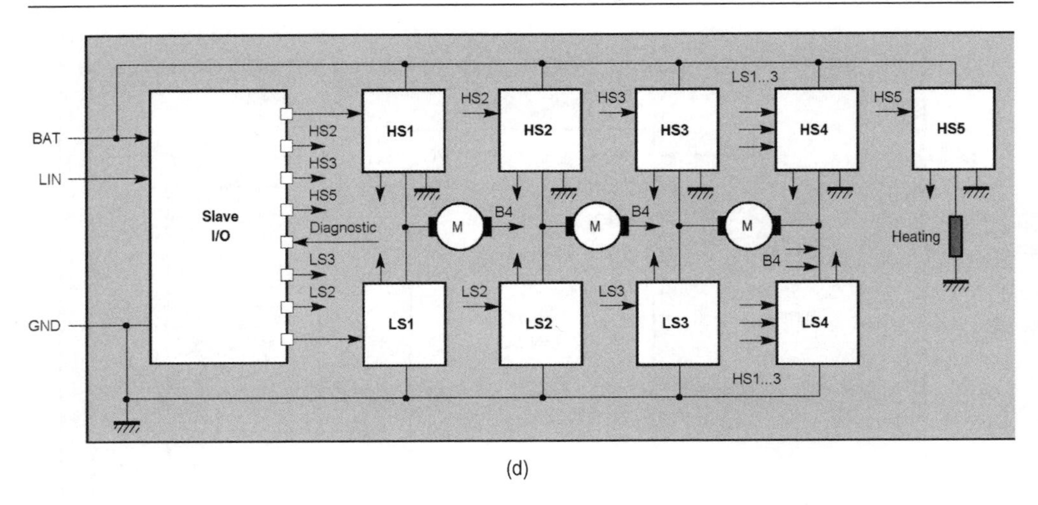


(d)

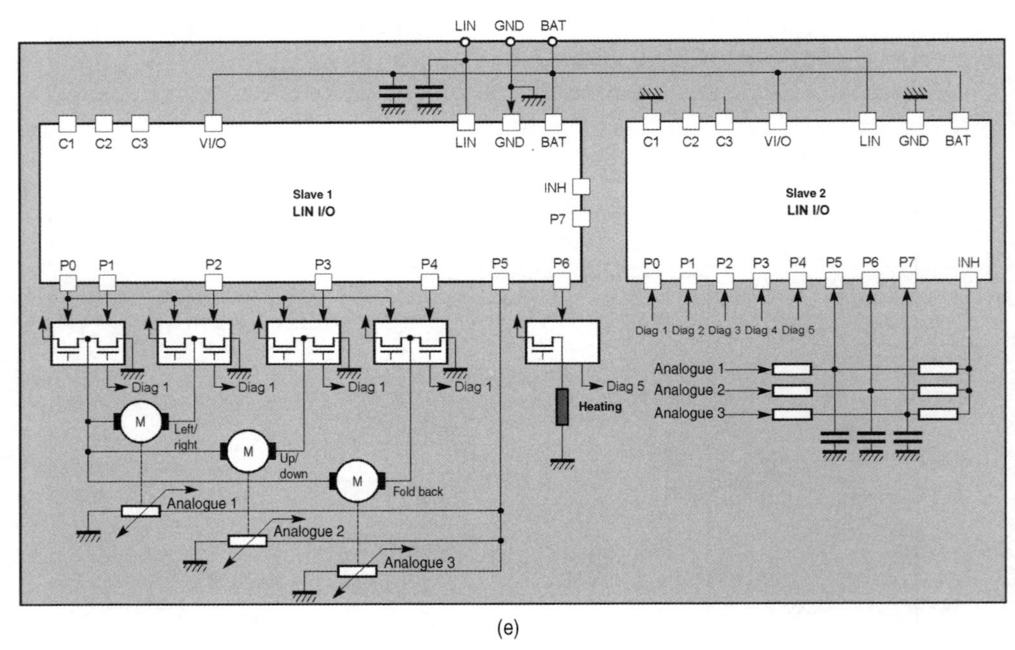


(e)

**Figure 7.20** *Continued.*

- electric motor control via an H-bridge circuit, with motor position return via analogue data element;
- control of multiple motors (e.g. for wing mirror motor controls);
- serial connection of circuits;
- etc.

## Demo board

Figure 7.21 shows the demo board associated with this circuit. It enables us to master this circuit, by providing an understanding of the details of its functionality and assisting with the development of a module equipped with this circuit, and also by incorporating all the latest subtleties of LIN 2.0, by directly displaying the contents of the frames sent and received (Figure 7.22).

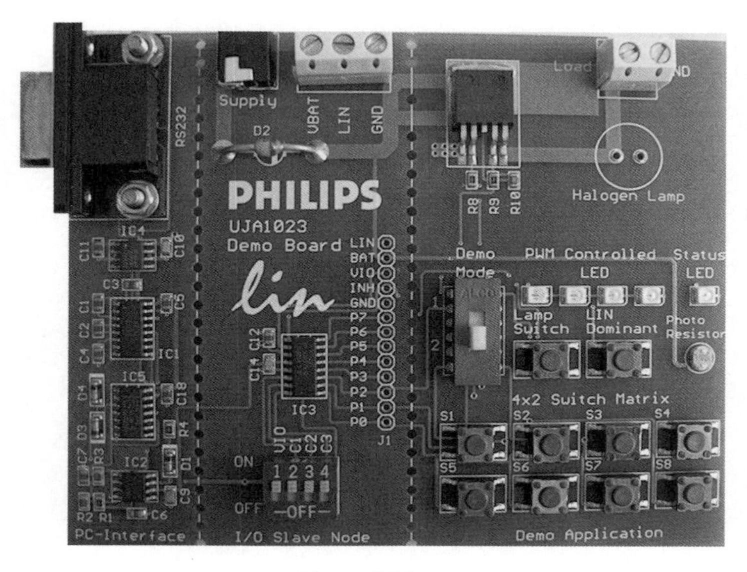


Figure 7.21

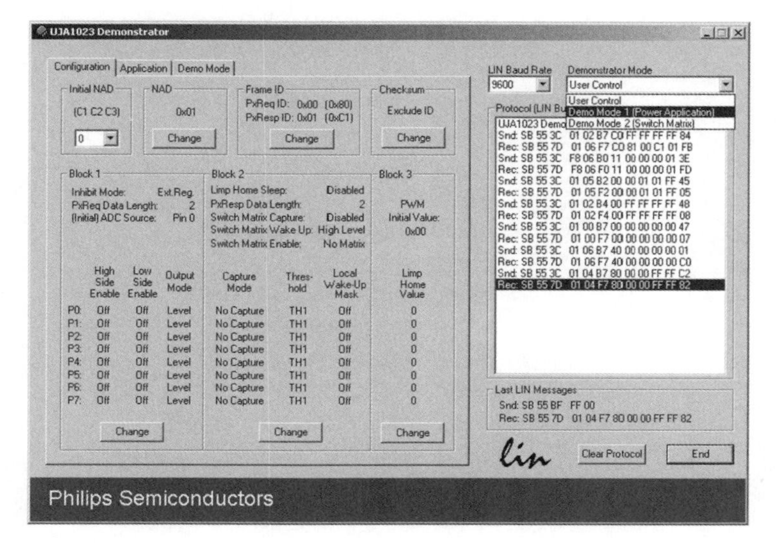


Figure 7.22

# 8

# Think 'Bus', Think 'Fail-safe SBC', 'Gateways' ...

I have now briefly described a large number of 'serial' communication systems. In the previous chapters, you would no doubt have noticed that each system has its own characteristics which make it preferable for certain proposed applications (protocol power, cost, dedicated functionality, etc.).

In an industrial, motor vehicle or aeronautical project, many of these applications are present and coexist in the final system. Moreover, some messages in any communication system must, for many reasons, be carried on another system as well. For example, the engine rpm (at high speed) must be displayable on the dashboard (at lower speed). We therefore need to design gateways and repeaters between networks and buses (see Figure 8.1) for various transfers:

- from one speed to another, for example
  - from high-speed CAN (controller area network) 1 to high-speed CAN 2;
  - from low-speed fault-tolerant CAN to high-speed CAN;
- from one medium to another, for example
  - from one-wire CAN to differential pair CAN;
  - from differential pair CAN to optical fibre CAN;
- from one bus to another, for example
  - from CAN xxx to LIN (local interconnect network),
  - from CAN xxx to D2B or MOST or IEEE/IDB 1394;
- etc.

Very commonly, a minor or major operation is also performed on the transferred message as it passes through the gateway; this may involve a change or readdressing of the destination address(es), calculations using the message content, etc. This tends to occupy a lot of the capacity of the system CPU, and therefore a specific CPU is often dedicated to the gateway

*Multiplexed Networks for Embedded Systems: CAN, LIN, Flexray, Safe-by-Wire...* D. Paret

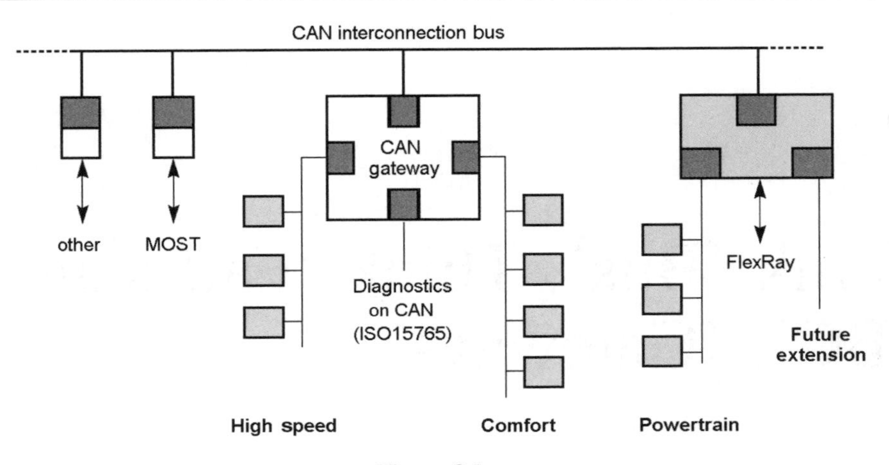


**Figure 8.1**

functions for these applications, as described at the end of this chapter. Given the wide range of protocols, buses, etc., the market for gateways looks promising! But before entering the domain of high-level gateways, let us first take a detour via another gateway concept found in many on-board systems: fail-safe SBC (system basis chip).

## 8.1 Fail-safe SBCs: Their Multiple Aspects and Reasons for Using Them

The electronic systems and networks of motor vehicles or industrial systems consist of numerous nodes or modules of varying complexity. There are different ways of designing them with simple levels of complexity and functionality, on different scales.

### *8.1.1 The conventional approach*

A first approach (now obsolete) was to design each module independently, so that it was dedicated to its local functionality. By way of example, Figure 8.2 shows three variant designs of electronic circuits for car doors, developed by different manufacturers for different models, or for optional functions. This means that

- each module is physically different from its associated modules;
- manufacturers use very specific integrated circuits or the ones specially designed for the purpose (ASIC), requiring long development times, high risk and costs;
- large quantities are desirable, in order to write off the development costs of circuits;
- there are many variant modules, and consequently
  - the quantities of each module are low;
  - the life of each is short;
  - approval must be obtained for each module in turn, a lengthy and expensive process.

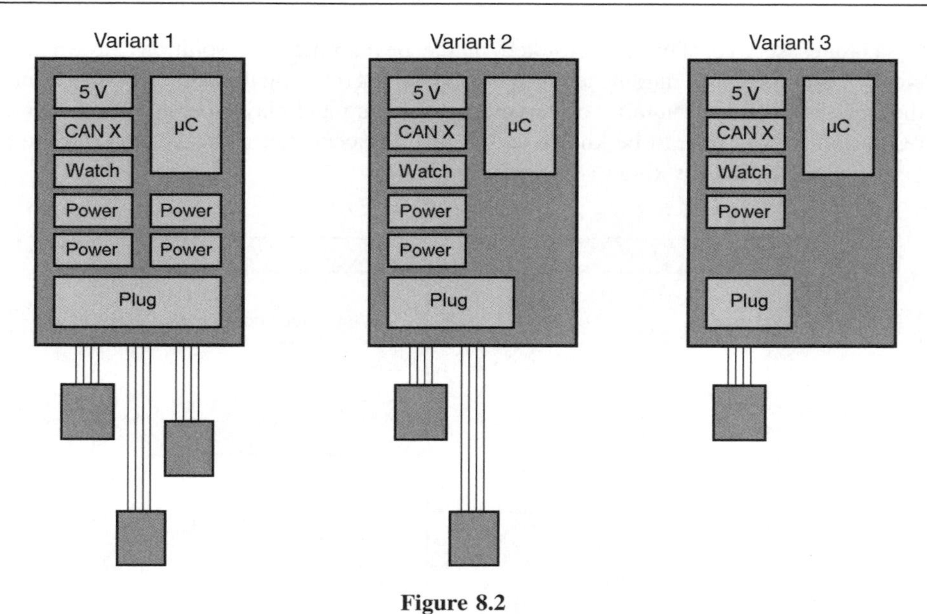


**Figure 8.2**

### *8.1.2 The SBC approach*

A second approach, in common use but already becoming outmoded (Figure 8.3), has developed in response to the growing demand for a reduction in the cost and the mechanical dimensions of modules and the large number of variants, for reasons of production, storage

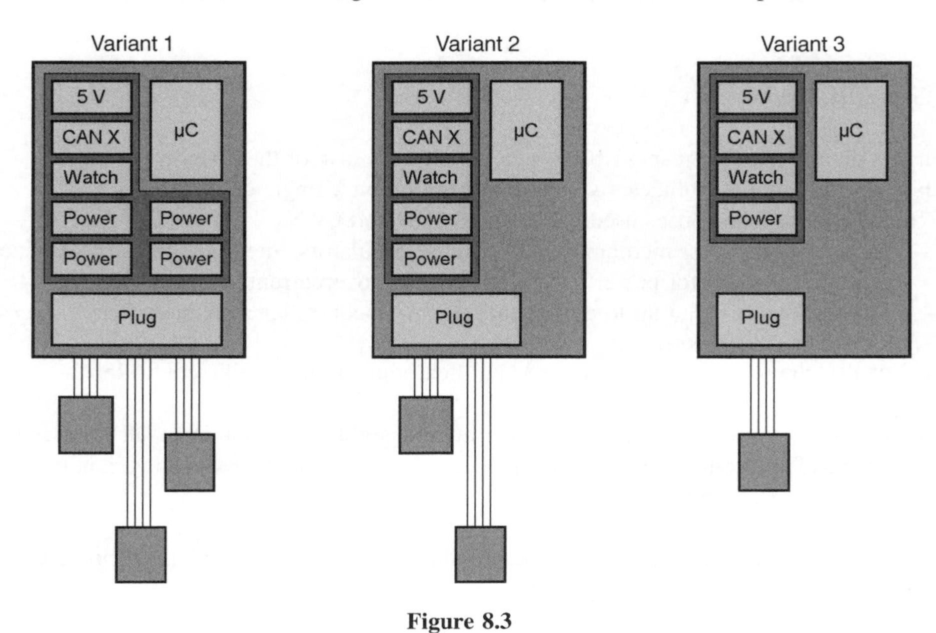


**Figure 8.3**

and after-sales service. This has resulted in the development of solutions in which the efficiency/cost ratio of the module tends to be maximized by using the same modules (almost) all the time, with a small number of variants, of the plug and play type, etc. This approach represents what has come to be known as the SBC concept. Let us analyse a conventional module equipped with this kind of system (Figure 8.4).

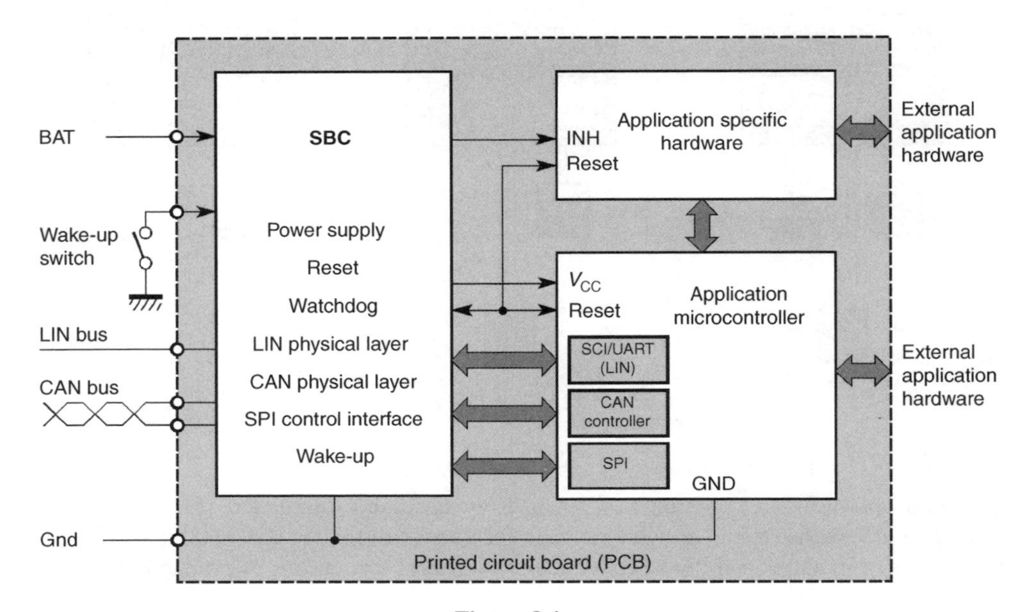


**Figure 8.4**

Generally, it comprises

- first, a microcontroller responsible for the application task of the node in question;
- then, an SBC which includes, by simple integration on a single chip:
  - line drivers for the buses used, for example LIN and CAN;
  - one or two low- or medium-power voltage regulators, including if possible some monitoring devices for power losses, power blips, overcurrent, etc.;
  - a conventional watchdog, to ensure that the microcontroller is actually performing its task and has not crashed;
- and finally, the relay or power controls and their application-specific controllers.

Note that this solution, although attractive at first sight, does not meet all the security requirements of users; in other words, the problems of reduction of costs and variant modules are not resolved because

- there are still may different SBCs and modules for the various planned applications, resulting in low volumes and high prices;

- the development times are long and costly for module and chip manufacturers;
- the life of an SBC is short, posing problems of future proofing;
- the number of connectors is still large and numerous connecting cables are required;
- the number of variant modules is not reduced.

### *8.1.3 The fail-safe SBC approach*

Let us take another step upwards! A third approach (recent and promising) is the fail-safe SBC concept – see Figure 8.5. This is based on the same technical and economic demands and on a functional division apparently similar to the previous one; however, it is implemented in a radically different way, as it resolves the problems of cost and variant modules mentioned above, while still using the same type of circuit (thus allowing larger scale production at lower cost). This concept also aims to provide the module with a very high level of operational security and reliability, as implied by the term 'fail-safe', eliminating the numerous deficiencies of the simple SBC concept described previously.

**Note:** As shown by the example in Figure 8.5, this concept benefits from the LIN bus by reducing the size of the connectors and the number of connecting wires to the various slave units connected to this sub-bus.

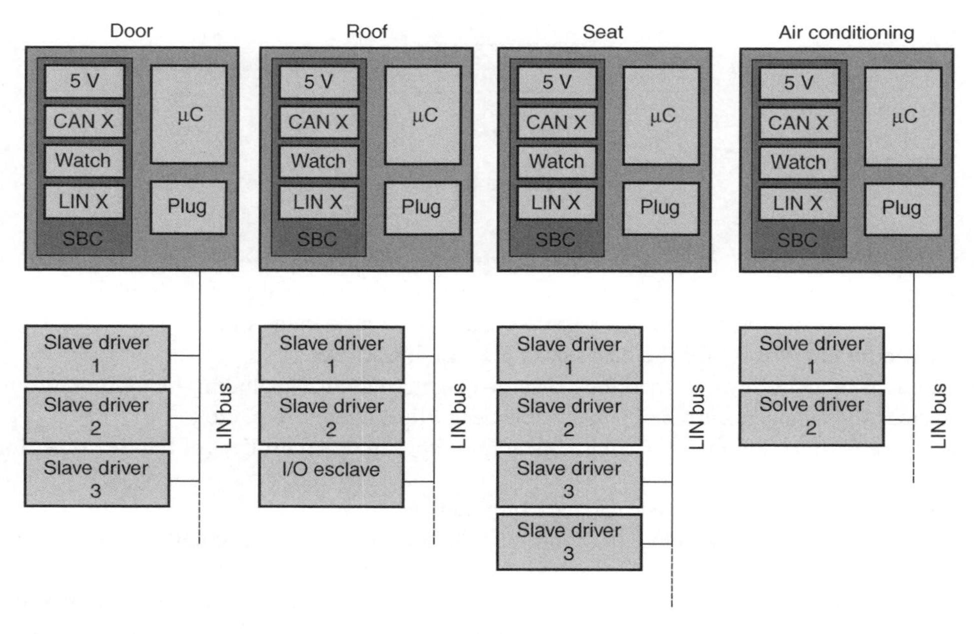


**Figure 8.5**

For this purpose, a fail-safe SBC node currently comprises a microcontroller responsible for executing the application task of the node, but, unlike the simple SBC node described above, it also includes a sophisticated fail-safe control device consisting of different functional units. These can be listed, in no particular order, as follows (Figure 8.6):

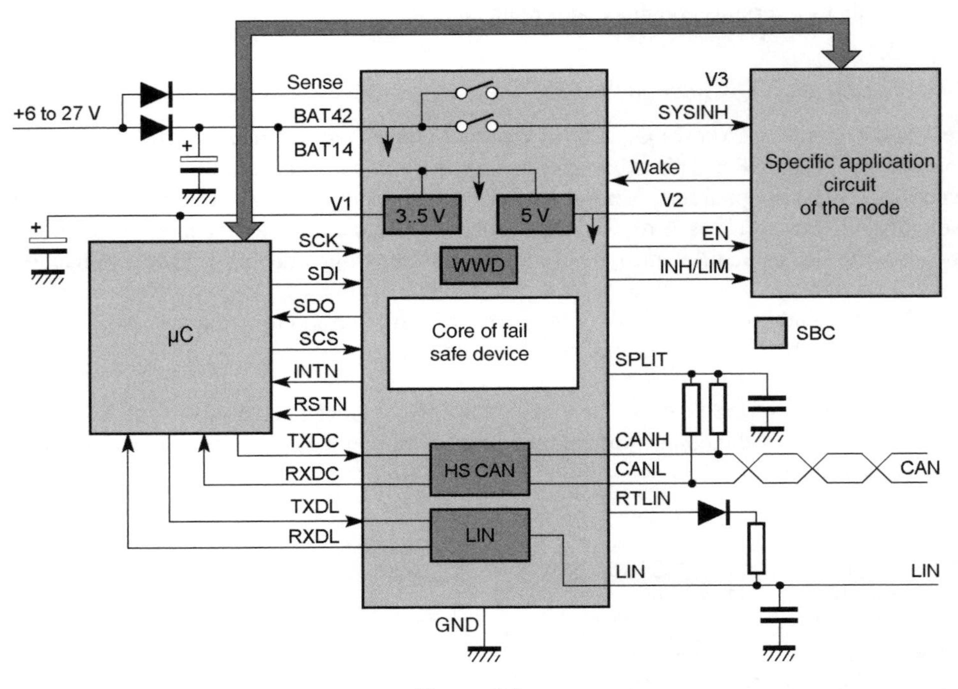


**Figure 8.6**

- a conventional SBC part, with all the properties described above, implemented in a semiconductor technology which can also support a wide range of operating voltages (18/27 V), to support the voltage regulator and LIN control stage functions;
- the secure line drivers for the buses used (e.g. LIN, CAN, etc.);
- one, two or more low- or medium-power voltage regulators, including many monitoring devices for voltages, currents, power losses and blips, overcurrent, etc.;
- a secure (serial) interface for communication with a microcontroller (of the SPI type, for example);
- a local oscillator, for reasons explained fully below;
- a high-performance watchdog, to ensure that the microcontroller is actually performing its task and has not crashed, and that it has not made the whole network crash;
- a whole sophisticated electronic system forming the core of the system, ensuring that the node concerned is kept secure and healthy in case of malfunction of one or more internal or external components and reliably providing a known 'limp home' function in order to escape from any 'deadlock' not listed on the menu;
- many other additional security functions, which are impractical to handle when the system is formed from discrete elements, for example a high-performance FMEA design (see below);

- and finally, as usual, the relay or power unit controls and their application-specific controllers.

Moreover, these on-board systems must be economical in terms of cost and area occupied by the components and must have low power consumption! We must therefore revise our earlier ideas, as you will now see.

**Economy in cost and area**

These concepts of economy in cost and surface area lead directly to a higher level of integration of components on the chip and a redistribution of the many functions used in this design. As we can now fit 100,000 gates or more into each square millimetre (about 10 million transistors per square centimetre) by using a '90 nm' technology and as '30 nm' technologies are on their way, we should be successful in this.

**A new philosophy of nodes and fail-safe functionality**

To provide a clearer picture of the fail-safe concept, I will try to demonstrate the hidden implications of this term with the aid of a few examples. Prepare for some hard work: After reading this section, I hope you will agree that it has been well worth the effort. First, you should know that many hours of discussions with network designers and users (motor and aeronautical manufacturers, equipment makers, industrialists, designers, OEMs, after-sales services, etc.) were required for the purposes of the following section, in order to provide a detailed description of the origins of the problems encountered at each stage (design, development, production, on-site, after-sales, etc.). So here is a non-exhaustive list (simplified slightly to meet the requirements of publication) of the problems that a system must be able to resolve in order to be a 'fail-safe SBC', rather than a 'simple SBC' in which several functions are integrated without any concept of operating security, such systems being abundant in the market.

### *8.1.4 The basic axiom for the design of a fail-safe SBC*

All the designers and users we have met would agree that '... the number of CPUs in an on-board (vehicle) system is constantly increasing ... a local failure must never affect another node ... also the battery must never be discharged accidentally when the vehicle is stationary ... it must always be ready to act ... and must always be in a specified and known state'! So the general framework is established: 'Just do it!'

Here is a classic example. Although it has been included in all systems for many years, the 'switched off' function is now introducing more and more new risks into on-board systems. This is because many elements have to remain on standby in order to provide new functionality required by the application. As I will show in the following section, this makes it necessary to make very complicated provision for secure operation, in the on-board hardware and software, if we are to avoid fast battery discharge and other problems with serious consequences. To make this clearer, let us take the example of sending a command for a switch to low-power, standby or sleep mode which is unsuccessful, a few instants before switching off. This causes severe problems for the system and its multiplexed network because it results in high current consumption, leading to a more or less speedy discharge of the battery with all the tiresome effects of this (a loss of charge over a few days or weeks is a nuisance not only for individual users like you or me, but also for manufacturers who sometimes keep their new vehicles parked for many weeks before delivery).

On the basis of this axiom, we can immediately draw two very important conclusions:

- The control system of the fail-safe device must be connected to all the functions of the SBC without exception, in order to ensure and guarantee the overall functional behaviour of the fail-safe part of the SBC.
- Because of the performance requirements, a fail-safe SBC cannot be implemented as a simple integration or combination of pre-existing functions, either in the form of discrete components or in the form of existing integrated circuits. A high-performance fail-safe device can only be implemented in the form of a fully integrated solution.

We need to create a more intelligent integrated system, ensuring correct operation at all times without dead ends, to base the chip and casing design on an FMEA structure (see the end of this chapter), and to develop an approach using component families rather than isolated circuits, to benefit from the effects of re-use and accumulated know-how.

*All the states of a fail-safe SBC.* Now let us take a closer look at the main problems to be resolved by the component units of a fail-safe SBC. Briefly, we need to monitor and resolve problems which may arise

- on the CAN or LIN lines;
- in relation to the correct operation of the application handled by the local microcontroller;
- in relation to the correct operation of the microcontroller itself and its peripherals;
- in the power supplies of all kinds;
- in the protection systems of all kinds;
- etc.

Let us begin with the communication lines.

### Monitoring and controlling the communication lines (CAN, LIN, etc.)

Obviously, a node equipped with a fail-safe SBC circuit communicates with the outside world via the buses mentioned above, mainly CAN and LIN at the present time. But do not worry, CAN/FlexRay fail-safe SBCs will appear sooner or later! As I have already mentioned, the communication lines to other nodes must not be affected by a local fault or by a fault in any other specific node of the network. To meet this requirement, we need to examine and resolve a number of problems.

*Power supply to the line drivers (CAN, LIN, etc.)*
The power supplies of the line driver stages (CAN, LIN, etc.) and that of the microcontroller outside the fail-safe SBC controlling the node's functions must be completely isolated from each other, to ensure that the network is not affected by any incidents occurring either physically in the CAN or LIN lines or because the microcontroller has not started up for any reason (for

example, following reset or latch-up problems, or any other stoppages). In a conventional or traditional system, the verification of the last-mentioned faults cannot be handled if the line driver is physically separated from the microcontroller because they would then have to be activated via the microcontroller standby/enable (STB/EN) pin. Moreover, although the output stages of the line drivers are always physically connected to their individual power supplies, the nodes must be 'transparent' (i.e. must not disturb the operation of the bus) if they are not powered, or if they cease to be powered. This means that the operation of the nodes can be stopped in turn, one after the other, for functional reasons, instead of being stopped all together. In fact, in a one-by-one stop procedure, it is easy to request a formal functional acknowledgement of the current procedure from each node individually, enabling it to be stopped unequivocally (see the section below on partial networks).

*Monitoring the I/O pins of the Rx and Tx logical signals of the line drivers*
Faults in transmission lines (Tx) and receiving lines (Rx) must be detected and dealt with.

To make this clearer, let us look at the examples which arise

- when the Tx line is always kept at the dominant level;
- when Rx is always recessive;
- in the presence of short circuits between the Rx and Tx command lines (these pins are often physically close);
- with hardware faults (e.g. drops of solder between the pins);
- or in case of software errors in the program of the microcontroller controlling these lines, which may cause the whole line control interface to become inoperative.

*Monitoring the physical communication lines (the buses)*
Even if physical communication lines of the CAN, LIN or other types are short-circuited either to + or to ground, they must not cause a fast discharge of the battery in any circumstances. Note that a special pin RTLIN is dedicated to this function for the LIN driver.

*Withstanding ground disconnections of modules and ground differences*
The familiar problem of poor ground connections (for example, when rust is present on and around the module fixing screws, etc.) creates different potential variations in a system, according to the relative positions and arrangements of the nodes on the chassis. It also causes significant ground potential variations (sometimes as much as 12 V in braking operating the regulator (e.g. the 'Telma') in trucks powered by 24 V batteries): these are known as 'ground differences' or 'ground shifts' (GND shifts). Clearly, these must not be allowed to interfere with the system or stop it, and their detection must not be affected by errors due to radio interference (electromagnetic compatibility or EMC). For this purpose, a reliable ground potential difference detection device must be provided. To provide different operating options, two possible thresholds ($-0.75$ and $-1.5$ V after filtering) must be provided and must trigger a warning to the microcontroller handling the application. This makes it possible, for example, to analyse the activity of certain nodes whose local ground connections are uncertain because of the possible local presence of rust.

*Selective power-down to support partial network topologies*
To keep only a few specific elements of a network in operation (for example, the monitoring of the passive keyless entry function of a vehicle) while keeping some nodes in sleep mode or

power-down when the vehicle is stopped, it is sometimes necessary to construct a temporary 'partial network' (Figure 8.7).

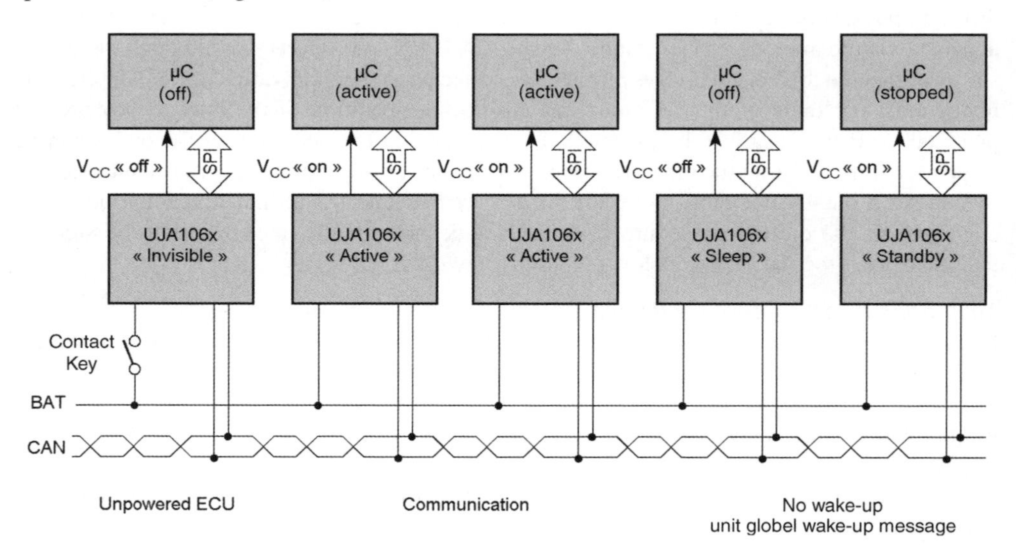


**Figure 8.7**

Thus, some participants of a main network have to be put into sleep mode, after which all the participants of the same network have to be woken globally. Because of the characteristics described above, we can now put different participants to sleep one by one (for example, by sending a series of messages such as 'node 3', *prepare to sleep*, *please confirm* and finally *go to bed*. Following this, even though the line driver of the node in question sees messages passing on the bus, it does not wake up its CPU(s). All the network participants can then be woken up by sending a specific message containing a frame whose configuration (pattern) contains symmetrical data. This prevents any incorrect wake-up, due for example to another node whose CPU may be faulty and therefore remains active and sends messages constantly. Note that, to avoid probable collisions with messages in other CAN frames which may arrive from other nodes, this 'local go to sleep/wake-up' instruction is sent in two different messages.

### Cyclic WAKE port supply

A fail-safe SBC must offer the possibility of cyclic supply for running tests of external wake-up switches. For this purpose, the local wake-up port must operate in synchronization with its cyclic power supply. Consequently, a wake-up switch, connected permanently or intermittently to ground, will not be able to discharge the battery unduly.

### Units specific to the global fail-safe function

In order to have a known level of reliability, the core of the fail-safe controller (Figure 8.8) must be based on an asynchronous state machine system, rather than a microcontroller. We shall now examine the behaviour and performance of the fail-safe fall-back unit in case of

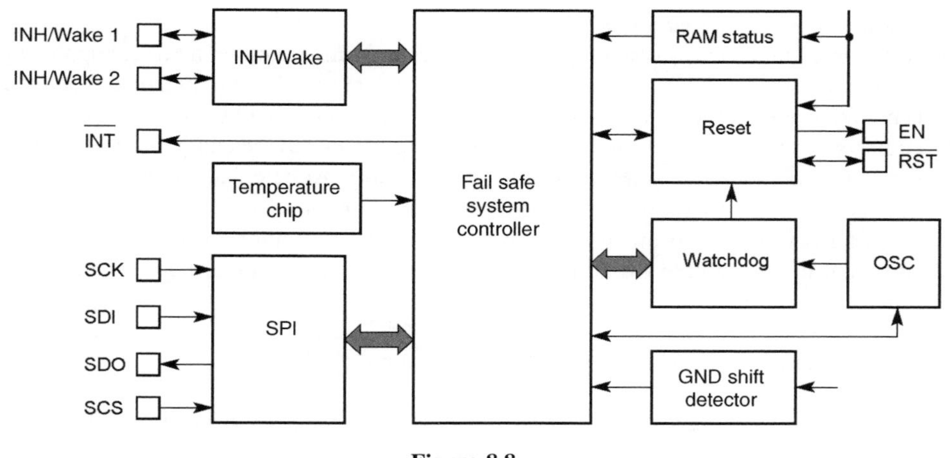


**Figure 8.8**

serious malfunction of the ECU (the complete node). We shall start with the two most important elements.

*The need to provide a 'limp home output signal'*
If the module suffers a serious malfunction, the fail-safe SBC must switch unambiguously into the fail-safe state and must supply a special fall-back control signal, called the 'limp home signal', to the exterior, via a pin specially dedicated to this information and this use, which becomes active (in the high state) at this instant. The function of this signal is to activate all the special hardware designed to support the chosen fall-back functions, for example the illumination of alarm lamps.

*Global enable signal*
The critical hardware specific to certain functions must be switched off immediately if there is a CPU malfunction. For this purpose, the fail-safe SBC provides a dedicated global enable pin, which can be used to control the critical hardware components (Figure 8.9). The signal present on this pin immediately changes to the low state if there is a serious malfunction of the microcontroller, and it can be used to stop the raising of a window, for example. Note that this function is active only if the watchdog has been correctly put into operation.

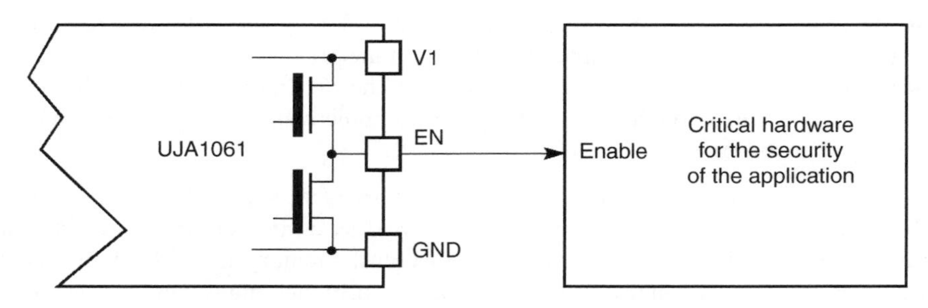


**Figure 8.9**

*Dealing with the moods and fantasies of the reset pin*

The reset pin and its control system are always a very sensitive point in a critical application using a microcontroller. It is usually controlled by a conventional RC circuit, but unfortunately this is not advisable in secure applications. To avoid any doubt about the quality (tolerances, deviations, parasitic signals, etc.) of the (analogue) signals due to the charging or discharging of a capacitor connected to the reset pin, the reset signal must be created digitally.

Here again, the battery supplying the whole system must not be subject to fast discharge by a corrupted reset signal. To prevent this, it is necessary to monitor the reset pin itself, for detection and action in case the link carrying the reset control signal happens to become connected

- either constantly to the positive power supply terminal ('+'),
- or to ground ('−'),
- or if the 'wire' carrying the reset control signal is broken (open), for example if a printed circuit track is interrupted.

This is because a control failure on the reset line will always have drastic consequences for the system, for example

- if a line is cut off, the watchdog can no longer 'reset' the microcontroller;
- if there is a short circuit, the microcontroller is 'reset' permanently, which is not always desirable where the rate of battery discharge is concerned!

To avoid this kind of situation, if a reset command is sent to the corresponding pin of the circuit, the fail-safe SBC sends a frame via the SPI bus to obtain confirmation. This confirmation can only be received once, solely when the program is first started after the reset. Figure 8.10 shows the complete state diagram of the reset function.

*Flash memories and their problems*

*Procedure for programming protected Flash memory.* It is customary for module CPUs to support a (re)programming mode for protected and secure Flash memories, if only for the purpose of updating the vehicle software in the course of regular servicing.

To avoid making any mistakes (and to prevent others from making them), a fail-safe SBC must provide a totally secure input device in the Flash memory reconfiguration mode. This is because there has been no suitable tool for reprogramming/reconfiguration (often for reasons of cost) for many years, and therefore skilled 'hackers' often bypass accesses with little or no protection in order to modify the contents of the Flash memory. To prevent all these problems and the resulting disasters, a special trigger function of the watchdog, combined with a specific reset signal and the reset source data, can be used to implement a highly reliable system for updating Flash memories.

*Managing the problems and failures of the Flash memory of the external microcontroller.* The 'intelligence' of most modules is currently controlled by a microcontroller which runs its program and which is largely or wholly written to a Flash memory, instead of being stored in ROM as was the case for many years. Clearly, this provides many benefits, such as the possibility of customizing the module in the final stage of production of the vehicle or

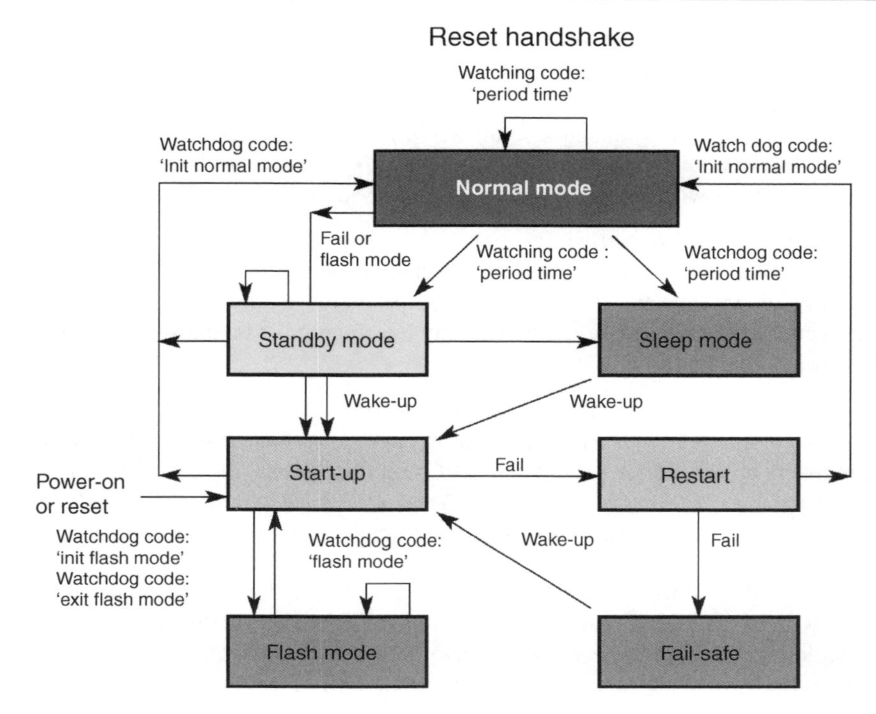


**Figure 8.10**

system. This also enables it to be reprogrammed if the software has to be updated during servicing, in the after-sales period or when a system is recalled to the factory. In fact, a Flash memory is the best of all! Or almost – in fact, as a Flash memory can be deliberately programmed, then it can also be unexpectedly or accidentally deprogrammed. Everyone will tell you that no provision is made for this, but it only happens to other people. In theory, though, it could be very costly! For information, it is possible for a Flash memory containing the essential application program and its various updates (releases) to be damaged. For example, the value of one or more bits may be modified by an electrostatic discharge (ESD) when the diagnostic equipment probe is connected to a vehicle during servicing. Curious!

Briefly, the Flash ROM program store can be subject to faults, and these must never be allowed to make programs crash or enter risk zones, stop ECUs and/or cause a fast battery discharge. The following text deals with the resulting software problems that may be caused by faults of this kind.

Normally, faults caused by problems in the Flash memory, including the failure of the application program, force the watchdog to 'reset' the CPU. Unfortunately, this does not solve the problem because the CPU, after normal restarting, inevitably falls back into the same loop sooner or later, causing a new CPU reset, and so on forever. Again, this leads to a progressive discharge of the battery. To be sure of detecting this kind of fault, we must provide a cyclic detection of these cyclic faults.

In order to distinguish these special problems from the routine ones and handle them correctly, we must implement a register storing the details of the origins of the resets concerned and protect the memory with the aid of statistical fault data. This can be done quite simply, for example by using a small RAM (always supplied directly on $V_{\text{bat}}$ before contact is made) which stores the number of unnecessary tests and holds these data in memory even after the contact with the vehicle or system has been switched off.

*Special state of the fail-safe system*
Fail-safe SBC provides a special low-power state to which it can be switched directly, independently of any system failure or serious problems in the application. To provide complete security, no clamping condition of any interface must be allowed to stop the fail-safe SBC from switching to this state.

*Reliability of commands transmitted to the fail-safe SBC*
The transmission of commands from the external microcontroller, in order to add them to the internal registers in the core of the fail-safe SBC and thus control its functions, must be very reliable and must include many protective devices relating to the integrity of the message contents (bits) carried, to avoid the presence of incorrect commands or, worse still, the absence of any commands.

In a fail-safe SBC worthy of the name, it is therefore essential to control each command transmitted with the aid of a unique access, for example on 16 bits via the SPI bus. Why is this?

This method is chosen because we want to avoid the formation of 'zombie' messages which would be created by the incorrect assembly (concatenation) of function command messages of the fail-safe SBC, consisting of multiple bytes, whose global transmission could be interrupted by the external interrupt handling taken into account by the microcontroller. This would create a mixture of unrelated byte sequences on the SPI bus, causing civil war to break out in the fail-safe SBC system!

Additionally, all critical commands are encoded with redundant bits, for the detection of faults in the shifts of bits in transmission, and finally a check is made to ensure that the number of clock pulses is correct for the command which is sent.

Note that a fail-safe SBC worthy of the name should also offer, for simple reasons of security, a facility for re-reading its configuration and command registers easily, without necessarily having to write them beforehand, so that the correctness of the system configuration can be checked at any time.

A special pin is also dedicated to the chip select function of the fail-safe SBC.

*Permanent synchronization between software and hardware*
To ensure that the software being run is not inappropriate for the current operating mode (e.g. Flash memory programming instead of running the main operating program), a handshake mechanism has been installed to provide mode changes including third parties, specific commands to the watchdog and instructions for commands for specific time slots and/or specific operating modes. This ensures that the software and hardware always operate in a perfectly synchronized way.

*Unique identification number ('unique ID')*
Each fail-safe component must include a unique identification number, including in particular detailed information about its type (reference), its version (release) number, its date of

manufacture, its serial number, etc., especially if it is decided, for reasons of cost (longer production runs, reduction of approval costs, etc.), that all circuits of the same family should be pin-to-pin compatible. This does indeed provide a high level of traceability of the component or module to which it is applied, but above all it has enormous functional benefits.

This is because the microcontroller software can automatically check, on each start-up, the pairing and consistency that exists (or should exist) between the existing software version and the type and version of the component actually implemented in the printed circuit of the module. This avoids all those problems which might be risky for the user (think of modules taken hurriedly off the shelf to fix a customer's system: 'Oh yes, I'm quite sure it's the same model!'). Changing one module of a fail-safe SBC will no longer cause an application or system to behave unpredictably. How is this done? ! Not during operation (clearly), but, in order to reduce production costs, this makes it possible to produce only a single type of module, including a single and unique type of microcontroller incorporating in its ROM a program suitable for several applications. Depending on the product to be delivered, it is possible to install a fail-safe SBC of the HS CAN or FT LS CAN + LIN type. When the fail-safe SBC is installed in the module, the software can identify the environment and the product, using the unique circuit number, and configure itself accordingly, thus achieving the best possible outcome.

*Local wake-up*

Local wake-up can be provided, using a special 'wake' pin, with or without the presence of short-circuit detection on the pin.

*Operating modes*

Because of its structure, a circuit such as a fail-safe SBC must be capable of supporting various operating modes, and these modes must also coexist. The principal modes are

- normal operating mode;
- fail-safe mode;
- start-up mode;
- restart mode;
- standby mode;
- sleep mode;
- flash mode;
- development mode;
- test mode.

Of course, the normal operating mode is usually predominant during the active phase of the system. But this stage must be preceded by successful development, testing, production, configuration, checking, etc. of the product, in which the fail-safe mode has a part to play, and this requires some special provision to be made to ensure that none of these modes interferes with the others.

Let us re-examine this by means of one simple example, namely the development mode which is usually the first one in a design process. The same principles apply to the other modes and to mode changes.

*Development mode and an example of mode change*
To be able to develop a hardware module and its associated software and then move on to emulation, we have to partially disable the watchdog function of the fail-safe system, which would otherwise constantly interrupt the emulation of the functional part (excluding the watchdog) of the application. Once this development phase is completed, we must be certain, for simple reasons of security, that the system can never return to this operating mode which is specifically dedicated to development; this again requires a number of functional security devices. This nature of this operating mode is indicated by the fact that it is only possible to return to the development mode by carrying out a very specific action on an external (test) pin and protective software:

- At the hardware level, the test pin must be connected to the '+' of the supply before power is supplied to the ECU, and, on the contrary, this pin is tested only during the rise in power supply voltage.
- At the software level, the watchdog must not be triggered (optionally, an interruption can be forced to avoid an overflow or triggering at incorrect times). The reset is then forced, i.e. clamped forever (problem of development tool, emulator), and its period of 256 ms of start-up time is ignored.

Clearly, if a parasitic signal were to switch the system to this operating mode, it could only do so with a considerable effort!

Furthermore, there must be many other operating modes such as the watchdog mode in addition to the normal mode and software development mode. Note that a mode change is always simultaneously initialized with a special watchdog trigger code, this being authorized and valid once only, during the mode change. In the hypothetical case in which the application software is not diverted towards another software routine to which the new watchdog code now corresponds, this is immediately detected by the fail-safe SBC and dealt with in conjunction with the reset system. This ensures that the operating mode change is always carried out in perfect agreement with the corresponding watchdogs. Thus, we can be sure that the new software task will always start with identical watchdog conditions, and this will also ensure a completely predictable behaviour of the task, totally independently of the previous history.

## Watchdog

*A watchdog – yes, but with a pedigree!*
To prevent any interaction in the microcontroller hardware and software between all the events or disturbances that may occur in the power supply, the watchdog must be 100% independent (including power supply) of any other component, so that it can observe the host microcontroller of the application, and it must also have its own clock (RC oscillator) without external components. In this case, even if the internal oscillator is faulty, this must be detected, and the fail-safe SBC switches the watchdog to fail-safe low-power mode.

In addition to the conventional operation of the watchdog to prevent overflow, the implementation of a fail-safe SBC watchdog must be able to detect whether or not the microcontroller is operating correctly in the application within a minimum–maximum time slot (watchdog window).

*Quality of the triggering instant of the watchdog and its triggering*
The form and quality of the triggering of the watchdog must be perfect, to ensure that the triggering instant of the watchdog timer is correct. For this purpose, the commands sent to it by the microcontroller (via the SPI bus, for example) must be designed with a bit redundancy (CRC) to allow detection of the faults due to message shifts (using a 16-bit train, for example, instead of $2 \times 8$ bits transmitted twice), or simple and conventional faults at the bit level (parasitic signals, EMC, etc.). During the start-up phase of the system or module (when the module is switched on or rather when it is first switched on, as it is considered to be permanently switched on after this point), because of the time taken for the start-ups of the different oscillators (the microcontroller's quartz oscillator, etc.), the watchdog period has to be longer (256 ms) than in normal operation. Following this start-up phase, the system changes to normal operating mode, and the watchdog operates according to the execution of the desired program and the 'normal' application tasks (standby/flash, sleep) to be performed, in a wide range of programmable periods from 4 ms to 28 s and in different operating modes, as follows:

- a cyclic wake-up system using interrupts or resets;
- fail-safe OFF mode;
- optional automatic disabling in cases where the external current $I_{cc} < I_{cc\ min}$;
- automatic activation of the watchdog if $I_{cc} > I^{cc\ min}$ (after 256 ms);
- software development mode with partial disabling of the watchdog;
- possible switch to fail-safe mode only after the power-up (BAT on) phase;
- fail-safe mode linked to the watchdog access codes, via the SPI bus;
- support for the Flash memory loading mode;
- possibility of entering fail-safe mode constantly in the absence of a hardware signal;
- possibility of loading new software for the CPU without the use of the watchdog.

## Managing interrupts

Interrupts must be monitored, recognized and dealt with. The fail-safe system checks whether forced interrupts are served in a reasonable time. If an interrupt is not served, this is signalled either in the form of a broken (open) interrupt line or by the fact that the microcontroller is out of service.

The fail-safe system must be designed to avoid the overloading of the microcontroller by too many interrupts (possibly at the wrong times) and by their management. The interrupts are therefore synchronized with the watchdog period and can only occur once in each of these periods. This ensures that the software is not run in a dangerous way, for example after an interrupt due to the failure of a 'chattering' CAN line.

## General power supply

A fail-safe SBC can have several voltage regulators, but their implementation needs to be thought through. The choice of their implementation architecture is fundamental. For example, the voltage regulators of the line drivers and power supply of the microcontroller will be located on the same integrated circuit, but, as I will demonstrate, they must be completely isolated.

*Microcontroller regulators*
A conventional linear current–voltage regulator can be used for the power supply to the external microcontroller of the module. The circuit automatically adjusts its maximum current as a function of the input voltage to be controlled.

*Regulators for line drivers (e.g. the CAN line driver)*
As the electrical levels of CAN lines are directly based on the 5 V power supply, a separate 5 V regulator must be provided for these stages. This regulator must be completely independent of the other controllers because no faults, EMC pollution, noise, etc. that may occur on the CAN bus must be allowed to affect the operation of the microcontroller, and it must be completely independent of the microcontroller. For this purpose

- the physical line determines its requirement at 5 V;
- even if the microcontroller crashes, the physical layer remains active and secure.

It is also necessary to be able to manage the consequences of poor contacts that may occur on the battery terminals and to provide extensions for controlling external controllers (via the INH pin).

*Controlling overloads*
A fail-safe SBC must also ensure that current and temperature overloads do not destroy the system. For this purpose, a device must be provided to monitor the temperature of the chip and send a warning if the maximum temperature is exceeded. If the temperature is too high, a special interrupt is generated, to cut certain loads at the command of the software. However, it must not cut everything (as occurs all too often) because this could damage the system.

## FMEA – failure mode and effect analysis

We cannot leave the subject of 'secure operation' without mentioning FMEA. Meaning what, exactly? This is the acronym for failure mode and effect analysis. In response to this type of analysis and its results, certain actions are taken at different levels.

*The FMEA concept of pin assignment*
In order to make the operation of a system secure by using an integrated circuit, it is always important to provide a precise specification of the arrangement and order of the pins of the integrated circuit (or connector), so that undesired short circuits between pins cannot block or damage the system under any circumstances. Figure 8.11 shows an implementation of this approach, in the context of the fail-safe SBC family of Philips/NXP Semiconductors. This shows the physical separation of the strong signals on the one side of the integrated circuit from the weak control signals on the other side.

Also, still using the FMEA approach, Figure 8.12 shows how this does not disturb the system and indicates the states resulting from these examples. It also provides all the information required by the designer to manage the case of perturbation in question if he wishes.

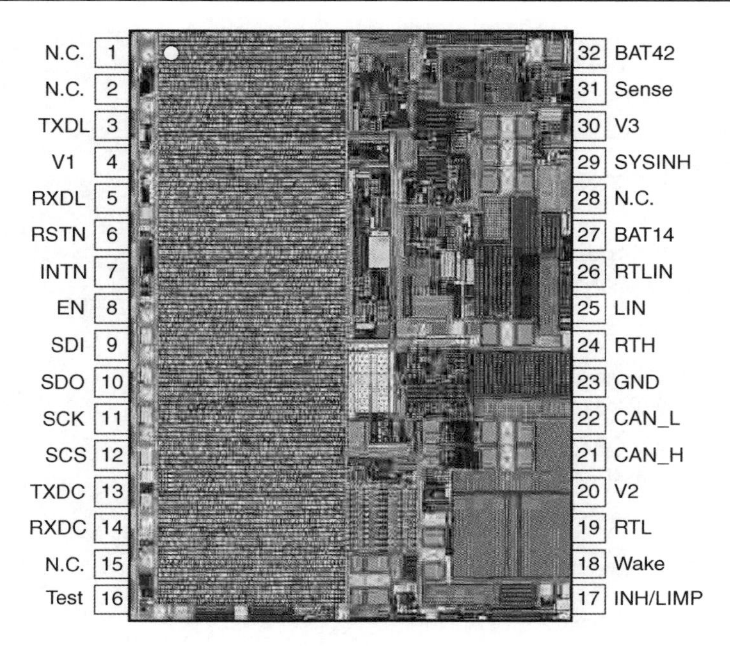


**Figure 8.11**

*The FMEA concept of CPU signals*
The concept of the fail-safe SBC inherently ensures that the operation of the CPU does not subject the system to unpredictable conditions or battery discharge. This is important, for example, for the TXD/RXD, RST and INT pins, and all power supplies.

*The FMEA technological design concept*
The choice of technology is also very important. For example, using SOI (silicon on insulator) enables the functional units of the fail-safe SBC to be isolated, so that a function that has failed completely cannot destroy other units in the integrated circuit. At the same time, in order to achieve the best operating reliability of the integrated circuit, the use of shared resources in its design must be avoided. All the units therefore generate their own references, thus reducing the failure rate per function and allowing designers to work on a family principle, using a 'toolbox' structure. Similarly, the shared resources and signals that cannot be tested easily in the production phase must be designed with additional safety margins.

Finally, as in circuits for secure chips, the use of asynchronous handshake technology and hardware makes the operation of the system secure, even if the local oscillator fails. In this particular case, the fail-safe SBC stops in a very precisely specified secure state, which is also a very low power state.

*The concept of covering production tests*
The design, development, production, etc. of integrated circuits is a simple matter, if we disregard the simultaneous requirements for effective tests and test schedules

| Pin name | Failure | Remark | Severity class |
|---|---|---|---|
| V1MODE | Shorted to neighbour (nc) | – | – |
| | Left open erroneously instead connected to GND | 3 V operating mode of V1 as expected | – |
| | Left open erroneously instead connected to V1 | 3 V operating mode of V1, 5 V operation expected, microcontroller might not operate correctly | C/D |
| | Shorted to V1 erroneously, is also connected to GND | V1 is shorted to GND, system will enter fail-safe | C |
| | Shorted to GND erroneously, is also connected to V1 | V1 is shorted to GND, system will enter fail-safe | C |
| TXDL | Shorted to neighbour (nc) | – | – |
| | Shorted to neighbour (V1) | LIN stays recessive all time | D |
| | Shorted to GND | LIN stays recessive all time due to TXDL dominant timer protection | D |
| | Left open | LIN stays recessive all time | D |
| V1 | Shorted to neighbour (RXDL) | LIN locally clamped recessive | D |
| | Shorted to GND | V1 off, system will enter fail-safe | C |
| | Open circuit | Microcontroller un-powered, system will enter fail | |

**Figure 8.12**

throughout the production sequence. This can introduce a number of complications, for example

- For secure testing, all the equipment testing digital functions must include timer functions, which slightly lengthen the test period but enable the quality of the product to be assured.
- Additional hardware must be provided to enable all the components and all the interconnections to be checked. This significantly increases the surface area of the chip, but also improves the quality and reliability.
- Everything else, including all the industrial secrets of all component manufacturers!

| | Some problems to be resolved | Examples of fail-safe SBC solutions |
|---|---|---|
| a | Any local failure must not affect communication between the other nodes of the network<br>Local failures of a particular ECU must remain local and must not affect other nodes | Autonomous power supply protected against short circuits of the CAN driver power supplies<br>Failures of TXD/RXD interfaces are detected and managed |
| b | Fail-safe performance in case of malfunction of a CPU | LIMP Home output signal |
| c | The software version and the hardware version do not match | Identification of the fail-safe SBC component including the chip references |
| d | Ground shifts between ECUs must not cause the system to stop | Detection of ground shifts and EMC immunity |
| e | Curruption of the RESET signal must not discharge the battery | Detection of short circuits and RESET power cut and management of these |
| f | Failures of the Flash memory must not block an ECU of discharge the battery | Cyclic failure detection devices |
| g | The critical hardware must be disconnected immediately in case of malfunction in an ECU | Global Enable signal device |
| h | Switching to low-power mode must be independent of any failure in the system | Specific state of fail-safe system |
| i | Communication via the SPI must never allow corrupted commands to be present | Device for 16-bit access to the SPI with protection |
| j | The system configuration must be checkable by software | Read only access device |
| k | Interrupts may overload the microcontroller and cause unpredictable software performance | Devices for limiting SBC interrupts – predictable software performance |
| l | Interrupts must not overload the microcontroller. A system with predictable performance is a 'must' | As above |
| m | The system must with stand safe reprogramming of Flash memories on site | Special protected system for entry into Flash programming mode |
| n | Software development must not be complicated by the fail-safe SBC functions | Special protected system for entry into software development fail-safe mode |
| o | Software and hardware may lose their synchronization and make the system subject to unknown factors | Fail-safe SBC provides a handshake for mode changes with synchronization of the watchdog<br>Each mode change must reload the watchdog content<br>Watchdog synchronized with the operating mode change |
| p | Short-circuited lines must not discharge the battery | Autonomous disabling of the CAN and LIN terminations cyclic supply of port wake-up functions |
| q | A 100% independent watchdog is required to monitor the microcontroller of the application | Integrated RC oscillator without external components<br>'window' watchdog with code protection relation between watchdog-off during the low-power phase and the current ICC used |
| r | The system must not be destroyed by overloads | Excess temperature alarm<br>SOI high-temperature silicon technology |
| s | The various types of interrupts must be recognized and dealt with | Interrupt monitoring device |
| t | A wake-up solution must always be possible | Wake-up on leaving sleep mode<br>CAN and LIN wake-ups are always ensured for CAN and LIN derivatives |
| u | No partial power supply to the microcontroller | Provide full discharge of the power supply in question |
| v | Failure situations must be transparent for the system | Devices for detailed information on failure analysis and diagnostics |
| w | The CPU must not lose data which are important for the application | 'Sense' devices with interrupts in case of power loss |
| x | The pin assignment of the circuit must follow FMEA rules | Clear separation between the different voltages of the different modes to be powered |
| y | Support numerous low-power modes | Standby mode with permanent supply to the microcontroller<br>Sleep mode with microcontroller unpowered partial CAN networks<br>Cyclic wake-up using the watchdog during the standby and sleep phases |

**Figure 8.13**

### *8.1.5 Summary*

Figure 8.13 summarizes the major real-life problems encountered in the application of on-board systems and the solutions that must be provided by a fail-safe SBC. Note that I have not classified these items in order of importance because an on-board system, if it is to be secure and reliable, must meet each of these requirements.

This brings us at last to the end of this 'brief' description of the functions to be supported by any fail-safe SBC worthy of the name! I hope you have found this exhaustive list useful and will make good use of it when designing your future systems.

## 8.2 The Strategy and Principles of Re-Use

When all these elementary units have been implemented, they naturally become 'standard issue' for all on-board applications of this type and therefore join the great family of re-usable components. This brings a whole family of fail-safe SBCs within your reach.

| | CAN FTLS | CAN HS | LIN 2.0 | Enhanced LIN |
|---|---|---|---|---|
| UJA1061 | X | | X | |
| UJA1062 | X | | | |
| UJA1065 | | X | X | |
| UJA1066 | | X | | |
| UJA1068 | | | | X |
| UJA1069 | | | X | |

**Figure 8.14**

Let us take the characteristic example of the Philips/NXP Semiconductors UJA 106x fail-safe SBC circuits. Why choose these, rather than the others? The answer is simple: First, why not? Second, as a way of examining the 're-use' aspect of the story. This is because this family (Figure 8.14), the first of a large series in the market, is totally pin-to-pin compatible and offers many advantages to designers or users, namely

- re-use of the same printed circuit tracks, providing high flexibility for 'off the shelf' modules and fast delivery in after-sales service; the same low-level libraries are used for system processing, etc., resulting in
  - a significant reduction in, and maximum return on, system development costs;
  - a shorter 'time to market';
  - a much easier passage (in terms of time and money) through the tortuous stages of product approval.

Figure 8.15 shows specific examples of all these matters.

## 8.3 Demo Board

To conclude this section, Figures 8.16 and 8.17 are photographs of a demo board for handling and evaluating all the functions of a fail-safe SBC integrated circuit of the

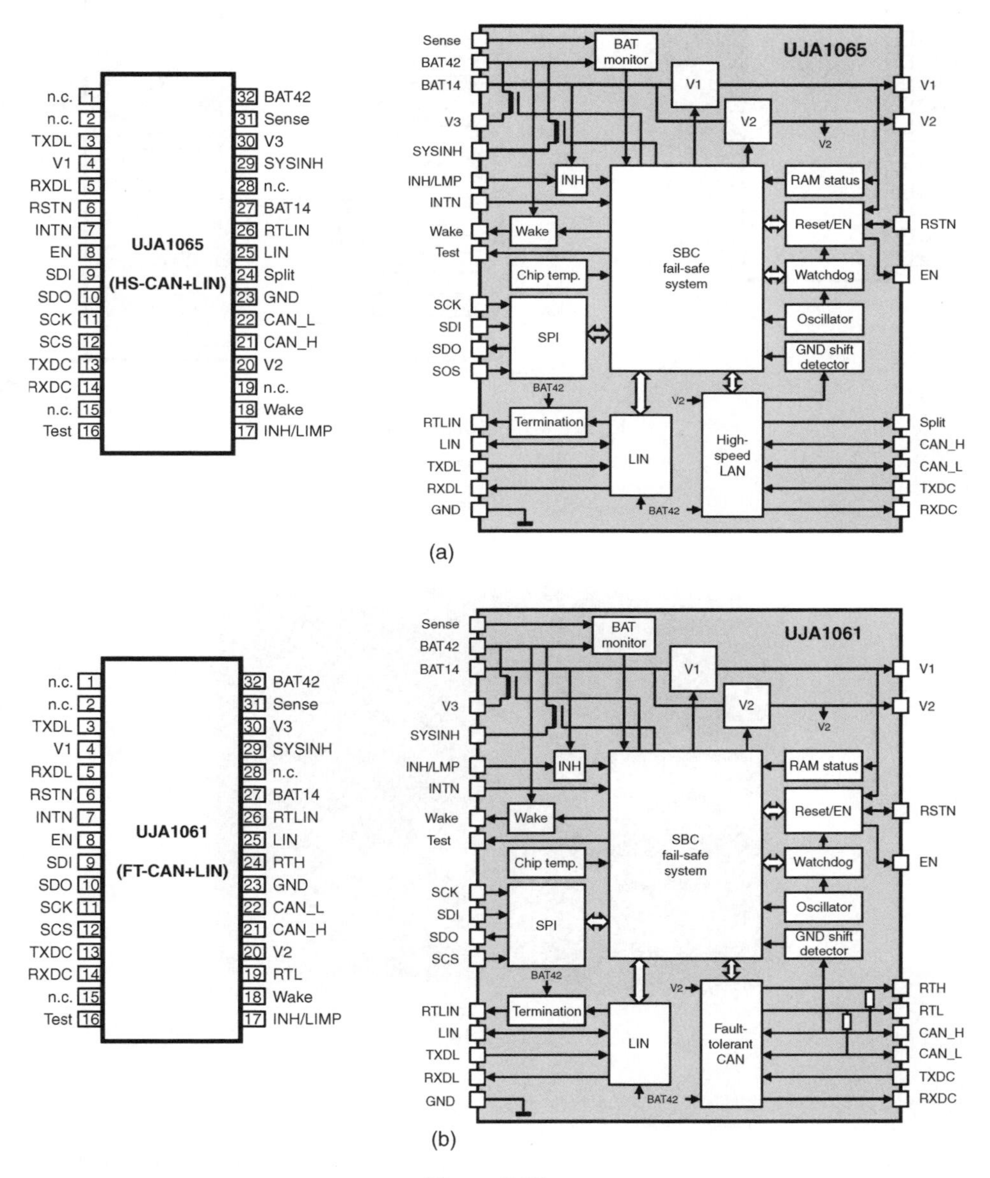


**Figure 8.15**

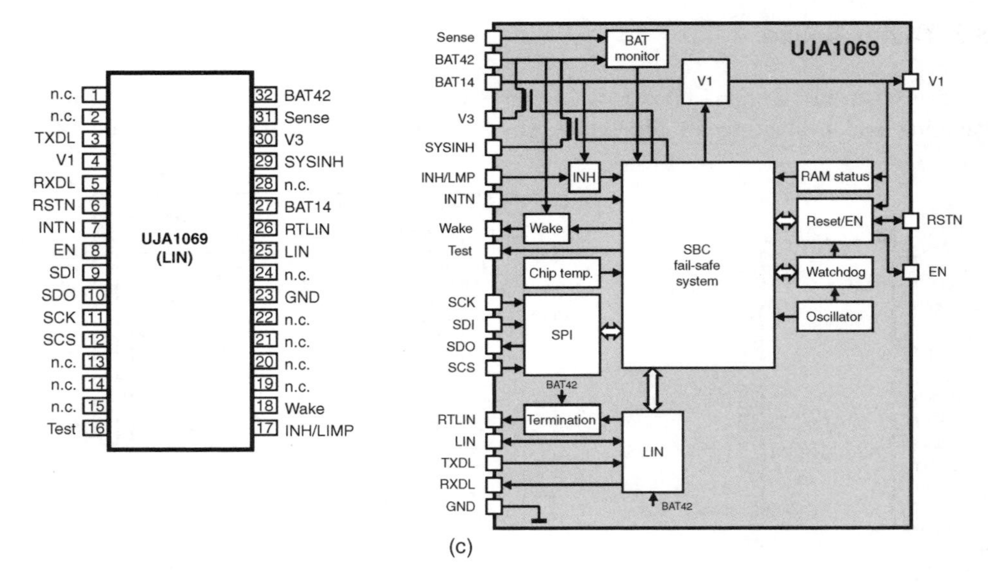


(c)

**Figure 8.15** *(Continued)*

**Figure 8.16**

component family described above, together with a screen shot of the results relating to these functions.

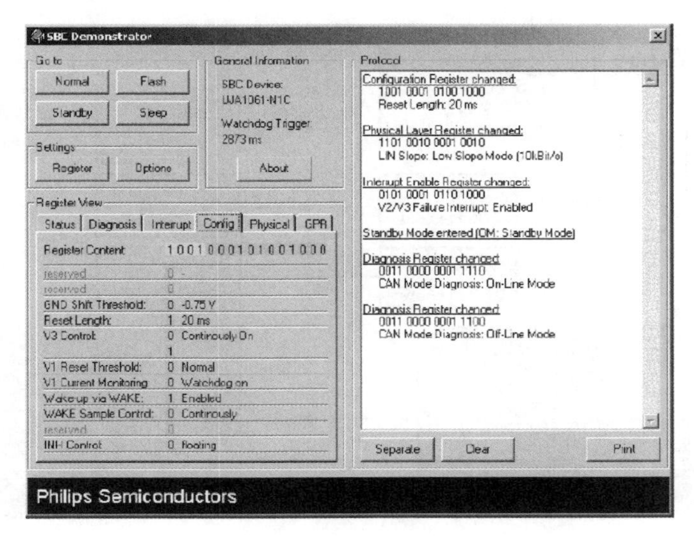


**Figure 8.17**

## 8.4 Gateways

To conclude this chapter, let us return to the question of gateways between networks (which can, or rather must, be fail-safe).

We often need to manage several networks of the same type (CAN, for example) or of different types (CAN/LIN) at the same time. For numerous reasons, we frequently have to provide several networks in a system, for example in order to make certain functions secure (e.g. if a door is severely bent by a collision, this must not prevent the starting of the vehicle, etc.). Clearly, this all has a cost, in terms of components and also in the CPU workload for the control and monitoring of all these features. This is why new circuits, called 'gateways', have been provided between buses, with the aim of decreasing the workload on the main CPU.

### *8.4.1 Example*

Figure 8.18 shows a conventional architecture of this kind of functional division in a motor vehicle, but it can easily be adapted for industrial application architectures.

These gateways provide, for example, functions between buses such as speed changing, physical medium changing, modification of the data carried, protocol changing, etc. at the highest possible speed to meet the requirements of applications. This makes it necessary to provide CPUs with high performance, mainly in terms of Mips, implemented for example on ARM 7, 9 or 11 platforms or competitors at the same level. The example in Figure 8.19 shows the conventional schematic architecture of this type, for a gateway application between five CAN networks which can equally well be HS CAN or FT LS CAN, on differential pairs or on optical fibres.

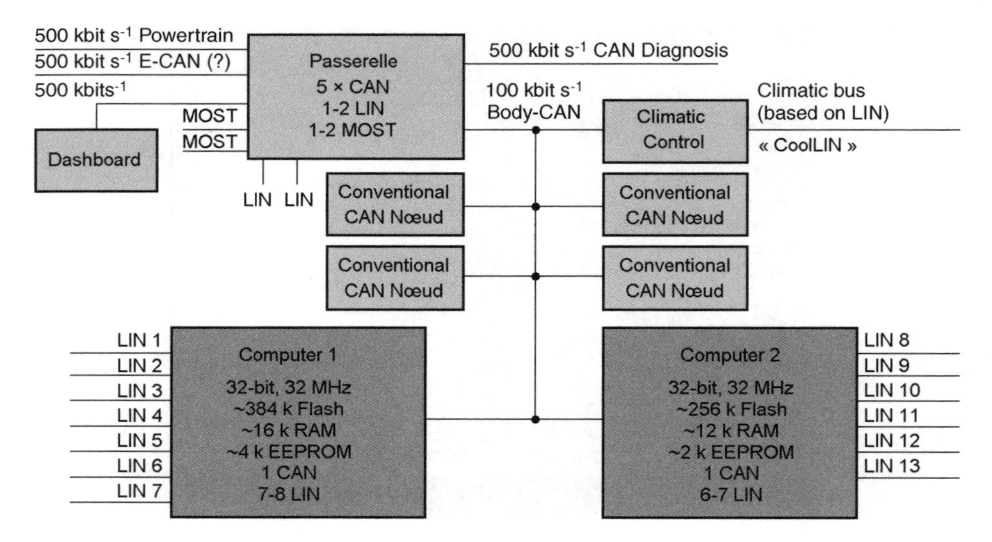


**Figure 8.18**

### *8.4.2 Components for gateways*

Several component suppliers are currently offering more or less application-oriented gateways of this type. A specific example is illustrated in Figures 8.20–8.22. This consists of the UJA 2010, 2020 and 25xx circuits, based on ARM 7 and ARM 9 processors, with numerous interfaces (SPI bus, AD/DA converters, watchdogs, etc.), and above all provides control of four to six CAN buses, four to eight LIN buses and two FlexRay channels for the UJA 25xx. (Yes, these systems really do exist: some top-range vehicles have up to five different CAN buses and four to six LIN buses.) Consequently, these kinds of circuits are used in a large electronic module which simultaneously provides separate control of several buses and numerous gateway functions between buses. To illustrate this, here is an example of what is known as the 'body controller' board of a car, which has the task of controlling and monitoring the functions regulating the passenger compartment of a vehicle (air conditioning, dashboard, control of opening parts – doors, boot, sunroof, etc.). As mentioned throughout this book, these functions have very different requirements in terms of speed, data consumption, etc.

## 8.5 Managing the Application Layers

Now for the last important point which has not been emphasized in this book. This concerns the application layers, i.e. level 7 of the OSI/ISO (International Standardization Organization/ Open Systems Interconnect) model which controls the operation of applications. These layers can be based on CAL, CANopen, DeviceNet, SDS, OSEK or other application layers, or any of those being developed by the AUTOSAR group. Anyone wishing to find more information on these matters should read my earlier book, *Le bus CAN – Applications*, published by Dunod, which is specifically concerned with them.

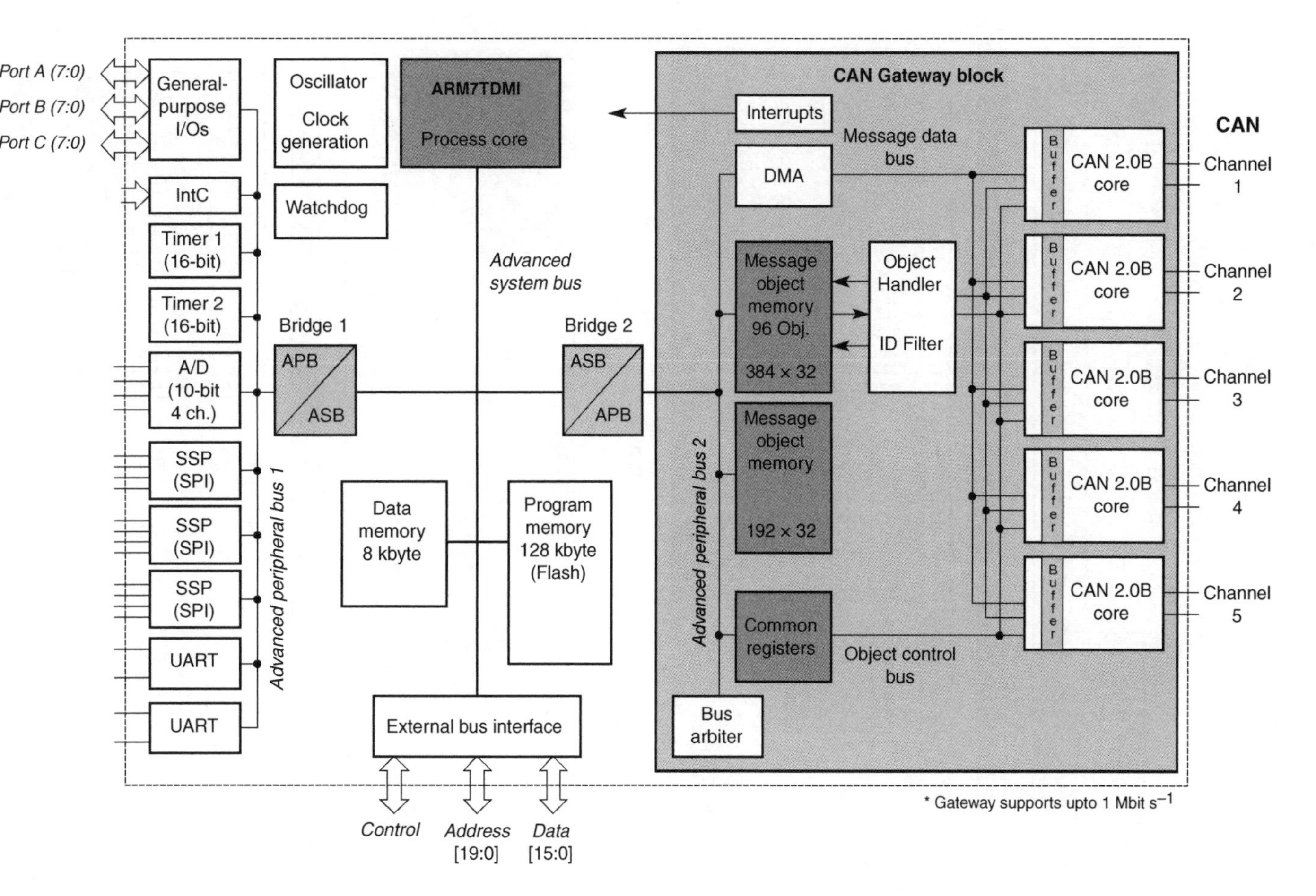

Port A (7:0)
Port B (7:0)
Port C (7:0)
General-purpose I/Os
IntC
Timer 1 (16-bit)
Timer 2 (16-bit)
A/D (10-bit 4 ch.)
SSP (SPI)
SSP (SPI)
SSP (SPI)
UART
UART
Advanced peripheral bus 1
Oscillator
Clock generation
Watchdog
ARM7TDMI
Process core
Advanced system bus
Bridge 1
APB
ASB
Bridge 2
ASB
APB
Data memory 8 kbyte
Program memory 128 kbyte (Flash)
External bus interface
Control
Address [19:0]
Data [15:0]
CAN Gateway block
Interrupts
Message data bus
DMA
Message object memory 96 Obj.
384 × 32
Object Handler
ID Filter
Message object memory
192 × 32
Advanced peripheral bus 2
Common registers
Object control bus
Bus arbiter
CAN
Buffer
CAN 2.0B core
Channel 1
Channel 2
Channel 3
Channel 4
Channel 5
* Gateway supports upto 1 Mbit s−1


Figure 8.19

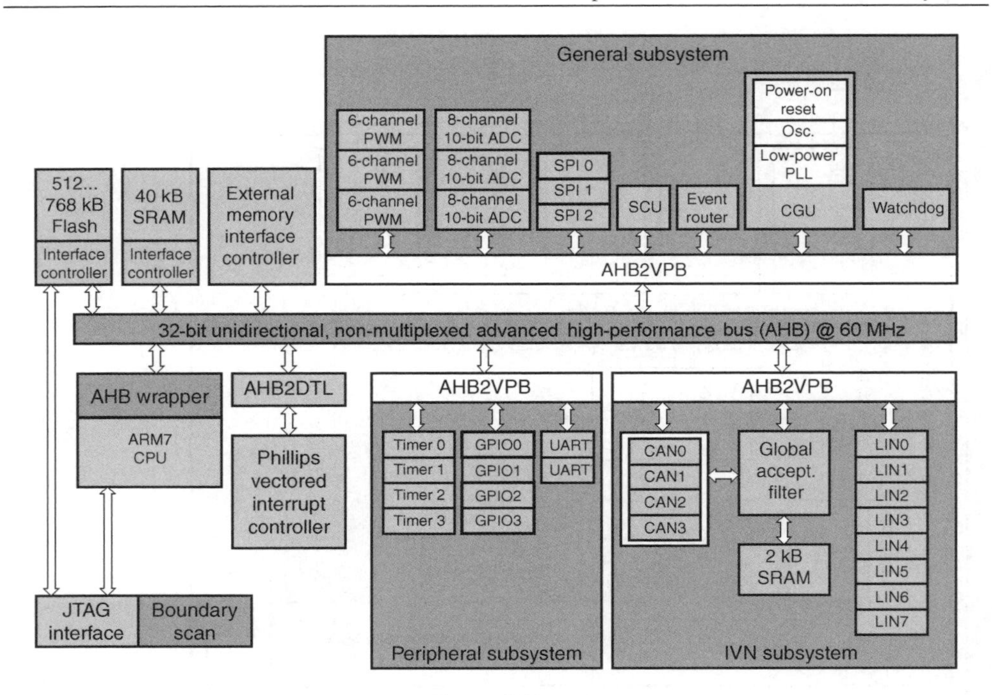


**Figure 8.20**

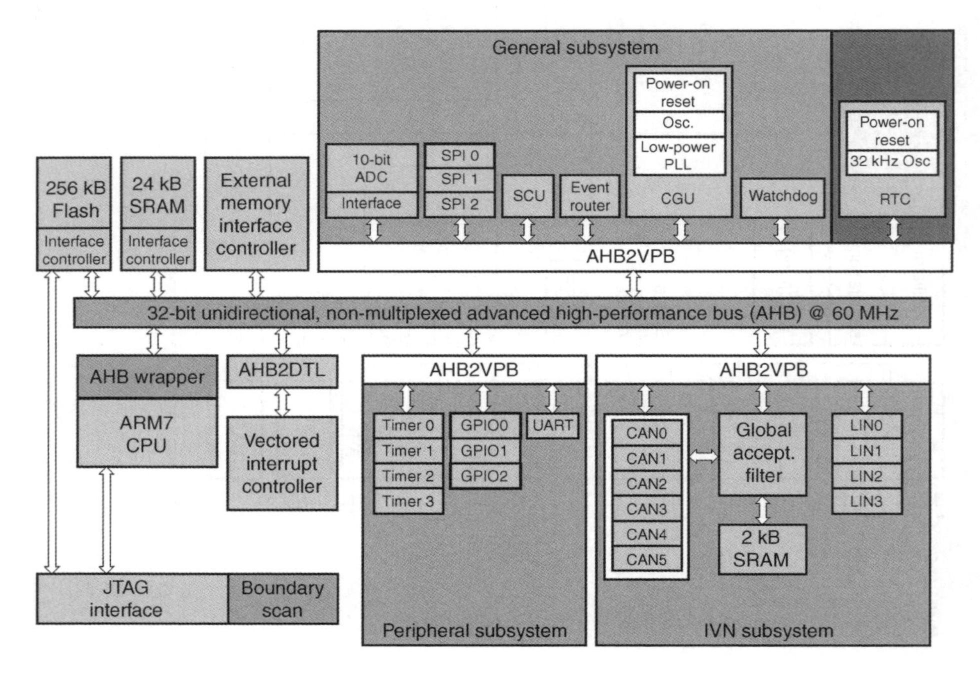


**Figure 8.21**

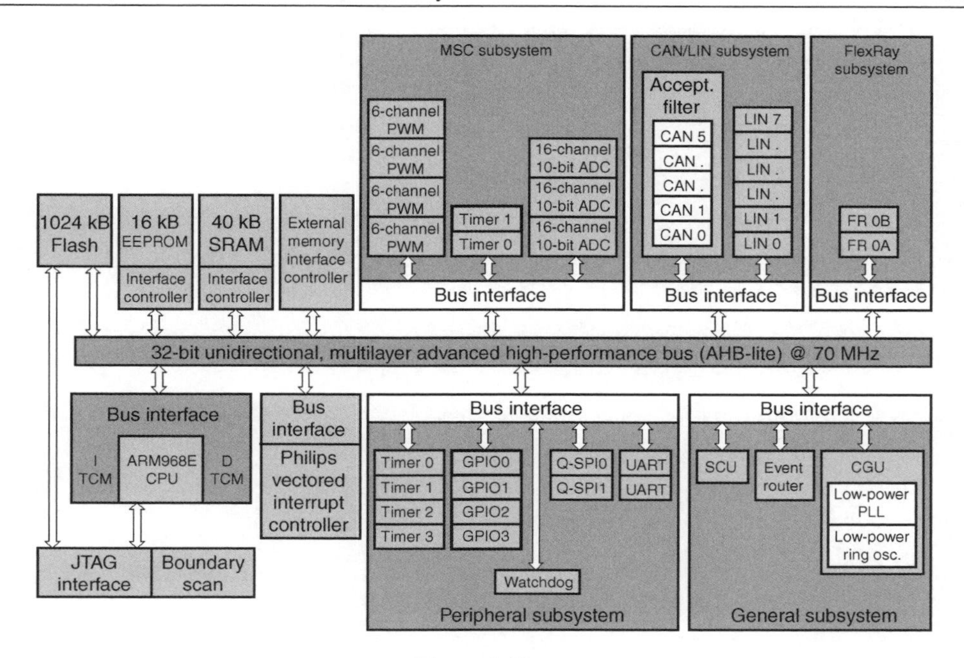
MSC subsystem
6-channel PWM
6-channel PWM
6-channel PWM
6-channel PWM
Timer 1
Timer 0
16-channel 10-bit ADC
16-channel 10-bit ADC
16-channel 10-bit ADC
Bus interface
CAN/LIN subsystem
Accept. filter
CAN 5
CAN .
CAN .
CAN 1
CAN 0
LIN 7
LIN .
LIN .
LIN .
LIN 1
LIN 0
Bus interface
FlexRay subsystem
FR 0B
FR 0A
Bus interface
1024 kB Flash
16 kB EEPROM
Interface controller
40 kB SRAM
Interface controller
External memory interface controller
32-bit unidirectional, multilayer advanced high-performance bus (AHB-lite) @ 70 MHz
Bus interface
I TCM
ARM968E CPU
D TCM
Bus interface
Philips vectored interrupt controller
JTAG interface
Boundary scan
Bus interface
Timer 0
Timer 1
Timer 2
Timer 3
GPIO0
GPIO1
GPIO2
GPIO3
Q-SPI0
Q-SPI1
UART
UART
Watchdog
Peripheral subsystem
Bus interface
SCU
Event router
CGU
Low-power PLL
Low-power ring osc.
General subsystem

**Figure 8.22**

# 9

# Safe-by-Wire

Let us now examine a particularly good example of a communication bus, namely the one used to control airbags.

## 9.1 A Little History

The airbag system of a vehicle is highly complex, and a detailed description would require more space than this book can provide. For anybody unfamiliar with the conventional structure of this system, it can be divided into three major parts:

- everything relating to the sensors (accelerometers, inertial unit, etc.) and the information that they deliver;
- everything relating to the triggering of the actuators, igniters (squibs), safety belt pre-tensioning devices, etc.
- the intelligent system that controls everything else.

For ordinary vehicles, because of the large number of interconnections required between these three units, the members of the system are generally all connected to each other by a 'point-to-point' architecture. Figure 9.1 provides an idea of this kind of structure.

However, it should be borne in mind that security architectures can be complicated to suit any requirements, and top-range vehicles can have up to 20 airbags at least (in front, at the sides, for the head, for the knees, under the knees, to prevent sliding in a collision, etc.) and a large number of sensors. In this case, the processing unit must be centralized in order to limit the lengths of the connecting cables; indeed, we can soon reach the mechanical limits on the size of the connector(s) which may include up to 100 interconnection points and which end up looking more like mouth organs than connectors (Figure 9.2).

In order to open up the architecture and topology of the implementation of safety restraint systems, attempts were made some years ago to tackle the problem of specifying a network and bus structure with the aim of permitting a modular, variable-geometry architecture.

*Multiplexed Networks for Embedded Systems: CAN, LIN, Flexray, Safe-by-Wire...* D. Paret

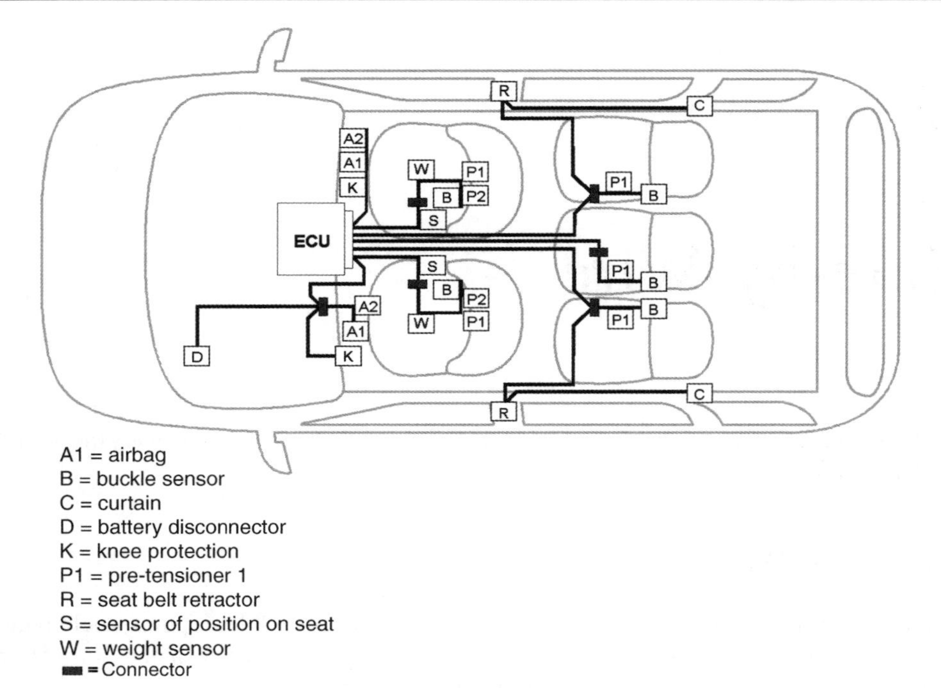


**Figure 9.1**

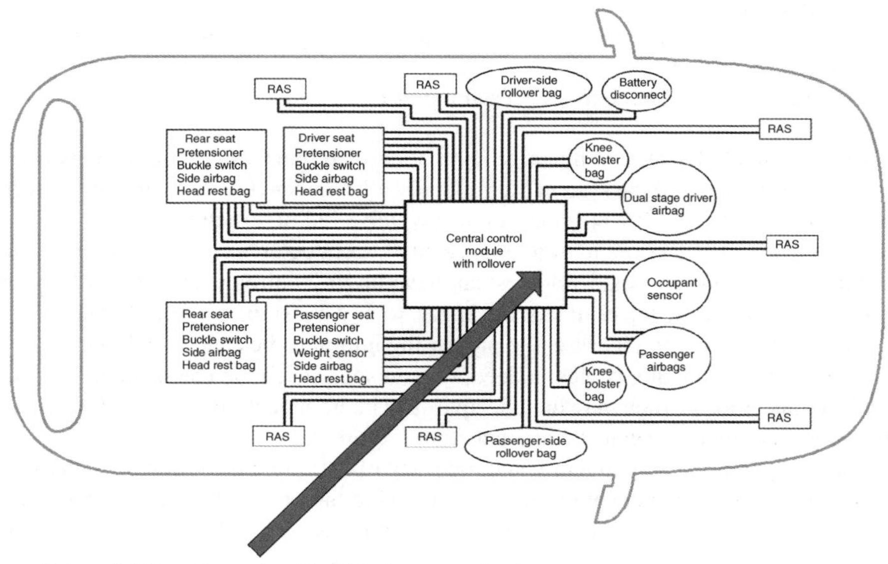


**Figure 9.2**

Clearly, this entails a degree of expense, and, if the problem is viewed from this angle, one of the main questions is how many nodes (sensors, actuators, etc.) are required to make the whole process worthwhile. Detailed economic calculations have shown that as the number of nodes in a network approaches twenty it becomes preferable to consider a bus structure for interconnecting all the network participants (Figure 9.3).

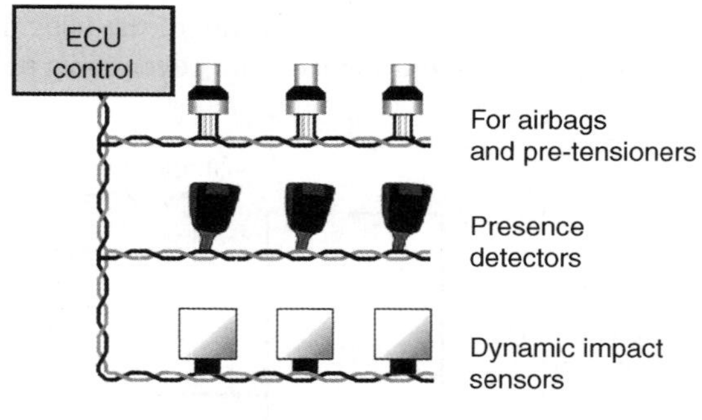


**Figure 9.3**

## 9.2 Safe-by-Wire Plus

Various designs have been proposed since 1999, particularly these two: on the one hand, a design developed by the 'BST' group, consisting of Robert Bosch GmbH, Siemens VDO Automotive AG and Conti Temic Microelectronic, and on the other hand one developed by Autoliv, TRW Inc., Philips Electronics N. V., Analog Devices Inc, Delphi Electronics Systems, Key Safety Systems Inc. and Special Devices Inc. As usual, after several years of ferocious arguments, concessions and counter-proposals, everyone joined a single consortium in 2001, known initially as Safe-by-Wire. Subsequently, under the aegis of USCAR (United States Council of Automotive Research), this consortium was rechristened 'Safe-by-Wire Plus' (see the logo in Figure 9.4), with the aim of developing an open standard for safety

**Safe-by-Wire Plus**

**Automotive Safety Restraints Bus Specification**

**Version 2.0**

**Figure 9.4**

restraint specifications for the motor vehicle market. The end result of all this work was the public appearance of the 'automotive safety restraints bus – ASBR 2.0' in late September 2004, which was to lead in the usual way, and at the usual pace, to an ISO standard, ISO 22 896.

The proposed network architecture is illustrated in Figure 9.5, which shows two different areas: on one hand, the very fast domain of impact (crash) sensors, and on the other hand the area controlling the firing of the various safety elements of the network (control of airbag igniters ('squibs'), safety belt pre-tensioning devices, etc.), everything being orchestrated by a central CPU.

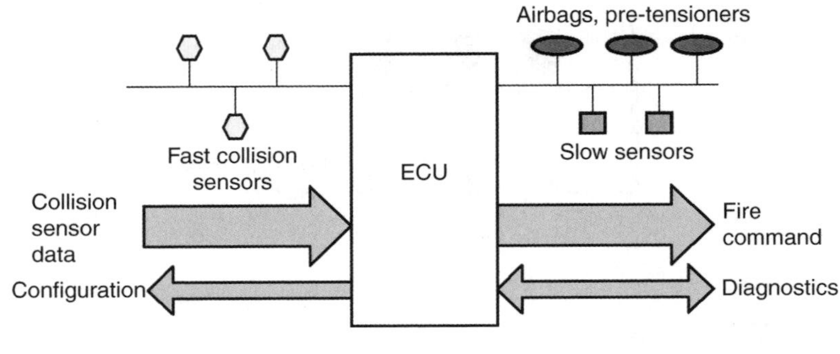


**Figure 9.5**

Compared with the conventional architectures of present-day vehicles, this is much simpler and, 'on paper', takes the form shown in Figure 9.6.

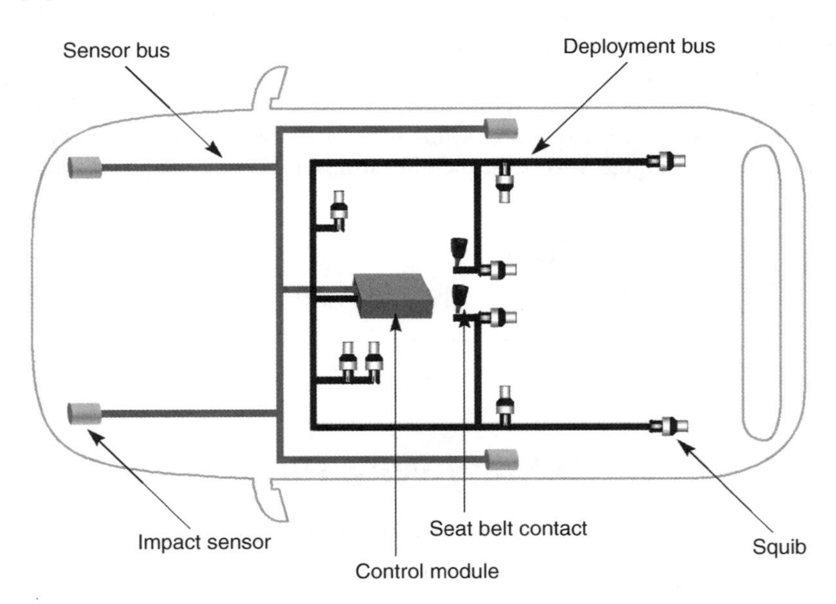


**Figure 9.6**

To meet its functional requirements, the Safe-by-Wire Plus concept must have safety features related to a very specific operating mode, itself divided into two very separate phases:

- The first is when the system appears to be inactive (on standby – but in fact it is carrying out diagnosis), i.e. most of the time (at least we hope so!). It must be possible to test the operating potential of the elements on the bus on a cyclic basis, to ensure that they are ready for operation when needed. This form of cyclic polling of the network elements must not disturb other devices in the vehicle (radio receiver, etc.) or create interference, and therefore slow communication speeds (about 20 kbit $s^{-1}$) are used during this phase.
- The second is when the system is active, in a collision. The system must be able to communicate immediately, at high speed (about 250–500 kbit $s^{-1}$). It must be able to operate instantaneously to trigger the squibs (igniters). Communication must be fast and secure, to control all the elements to be brought into action. The chosen network operating principle must allow for independent power supplies to the participants, so that each element in the system can carry out its function even in case of an incident or battery loss during the collision, and even a few moments later. The system must not be sensitive to short circuits or connection of the bus wires to the positive battery terminal or to ground.

All these operational characteristics require a distinctive architecture and a special design, in terms of the principle followed, the protocol and the topology – all being linked sooner or later to the CAN (controller area network) bus, if only as a way of indicating on the dashboard that your seatbelt is unfastened!

## 9.3 Some Words of Technology

The above requirements have resulted in the development of some very special hardware for the physical layer.

### *9.3.1 Operating principle*

The operating principle is shown in Figure 9.7.

First of all, we must provide a fully floating sinusoidal voltage generator, with no reference to a ground potential of the system.

During the positive half-waves delivered by the sinusoidal voltage generator, the two diodes shown in the figure are conducting, and because of the capacitors located in the nodes present on the network, half-wave rectification and filtering is carried out locally in each node, thus creating a continuous voltage for supplying the node in question.

The node is now correctly supplied and its local electronic circuit can be synchronized with the frequency of the incoming signal (from the generator) and can operate and perform the specified task.

In order to communicate correctly via the network, the node modulates the load which it represents only during the negative halves of the sine waves of the sinusoidal voltage supplied by the voltage generator. To achieve this during the negative half-waves, a resistor with a fairly

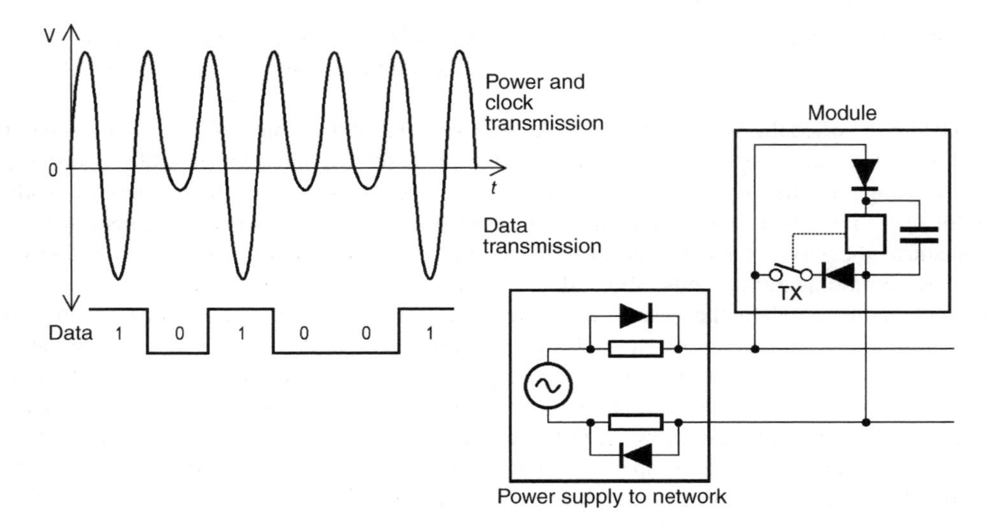


**Figure 9.7**

high value must be connected in parallel with the rectifier diodes, to provide a significant pd in the network during this modulation. This makes it possible to obtain one bit per half-wave, directly related to the generator frequency, while being totally independent of the value of this frequency.

The proposed principle allows remote supply to the nodes and communication between nodes, and offers

- the possibility of instantaneous speed changes (to the nearest half-wave, with the need to provide automatic bit rate detection systems for the generator frequency). The theoretical and actual oscillograms of this important feature are shown in Figures 9.8;
- less RF pollution, as the wave carried on the bus is always sinusoidal;
- highly secure operation in case of short circuits of the power supply wires to the positive terminal or to ground, as the sinusoidal generator is floating;
- simple bit recognition;
- no interference between the power supply and the data signal;
- a high degree of flexibility and simplicity of design in terms of implementation and network topology;
- an economical solution, as the power supply and the clock signal are supplied via the physical medium of the bus.

Furthermore, because of the chosen principle of load modulation, the bit value is formed by creating a dominant bit on the network (it is a load modulation *and* wired circuit), making it possible to apply the same arbitration principle as that of the CAN bus and therefore makes the network to operate in multi-master mode. Because of the protocol used and the authorized range of identifiers, this type of bus can accept up to 64 participants.

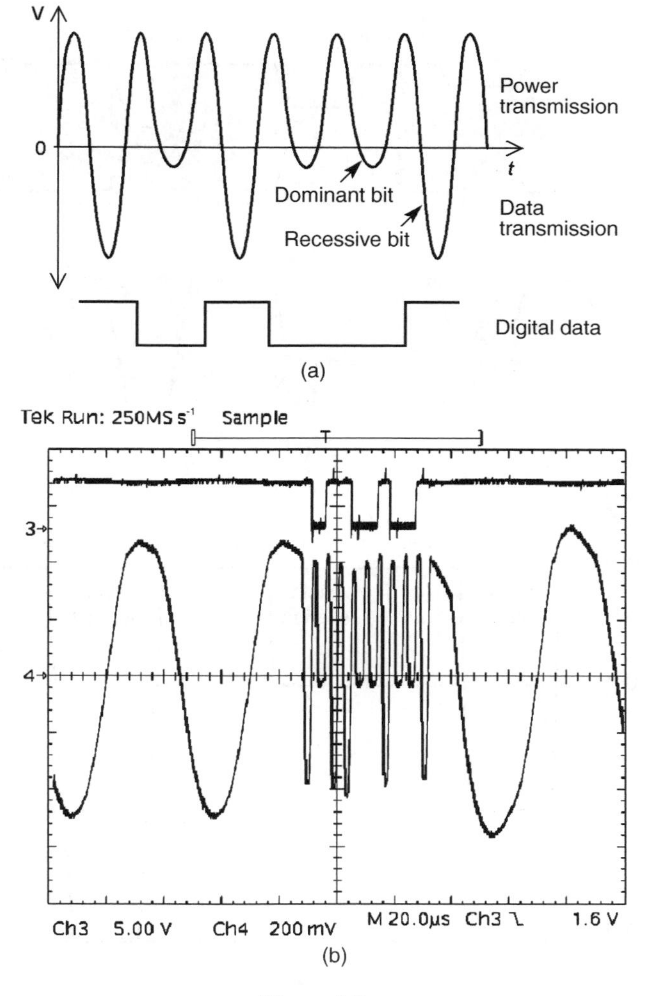


**Figure 9.8**

### *9.3.2 Coding and physical representation of the bit*

The bit coding is NRZ and its physical representation is shown in Figure 9.9. This clearly shows (in the upper part of the figure) the different phases of carrier modulation corresponding to the binary levels '1' and '0', together with their electrical values in differential mode (in the lower part of the figure).

### *9.3.3 Safe-by-Wire network configurations and topologies*

Depending on the specified safety requirements, the network topology can be of the bus, parallel, ring, daisy chain bus, daisy chain ring or other type (Figures 9.10).

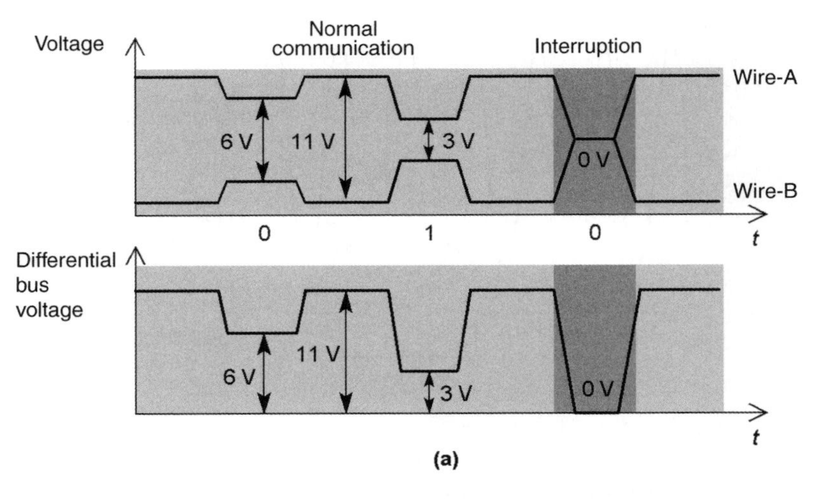


(a)

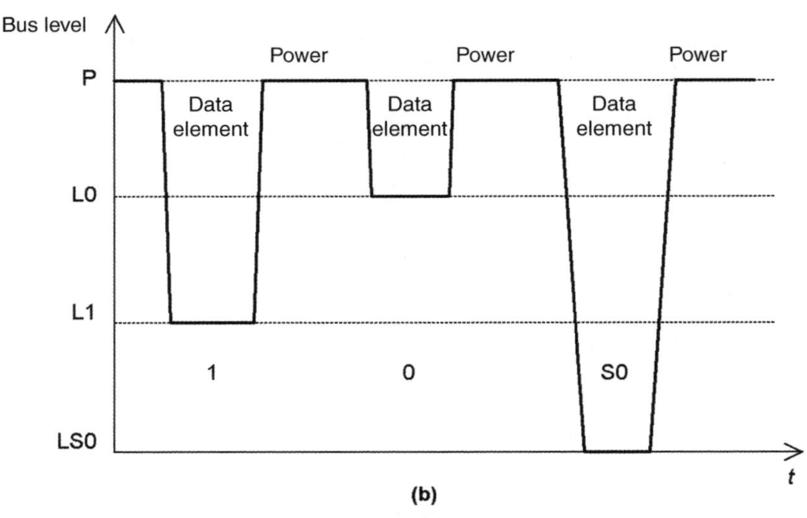


(b)

**Figure 9.9**

Very often, the topology used and implemented in a vehicle is a happy mixture of all these possibilities. This is because, for many reasons, for example in a collision, there may be a power loss that disconnects part of a bus-type network. For example, one of the wires may be short-circuited to ground after an impact on the right-hand side. This must not endanger the rest of the network if, following this impact, the vehicle, being pushed from one side, suffers a further impact on its left-hand side a few tenths of a second later. To avoid this, a ring or mixed configuration makes it possible to go on serving all the network participants. The same applies if the bus wires are accidentally short-circuited in a collision. The following section provides more details on this point.

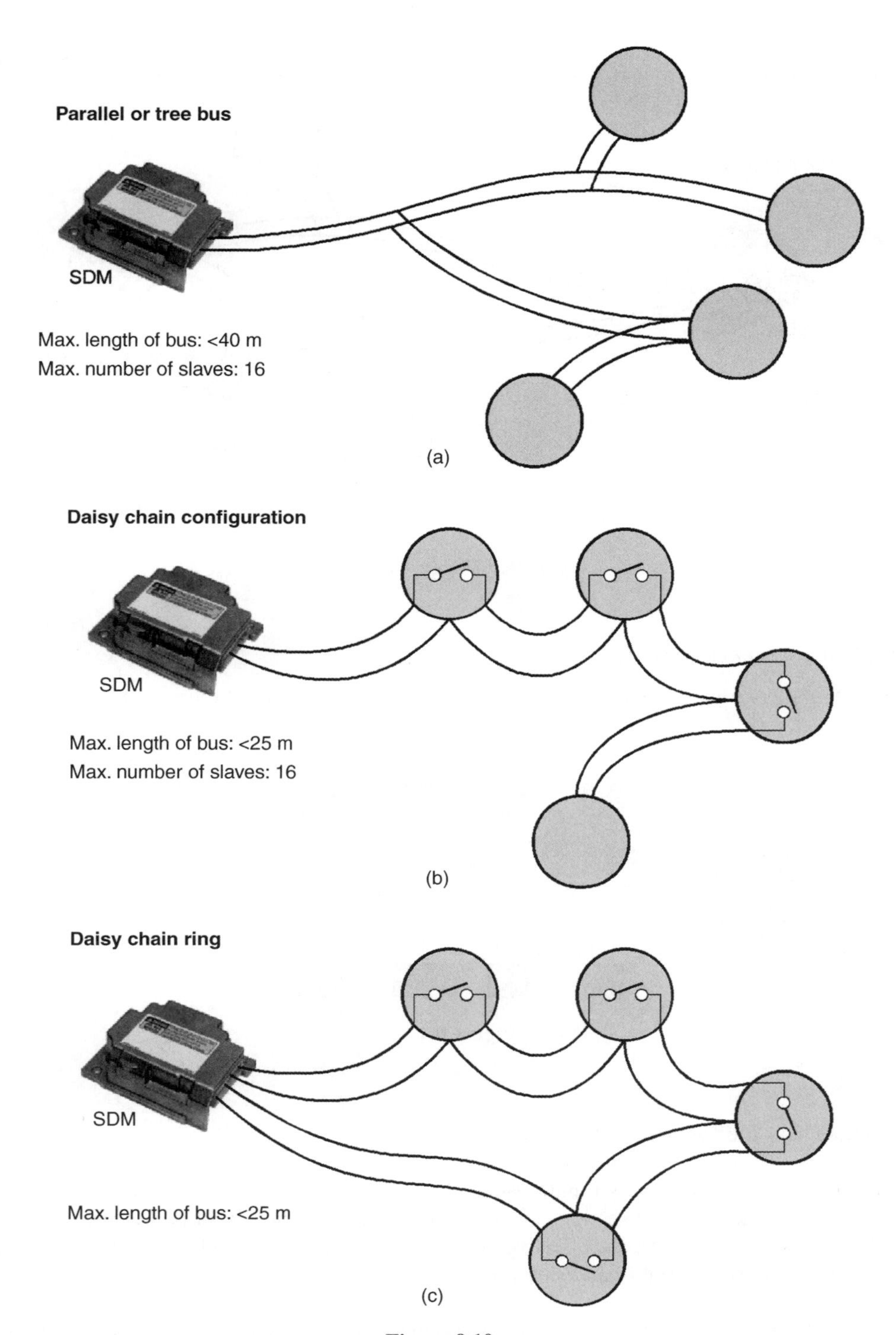

Parallel or tree bus
SDM
Max. length of bus: <40 m
Max. number of slaves: 16
(a)
Daisy chain configuration
SDM
Max. length of bus: <25 m
Max. number of slaves: 16
(b)
Daisy chain ring
SDM
Max. length of bus: <25 m
(c)


**Figure 9.10**

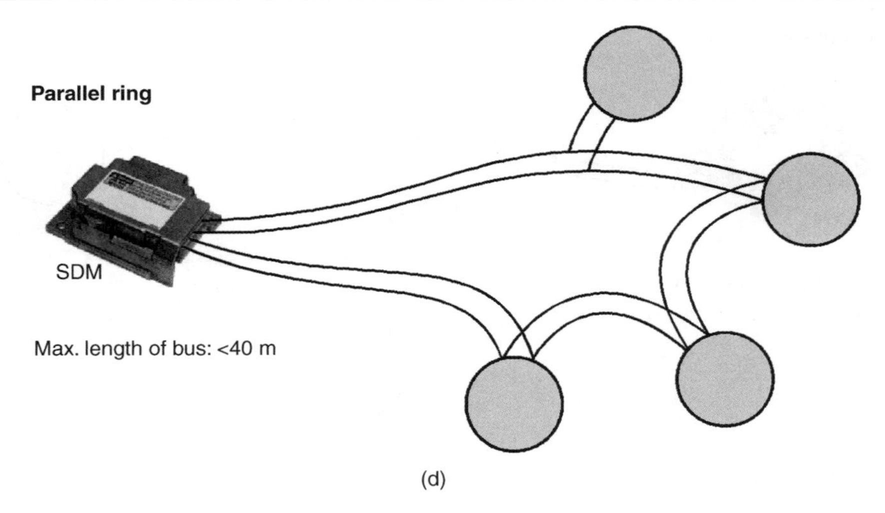


**Figure 9.10** (*Continued*)

### *9.3.4 Bus section separators*

In the latter case, bus section separators (BSS), like those shown in Figure 9.11, are used.

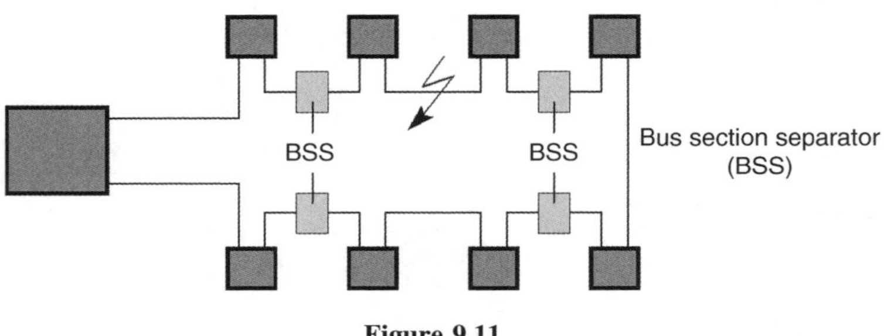


**Figure 9.11**

These have the following advantages:

- the damaged section is very rapidly isolated;
- the squibs of the isolated section can still be ignited, using their local power reserves;
- the system remains active and effective even if there is a secondary collision (for example an initial impact on the right-hand side followed by one on the left-hand side).

Figure 9.12 shows an example of a topology combining a ring, a conventional bus and BSS, to optimize the safety aspect of a given specification.

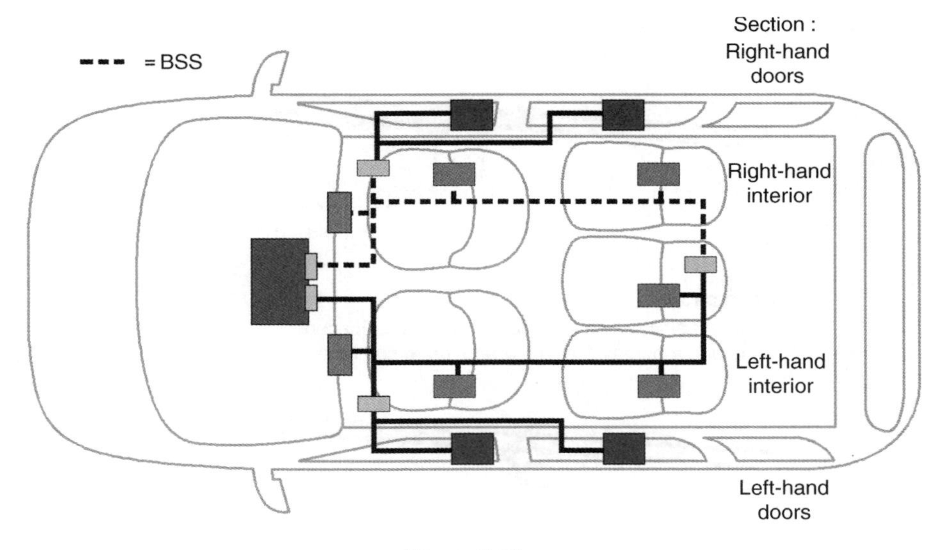


**Figure 9.12**

To allow the wires of the bus in the vehicle to be connected accidentally (in a collision or due to error) to the positive battery terminal or to ground (Figure 9.13), or very often the circuit of the sinusoidal generator, which itself has a ground reference, galvanic isolation of the transmission line is provided, using a transformer (Figure 9.14), so that short circuits of the communication line to the battery positive or to ground can be tolerated.

This transformer, which is not very large because it operates at HF, must of course pass the lowest frequency of the band (approximately 20 kHz) and is generally followed by a power

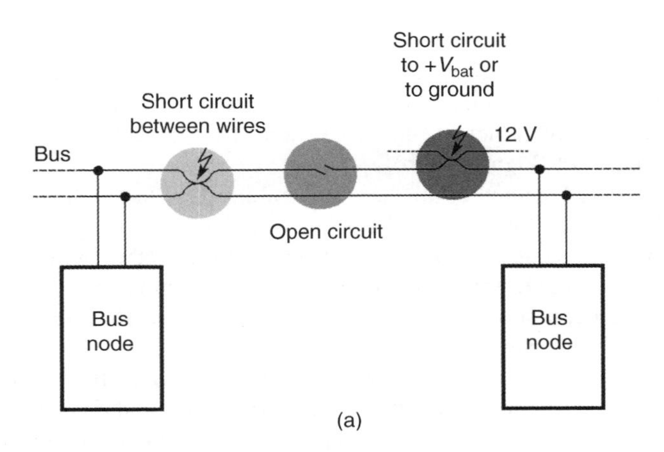


**Figure 9.13**

**Short circuit of one wire:**
- Capacitive coupling

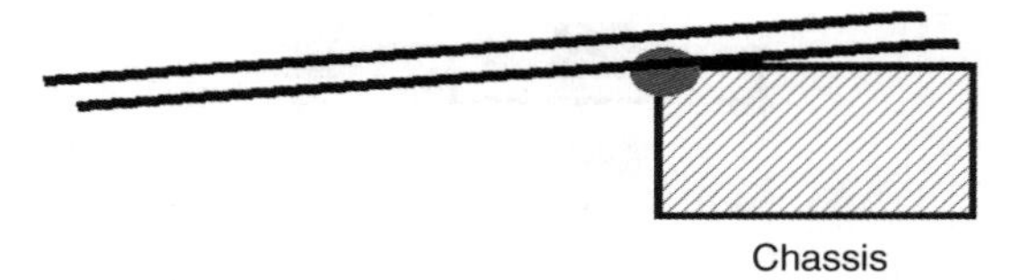


**Open circuit:**
- The remaining wire maintains the communication
- The ring topology enables operation to continue

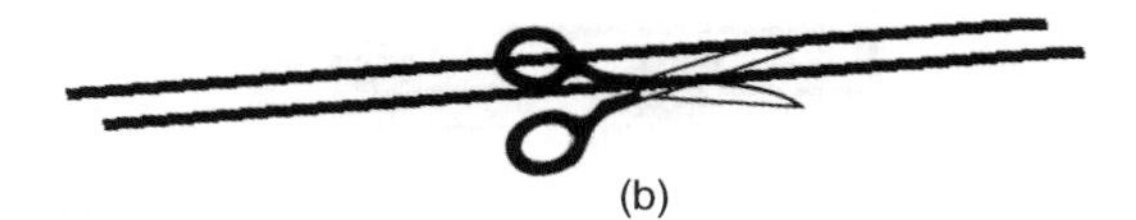


**Figure 9.13** (*Continued*)

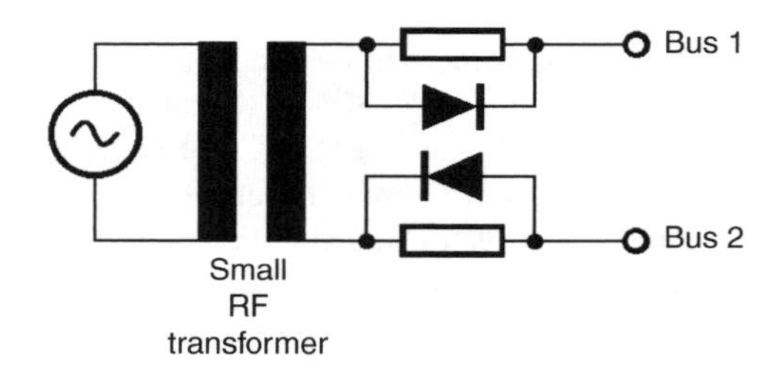


**Figure 9.14**

shuttle whose task is to create ‘galvanic isolation which can pass continuous current’. In the first phase, the double switch on the left allows the capacitor to charge (Figure 9.15a); in the second phase, the capacitor, if there are no connections to the generator, transfers and supplies this power to the network (Figure 9.15b).

We have now reached the end of this brief general introduction to the new concept of the Safe-by-Wire Plus bus and the specific characteristics chosen for the physical layer. As you will certainly have noticed, it has nothing in common with the other systems in a vehicle (in terms of either the protocol or the physical layer), but it needs to communicate with them, if only to provide an indication of correct operation on the dashboard of the vehicle when it is started – so we need a gateway.

For details of the content of the data transfer frame, its various fields and the associated specific commands, you should consult the Safe-by-Wire Plus standard on this occasion, even though its contents (although very instructive) go beyond the intended scope of this book, in terms of information but not in respect of the technical level.

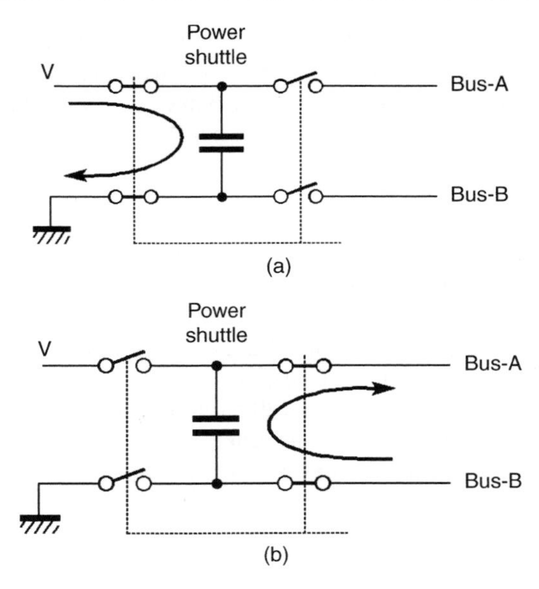


**Figure 9.15**

To conclude this chapter, I will now mention some industrial applications of this concept which are very unexpected, but which are also responsible for high volumes of products (it is not only in the automotive or telephone markets where we find high volumes, after all). I will then provide a final description of the existing electronic components dedicated to this concept.

### *9.3.5 Industrial applications*

Although I have provided many examples of motor vehicle applications, this type of bus is also used in industrial applications having many technical characteristics in common with airbag controllers. An airbag control procedure consists of remotely commanding, at a precise moment, an igniter which only operates once – and hopefully never! However, if it does happen, it must happen in 100% of cases. You should be aware that there are other applications of this principle. Here are two of them.

#### Mineral extraction from opencast quarries

You may say that a data bus is a very expensive way of making some fragments of rock, because theoretically all you have to do is to place a few sticks of dynamite, light the fuse, retire to a safe distance and wait. However, it is more complicated than that . . . Let us take the example of mineral extraction from opencast quarries. If we want to avoid making too much noise and dust (environmental considerations), and if we want to have small pieces of rock conforming to a desired gauge without passing the larger pieces through a noisy, expensive crusher, it is simply necessary to set off the dynamite sticks (using special squibs) not all at once, but in a very precise sequence, at a given time which can be programmed according to the location, the type of rock, etc. In this way the sound wave will have a subsonic frequency, the dust will remain in place and the size of the rock pieces will meet your requirements. By doing this, you have reinvented the 'airbag controller'!

### Controlling fireworks

Same reason, same problem. To achieve a beautiful, musical effect, each firework display must be programmed very precisely. Here again, to avoid the need to run from one firing base to another, a good squib firing bus is the perfect answer.

A final detail: the market for squibs for mineral extraction and fireworks is much larger than the car airbag market. The world output of cars is about 60 million per year, with an average of two airbags per car worldwide, that is to say 120 million, as against several million squibs per week for the industrial market – and, to the delight of electronic component vendors, the squibs for mineral extraction and fireworks can only be used once – proper consumables!

### *9.3.6 Examples of components*

The first components are emerging from the diffusion furnaces. Some examples are the UJA 6203 sensor bus slave (see the functional diagram in Figure 9.16), the UJA 6402 sensor bus master (see Figure 9.17), the squib driver (Figure 9.18) and an example in Figure 9.19.

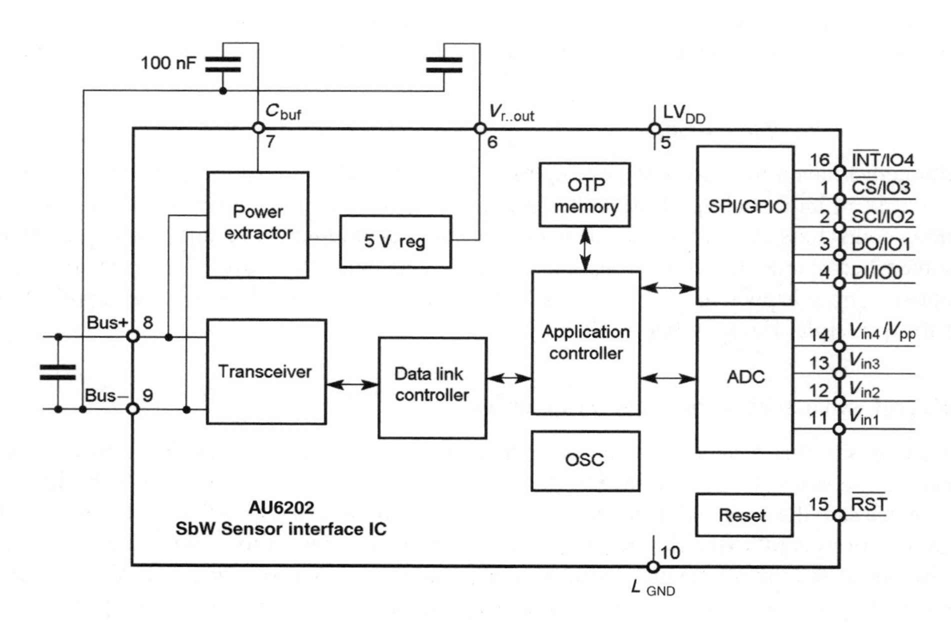


**Figure 9.16**

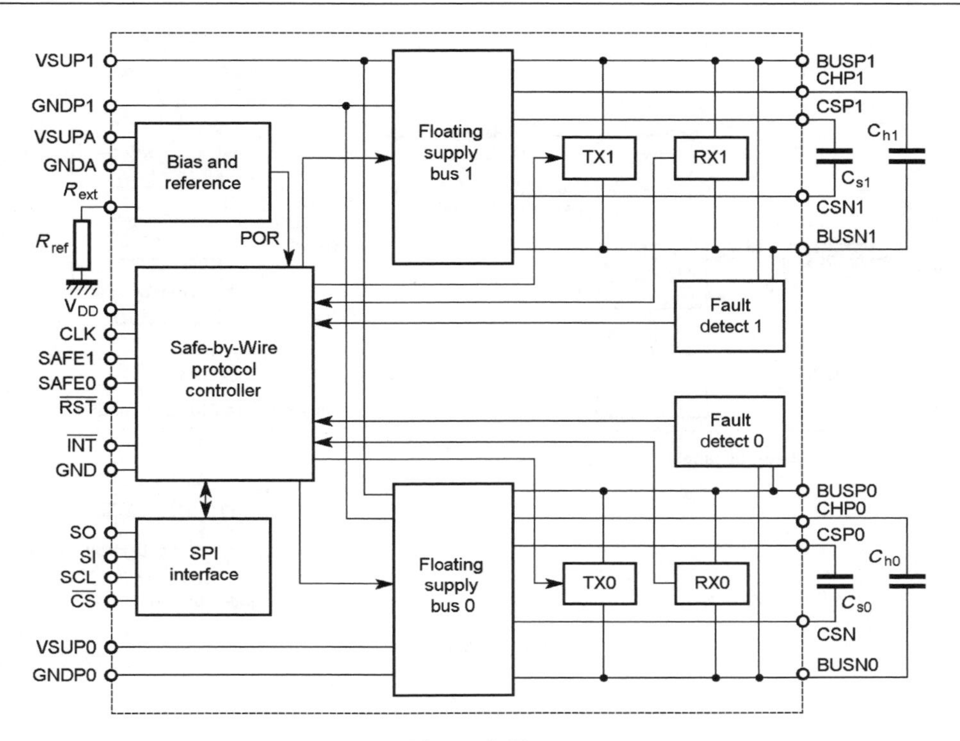


Figure 9.17

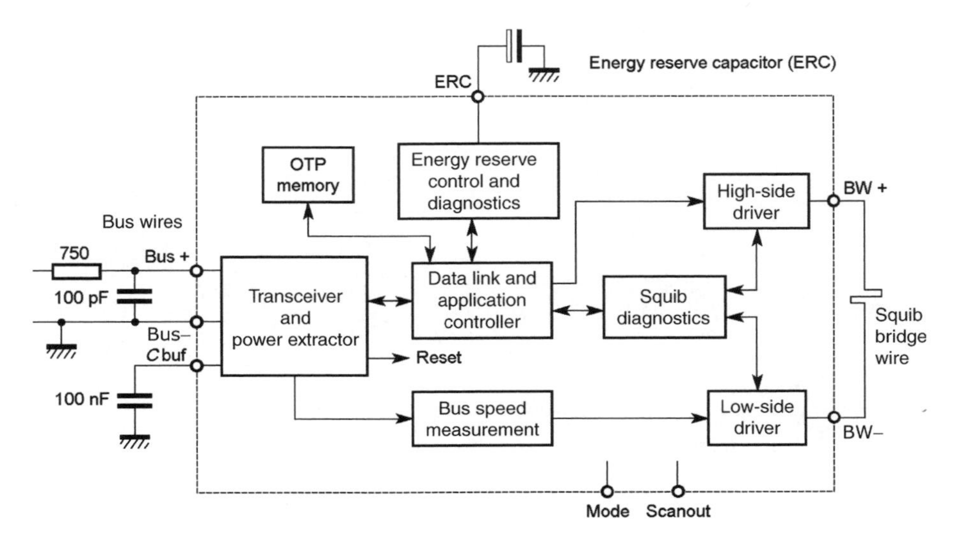


Figure 9.18

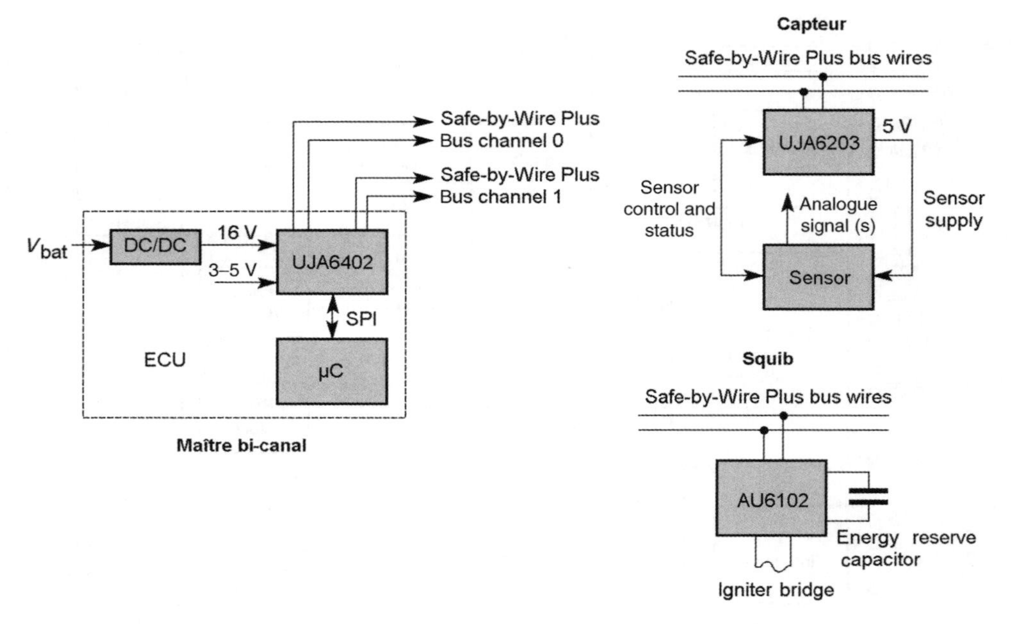
Safe-by-Wire Plus
Bus channel 0
Safe-by-Wire Plus
Bus channel 1
$V_{bat}$
DC/DC
16 V
3–5 V
UJA6402
SPI
ECU
μC
Maître bi-canal
Capteur
Safe-by-Wire Plus bus wires
UJA6203
5 V
Sensor
control and
status
Analogue
signal (s)
Sensor
supply
Sensor
Squib
Safe-by-Wire Plus bus wires
AU6102
Energy reserve
capacitor
Igniter bridge

**Figure 9.19**

# 10

# Audio–Video Buses

This chapter will deal with the main communication networks and buses used in industrial and motor vehicle on-board systems, etc., for audio and video applications.

These applications are structured around conventional digital audio CD players, MP3 compressed digital audio CDs, CD-ROMs containing, for example, data representing road maps for assistance in a navigation system, video games, conventional DVDs, etc. They also include links carrying pure video, in other words images captured by a camera located at the rear of a vehicle, for example, to send information to the instrument panel (reversing camera), reception of coded OFDM digital terrestrial television (DTT) channels, and multiplexing MPEG2 and MPEG4 digital streams from multiple video sources for distribution to different points (e.g. the seats and instrument panel of special vehicles, coaches, aircraft, etc.).

## 10.1 I2C Bus

### *10.1.1 General*

The I2C bus, developed by Philips Semiconductors, is well established and is still a dominant feature of on-board systems. A vast number of these integrated circuits are produced every year for all kinds of applications. We should remember that this is a synchronous (two communication wires, the first carrying data SDA and second carrying the clock, SCL), multimaster bus and that its initial speed of 100 $\text{kbit s}^{-1}$ has increased progressively to 3.4 $\text{Mbit s}^{-1}$. For further information, I would refer you to many books that I have written on this subject, available from the same publisher. The applications of this bus are often limited to short communication distances (for example, within an instrument panel, for controlling the central display unit).

### *10.1.2 Buffered I2C and optical I2C*

Many variants of the physical layer have been developed to increase the communication distance. The asymmetric ‘buffered’ I2C and its ‘optical’ version have been, and are still,

*Multiplexed Networks for Embedded Systems: CAN, LIN, Flexray, Safe-by-Wire...* D. Paret

frequently used as series links for conveying commands between an instrument panel and a CD changer, when the latter is located in the boot (for changing discs or selection ranges, for example), as well as for data exchange.

At present, for simple reasons of network uniformity, it should be noted that CAN (controller area network) is used more and more frequently for transmitting commands or short numeric messages, particularly for display commands and data communications applications.

**Note:** The well-known SPI bus (developed by Motorola) which is also widely used in motor vehicles (for short links between integrated circuits on a single card, or between engine control elements, for example) is not dealt within this chapter, as it is rarely used, if at all, in audio–video applications.

## 10.2 The D2B (Domestic Digital) Bus

Let us begin with a brief history of this subject.

### *10.2.1 A short history*

The D2B bus was also developed by Philips (again!) at the beginning of the 1980s, for applications of the system-to-system communication type, mainly domestic and especially audio–video. At that time, the bus appeared to be rather ahead of its time, and it aroused considerable jealousy among other corporate structures (the post office, EDF, the construction industry, etc.) because of its very considerable audiovisual potential. Finally, after much work, and the signing of partnership agreements between Philips and Matsushita initially, and then with Thomson and Sony (which themselves represented only 60% of the world market for domestic and audio–video applications), the project reached its final stage as the European standard IEC EN 1030, which describes in detail, like any standard, the protocol, the codes assigned, etc.

The main purpose of D2B is to interconnect different devices having functions which can be completely different (linking one brand of car radio to another brand of CD player and changer, for example). Clearly, the preferred applications of this bus, from the outset, have been associated with audio and video. However, there is no reason why it cannot be used for other applications.

### *10.2.3 The D2B protocol*

This section is not intended to describe D2B in the same way as I described CAN, but simply to summarize in a few words the contents of a D2B frame.

This multimaster bus operates in asynchronous mode on a differential pair (with ground return), and its frame format has a maximum duration.

Figure 10.1 shows the composition of a D2B frame. If you examine it closely, you will find that it differs radically from the CAN frame, mainly in the following ways:

- the bit structure is completely different;
- the address field (master or slave) is more important than in I2C;

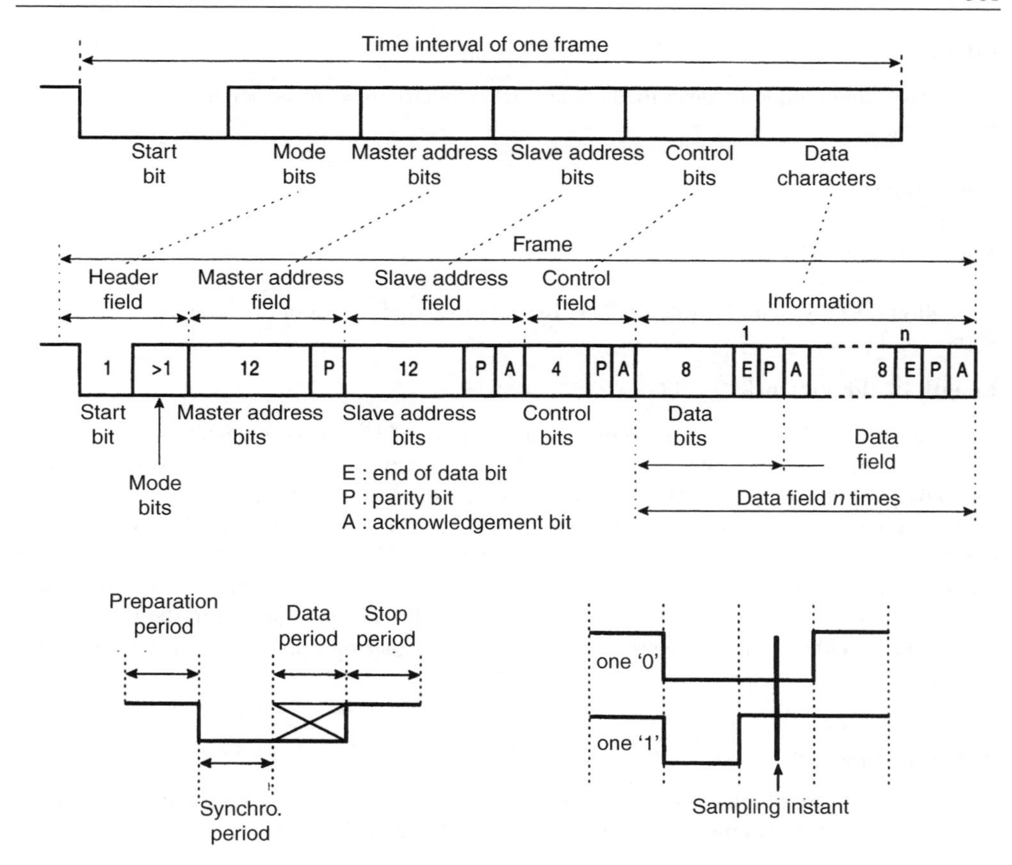


**Figure 10.1**

- the master declares its identity;
- arbitration takes place only at the start of a frame;
- a continuity bit has appeared.

Before describing how to form a gateway between D2B/I2C and CAN, I will briefly summarize the main characteristics of D2B to provide a clear context for its use, instead of going through a lengthy list of functions of each bit, which would be largely outside the scope of this book.

### *10.2.4 The network aspect of D2B*

D2B is one of the very large family of local area networks (LANs). For information, the associated I2C bus also belongs to this family. The local aspect relates mainly to the distance over which the data can travel, and the quality of these data.

**Bit rate**

Three operating modes are described in the D2B standard; they are associated with three clock frequencies:

- in mode 0: 750 kHz,
- in mode 1: 3 MHz,
- in mode 2: 6 MHz.

This allows a maximum bit rate of 7760 bytes per second, i.e. about 62 kbit $s^{-1}$.

**Length of the network**

D2B was designed to operate over a maximum distance of 150 m. This is because research has shown that this distance is the maximum point-to-point wiring length of a standard installation for individual residences or private cars (hence the 'domestic' name tag).

The maximum bit rate shown in the preceding section and the length stated above provide a practical definition of the types of application in which D2B can be used. In the motor vehicle market, for example, one of its main applications is for the control of standard electronic systems (audio CD players, video, road maps on CD-ROM, etc.) located in the boot, where control is provided from the instrument panel or the steering column (or at the top of the column).

**Network topology**

There are many topologies for the construction of networks. The chosen topology is a daisy chain serial bus type (Figure 10.2), in order to meet the criteria of length, bit rate, installation cost, etc.

### *10.2.5 The line aspect of D2B*

**Symmetry/asymmetry**

The I2C bus described above has two wires (SDA for data and SCL for the clock), both asymmetric with respect to ground. D2B, on the contrary, was designed for symmetrical transmission, using a twisted differential pair, with potential recovery acting as a screen. This was done in order to provide protection against any parasitic signals (radio frequency, electromagnetic and/or electrostatic). This combination of characteristics provides a line with excellent electrical properties.

**Impedance**

Any line has electrical characteristics, especially an intrinsic impedance, which in most cases is where the problems really start.

The purpose of a network is to communicate, and therefore the signals must travel, and they must be propagated along these lines. However, you will be well aware that lines have a

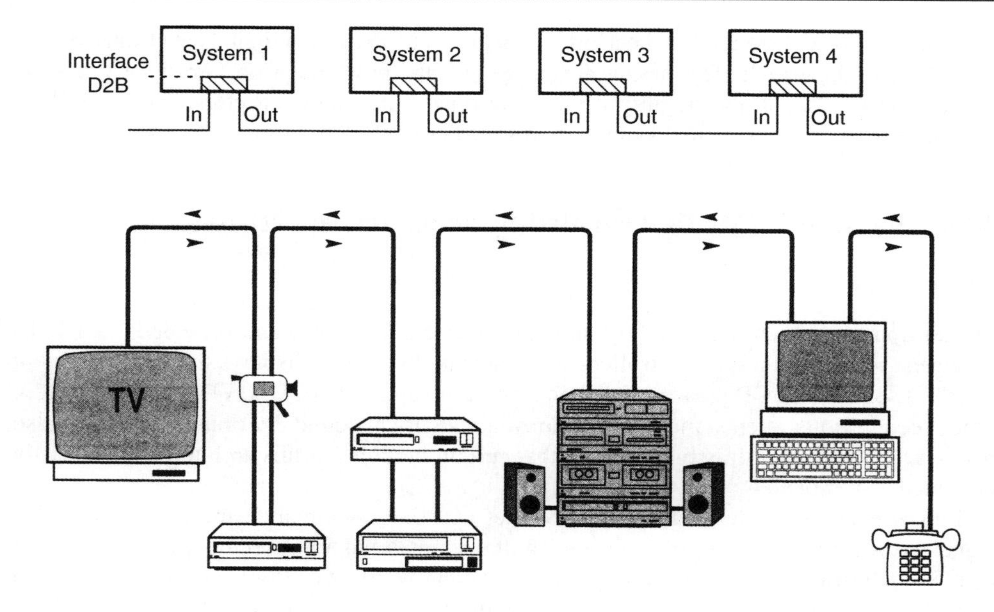


**Figure 10.2**

troublesome habit of reflecting signals to their terminations if care is not taken to prevent this. This can of course be entertaining at first – but it quickly palls!

Unfortunately it also causes problems in applications. In the context of D2B (and in many other cases as well), when a message is sent, the 'caller' often waits to receive an electrical 'acknowledgement' signal from the 'called party'. Nothing so unusual about that. But suppose that the 'called party' has gone away on holiday, for example, the 'caller' very politely sends his acknowledgement request message, which, when it is propagated, travels to the end of the line, is partially reflected at the end of the line and comes bouncing back. At this point, two problems can arise:

- the returning signal is interpreted as the one which should have been sent by the 'called party' (although of course the called party is not there);
- or the signal, returning too soon (or too late), causes a head-on collision with a new signal sent by the 'caller' (call the insurance company...).

This results in either an error or a bus conflict, which is hardly any better.

## D2B on an optical medium

In this case of conventional uncompressed digital audio communication (audio CD), the bit rate of the digital data is about 1.15 Mbit $s^{-1}$, and it is difficult to carry these data without protection against parasitic signals in a motor vehicle. Generally, the upper limit of wire operation on a differential pair is reached, unless special precautions are taken. An 'improved' D2B on a fibre optic physical layer has therefore been used for a long time to transfer digital data from the back to the front of the vehicle. This has the advantage of better protection from parasitic signals and the drawbacks of cost, the fragility of the fibre optic links and connectors, and the lack of flexibility when creating a bus topology in this medium, other than point-to-point links.

## 10.3 The MOST (Media Oriented Systems Transport) Bus

### *10.3.1 General*

In addition to conventional audio applications, a car must now be provided with links between radios, navigation controllers and associated systems, displays (on the instrument panel, at seats, etc.), CD players and changers (audio and video CD, DVD, CD-ROM, etc.), voice recognition systems, mobile telephony, active in-car sound distributors, etc. It is also necessary to provide infrastructure such that certain modules of this architecture can easily be added or removed.

Before examining the details of MOST, let us briefly summarize the audio and video signals to be carried, in terms of associated digital bit rates and transport modes. Some of these signals are transmitted in asynchronous mode, in other words at a closely specified rate which does not allow any freedom of design (as with an audio CD, for example). Others have essentially variable digital streams (for example, a video sequence compressed by the MPEG2 source coding). Figure 10.3 summarizes the main properties of most of the signals that can travel on a MOST bus.

| | | Bit rate | Format |
|---|---|---|---|
| Control signals | | 125/250 kbit $s^{-1}$ | Asynchronous |
| Data signals | Digital audio:<br>– uncompressed audio CD<br>– MPEG compressed audio | <br>1.41 Mbit $s^{-1}$<br>128/384 kbit $s^{-1}$ | <br>Synchronous<br>Asynchronous |
| | Digital video:<br>– uncompressed CCIR 601/4:2:0<br>– compressed MPEG1<br>– compressed MPEG2 | <br>249 Mbit $s^{-1}$<br>1.86 Mbit $s^{-1}$<br>2/15 Mbit $s^{-1}$ | <br>Synchronous<br>Asynchronous<br>Asynchronous |
| | navigation:<br>– data carrier<br>– MPEG1 video<br>– voice | <br>250 Kbit $s^{-1}$<br>1.4 Mbit $s^{-1}$<br>1.4 Mbit $s^{-1}$ | <br>Asynchronous<br>Synchronous<br>Synchronous |
| | Data communications | Several bytes | Asynchronous |

**Figure 10.3**

Figure 10.4 shows an example of a conventional configuration for audio and video signal distributions in a motor vehicle.

In the example shown in this figure, the most unfavourable situation is where the following bit rates are simultaneously present on the bus:

| | | |
|---|---|---|
| Four-channel stereo audio | $(4 \times 2) \times 1.4$ Mbit s$^{-1}$ | 11.2 Mbit s$^{-1}$ |
| Multiplexed video | + from 2.8 to 11.0 Mbit s$^{-1}$ | + 2.8–11.0 Mbit s$^{-1}$ |
| | + a reserve of 4.0 Mbit s$^{-1}$ | + 4.0 Mbit s$^{-1}$ |
| **Making a total of:** | | = **18–26.2 Mbit s$^{-1}$** |

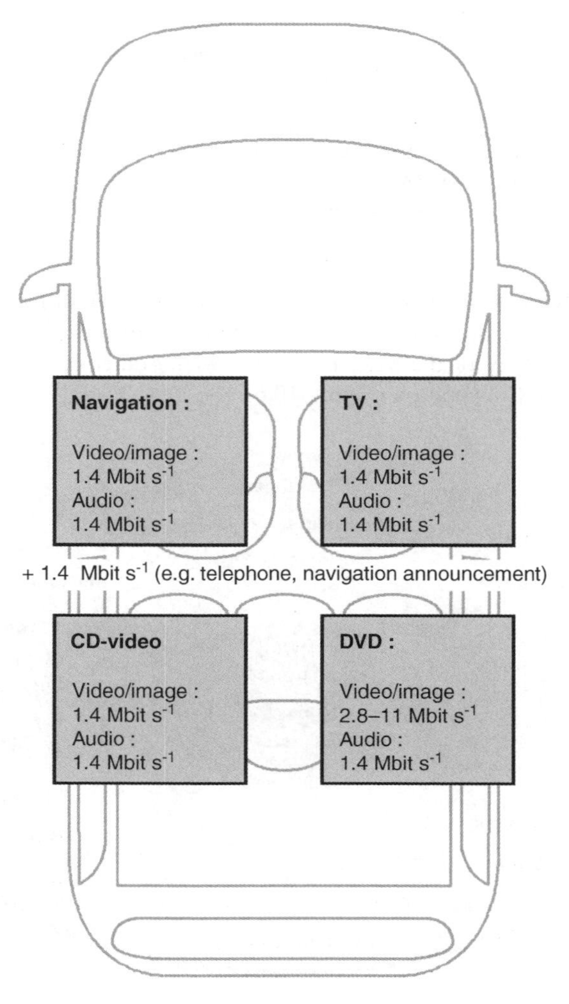


**Figure 10.4**

### 10.3.2 The MOST concept

The aim of the MOST concept (Figure 10.5) is to distribute audio, video and multimedia signals in serial and digital forms in a motor vehicle.

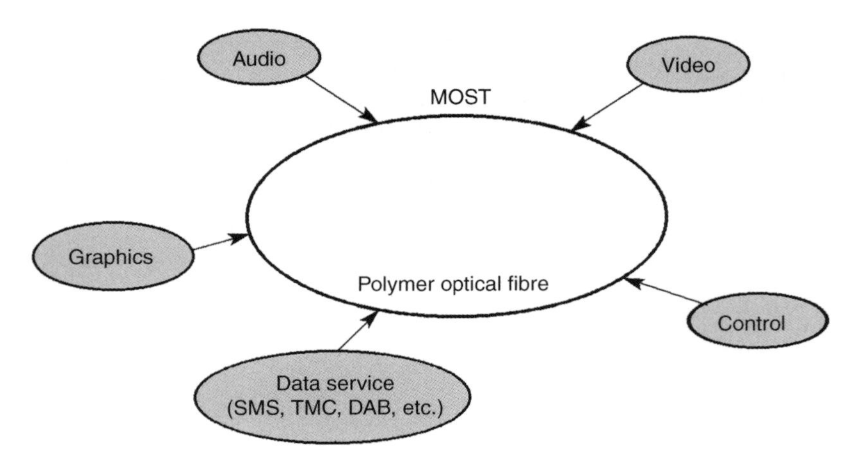


**Figure 10.5**

Based on the D2B solution, the MOST Cooperation was set up in 1998 by BMW, Daimler Chrysler, Harman/Becker and OASIS SiliconSystems, with the aim of standardizing the communication technology of the MOST concept. Many other companies are 'Associated Partners' of the MOST Cooperation (Figure 10.6).

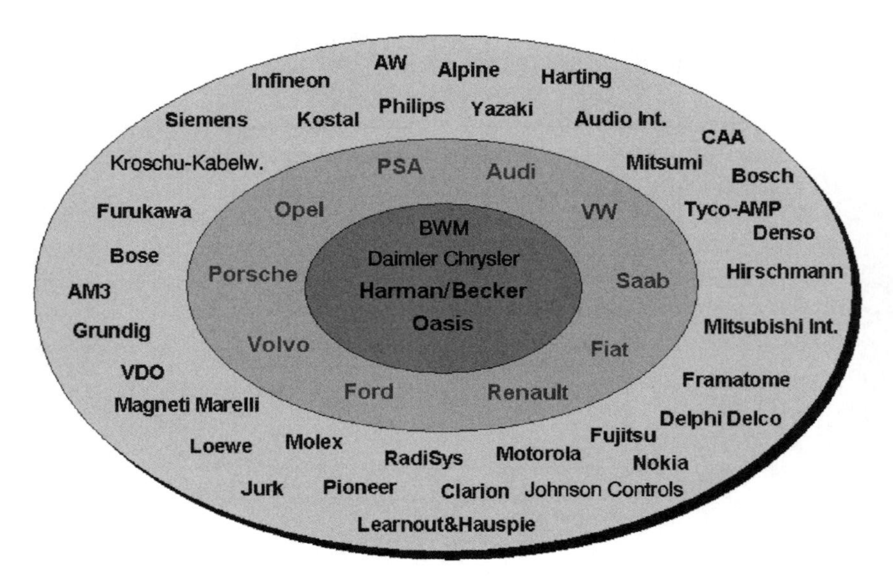


**Figure 10.6**

I shall now briefly survey the major technical features of this concept, starting with the physical layer for once.

## Physical layer and medium

The physical layer, originally designed around a 'copper' twisted pair wire link, has now evolved to its present form, supported by a fibre optic medium, to provide both a wider range of applications and greater immunity to external parasitic signals, while avoiding interference by radiation with the immediate environment. Generally, the optical fibre is a 1-mm diameter fibre with a variable index (of the 'step index' type), made from a polymer, i.e. plastic, and known as POF (plastic optical fibre). Figure 10.7 is a block diagram of the interface between the digital data input and the MOST physical layer.

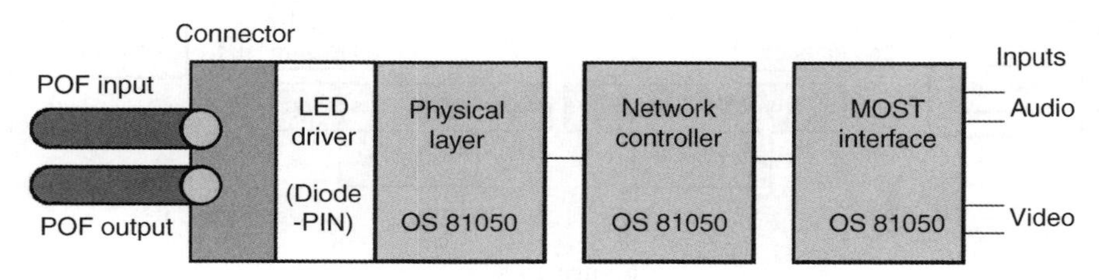


**Figure 10.7**

As shown in the figure, it consists of the following, in this order:

- special integrated circuits made by Oasis, with the functions and references shown in the figure itself;
- a transmitter comprising a PIN diode, an LED and a physical interface with the POF;
- a specially designed (sealed) connector which prevents the ingress of dust and moisture, and can withstand many connections and disconnections.

## Topology

The application topology of this bus is often in the form of a ring (Figure 10.8).

## Data transfer and associated bit rates

Digital data can be transferred in two operating modes: synchronous and asynchronous.

*Synchronous operation*
As MOST was initially designed to carry digital audio CD data at a fixed bit rate, the 'synchronous' mode is still its main operating mode. In this case, a master supplies a clock signal to all other network participants, which synchronize themselves to this clock, thus avoiding the need for buffers and devices for sampling the transmitted data. In this operating

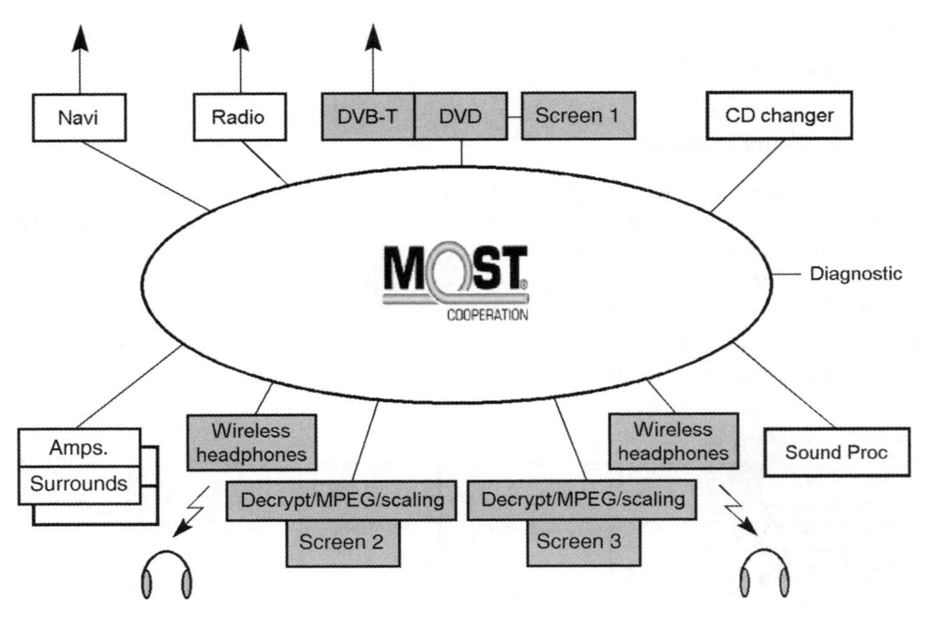


**Figure 10.8**

mode, the digital data are transferred in frames with a bit rate of 44.1 kHz (the rate of digital audio CD), each consisting of 60 channels of 1 byte each. Theoretically, the maximum bit rate of the bus is therefore

$$(60 \times 8) \times 44,100 = 21.168\,\mathrm{Mbit\,s^{-1}}$$

In fact, because of the format of the byte transmitted (in 10 bits, 8N1) in this mode, the maximum gross data transfer rate is

$$(60 \times 10) \times 44,100 = 25.46\,\mathrm{Mbit\,s^{-1}}$$

but the MOST networks currently operate at around 8–10 Mbit $\mathrm{s^{-1}}$ gross for video, audio, subtitling and similar applications, or with an average working rate of 4–5 Mbit $\mathrm{s^{-1}}$.

Figure 10.9 shows an example of a synchronous frame application. As shown in this figure, the frame consists of a number of channels, each having a width of 1 byte. When the meaning of the position of each channel (service channel, data channel) has been defined, the hard work is done. It is then simply necessary to have recurrent frames in the same format to enable each network participant to find the bytes (channels) that relate to it. This communication frame format makes it easy to determine many functional configurations of nodes in advance.

In the example shown above, six channels are reserved for audio channel transmission:

$$(6 \times 8) \times 44{,}100 = 2.1168\,\mathrm{Mbit\,s^{-1}}$$

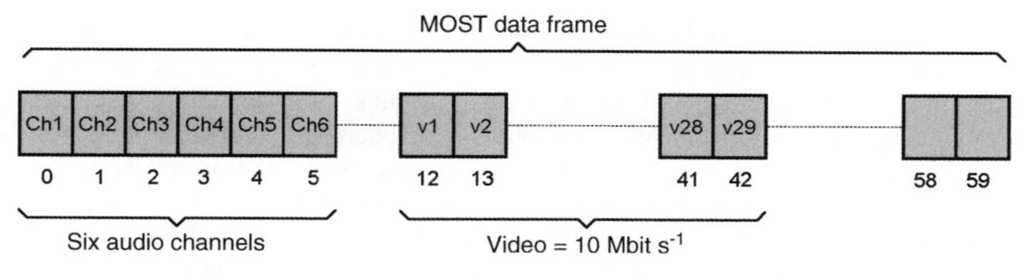


**Figure 10.9**

and 29 (channels 12–42) are reserved for carrying video signals:

$$29 \times 8 \times 44{,}100 = 10.231{,}200\,\mathrm{Mbit\,s^{-1}}$$

Note that the last value is sufficient for multiplexing several video sources with acceptable qualities (resolution) in a single medium, on small on-board screens, provided that the observation distances are relatively short.

MOST also supports the presence of several masters in a single network and the maximum number of participants is 64.

***Asynchronous operation***

In this operating mode (without a clock signal supplied by one of the participants), the maximum bit rate is 14.4 Mbit $\mathrm{s^{-1}}$, and the associated dedicated channels run at 700 kbit $\mathrm{s^{-1}}$. This mode can be used, for example, to transmit short bursts of signals such as those corresponding to voice signals for navigation assistance and other driver assistance messages.

**Messaging system**

The type of messaging system offered by MOST makes it possible to use communication channels dedicated to data transfer, on the one hand, and on the other hand communication channels dedicated to the transfer of commands, which indicate for example the type of data sent by the transmitter. Consequently, when the connection has been established, the digital data stream can be transmitted without being formed into packets as in other transmission systems of this type.

### *10.3.3 Some notes on applications*

MOST was introduced in on-board applications for audio links. Originally, the I2C was used to control the whole infrastructure of the 'car radio' component, and then, as I2C became rather more independent, it was 'buffered' and 'screened' so that it could control disc changing in a CD player located in the boot. It was then replaced by the D2B 'copper wire' bus on a symmetrical pair, which improved matters, and subsequently by the fibre optic version (expensive for the application concerned, but providing two benefits: no radiation and very low susceptibility). In all these examples, the audio content that was carried was analogue. Then MOST came on the scene, making it possible to transfer the digital frames of a 'normal' (uncompressed) audio CD from the back to the front of the vehicle. Here are some more details, with figures.

In conventional digital audio systems (audio CD, digital audio, uncompressed), the analogue audio signal is sampled at a constant rate of 44.1 kHz on $2 \times 16$ bits or 24 bits, resulting in a global digital data bit rate of about 1.5 Mbit $s^{-1}$ for an audio CD system. This is the beginning of the limit of conventional copper wire links (without special protection from the parasitic signals present in a motor vehicle). Consequently, MOST originally appeared in the market in the form of an asynchronous mode link (frame format transmitted at 44.1 kHz), without the need for clock transmission, as the principle of digital audio was based on a constant bit rate system. MOST was therefore very suitable for the transmission of uncompressed digital audio at a constant rate. However, it entailed some problems (which have since been resolved) when transmitting variable-rate signals, such as compressed video signals (MPEG1, MPEG2 or MPEG4, for example). This is because, in spite of speed controllers, the structural principle of the source coding of the video signal is such that these signals have digital flow rates that are essentially variable (about 1–5 Mbit $s^{-1}$). Moreover, in order to distribute several different programmes simultaneously in a vehicle (video games for the little boy in the back seat, cartoons on DVD for the little girl, digital terrestrial TV in the front to give Mum the latest information and finally a navigation map or rear camera video for Dad who is driving this miniature world), a 'system layer' of MPEG2 must be added to assemble and multiplex several data streams from different video, audio, data and other sources. This bit rate can vary from a few megabits per second to several tens of megabits per second, requiring the use of the 29 MOST channels as mentioned above to carry all this information ($44{,}100 \times 8 \times 29 = 10.231{,}200$ Mbit $s^{-1}$).

If we want to combine conventional digital audio data (uncompressed) and digital video data (compressed) on the same MOST network, we must be able to resolve the problem of 'synchronous asynchronism' or 'asynchronous synchronism'. What we must actually do in

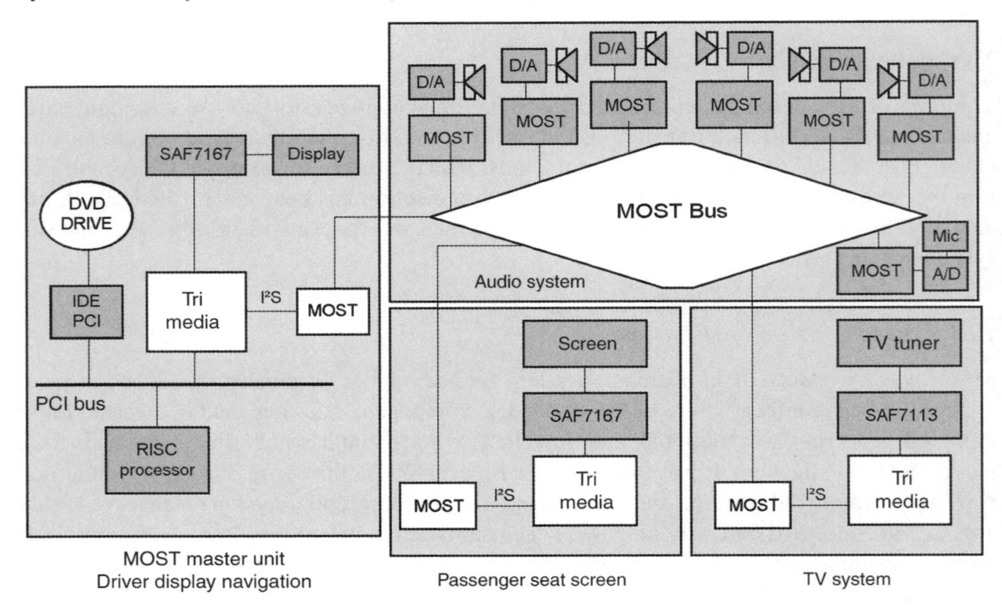


**Figure 10.10**

most cases is to take all the variable speed material, cut it into time 'slices' (packets), and encapsulate it in frames with a fixed time format, using a powerful real-time processor (such as a Tri-Media, TM, processor or a Nexperia platform), in order to transmit it in asynchronous MOST format at constant speed. To conclude this discussion of MOST, here is an example of the usual sequence of operations carried out to transfer the data from a DVD (in the boot) to a screen using MOST (Figure 10.10).

- the TM acts as a controller, receiving the data from the DVD player;
- the software module of the TM demultiplexes the digital stream from the DVD and extracts the programme chosen by the user;
- the TM decodes the AC-3 audio and distributes it via MOST to the loudspeakers;
- the TM converts the VBR MPEG2 data to CBR data with MPEG2 stuffing;
- the TM sends the CBR MPEG2 data stream via MOST, using the I2S audio interface;
- the TM reduces the bit rate to less than 5 Mbit $s^{-1}$ by transcoding the MPEG2 stream, enabling more than one video channel to be transferred via MOST;
- the TM in the screen decompresses the MPEG2 stream and adapts the image resolution to match that of the display screen.

## 10.4 The IEEE 1394 Bus or 'FireWire'

I will end this chapter with a few words on the IEEE 1394 protocol which is a direct competitor of MOST for audio and video applications in on-board systems (Figure 10.11).

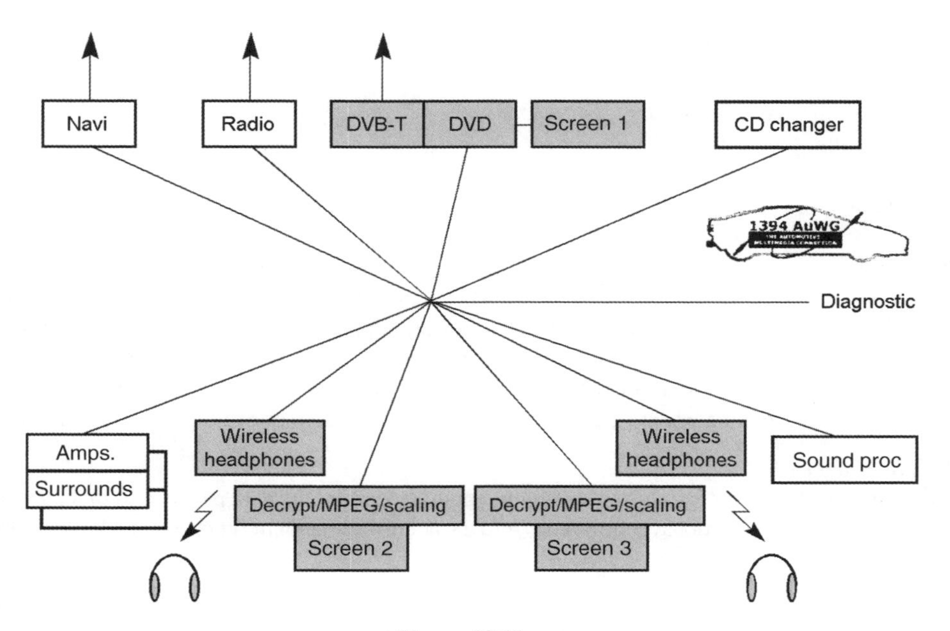


**Figure 10.11**

This communication protocol has given grandiose dreams to many system designers in their laboratories. Indeed, its performance is highly suited to high-speed digital data transfer (100, 200 and 400 Mbit $s^{-1}$, or even 800 Mbit $s^{-1}$). However, one of the main obstacles to its use at present is, on the one hand, the fact that it has appeared after MOST, and, on the other hand, its cost and the availability of components which can withstand the temperature ranges encountered in motor vehicle applications.

### *10.4.1 History*

IEEE 1394 made its first appearance in the early 1980s. Its original purpose was to meet the requirements of high-speed applications, up to 400 Mbit $s^{-1}$, capable of carrying, for example, MPEG 2 digital data streams (in the range from 1.5 to 70 Mbit $s^{-1}$) and DVC (25 Mbit $s^{-1}$). Because of its intrinsic potential, it was subsequently given different names, particularly 'FireWire' or 'FireWire 400'.

#### IEEE 1394–1995

The first published version was that of 1995, subsequently called IEEE 1394–1995 to avoid confusion with later updates.

This protocol is based on what is called 'packet' communication, supporting speeds of 98.304, 196.608 and 393.216 Mbit $s^{-1}$, in a transport mode which may be isochronous (and therefore with a known and guaranteed bandwidth) or asynchronous (and therefore with a degree of flexibility in the bus access).

The control of IEEE 1394 is based on a communication model divided into four layers, as follows, starting from the lowest:

- the physical layer,
- the data link layer,
- the transaction layer,
- and finally the application layer itself.

*Physical layer*
The physical layer controls the analogue interface (in terms of the electrical levels) when the bus is implemented in wire form. The network can support up to 63 nodes, each having a bidirectional port allowing operation with a bus topology or in the daisy chain mode as described in the chapter on the LIN bus. Its design is such that it allows for the provision of repeaters and the management of the arbitration that may arise after collisions on the bus.

*Data link layer*
The purpose of this layer is to format (assemble or disassemble) the data packets, so that they can be routed synchronously or asynchronously and the acknowledgement functions can be provided.

*Host controller*
The higher layers (session, transport, etc. and application) are implemented within the host controller. The structure formed in this way (Figure 10.12) provides the system with

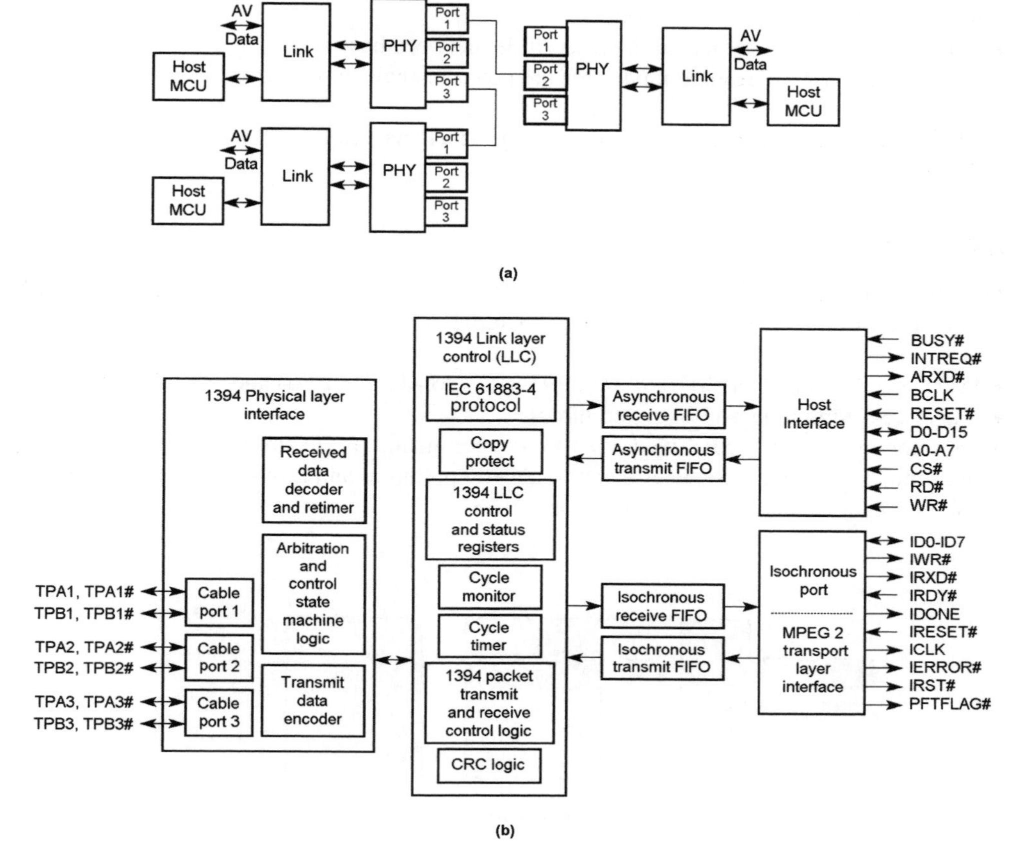


**Figure 10.12**

- a true 'peer-to-peer' bus architecture;
- possibilities for implementing true 'plug and play' hot connections;
- the possibility of dispensing with a server or central element for transactions.

In fact, this is really neither a master–slave nor a true multimaster architecture.

The initial version has been followed by different versions and enhancements.

## IEEE 1394a

The year 2000 saw the appearance of the IEE1394a, or IEEE 1394–2000, version, with the following aims:

- to clarify the ambiguities in the IEEE 1394–1995 version;
- to permit new equity ranges;

- to create fast arbitration after reset;
- to provide new rules for use in the reset of the network bus;
- to introduce suspend/resume modes for the power management part.

IEEE 1394a is also known in the form of a product marketed by Sony under the names 'i. LINK®' and 'FireWire'.

**IEEE 1394b**

More recently, the market has seen the arrival of the IEEE 1394b version, downward compatible with the previous version, but providing

- faster speeds, namely 98.304, 196.608, 393.216, 786.432, 1572.864 and 3145.728 Mbit $s^{-1}$;
- longer operating ranges (100 m and more);
- new communication media in addition to 'copper media', for example POF operating as a multimode fibre, which reduces electromagnetic pollution and provides a slightly cheaper solution than an analogue system (Figure 10.13).

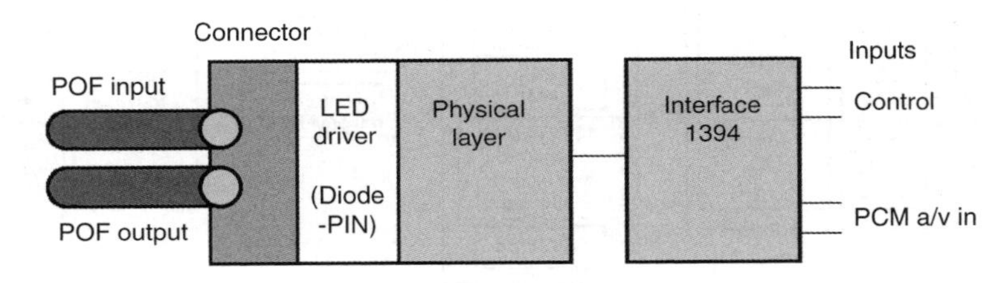


**Figure 10.13**

*10.4.2 Applications*

Because of the high speeds that it can support with either analogue or digital signals, IEEE 1394 is very well positioned as a common hardware platform for use with electronic consumer goods such as PCs and accessories such as external hard drives, audio and video equipment, digital camera encoders and DVD engravers, and for home networks.

Many application protocol infrastructures are already supported by IEEE 1394. We may mention those relating to

- PC peripherals;
- point-to-point links for audio and video transmission systems and their controls;
- point-to-point interface links for EIA 4.1 for digital TV and entertainment systems;
- Internet protocols;
- DVB interfaces for digital television;
- interfaces for VESA local home networks;
- HAVi (Home AV interoperability).

### *10.4.3 IEEE 1394 in motor vehicle networks – IDB-1394*

IDB-1394, *intelligent transportation systems data bus using IEEE1394 technology*, is a network designed to support entertainment applications for public use, as well as data communications. As IDB-1394 can operate at bit rates of up to 400 Mbit $s^{-1}$, it can carry high-quality video signals (HDTV) even if these are uncompressed (for blind spot and blind angle detection applications, and other ITS (intelligent transportation system) real-time applications), several audio channels and high-speed data for vehicle-specific applications. This network can also be used to interconnect IEEE 1394 compatible portable equipment, such as the Apple iPod, using a 'consumer convenience port'. The aim is to allow integration as a plug and play element in a digital network to form a multimedia local network (LAN) in the vehicle.

The IDB Forum, an international association of motor and consumer goods manufacturers, based at Washington DC, has drawn up specifications over the last 5 years, by agreement with the IEEE 1394 Trade Association, which are essentially oriented towards power management problems, FOT (fibre optical transmitter) standardization and the following technical matters:

- IDB-1394/1 *audio/video profile for DVD*,
- IDB-1394/2 *audio/video profile for DVB-T*,
- IDB-1394/3 *audio/video profile for camera*,
- IDB-1394/4 *audio/video streaming and control profile*,
- IDB-1394/5 *IDB power specification, physical media dependent for wake-up on LAN.*

By way of example, to show how each passenger is completely free to choose his own audio/video programme, implementations using POF in vehicles have already proved at this date (2005) that the following can be achieved:

- simultaneous transport of three different video information sources:
  - DVD player,
  - digital TV reception,
  - uncompressed digital images from a camera at the rear of the vehicle, used for parking assistance. Access to the image is permitted (via a CAN – IDB-1394 gateway) only when reverse gear is engaged.
- connection of the navigation system and plug-and-play access for an Apple iPOD,
- access to three touch display screens (one for the front-seat passenger and the other two for passengers in the back seats of the vehicle).

The IDB Forum IDB-1394, already used to carry data for navigation cards and multi-channel audio applications including high-definition super audio CD, is also the first digital network technology to be approved by the DVD Copy Control Association (DVD-CCA) for the distribution of protected copies of digital video content (CSS), from a DVD player for example, over a 'localized' digital network (in a motor vehicle, for example). The DVD-CCA has approved the use of IDB-1394 with DTCP/IDB-1394 (*digital transport content protection on IDB-1394*) which was developed by the Digital Transmission Licensing Administration (DTLA).

# 11

# RF Communication and Wireless Mini-Networks

To conclude this book and look ahead to the near future, I will briefly describe the radio-frequency links and mini-networks mentioned in Chapter 7, better known under the term 'wireless'. The 'internal' and 'external' aspects of these applications will be considered. As mentioned above, the most popular ones have names such as Bluetooth, Zigbee, IEEE 802.11, NFC, RKE, PKE, passive go, TPMS, TiD, etc.

## 11.1 Radio-Frequency Communication: Internal

Many internal applications in vehicles also use radio-frequency media for digital communications and for connection at some point to other networks (controller area network (CAN), local interconnect network (LIN), etc.) of a vehicle or other industrial system.

### *11.1.1 Radio*

Let us start with the most conventional kind of radio-frequency receiver, namely the AM/FM/digital radio, which may or may not have antennae incorporated in the windscreen or in the rear screen de-icing system. These systems are entirely conventional, except for those using voice commands (voice synthesis and/or recognition) for tuning and volume, with control via I2C, LIN or CAN buses, etc.

### *11.1.2 Immobilizer or engine start prevention*

The immobilizer function consists in preventing the starting of the engine of a vehicle (not to be confused with an anti-theft function), using RF transponders (usually operating on an low-frequency (LF) carrier at 125 kHz). These systems have been well known for several years.

*Multiplexed Networks for Embedded Systems: CAN, LIN, Flexray, Safe-by-Wire...* D. Paret

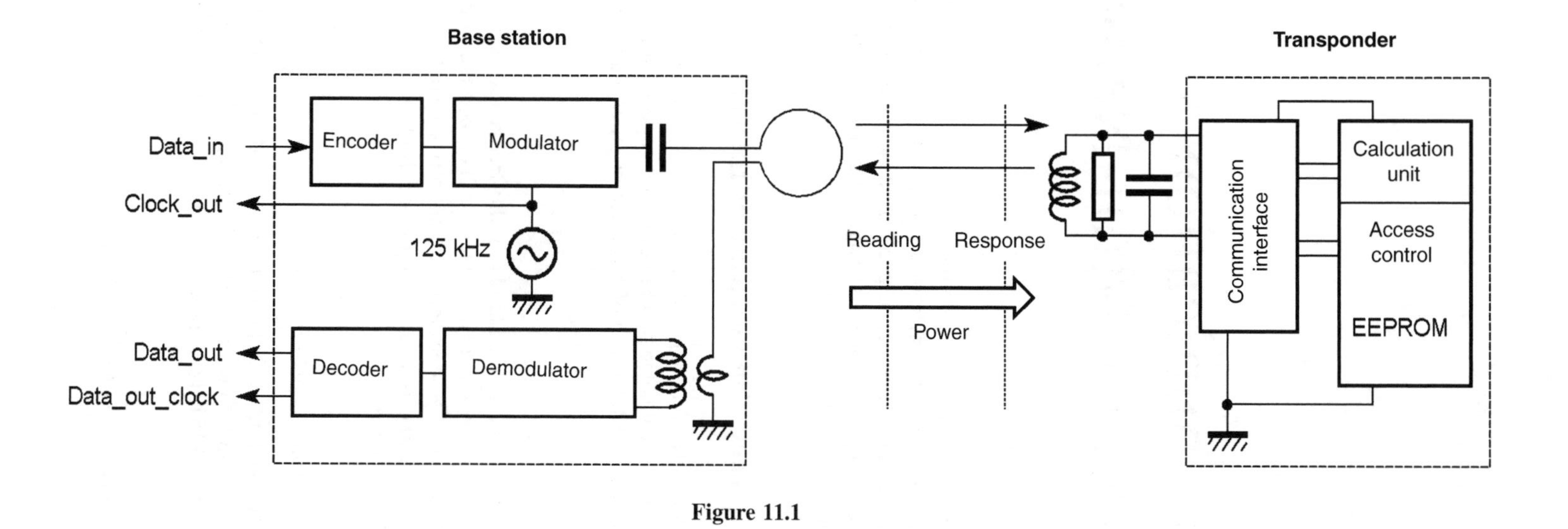
Base station
Transponder
Data_in
Clock_out
Data_out
Data_out_clock
Encoder
Modulator
125 kHz
Decoder
Demodulator
Reading
Response
Power
Communication interface
Calculation unit
Access control
EEPROM

Figure 11.1

Their operating principle is shown in Figure 11.1, and an example of a specific circuit is given in Figure 11.2.

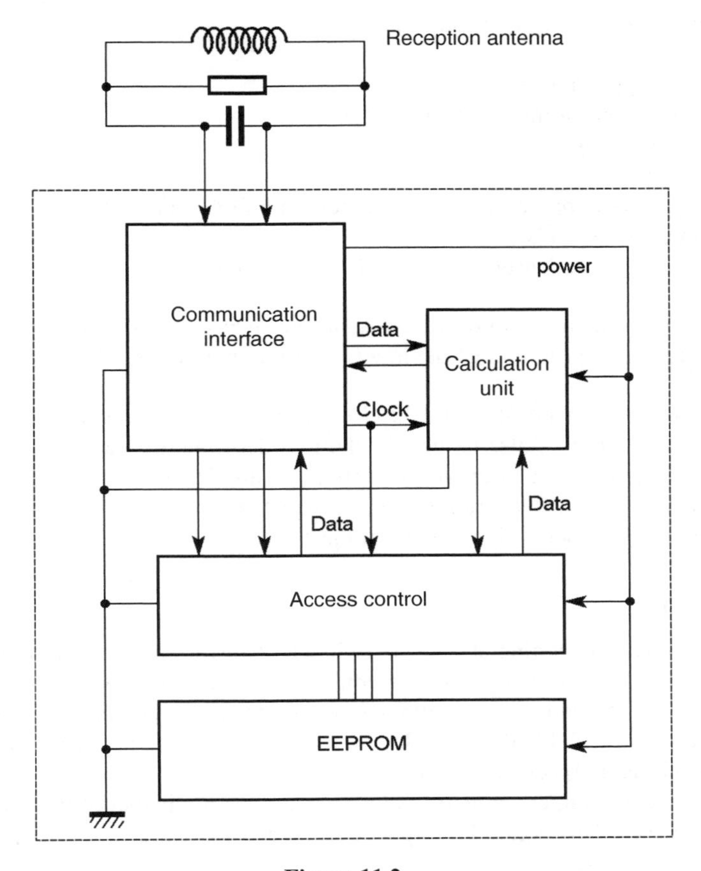


**Figure 11.2**

The aim of this immobilizer function is to enable an element operating in contactless mode (transponder), located in the head of the contact key (or in a badge), to communicate in encrypted mode at several kilobits per second with a fixed base station, located in the vehicle (often near the steering lock barrel). The purpose of this communication is to enable the operation of the ignition/injection part of the vehicle in which the decryption system is located (as a buried layer), allowing the engine to be started.

One of the main problems of these systems is that of starting the engine in the very short time allowed (less than a hundred milliseconds). In this brief time interval, it is necessary to remotely supply the transponder, which has to carry out its internal reset, execute the authentication phase between the vehicle and the key, manage the digital data exchange between the transponder and the vehicle, etc. and communicate with the injection control via a LIN gateway, then a CAN one and so on. At the same time, the engine must be checked to see if it is ready for starting. A considerable problem!

### *11.1.3 GSM*

In addition to the well-known simple mobile telephone link, this radio-frequency link provides access to internal/external services such as

- emergency call applications;
- communications controlling the panic button;
- service calls, including
  - hotel reservations;
  - breakdown service, operating in conjunction with GPS location;
- telephone links of all kinds;
- dynamic updating of data used for the navigation systems.

If we take the simple example of breakdown assistance using a GSM link assisted by GPS, we can immediately see that there must be, in terms of the internal aspect at least,

- a direction link between the GSM and the CAN, LIN and other buses, which can inspect and diagnose the different modules in the vehicle, in order to notify the breakdown assistance company of the functions that are most likely to be implicated in the operating failure;
- a direct link with the GPS location system, connected to a greater or lesser degree to the vehicle's audio–video network (MOST or other).

### *11.1.4 Bluetooth*

The Bluetooth system (operating at 2.45 GHz) is well known at the present time and is very commonly used as a local accessory for, or in combination with, the GSM link mentioned above.

Even where local regulations vary from one country to another, making it impossible to use the same bandwidth at 2.45 GHz and the same radiated power (EIRP), it is easy to use Bluetooth for internal purposes in a vehicle.

The first, very common, example of an internal application is that of the use of earpieces which enable a driver to use a mobile telephone while driving, when he must legally keep both hands on the steering wheel.

Also, as mentioned above, it is easy to produce very small local networks, or 'pico-networks', to interconnect on-board IT devices or office technology (e-mail printer, etc.) within limited perimeters, such as the passenger compartment of a vehicle. This forms the 'internal' aspect of Bluetooth applications.

Some applications are of the internal/external type, but more internal in nature. They include

- supplementary applications for providing 'hands-free' entry systems;
- anti-theft system applications based on the principles of relay attacks;
- passenger location in vehicles, using triangulation, for triggering airbags according to their instantaneous postures;
- etc.

Other Bluetooth applications will be discussed in the 'external' part. After this brief list, let use move on to the truly 'external' aspect.

## 11.2 Radio-Frequency Communication: External

We shall now briefly examine the radio-frequency applications that provide a link between the interior of the vehicle (or of an industrial system) and the world outside it. Let us start by looking at what is generally called 'passenger compartment safety management' or 'secure access control', depending on the industry for which it is developed.

### *11.2.1 Remote control of opening parts*

This includes all remote controls which open or close access to private places, such as private houses, apartments, buildings or motor vehicles.

Although optical infrared links have been used for many years, conventional remote control systems now aim to provide more user-friendly operation, mainly by providing less directionality and longer operating ranges, using ASK or FSK modulated UHF radio-frequency links, with operating frequencies mainly in the ISM bands of 433 and 866 MHz in Europe, around 915 MHz in the United States and 315 MHz in the United States and Japan (Figure 11.3).

#### One-way remote control

By far the greatest number of these remote control systems are purely one-way, with communication from the remote control module to the vehicle. The bit rates for these links are usually about several (tens of) kilobauds, and the associated protocols are generally proprietary ones. The messages sent by these remote control units are also encrypted by the same method as that used by the system for the immobilizer function. When received, they are transcoded into CAN frames, to be forwarded to the descramblers (the same electronic system is often used for decrypting the signals from the immobilizer function and those of the remote control unit in the key or badge).

*Doors*

Often implemented mechanically in the immobilizer key or badge, the remote door opening function is more complicated. This is because, when the command is received in RF and validated to check that this remote command is the right one for the vehicle, it is used to wake up the CAN network (or part of it), after which its content is sent across fast CAN buses and then low-speed ones. It must then be decoded and validated, then resent via CAN and LIN to the door opening striking plate, all in a very short time (less than 100 ms), to give the impression that the action is instantaneous, which sometimes gives rise to challenging problems of gateway design. Usually, the desired operating ranges of these remote controls vary from several metres to about ten metres – although care must be taken to ensure that no person other than the user can enter the vehicle before the user reaches it!

*Boot*

Opening a car boot with a remote controller is easy enough, but it is quite a different matter when you approach the boot with your arms full of parcels and no access to the remote

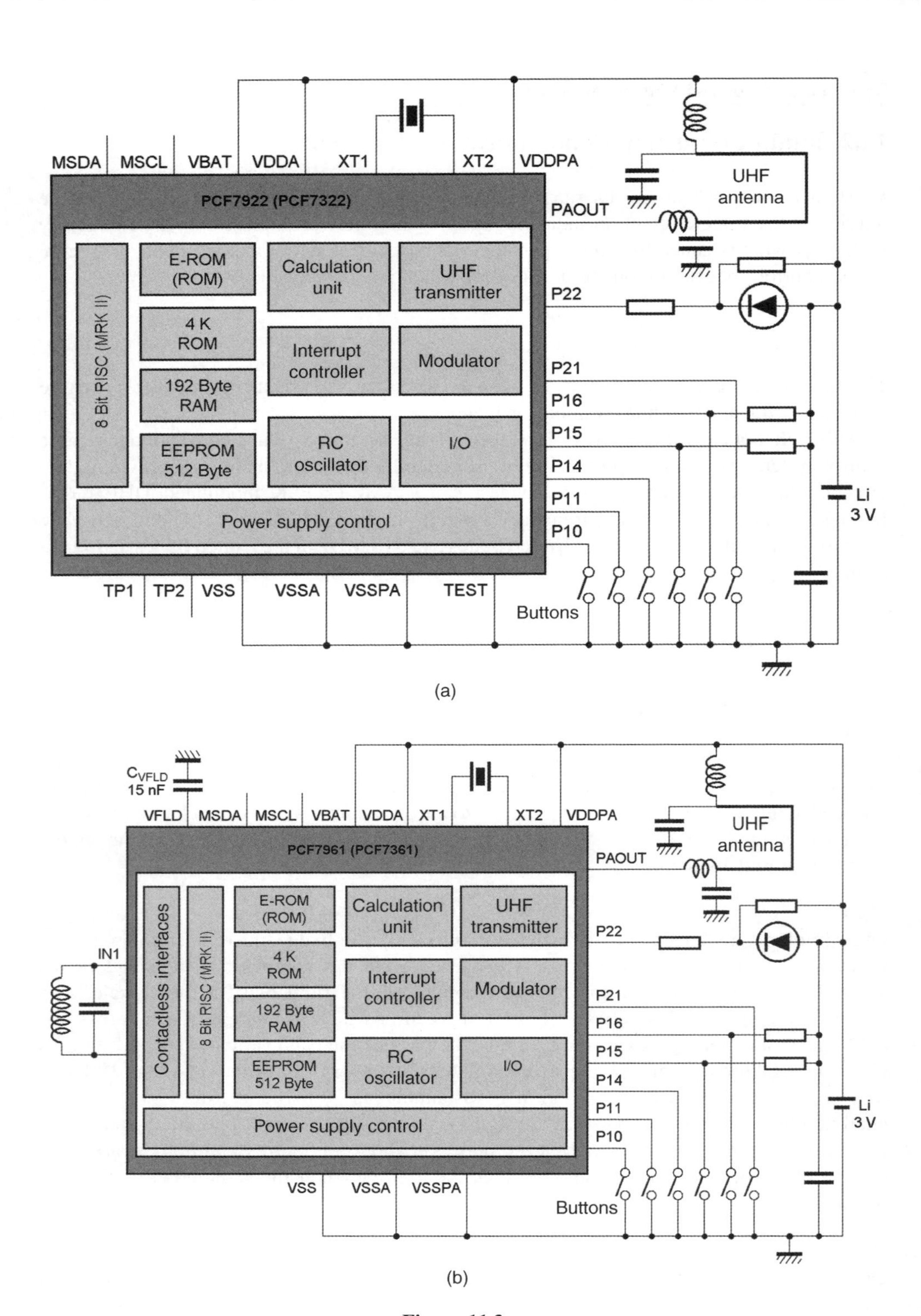
MSDA
MSCL
VBAT
VDDA
XT1
XT2
VDDPA
PCF7922 (PCF7322)
UHF antenna
PAOUT
8 Bit RISC (MRK II)
E-ROM (ROM)
4 K ROM
192 Byte RAM
EEPROM 512 Byte
Calculation unit
UHF transmitter
Interrupt controller
Modulator
RC oscillator
I/O
Power supply control
P22
P21
P16
P15
P14
P11
P10
Li 3 V
TP1
TP2
VSS
VSSA
VSSPA
TEST
Buttons
(a)
CVFLD 15 nF
VFLD
MSDA
MSCL
VBAT
VDDA
XT1
XT2
VDDPA
PCF7961 (PCF7361)
UHF antenna
PAOUT
IN1
Contactless interfaces
8 Bit RISC (MRK II)
E-ROM (ROM)
4 K ROM
192 Byte RAM
EEPROM 512 Byte
Calculation unit
UHF transmitter
Interrupt controller
Modulator
RC oscillator
I/O
Power supply control
P22
P21
P16
P15
P14
P11
P10
Li 3 V
VSS
VSSA
VSSPA
Buttons
(b)

**Figure 11.3**

controller which is still in your pocket! The answer to this predicament is given farther on in this chapter, in the section on 'hands-free' devices.

*Sun roof*
Why have a section devoted to this particular remote control system? The answer is simple. Suppose that you are inside the house, in summer, with your car parked on the drive with the sun roof open because the weather is so hot – and all at once a thunderstorm breaks out. To stop your car turning into a swimming pool, you need to close the roof with your remote controller, at 'long range', i.e. 20–30 m. Here again, the transmission sequence is via radio frequency, then LS CAN, then HS CAN, and finally LIN.

### Two-way remote control

Another newcomer in the field of radio-frequency links and their relations with other buses. We need to move on from the examples above – and yet stay with them for a while. We have just used the remote control to close the sun roof 'at long range', to avoid getting wet. Fine – but we do not want to get wet by going to check that the roof has actually closed! The finishing touch is to make the vehicle send back to the user's remote controller (hence the term 'two-way') a message such as 'the roof is now closed'. In this case, we have a system with a double set of transceivers, one in the remote controller (and therefore having a very low power consumption) and the other in the vehicle (Figure 11.4).

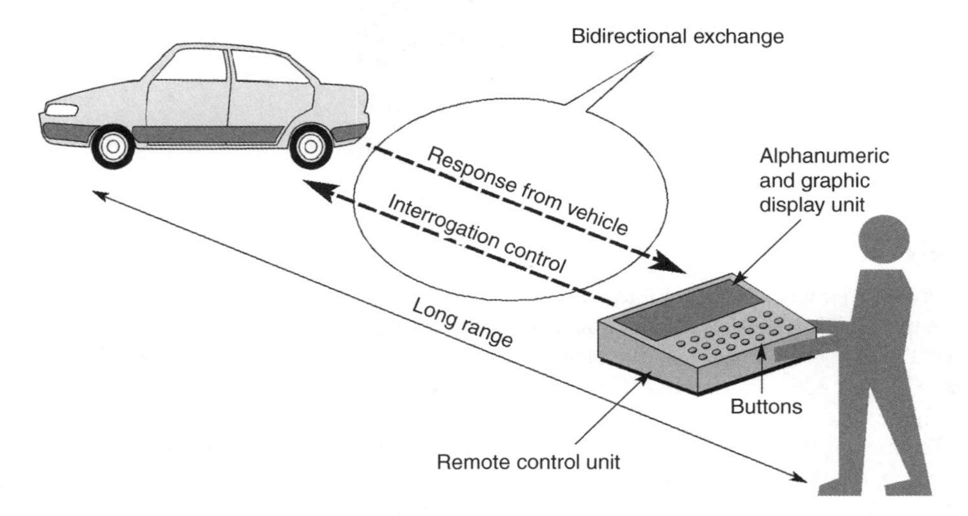


**Figure 11.4**

A simple command that illuminates a confirmation LED on the remote controller may be enough to indicate that the desired task has been carried out correctly. Sometimes, for applications with higher levels of security or convenience, we need to provide a small alphanumeric screen on the remote controller, using LCD or similar technology, so that certain information from the vehicle's on-board computer can be downloaded locally. For instance, in some countries that are especially cold (in Canada, for example) or hot (South Africa), the user can remain snugly at home and eat his breakfast in peace, while starting the engine to heat or cool the car which is still locked up outside. The remote controller displays the information on the desired operation in full, so that the whole procedure can be checked.

### *11.2.2 PKE (passive keyless entry) and passive go*

These terms are applied to the functions of 'hands-free entry' (passive keyless entry) and 'push-button starting' (passive go), without the use of any visible key (Figure 11.5).

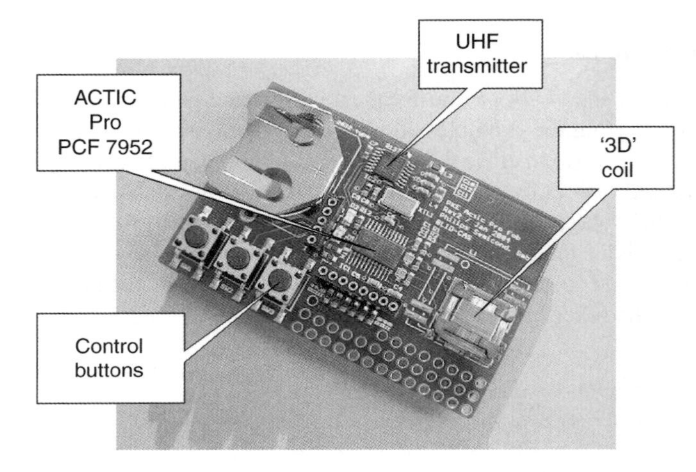


**Figure 11.5**

#### Passive keyless entry

The radio-frequency link providing this function is also of the two-way type, with the uplink from the vehicle to the key in your pocket (it operates on an LF carrier at 125 kHz at a speed of several kbit $s^{-1}$) and the downlink from the key or badge to the vehicle, using a UHF carrier in the ISM 315, 433, 868 or 915 MHz bands, depending on the country.

Figure 11.6 shows the operating principle of this kind of system.

To prevent the vehicle's LF transmitter from running down the battery by constantly scanning for the arrival or presence of the carrier in the immediate environment of the vehicle, the user's presence is normally detected when he grasps the door handle to open it, or even when his hand enters the handle housing (sooner or later you have to open a door manually; not all of them are motorized as yet!), either by a mechanical means in the first case (a contact) or by the breaking of an infrared beam or by capacitive coupling, for example, in the second case.

From this moment, the vehicle establishes an uplink at 125 kHz, with very weak inductive coupling, in 3D, to search for the key or badge that may be present. If its presence is confirmed, the key or badge sends a UHF (propagation) response to the vehicle, in order to establish a

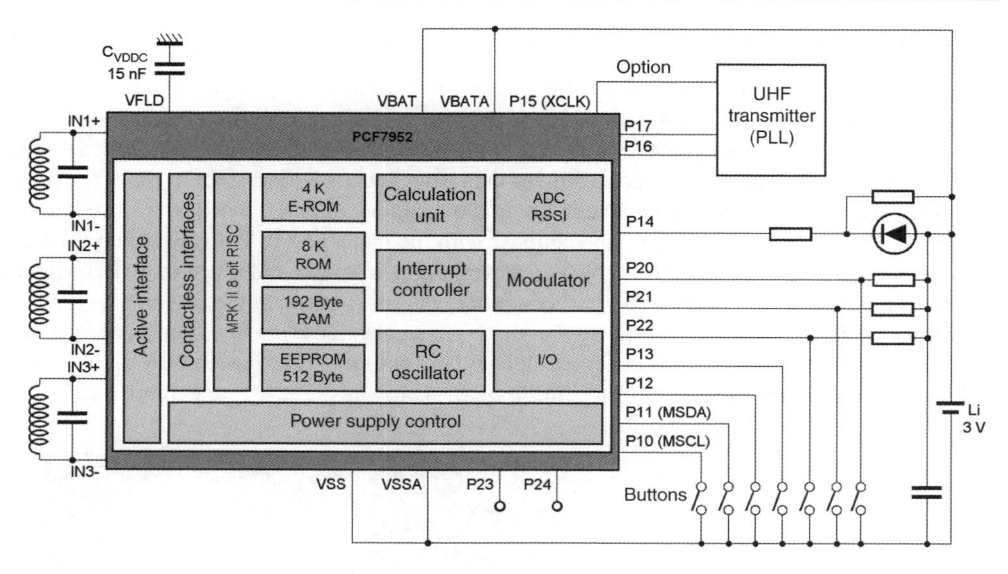


**Figure 11.6**

dialogue between the vehicle and the key or badge to authenticate or check the latter, and finally allow access to the vehicle via the familiar sequence of CAN, LIN, etc., as described above several times. If the temporal control of this exchange is fairly brief, it gives the authorized person the impression of instantaneous entry, without the need to take anything out of their pocket – hence the term 'hands-free'.

The communication protocol (generally proprietary) must be able to control the different antennae operating at 125 kHz and also process conflicts of messages and/or carriers because all operations are hands-free, and therefore more than one passenger with a badge (e.g. a husband and wife, etc.) may approach the vehicle simultaneously. Here again, the times required for wake-up, communication, protocol and collision management, decrypting, striking plate actuation, etc. are critical!

## Passive go

When the user has entered the vehicle, another sequence starts to enable (or not) the starting of the engine. For simple reasons of security, the system needs to know whether or not the badge and its owner are actually present inside the vehicle, so that it can enable (or prevent) the starting of the vehicle by means of the relevant button. But what if a child is in the car by himself and playing with all the buttons? Over to you, and the best of luck!

Everything depends on the ability to detect the actual presence of the badge in the vehicle, in the pocket of a jacket neatly folded up on the back seat, or in a handbag carefully placed in the boot of a saloon car. To put it more simply, it must be possible to distinguish the interior from the exterior of the vehicle, something that often depends on the thickness of a window, the windscreen, the sun visor and the thickness of a roof that may be made from steel, plastic or even glass.

### *11.2.3 TPMS – tyre (or tire) pressure monitoring systems*

When driving, we are only in contact with a very small area of the ground. In fact, the total contact surface is no more than $4 \times (4 \times 10) = 160\,\text{cm}^2$! As the safety parameters of a tyre are directly related to its pressure and temperature, TPMS systems have the function of measuring the tyre pressure and correcting it according to the instantaneous temperature of the tyre. They must then send this corrected value cyclically to the vehicle computer, by RF transmission (at 433, 868 or 915 MHz, depending on the country, with the metal part of the valve often used as the antenna), about once every 1 or 2 min when everything is correct and much more frequently if there is a problem, so that the on-board computer can inform or alert the driver.

Almost all TPMS systems measuring these two physical values (pressure and temperature) with special sensors[1] use electronic devices located in small modules placed next to the valve and positioned on the outer part of the rim, in other words, inside the tyre (Figure 11.7).

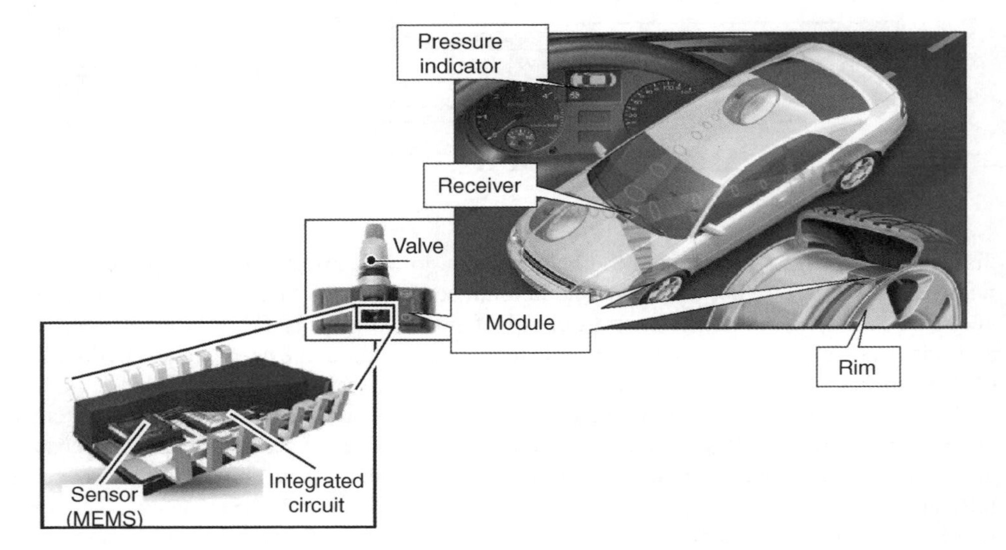


**Figure 11.7**

They are therefore inaccessible from the outside and must therefore have their own power supply devices in order to operate. (Note that some companies are researching remotely powered systems such as those used in RFID.)

Usually (if not always), TPMS systems are powered only by small batteries (with implications in terms of weight, unbalance, cost, etc.).

If we use batteries, we have to consider questions of service life: for example, a life of approximately 10 years ($\sim 3.2 \times 10^8$ s) with 300 mAh batteries ($\sim$110 μAs) requires a constant average power consumption of about 30 μA. This kind of challenge leads to several problems, but before considering how these two data elements are combined and sent to the driver, we must examine many other details.

[1] Some companies (Elmos, for example) offer 'combo sensors' including the pressure and acceleration sensors on a single chip.

### Movement detection

In order to optimize the management of this very low power consumption, which is essential to these applications, we must avoid any kind of wastage. The first idea that comes to mind will no doubt be to switch on the TPMS only when the wheel is rotating – but this is a whole new story, which I will tell in the following sections.

Movement detection or not movement detection, that is the question! So, once upon a time . . .

### TPMS without movement detection

For simple financial reasons (the cost of movement detection circuits), some TPMS systems do not detect whether or not the vehicle is moving. Thus, these systems are woken up cyclically by watchdog circuits and periodically read the pressure and temperature even when the vehicle is stationary and transmit their values, even when the vehicle is parked. This considerably reduces the life of the battery fitted in a TPMS.

*TPMS with movement detection*
The TPMS has a battery fitted in it. To extend the battery life, it is really preferable not to switch on the system unless the vehicle is moving. For this purpose, we need movement detection systems, based on either mechanical designs or electronic solutions.

*Mechanical.* By way of example, this movement detection can be carried out mechanically, using a metal ball associated with a contact (Figure 11.8) which switches on the TPMS when a

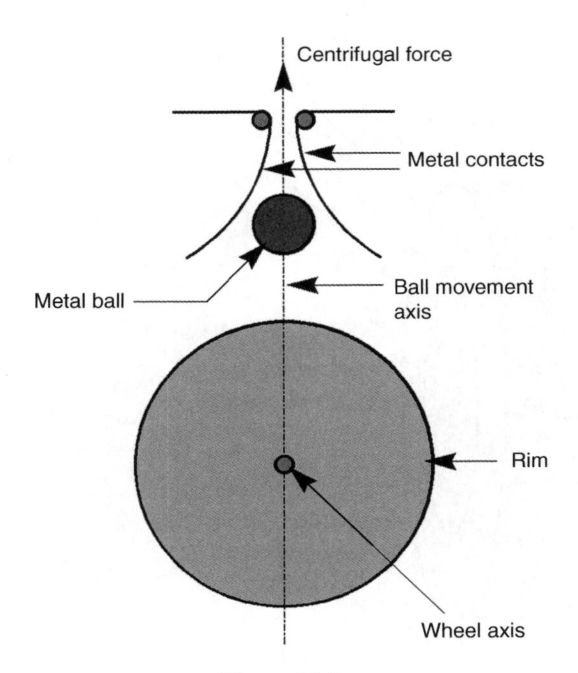


**Figure 11.8**

certain centrifugal force is reached along the radial axis of the wheel (in other words, when a certain linear velocity, and therefore an angular velocity, is reached).

These systems have (lower) limits which are directly dependent on the minimum angular velocities required to make this contact and are consequently dependent on the rim diameters, the reliability of the contacts, etc.

*Electronic*. If we wish to avoid the constraints mentioned above, and especially if we wish to avoid using a mechanical system for detecting movement, it is preferable to consider an electronic movement detection device to achieve a longer life for the battery and therefore for the whole system. Much work has been done in this area, and many clever patents have been filed. So there are now several possible solutions. By way of illustration, I have selected two of these, which are very representative and also demonstrate two very different ways of resolving the same problem.

**Example 1.**
In the first of these solutions, the rotation of the wheel is detected by detecting the synchronous magnetic flux of the movement of the wheel. This can be done by using an inductance $L_1$ placed on the rim, on the radial axis $Z$ of the wheel, cutting the magnetic flux $\varphi$ from a constant magnetic field $H$ whose vector $H$ is always substantially identical (in strength and direction), regardless of the geographic location of the vehicle (Figure 11.9). For example, this field can be created by means of a large permanent magnet (to provide a substantially uniform field) placed near the wheel. Incidentally, there is one magnetic field which, although weak, can perform this function very well, while being completely free: this is the Earth's magnetic field $H_t$ (induction $B_t$ of the order of 50 μT, with an angle of inclination $I$ of the order of 60–65°, see Figure 11.10).

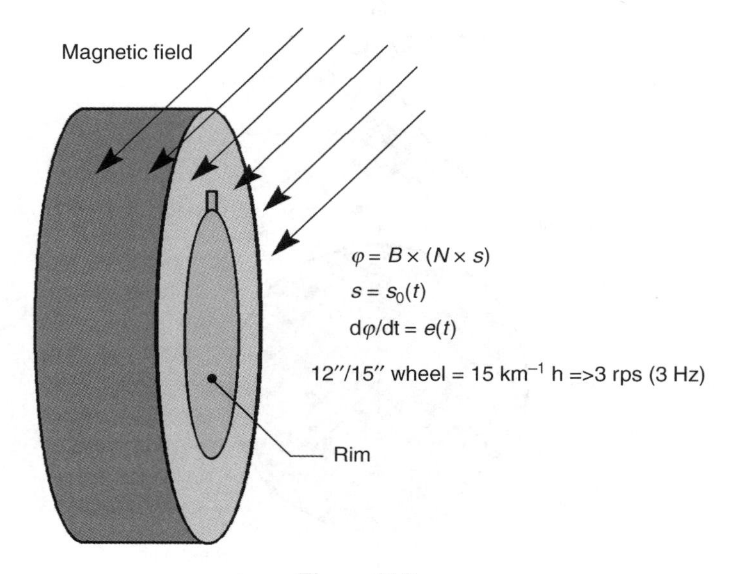


Figure 11.9

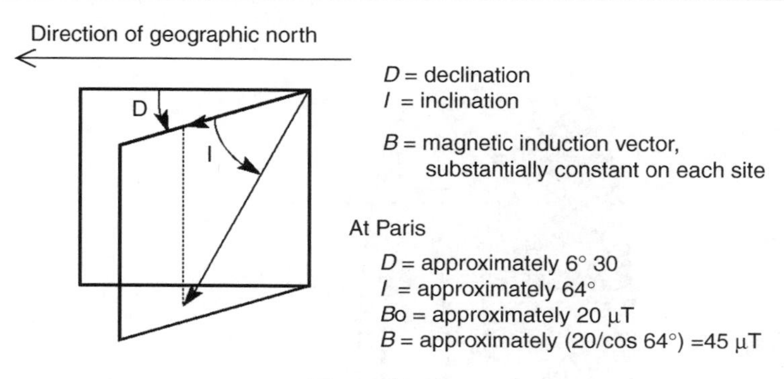


**Figure 11.10**

The use of this principle (inductance rotating with respect to a fixed magnetic field, or vice versa) requires the production of an induced voltage in the inductance $L_1$ (proportional to $d\varphi/dt$) varying sinusoidally at a very low frequency (VLF), at the same rate as the wheel rotation, in other words, about several hertz (for 13, 15 and 17 in. rims at 15 km h$^{-1}$ the frequency is about 2 Hz, and for a speed of 2 m s$^{-1}$ it is about 3 Hz). The voltage induced across the inductance $L_1$, being proportional to $d\varphi/dt$, is therefore small but detectable, and it indicates the movement of the vehicle. Clearly, when the vehicle runs at high speeds (40 m s$^{-1}$ = 144 km h$^{-1}$ or above), the frequency increases together with the induced voltage.

**Example 2.**
In another solution, an accelerometer is placed inside the TPMS, to measure the acceleration $\gamma$ at the position of the valve. If $v$ is the linear velocity of the valve fitted to the rim, $r$ is the radius of the rim and $f$ is the frequency of rotation of the wheel, any good physics textbook will tell you that

$$\gamma = \frac{v^2}{r} = \omega^2 r = (2\pi f)^2 \cdot r$$

The movement detection can then be carried out as follows:

- By measuring the acceleration $\gamma$ in m s$^{-2}$, along the $Z$-axis (Figure 11.11) (the radial axis of the wheel, along its radius). This is equivalent to measuring, as in the preceding case, the axial centrifugal force ($f = m\gamma = \omega^2 r$) present when the vehicle starts to move. If this acceleration information is to be used for other purposes, this method has the drawback that, when the vehicle moves fast at constant speed, this force is constant and the information from the sensor has a constant 'offset' whose value depends on the speed. The initial positioning of the sensor is also critical because a mechanical misalignment (tolerance) in the horizontal plane can cause large measurement errors.
- Or by measuring the acceleration ($\gamma$ in m s$^{-2}$) along the $X$-axis (the axis of the tyre tread; therefore, this value is related to the derivative $dv/dt$ of the instantaneous linear velocity of the vehicle). This is equivalent to measuring the acceleration of the vehicle in the direction of its movement when the vehicle starts to move. This has the drawback that this force is

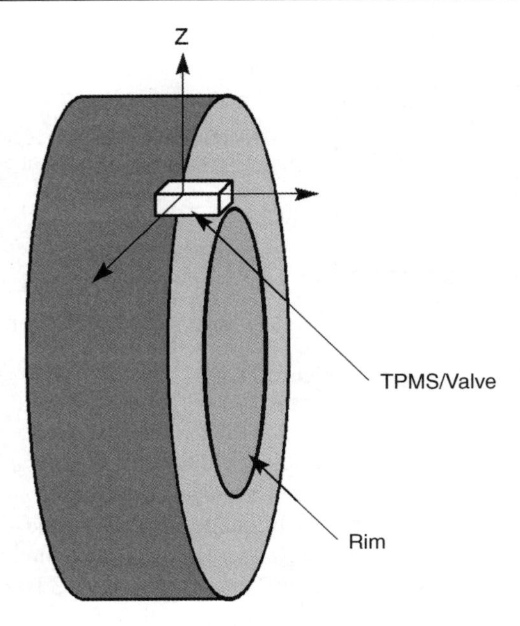


**Figure 11.11**

constant when the vehicle moves fast at high speed, and the information from the sensor has a structural offset. The initial positioning of the sensor is also critical because a mechanical misalignment in the horizontal plane causes large errors in measurement.

- Or by measuring the instantaneous variation of acceleration $d\gamma/dt$, so as to cancel out the static factor (offset) mentioned above, and thus enable the detection of vibrations, impacts and micro-impacts affecting the vehicle.

**Locating the wheels**

Measuring the pressure of a tyre is very good, measuring its temperature and performing a correlation is excellent and transmitting all this to the vehicle by an RF link is superb – but not very useful if we do not know which wheel this relates to! If the information is to be of any practical use for the driver, he must able to locate the wheel that it comes from (right-hand front, left-hand back, etc.), and this must be done

- either dynamically (the vehicle must move for a minimum period to enable the system to correctly initialize the location of the wheels with respect to the vehicle),
- or statically, in other words, without any need to move the vehicle in order to carry out this wheel location survey.

These solutions have some surprising features which will now be revealed.

*No location*

In order to avoid the complexity and cost of this electronic wheel location system, some companies make the economic choice to initialize the vehicle and its wheels in the factory, on the production line. This is done during a special learning phase, by creating an absolute one to-one mapping between a wheel and its location on the vehicle, everything being based on the fact that each integrated circuit in the TPMS has a unique serial number.

This is an economical solution, but is clearly rather limited. Whenever a wheel location is changed (when a wheel is replaced after a puncture, when a worn wheel is replaced with the spare wheel, when the wheels are swapped, when rims are changed for tuning, when wheels and/or tyres are changed between summer and winter, etc.), the initialization has to be repeated with the correct tool, which is often only available in specialist garages or branches. In short, using this system is not so simple!

*With location*

The above problems must be resolved if we wish to allow for the fact that a user can swap the wheels or carry out any of the other operations mentioned above on his vehicle, at his own home, at the weekend. Here again, various technical solutions can be considered.

*In dynamic mode (after the movement of the vehicle has been detected).*

**Example 1**.

A first possible solution is as follows. Seen from the valve, the directions of rotation of the front right-hand and front left-hand wheels are different (the same applies to the rear wheels): movement takes place in the trigonometric direction or in the opposite direction with respect to a fixed reference point, for example the direction of the Earth's magnetic field. A second inductance $L_2$ can be positioned on the $X$-axis (axis of the tread), in quadrature with the inductance $L_1$ mentioned previously (Figure 11.12).

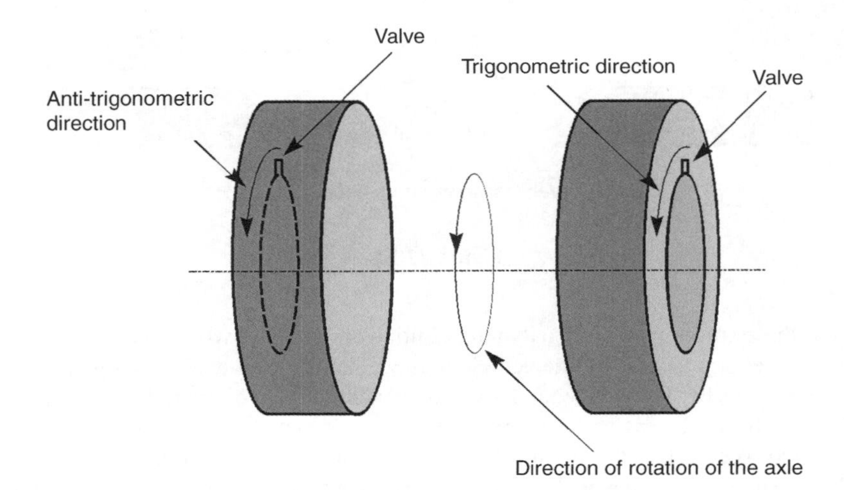


**Figure 11.12**

When the wheels rotate, the two inductances $L_1$ and $L_2$ deliver voltages which, because of their relative position (90°), are in quadrature, with $L_1$ leading $L_2$ for one wheel and $L_2$ leading $L_1$ for the other wheel. Thus, by making a relative measurement of the phases of the signals from $L_1$ and $L_2$, with the inductances kept in the same position in the TPMS, using the sequence of movement of the voltages from $L_1$ and $L_2$, it is easy to determine which is the right-hand wheel and which is the left-hand one. Because of the mechanical topology of the vehicle, the wheels of the front and rear trains can generally be differentiated by using the differences in level of the UHF signals received over the whole receiving assembly, from the front and rear axles. Note that this piece of information, when related to the movement detection information, can also be used to determine whether the vehicle is moving forwards or backwards.

The same applies to the VLF obtained from a slow rotation of the wheels at low speed.

**Example 2.**
Another system can be considered, for example the one detecting the different degrees of fading of the propagation of the UHF signals obtained from the wheels, due to the mechanical shape of the vehicle. This is equivalent to establishing an 'RF signature' for each vehicle (Figure 11.13).

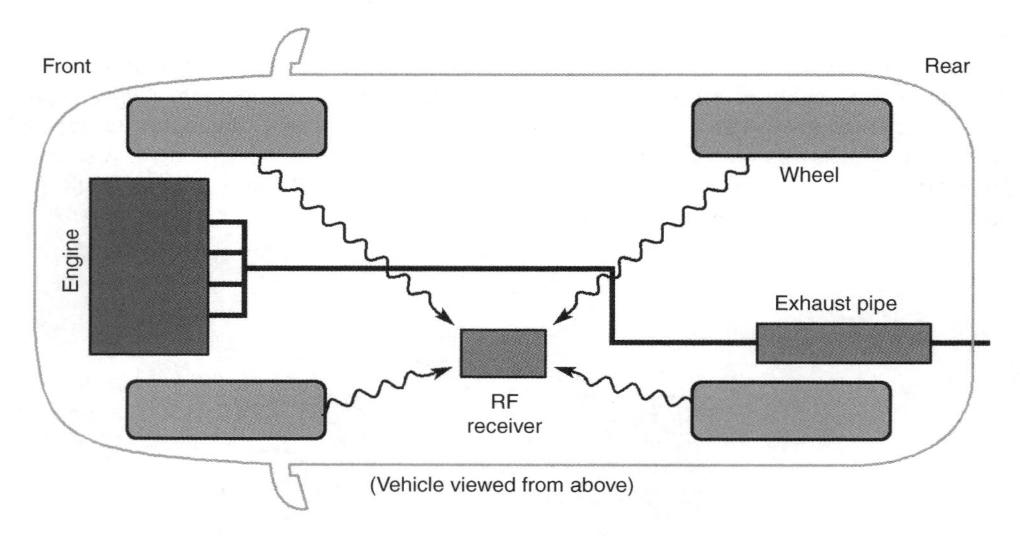


**Figure 11.13**

Clearly, these solutions require a dynamic initialization phase whenever the vehicle moves off, as it is necessary to wait for the vehicle to move before the wheel position can be located reliably. When this has been done, and given that each integrated circuit has a unique number (identifier), it becomes easier to 'mark' the messages sent by each wheel (including the identifier) and to correlate these with the wheel topologies marked during the initialization phase. It is assumed that the wheels do not usually change places while the car is moving... if they did, we would be in trouble!

*In static mode (when the vehicle is stationary).*
In this case, the TPMS must be operated without movement detection, regardless of the relative positions of the wheels with respect to the vehicle. These two conditions imply that

- after contact is made, the TPMS is woken up via an external signal, for example by sending a special message via one of the aforesaid inductances but at a higher frequency (for example in LF, at 125 kHz) (Figure 11.14a);
- a three-dimensional (3D) system is available, in order to make it possible to communicate with the TPMS at all times, even if the wheel has been taken completely out of its housing and the valve on the rim is at the opposite end from the base station controlling the communication (Figure 11.14b).

Clearly, this system also operates in dynamic mode.

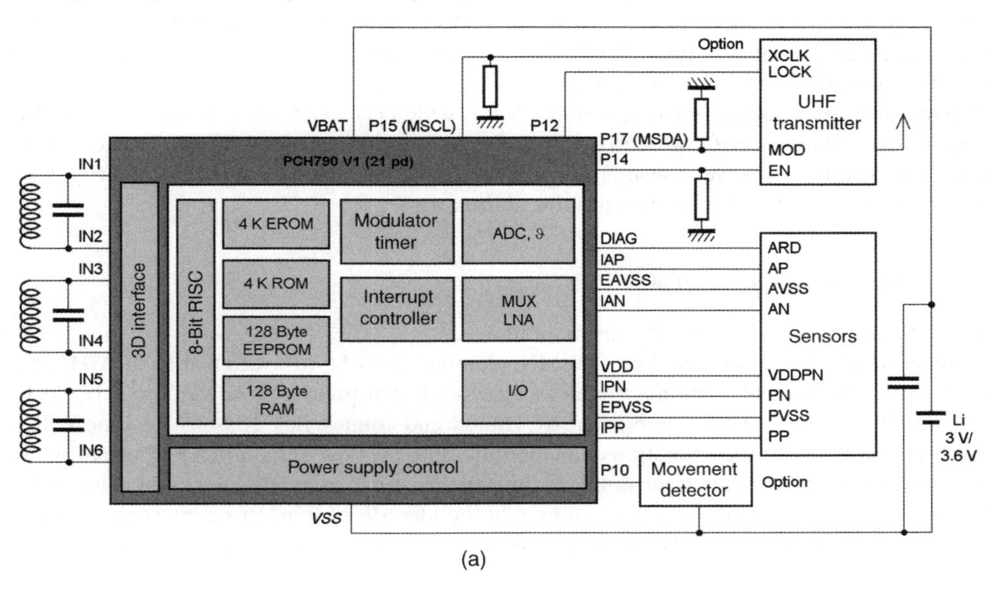


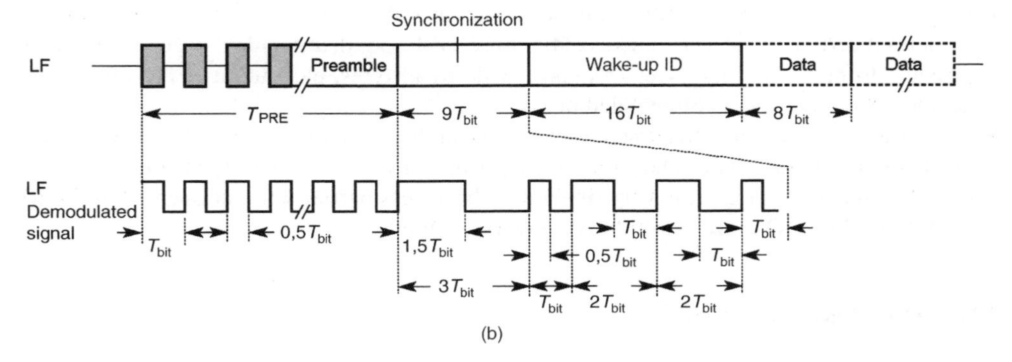


**Figure 11.14**

#### Digital communication

The main communication between the wheels and the receiver placed in the vehicle is usually based on a proprietary RF serial protocol. Generally, transmission takes place at a rate of about 20 ms once every 20 s, when there are no problems (a cyclic ratio of 1/1000, independent of the measurement rate), and can change to once per second if a problem occurs. Additionally, the final access to the RF medium is made pseudo-random by means of an algorithm including the value of the unique number of the TPMS integrated circuit, in order to avoid message collisions, arbitration and management in the receiver.

#### By way of conclusion

To complete this section on the TPMS, note that these systems, in spite of their apparent simplicity, are now included in the most advanced technology in this field. This is because it is not such an easy matter to provide physical measurements of pressure, temperature, acceleration, movement, etc. on a single silicon chip, together with a microcontroller, an HF transmission circuit operating at 434 or 868 MHz, which is capable of withstanding temperatures from −40 to +125°C, which is constantly subjected to acceleration and impact, which has a zero or very low power consumption and which operates with great precision through the life of the product in order to prevent false alarms, etc.

### *11.2.4 TiD – tyre identification*

Not to be confused with the previous system!

What we are concerned with here is the RF identification of tyres – often at 13.56 MHz or in the 860–960 MHz band – and not the measurement of their pressure (see numerous books by the present author on RFID, published by Dunod and John Wiley & Sons, Ltd/Inc.). This system is based on the incorporation of an identifier into the tyre. What is the purpose of such a system, apart from ensuring traceability of tyres in the course of manufacture and delivery? In fact, it is useful in several ways. Assuming that the identifier can itself be interrogated by an external base station located on-board the vehicle, here are two very typical examples:

- in the first case, a motor manufacturer states that he only guarantees the roadholding of his vehicles if certain models and makes of tyres are used, and, if an accident occurs, it would be possible to prove that the tyres (retreads, wide tread tyres, etc.) fitted at the time of the accident were not the recommended ones;
- to give another example, imagine that you arrive at the toll station of the ‘White Autoroute’ at Chamonix, France, in January for a skiing trip, and you are explicitly informed that you must use special fittings. It is very likely that the access barrier will not open, after the RF reading of the ‘TiD’, unless you have snow tyres fitted!

## 11.3 Wireless Networks

Now let us deal briefly with the wireless networks operating at radio frequency.

### *11.3.1 GSM*

In addition to the well-known simple mobile telephone link, the GSM radio-frequency link provides access to external services such as

- emergency call applications;
- communications controlling the panic button;
- telephone links of all kinds, used for example for the dynamic updating of information used by navigation systems;
- convenience or service calls:
  - reservations at the nearest hotel, carried out according to the exact location of the car as determined by its GPS;
  - a breakdown service, operating in conjunction with GPS location.

Looking at the simple example of breakdown assistance via a GSM link, assisted by GPS for the vehicle location, we can immediately see that there must be, at least as far as the internal system is concerned:

- a direct link between the GSM and the CAN, LIN and other buses of the vehicle, for inspecting and diagnosing the different modules in the vehicle, in order to notify the breakdown assistance company of the functions that are most likely to be implicated in the operating failure,
- a direct link with the GPS location system, connected to a greater or lesser degree to the vehicle's navigation and audio–video network (MOST or other).

### *11.3.2 Bluetooth*

Let us now return to the external applications of the Bluetooth link alone, leaving aside the ways in which it complements a GSM link. These differ slightly from one country to another. In fact, depending on the local regulations in force in different countries, it is permitted or feasible to transmit at different EIRP power levels (classes 1, 2 and 3), and consequently over different communication ranges (from several metres to several hundreds of metres). However, this is generally enough for the development of external applications. Here are some examples:

- RF links with beacons located along roads, for updating or re-updating road information.
- Applications for the reception departments of garages, branches, dealers, etc. These are generally designed to offer new facilities and more user-friendliness when vehicles are brought into branches or garages for repairs or servicing. This enables the receptionist to welcome the customer while remaining in his office, facing him and remotely interrogating the vehicle which has entered the service station, using a CAN–Bluetooth interface to read the mileage, the tyre pressure, the last service visits and repairs carried out, and certain characteristics of the vehicle.
- Downloading convenience data via a PC–Bluetooth link to a vehicle in a private garage.

These families of applications also include the Zigbee concept, a relative of the Bluetooth family, rather simplified, but of the same type.

### *11.3.3 IEEE 802.11x*

The IEEE 802.11x concept is generally known under the name of Wi-Fi. Admittedly, Wi-Fi is not aimed at the same applications as Bluetooth, although there may be many common groups of applications. However, it would take considerable foresight to be able to predict with any certainty the future share of applications of each of these two concepts. What are the prospects for this scheme of combining numerous participants in an RF network, by comparison with the pico-networks of Bluetooth?

In any case, the external links of on-board systems must always be separated into two major families:

- industrial applications, where there are no significant problems in choosing which systems to use because the consequences of a change or upset in the market, although sometimes annoying, can be corrected speedily;
- motor vehicle applications, where systems cannot be changed overnight because of the infrastructure that is already in place.

At present, therefore, manufacturers always take some time to think before making their final selection.

### *11.3.4 NFC (near-field communication)*

The last newcomer is NFC, i.e. a system using a communication range of approximately 10 cm. Can this concept be of any use in the external environment, when this tiny communication distance appears to place it firmly in the 'internal' category?

The answer is simple: this concept (Figure 11.15), which uses an HF carrier at 13.56 MHz, acts like a secure, contactless smart card in the equipment in which it is implemented (mobile phone, PC, etc.).

In this way, it enables a link to be established between different systems which apparently have nothing in common, for example between the car radio (internal) and your home computer (external) or a poster on a wall in the city centre (external, again). This may seem strange.

Let me explain it more clearly. I can use the NFC system to download some of the latest hits in MP3 format, in contactless mode, to my mobile phone. When I am in my car, I can transfer them via NFC, again in contactless mode, to my car radio and listen to them. Similarly, if I see something interesting on a poster while I am walking along the street, I go up to the poster and use the RF and NFC system to capture the desirable information about the web site address contained in it (a chip is integrated into the poster); later on I can connect to this site, possibly while driving, via the NFC link and then the GSM link of the car radio.

As you would have realized, the future lies open before us, once again! (You may reply that it is not very likely to be behind us, and of course you would be quite right.)

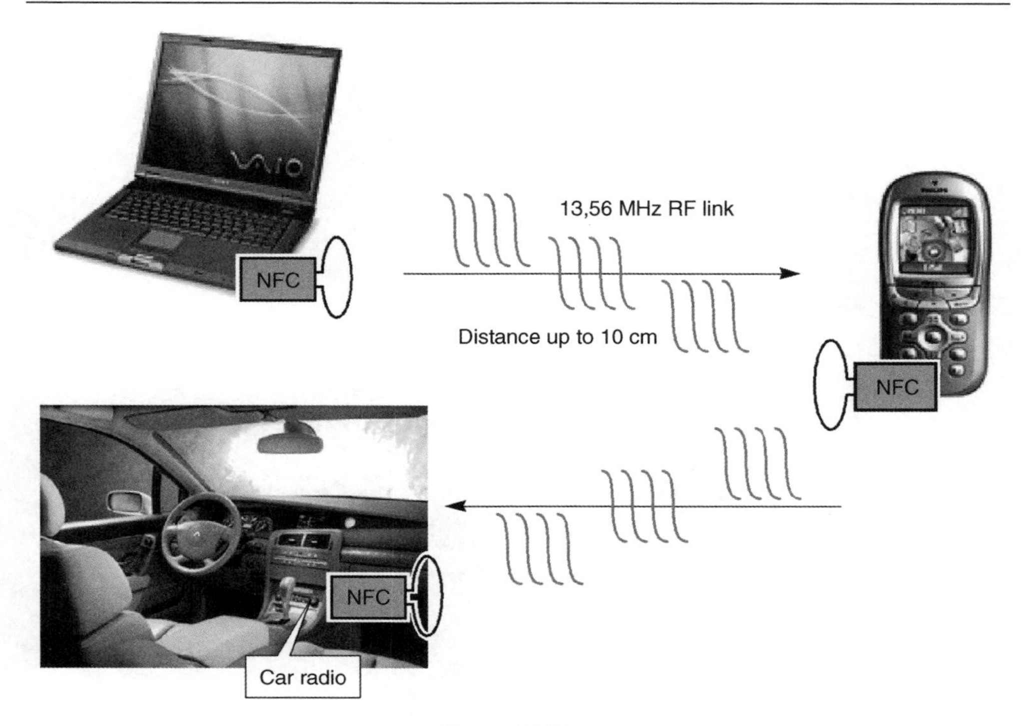


**Figure 11.15**

So much for this short introduction to these new communication protocols which form extensions and gateways for the protocols described previously (CAN, LIN, etc.). It is still too early to say how long any of these will last, but some of them at least will certainly be in use for many years, and their complementary functionality must be taken into account when producing intelligent on-board systems.

# Conclusion

This brings us to the end of the book. I hope that it has met all your expectations concerning the understanding and operation of the CAN (controller area network) protocol and some of its new associates such as LIN, FlexRay, *Safe-by-Wire Plus*, MOST, IEEE 1394, etc.

Throughout these pages, I have tried to cast some light on the characteristics, the peculiarities and the strengths of these concepts which, I hope, will have aroused your enthusiasm as much as the now familiar CAN.

Sadly, a shortage of time and space has prevented a detailed discussion of other applications using all the other components available in the market. But this is only a pleasure postponed! You should be aware that the principles described in this book will enable you to understand any other application without difficulty. So do not hesitate to enter the world of controller area networks and X-by-Wire!

Some of my colleagues and I intend to present new specific applications of these buses to you before too long.

See you then!

Dominique Paret
Meudon, June 2006

*Multiplexed Networks for Embedded Systems: CAN, LIN, Flexray, Safe-by-Wire...* D. Paret

# Part C

Appendices

# Appendix A

# CiA (CAN in Automation)

## A.1 Its Mission, Its Role, Its Operation

I have mentioned the CiA (CAN in Automation) user group several times in this book. Let us return briefly to this topic.

Founded in March 1992 by many users, vendors and manufacturers of CAN components, this fully independent international association is intended 'to provide technical, product and marketing intelligence information in order to enhance the prestige, improve the continuity and provide a path for advancement of the CAN network and promote the CAN protocol'.

At present (June 2006), the association has about 500 full class member companies, including all the major industrial businesses of the sector to which this subject relates, all the major electronic component manufacturers and a large number of SMEs, SMIs, universities, etc. of all nationalities. The open approach of this association, recognized by the whole profession, is structural, its main aim being to promote the CAN protocol. I should point out, by way of example, that the group supports all application layers (CAL, CANopen, DeviceNet, SDS, etc.) in the same way and with equal enthusiasm.

The general organization of CiA is shown in Figure A1. You will see from this that CiA has local organizations or delegations in each country.

## A.2 Its Activities

The main activities of this organization are as follows:

- at the international level
  - preparation and publication of technical recommendations to complement ISO, SAE and other standards, in the form of CiA Draft Standard (DS. . .) relating to the connectors, the application layers, etc., for all types and all fields of application of CAN;

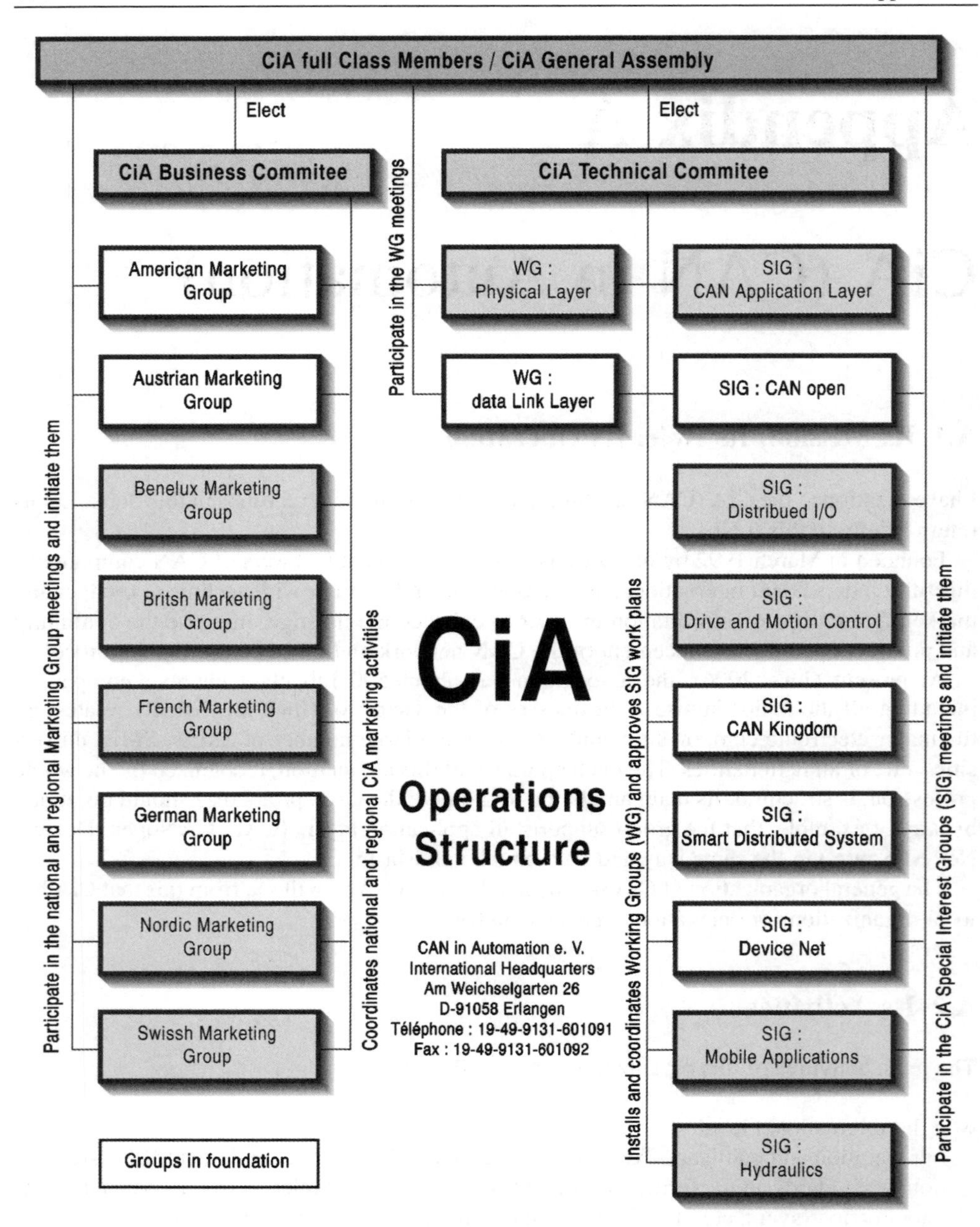
CiA full Class Members / CiA General Assembly
Elect
Elect
CiA Business Commitee
CiA Technical Commitee
Participate in the WG meetings
American Marketing Group
Austrian Marketing Group
Benelux Marketing Group
Britsh Marketing Group
French Marketing Group
German Marketing Group
Nordic Marketing Group
Swissh Marketing Group
Groups in foundation
Participate in the national and regional Marketing Group meetings and initiate them
Coordinates national and regional CiA marketing activities
WG : Physical Layer
WG : data Link Layer
SIG : CAN Application Layer
SIG : CAN open
SIG : Distribued I/O
SIG : Drive and Motion Control
SIG : CAN Kingdom
SIG : Smart Distributed System
SIG : Device Net
SIG : Mobile Applications
SIG : Hydraulics
Installs and coordinates Working Groups (WG) and approves SIG work plans
Participate in the CiA Special Interest Groups (SIG) meetings and initiate them
CiA
Operations Structure
CAN in Automation e. V.
International Headquarters
Am Weichselgarten 26
D-91058 Erlangen
Téléphone : 19-49-9131-601091
Fax : 19-49-9131-601092

**Figure A1**

  - organization of special days, seminars and technical conferences such as the very high level international CAN conferences (ICC) (dealing with the protocol, the components, the application layers, real-time CAN, approval, certification, etc.);
  - promotion events in Europe and for new markets, especially in the United States and the Far East;
  - periodical publication of newsletters and product guides;
- at the local level (in many countries)
  - organization of dedicated information forums, general training and special workshops on particular themes;
  - a presence in the specialist press, and at trade fairs and exhibitions.

For further information you can contact the group directly at

CAN in Automation
Am Weichselgarten 26
DE-91058 Erlangen
Tel.: +49 -9131 -69086-0
fax: +49-9131-69086-79
E-mail: *headquarters@can-cia.org*
Web site: *www.can-cia.org*

# Appendix B

# Essential References

## B.1 The ISO/OSI Model

ISO 7498 (1984)
*Information processing systems. Open systems interconnection – OSI*
Basic reference model, International Standard Organization

ISO/IEC 9646 – 1 to 4 (1991)
Information Technology – Open Systems Interconnection Conformance Testing methodology and framework, Part 1

## B.2 CAN Reference Document

CAN protocol specifications V 2.0 (A & B) - Robert Bosch (1991 & 1992)

## B.3 CAN and ISO

ISO 11898-1, -2, -3, -4 and -5 (1993 to 2006)
Road vehicles – interchange of digital information
Controller area network – CAN – ISO standard

ISO 11519-1 (1994)
Road vehicles – low speed serial data communications
Part 1 – general and definitions

ISO 11519-2 (1994)
Road vehicles – low speed serial data communication
Part 2 – low speed controller area network CAN

*Multiplexed Networks for Embedded Systems: CAN, LIN, Flexray, Safe-by-Wire...* D. Paret

## B.4 'Industrial' Application Layers

### CiA draft specifications (DS...)

| CAL (CAN application layer) | | | | |
|---|---|---|---|---|
| *CAN physical layer* | DS | 102 | 2.0 | 1994 |
| *CAN power management layer* | DS | 150 | 1.2 | |
| *CAN application layer for industrial applications* – CAL | DS | 201/207 | 1.1 | 1996 |
| CMS | *Service specification* | DS | 202-1 | |
| | *Protocol specification* | DS | 202-2 | |
| | *Data types and encoding rules* | DS | 202-3 | |
| NMT | *Service specification* | DS | 203-1 | |
| | *Protocol specification* | DS | 203-2 | |
| DBT | *Service specification* | DS | 204-1 | |
| | *Protocol specification* | DS | 204-2 | |
| LMT | *Service specification* | DS | 205-1 | |
| | *Protocol specification* | DS | 205-2 | |
| | *Application layer naming conventions* | DS | 207 | |

| CANopen | | | | |
|---|---|---|---|---|
| *Communication profile for industrial systems(EN 50 325 – 4)* | DS | 301 | 4.0 | 2003 |
| *Framework for programmable devices* | DS | 302 | 1.0 | 1997 |
| *Device profiles* | DS | 401 | 1.4 | 1996 |
| *for I/O modules* | DS | 402 | 1.0 | 1997 |
| *for drives and motion control* | DS | 403 | | |
| *for human machine interfaces* | DS | 404 | 1.0 | 1997 |
| *for measur. devices and closed-loop control* | | | | |
| *for encoders* | DS | 406 | 1.0 | 1997 |

| SDS (smart distributed system) specifications | | | | |
|---|---|---|---|---|
| *Application layer protocol* | GS 052 | 103 | 3.0 | 1996 |
| *Physical layer specification* | GS 052 | 104 | 1.0 | 1994 |
| *Interface guidelines specifications* | GS 052 | 105 | | |
| *Device guidelines specifications* | GS 052 | 106 | | |
| *Component modeling specification* | GS 052 | 107 | 1.0 | 1995 |
| *Verification test procedure spec. for I/O devices* | GS 052 | 108 | | |
| *Product selection guide* | 1996 | | | |
| *Conformance testing procedure* | | | | |

| DeviceNet specifications | | |
|---|---|---|
| DeviceNet specification – reference D 9240 – DNDOC | | |
| Volume 1: *Communication Model and Protocol* | 2.0 | 1997 |
| Volume 2: *Device Profiles and Object Library* | 2.0 | 1997 |
| Source code example – reference 9240 – DNEXP | | |

| CAN kingdom | | |
|---|---|---|
| *CAN kingdom specification* | Version 3.0 | |

| M3S – multiple master multiple slave | | |
|---|---|---|
| M3S – *multiple master multiple slave specification* | Version 2.0 | 1995 |

## B.5 Motor Vehicle Application Layers

| *OSEK/VDX specifications* | | |
|---|---|---|
| OSEK/VDX *operating system* | Version 2.0 | 1997 |
| OSEK/VDX *communication specification* | Version 1.2 | 1997 |
| OSEK/VDX *network management* | Version 2.0 | 1997 |

| SAE J1939/3 | | |
|---|---|---|
| Recommended practice for serial control and communication network (class C) for truck and bus application | | 1992 |
| J1939 | General standard document describing the general characteristics of the network, the OSI layer, etc. | |
| J1939/01 | Application document for truck and bus control and communication | |
| /02 | Application document for agricultural equipment control and communication | |
| /11 | Physical layer 250 kbit $s^{-1}$, shielded twisted pair | |
| /21 | CAN 29 bit identifier data link layer | |
| /31 | Network layer for trucks and buses | |
| /4X | Transfer layer | |
| /5X | Session layer | |
| /6X | Presentation layer | |
| /71 | Vehicle application layer | |
| /72 | Virtual terminal application layer | |
| /81 | J1939 network administration document | |

### ICC (International CAN Conferences) proceedings

Most of the documents listed above can be obtained from CiA. This independent organization, which has the mission of supporting CAN, also distributes and sells official documents relating to application layers.

# Appendix C

# Further Reading

## C.1 Journals, Magazines, Other Documentation

*CiA*
*Product guide*
*CAN newsletters* (published quarterly)

## C.2 Books in German

*CAN (controller area network) – Grundlagen, Protokolle, Bausteine, Anwendungen* – K. Etschberger, Éditeur Hanser.
*CAN – Grundlagen und Praxis* – W. Lawrentz, Éditeur Hüthig.

## C.3 Books in English

*CAN* – W. Lawrenz, John Wiley & Sons.

## C.4 Books in French

*Le bus CAN – Applications* – D. Paret, Dunod.

# Appendix D

# Useful Addresses

## D.1 ISO

In France, ISO standards can be obtained from
AFNOR
11, rue François de Pressencé
Saint-Denis
Tel: +33(1) 41 62 80 00
Fax: +33(1) 49 17 90 00

## D.2 CAL and CANopen

CiA – CAN in Automation
Am Weichselgarten 26
DE-91058 Erlangen
Tel: + 49 9131 69086 0
Fax: + 49 9131 69086 79
E-mail: *headquarters@can-cia.org*
Web site: *www.can-cia.org*

ODVA
8222 Wiles Road, suite 287
Coral Springs, FL 33067
United States
Tel: + (1) 954 340 5412
Fax: + (1) 954 340 5413

*Multiplexed Networks for Embedded Systems: CAN, LIN, Flexray, Safe-by-Wire...* D. Paret

## D.3 CAN Kingdom

Kvaser AB
P.O. Box 4076
51 104 Kinnahult
Sweden
Tel: + (46) 320 15 287
Fax: + (46) 320 15 284

## D.4 M3S

M3S dissemination office
TNO Institute of Applied Physics
PO box 155
2600 AD Delft
The Netherlands
Tel: + (31) 152 69 20 04
Fax: + (31) 152 69 21 11
Web site: 147 252 133 152/m3s

## D.5 OSEK

Institute of Industrial Information Systems – IIIS
University of Karlsruhe
Hertzstrasse 16
D - 76187 Karlsruhe
Germany
Tel: + (49) 721 608 4521
Fax: + (49) 721 755 788

**MOST Corporation**
Bannwaldallee 48
D - 76185 Karlsruhe
Germany
Tel: + (49) 721 966 50 00
Fax: + (49) 721 966 50 11

# Index



*Multiplexed Networks for Embedded Systems: CAN, LIN, Flexray, Safe-by-Wire...* D. Paret